Mechanical Engineering Series

Series Editor

Francis A. Kulacki
Department of Mechanical Engineering
University of Minnesota
Minneapolis, Minnesota, USA

The Mechanical Engineering Series presents advanced level treatment of topics on the cutting edge of mechanical engineering. Designed for use by students, researchers and practicing engineers, the series presents modern developments in mechanical engineering and its innovative applications in applied mechanics, bioengineering, dynamic systems and control, energy, energy conversion and energy systems, fluid mechanics and fluid machinery, heat and mass transfer, manufacturing science and technology, mechanical design, mechanics of materials, micro- and nano-science technology, thermal physics, tribology, and vibration and acoustics. The series features graduate-level texts, professional books, and research monographs in key engineering science concentrations.

More information about this series at http://www.springer.com/series/1161

Amit Agrawal • Hari Mohan Kushwaha
Ravi Sudam Jadhav

Microscale Flow and Heat Transfer

Mathematical Modelling and Flow Physics

Springer

Amit Agrawal
Department of Mechanical Engineering
Indian Institute of Technology, Bombay
Mumbai, Maharashtra, India

Hari Mohan Kushwaha
Department of Mechanical Engineering
Indian Institute of Technology, Bombay
Mumbai, Maharashtra, India

Ravi Sudam Jadhav
Department of Mechanical Engineering
Indian Institute of Technology, Bombay
Mumbai, Maharashtra, India

ISSN 0941-5122 ISSN 2192-063X (electronic)
Mechanical Engineering Series
ISBN 978-3-030-10661-4 ISBN 978-3-030-10662-1 (eBook)
https://doi.org/10.1007/978-3-030-10662-1

Library of Congress Control Number: 2019932162

This Springer imprint is published by the registered company Springer Nature Switzerland AG.
The registered company address is: Gewerbestrasse 11, 6330 Cham, Switzerland

This book is dedicated to
past and current students of "Rarefied Gas Dynamics" and "Microfluidics" laboratories, IIT Bombay,
and
to our families.

Preface

This book has been written with the objective of making the readers familiar with the exciting developments on gaseous slip flow and heat transfer in microchannel. A large amount of work is being currently undertaken worldwide in these areas, with numerous potential applications. The subject is therefore topical and also particularly significant as it leads to the question about the validity of the Navier-Stokes equations, which is usually considered sacrosanct. It obviously leads to a subsequent question: if the Navier-Stokes equations are themselves not valid, then how do we model the flow and heat transfer? Although the answer to this last question is still not clear, it is important for the fluids and thermal community to first appreciate the limitations of the continuum approach (which leads to the Navier-Stokes equations), to better appreciate the ongoing search for the "beyond Navier-Stokes equations," and to test some of the available equations for their accuracy, before the larger objective of finding beyond the Navier-Stokes equations can be practically met.

The book is therefore organized in two parts: the first part is on the gaseous slip flow and heat transfer in microchannel (Chaps. 2 and 3). In the second part, we examine beyond the Navier-Stokes equations (Chaps. 5–7). Chapter 1 summarizes the characteristics of microscale flows and provides an introduction to the various modelling approaches available. Chapter 4 is a transition chapter between the two parts, where it is shown that simple extensions of the Navier-Stokes equations are not adequate. Recognizing that only analytical tool will not be adequate for studying flows in the slip and transition regimes, a brief overview to relevant numerical and experimental techniques is provided in Chap. 8. Finally, Chap. 9 summarizes our current understanding and provides suggestions for future research in this subject.

The Knudsen number is the most important parameter, and several known solutions with Knudsen number as an additional parameter are compiled in the first part of the book. Interesting observations on Knudsen minima and flow separation are presented. However, it should become apparent that our understanding of heat transfer at the microscale is not that sound, as several additional effects such as axial conduction, pressure work, conduction in the substrate, viscous dissipation, etc. coexist, but it is virtually not possible to treat all these effects together and obtain an analytical solution for even the simplest problems. Only few experimental and direct simulation Monte Carlo (DSMC) data exist, and they do not agree well with that obtained from alternative approaches.

A good portion of the second part of the book is devoted to deriving and understanding the Burnett and Grad equations. These two sets of equations form the most important "beyond the Navier-Stokes" equations and are generally referred to as "higher-order continuum transport equations." The study of these equations is important for further development of the subject, as they involve several novel concepts and represent important breakthroughs in the subject. The derivation of these equations has not often been repeated, and it is expected that the stepwise derivation presented here will invoke wide readership. Special effort has been made to make the text readable through the insertion of a large number of figures. The hope is that with the help of this book, it should now be possible to derive the Burnett and Grad equations in a graduate class. A few problems are solved to illustrate the type of solution obtained from these equations. In the derivation of some solutions presented here, some minor error in the original source was noticed which has been corrected here.

The first part of the book (Chaps. 1–3 along with Chaps. 8 and 9) should appeal to readers interested in understanding fundamental aspects of microscale flows and heat transfer, while the second part (Chaps. 4–7) is primarily for slightly advanced readers interested in understanding equations beyond the Navier-Stokes equations.

Mumbai, India — Amit Agrawal
Mumbai, India — Hari Mohan Kushwaha
Mumbai, India — Ravi Sudam Jadhav
September 2018

Acknowledgments

We are grateful to all the students who have worked in the "Rarefied Gas Dynamics" and "Microfluidics" laboratories at IIT Bombay, including those who were involved in purely theoretical or numerical aspects of the problem. It is they who slowly and patiently uncovered the various fine points of the subject, which finally culminated in the form of this book. We are also grateful to Prof. K. Muralidhar (IIT Kanpur) and Prof. F. Kulacki (University of Minnesota), the editors of this series, for first placing their trust in us for writing this book and then patiently waiting for it while we went past several deadlines. The book would not have resulted without the coaxing of Prof. Muralidhar to write on this subject. While preparing the book, we gained from the comments of Prof. Atul Sharma (IIT Bombay).

Funding for this line of investigation has been provided by the Indian Institute of Technology Bombay (IITB), Indian Space Research Organisation (ISRO), Department of Science and Technology (DST), and recently by the Department of Atomic Energy (DAE) under its prestigious scheme—DAE-SRC Outstanding Investigator Award, awarded to the first author.

Contents

Nomenclature

Alphabetical

$\mathbf{c}$	molecular velocity, $\frac{m}{s}$
c_s	speed of sound, $\frac{m}{s}$
c_x, c_y, c_z	components of molecular velocity $\mathbf{c}$, $\frac{m}{s}$
c_p	specific heat at constant pressure, $\frac{J}{kg\ K}$
c_v	specific heat at constant volume, $\frac{J}{kg\ K}$
$\mathbf{C}$	peculiar velocity, $\frac{m}{s}$
C_1, C_2	Slip coefficients
C_x, C_y, C_z	components of peculiar velocity $\mathbf{C}$, $\frac{m}{s}$
d	molecular diameter, m
D_h	hydraulic diameter, m
e	specific energy, $\frac{J}{kg}$
$\mathbf{F}$	external force, $\frac{kg\ m}{s^2}$
f	particle distribution function, $\frac{s^3}{m^6}$
f_M	Maxwell-Boltzmann distribution, $\frac{s^3}{m^6}$
h	convective heat transfer coefficient, $\frac{W}{m^2\ K}$
k	thermal conductivity, $\frac{W}{m\ K}$
k_B	Boltzmann constant, $\frac{J}{K}$

L	characteristic length scale, m
m	mass of a molecule, kg
$\dot{m}$	mass flow rate, $\frac{kg}{s}$
n	number density, $\frac{1}{m^3}$
p	hydrostatic pressure, $\frac{N}{m^2}$
P_{ij}	pressure tensor, $\frac{N}{m^2}$
p_{ij}	divergence-free part of pressure tensor P_{ij}, $\frac{N}{m^2}$
$\mathbf{q}$	heat flux vector, $\frac{W}{m^2}$
R	specific gas constant ($= k_B/m$), $\frac{J}{kg.K}$
r, θ, z	cylindrical coordinate system
r, θ, ϕ	spherical coordinate system
t	time, s
T	absolute temperature, K
T_m	bulk mean temperature, K
T_w	wall temperature, K
$\mathbf{u}$	bulk velocity vector, $\frac{m}{s}$
u, v, w	components of bulk velocity $\mathbf{u}$, $\frac{m}{s}$
u_s	velocity of gas at the wall, $\frac{m}{s}$
u_w	wall velocity, $\frac{m}{s}$
$\bar{v}$	most probable speed, $\frac{m}{s}$
$\mathbf{x}$	position vector, m
x, y, z	Cartesian coordinates in physical space

Greek symbols

α	thermal diffusivity, $\frac{m^2}{s}$
γ	specific heat ratio, $\frac{c_p}{c_v}$
λ	mean free path, m
δ_{ij}	Kronecker delta

ϵ	specific internal energy, $\frac{J}{kg}$
μ	dynamic viscosity, $\frac{N\ s}{m^2}$
μ_2	second coefficient of viscosity, $\frac{N\ s}{m^2}$
μ_v	coefficient of bulk viscosity, $\frac{N\ s}{m^2}$
ν	kinematic viscosity, $\frac{m^2}{s}$
Φ	viscous dissipation, $\frac{kg}{m\ s^3}$
ρ	mass density, $\frac{kg}{m^3}$
σ_T	thermal accommodation coefficient
σ_v	tangential momentum accommodation coefficient
σ_{ij}	viscous stress tensor, $\frac{N}{m^2}$
τ_{ij}	stress tensor $(= -P_{ij})$, $\frac{N}{m^2}$

Non-dimensional numbers

Br	Brinkman number, $Br = \frac{\mu u^2}{k(T_w - T_m)}$
Ec	Eckert number, $Ec = \frac{u^2}{c_p(T_w - T_m)}$
Kn	Knudsen number, $Kn = \frac{\lambda}{L}$
Ma	Mach number, $Ma = \frac{u}{\sqrt{\gamma RT}}$
Nu	Nusselt number, $Nu = \frac{hD_h}{k}$
Pe	Peclet number, $Pe = \frac{\rho u c_p D_h}{k}$
Pr	Prandtl number, $Pr = \frac{\mu c_p}{k}$
Re	Reynolds number, $Re = \frac{\rho u D_h}{\mu}$

Abbreviations

CFD	computational fluid dynamics
DSMC	direct simulation Monte Carlo
MEMS	micro-electro-mechanical systems
TMAC	tangential momentum accommodation coefficient

Chapter 1
Introduction to Microscale Flows and Mathematical Modelling

The purpose of this book is to understand fluid flow and heat transfer at the microscale. The focus is on gases, as the mean free path of the gas can become comparable to the passage dimensions, and an additional non-dimensional number (Knudsen number) starts affecting the dynamics of the flow and heat transfer behavior. The effect of Knudsen number turns out to be non-trivial, as it leads to not only several new physical phenomena, but also exposes the limitations of the celebrated Navier–Stokes equations in modelling such flows. The second major purpose of this book is therefore to examine equations which can model such high Knudsen number flows. Again, we first show that simple modifications to the Navier–Stokes equations and boundary conditions is not sufficient to fulfill this objective. There is perhaps no alternative but to shun the conventional way of deriving transport equations, and resort to the Boltzmann equation for derivation of "higher-order continuum transport equations." The book presents a clear derivation of the Burnett and Grad equations starting from the Boltzmann equation, a discussion on the variants of the Burnett and Grad equations, and some known solutions of these higher-order continuum transport equations.

Microscale flows therefore help put the higher-order continuum transport equations in better perspective, and provide reasons to study these "beyond the Navier–Stokes" equations. The higher-order continuum transport equations may not just help understand high Knudsen number flows better, rather, being superset of the Navier–Stokes equations, help understand fluid flow and heat transfer phenomena in ways that we cannot even imagine as of now; thereby making the study of these higher-order equations extremely relevant. In this chapter, we provide an introduction to microscale flows and their characteristics, application of microscale flows, and challenges in modelling such flows.

A. Agrawal et al., *Microscale Flow and Heat Transfer*, Mechanical Engineering Series, https://doi.org/10.1007/978-3-030-10662-1_1

1.1 Introduction

Microscale flows are in vogue and the last three decades have seen substantial advancement in their understanding. This interest in microscale flows is because of counter-intuitive flow physics that can be encountered at the microscale along with several potential applications of microdevices that can be built using such an understanding. A large number of application of microdevices are in the domain of medical diagnostics, and lie beyond the scope of this book. Here, we rather focus on the physics of flow in microdevices, which is interesting because of slipping of the gas at the passage walls and substantial compressibility effect (or variation in density) as the flow occurs. Such *internal compressible* flows involving *slip* at the wall constitute a new class of problems, which have not been well studied in the past.

The reduction in length scale (small cross section of the passage across which the flow is occurring) means that a gas molecule experiences fewer inter-molecular collisions before hitting the opposite wall. This issue along with finite momentum at the wall (due to slip) is brought about due to rarefaction of the gas, with the amount of rarefaction being measured by a non-dimensional parameter—Knudsen number. The issue of relatively infrequent inter-molecular collisions leads to the gas not being in thermodynamic equilibrium close to the wall. The region of this thermodynamic non-equilibrium extends further into the bulk of the flow with an increase in the Knudsen number. The classical flow equations—the Navier–Stokes equations—are not applicable in the region of non-equilibrium, which makes modelling of the non-equilibrium region, and therefore flow at the microscale particularly challenging. Further complications arise as the typical no-slip boundary condition at the gas–wall interface is no longer applicable.

The equation that needs to be solved in place of the Navier–Stokes equations to describe the flow is not entirely settled, although several equations have been proposed in the literature. These higher-order continuum transport equations arise from the Boltzmann equation and are derived by making suitable assumptions regarding the nature of the particle distribution function. Most noteworthy of these higher-order continuum transport equations are by Burnett and Grad, and bear the name of these scientists. The derivation of the Burnett and Grad equations is intimately linked to concepts in kinetic theory—a subject about which engineers are ill-aware of. The awareness of the Burnett and Grad equations, even among senior practitioners in the field, is therefore limited. This deficiency in understanding is recognized and addressed in this book. We have further invoked minimal concepts from the kinetic theory in our discussion of the Burnett and Grad equations, and have taken the viewpoint that the "output" of the kinetic theory should serve as the "input" based on which higher-order continuum transport equations can be derived.

Currently, most rarefied gas flows are solved using the Direct Simulation Monte Carlo (DSMC) method of Bird. So can we not continue to work with the DSMC method, especially since the DSMC method is so well established? That is, why do we need higher-order continuum transport equations at all? DSMC is a numerical method and all numerical methods yield solution for the specific case solved, as opposed to a general solution possible with the analytical approach. Therefore, unless one solves a large number of cases, one does not get an idea of the trend with respect to variation of a given parameter. Of further interest is to have a scaling of the solution with respect to the various governing parameters. Solving multiple cases requires effort and is computationally expensive in terms of time and resources. On the other hand, understanding the variation or scaling with respect to various governing parameters is straightforward after having obtained an analytical solution. Knowledge of the governing equation that needs to be solved is however a prerequisite for obtaining an analytical solution, pointing towards the need for obtaining accurate transport equations.

Analytical solution from governing equations can further be applied to validate numerical results, and even experimental results especially when a new setup is being built. Further calibration of instruments can be undertaken in flows for which analytical solution is available and which are also simple and convenient to set up. In summary, having an alternative independent approach (knowledge of accurate higher-order continuum transport equations in the present case) is useful in many ways, and helps by enriching our understanding of the phenomena as it adds another dimension to the available tools (DSMC based simulations and experiments). The field of computational fluid dynamics (CFD), which solves the Navier–Stokes equations, is already well developed and its extension to solve the higher-order continuum transport equations should not be too difficult. Thus, the range of problems that can be solved using CFD can be substantially increased.

In Chaps. 2 and 3 of this book, we have discussed the various fluid flow and heat transfer solutions of microscale flows. These examples bring out the difficulty with respect to employing the Navier–Stokes equations as discussed further in Chap. 4. The Burnett and Grad equations and few known solutions of these equations are discussed in Chaps. 5 and 6, respectively. The variants of the Burnett and Grad equations along with some other recent works on higher-order continuum transport equations are covered in Chap. 7. It is further recognized that the analytical approach is not sufficient to understand flows at the microscale. Therefore, a brief introduction to the relevant numerical and experimental techniques is provided in Chap. 8. Finally, scope for further work is identified in the last chapter.

1.2 Applications of Microscale Flows

We begin by examining a few interesting applications of microscale flows. The ability to first design a microdevice towards its intended objective, followed by the fabrication of the micron size passages involved and their sealing, can lead to the development of numerous interesting microdevices. A good understanding of microscale flows is important for design of microdevices as most practical microdevices will involve movement of some fluid through them.

Although numerous potential applications can be quoted, here we restrict ourselves to four applications from distinct domains. Some of the applications take advantage of physical effects which are specifically present at the microscale. While other applications exploit typical features of microdevices—large surface area to volume ratio, small amount of sample and reagents required, requirement of lower amount of power which leads to a solution at a reduced cost, and the possibility of having a microdevice which is portable and disposable. These advantages however come at a cost as the microdevices suffer from the requirement of relatively large pumping power, low rate of mixing of fluids, low throughput, among others. Several issues related to selection of material and cost of fabrication also show up in making of practical microdevices.

1.2.1 Cooling of Electronic Devices

The electronics industry has seen remarkable advances in the last few decades because of which it is now possible to pack numerous advanced electronics circuits on a single chip. As the functionality of the electronics chip increases, the amount of heat being generated by the electronic circuit increases, and consequently the heat flux increases. The heat generated in the chip needs to be removed, so that the temperature stays below a threshold value and the functioning of the chip is not adversely affected. Although the amount of power that needs to be dissipated is only a few 100s of watt, the area over which this heat is generated is small ($<1\,cm^2$), which makes the heat flux (power per unit area) very large ($>250\,W/cm^2$). The conventional cooling techniques are not capable of handling such intense heat fluxes.

In this context, flow in microchannel, with or without phase change, has been proposed as a potential cooling solution in the literature. Figure 1.1a,b shows a microchannel with a microheater on the other side of the wafer to mimic the heat generating chip. A simple calculation shows that the heat transfer coefficient with flow of water in a $100\,\mu m$ passage can easily be greater than $10\,kW/m^2\,K$ without phase change, and much greater than this with boiling (Fig. 1.1c). Such a high value of heat transfer coefficient is obtained only with phase change at the conventional scale, illustrating the advantage of cooling electronic devices (and other high heat flux generating components) using an array of microchannels, with water (or another appropriate fluid) flowing through the array of microchannels.

1.2.2 Micro-nozzles and Micro-thruster

Micro/nano satellite technology has attracted increasing attention worldwide due to its desirable features of small size, light weight, low cost, short development cycle, excellent performance, and rapid launch. The existing propulsion systems

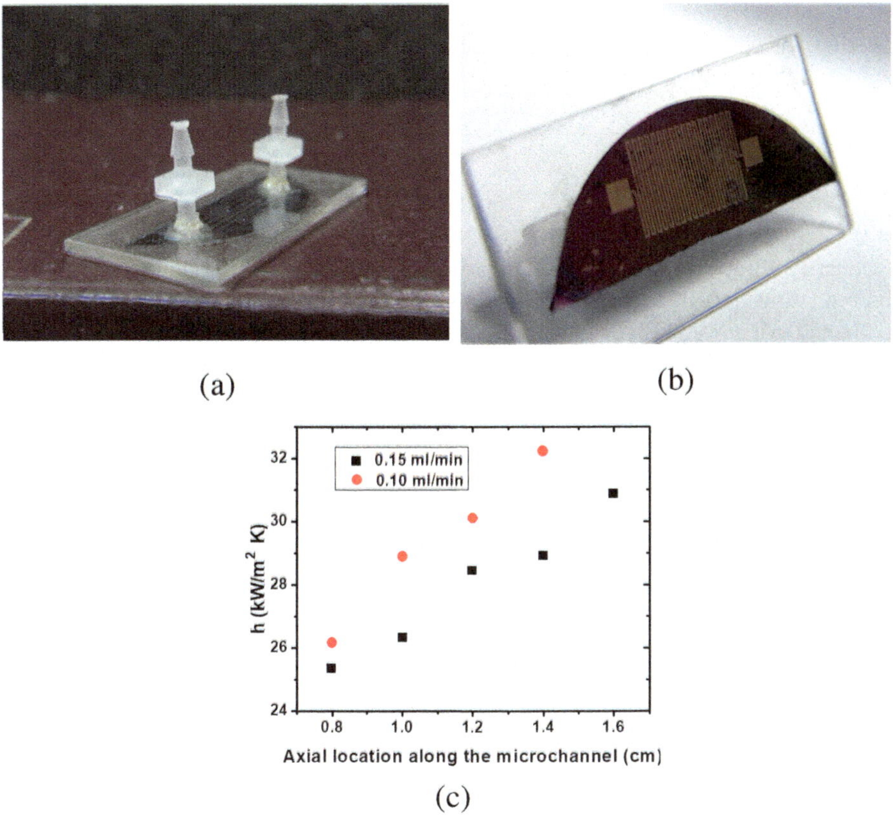

Fig. 1.1 (**a**) Microchannel made on silicon wafer and bonded with quartz. There are four parallel trapezoidal cross-section microchannels with dimensions of 78 μm (depth) by 273 μm (upper width) (159 μm lower width) by 20 mm (length), yielding a hydraulic diameter of 108.8 μm. Connectors for bringing in and taking out fluid from the microchannel are also seen in the figure [135]. (**b**) Microheater on the other side of the silicon wafer to supply a known amount of heat to the flowing fluid [135]. (**c**) Variation of heat transfer coefficient with position in the microchannel at two different flow rates of water [134]

and thrusters cannot meet the demands of micro-satellites due to the limitations in their size, mass, and thrust. Therefore, there is a need to develop micro-thrusters suitable for micro-satellites having features of low mass, small size, precise thrust, and low power consumption. Micro-thrusters are generally used in spacecraft that require small amounts of thrust, and can be classified as [10]: (1) mesoscopic dimensions (2) microscopic segments, and (3) nanoscopic structures. A micro-thruster generates 1–10 mN force and can be used for propulsion and attitude control in small space satellites. A smaller micro-thruster generating 10–1000 μN force is used for dynamic suppression/damping of vibrations in extended space structures. In the operation of the micro-thruster, micro-nozzle forms the basic component and therefore, fluid flow in micro-nozzles is of interest. Further, it is expected that the basic understanding of fluid flow characteristics in these relatively complex geometries will be beneficial in the design of other micro-devices for scientific applications.

1.2.3 Breath Analyser

Generally, three types of biological media are used to assess the human health, i.e. breath, urine, and blood. Working with breath is obviously advantageous as it can be collected in a non-invasive manner obviating the need of a trained clinical personnel, while also being unlimited in timing and volume. Breath analyser can be used to determine the human metabolic rate, estimate the blood alcohol content, diagnose diseases (such as lung cancer), and provide information on physical fitness. Usually three types of breath matrices are used to assess the human health: exhaled breath, exhaled breath condensate, and exhaled breath aerosol. Exhaled breath analyzer has received much attention in the field of medical diagnostic technology due to the simple procedure in sampling analytes with sufficient information for further diagnosis [163].

In order to assess the human health, a wide variety of sample collection and detection methods are available. These sample collection methods include polymer bags, canisters, sorbent tubes, exhaled breath condensate (EBC) collection and exhaled breath aerosol (EBA) collection, while detection methods include sensors, mass spectrometry, and immunoassay instrumentation. Currently, different types of sensors and mass spectrometry instruments are available for breath analysis. Extensive research is currently being done on several breath-relevant topics including security and airport surveillance, cellular respiration, and canine olfaction. In this context, several breath-relevant microdevices have already been developed [106]; such as for detection of volatile organic compounds in breath; for detection of nitric oxide in exhaled breath (a marker of asthma), and halitosis (or bad breath) caused by ammonia, amines, and certain sulfur gases.

1.2.4 Microdevice for Conducting Blood Test

Blood test is one of the most routinely undertaken procedures when a person is unwell. This is because the physiological condition of a person can be easily assessed by examining the concentration level of various analytes present in blood (particularly in plasma). The ability to perform blood test at the site of the patient itself—rather than the patient going to the hospital for the test—would clearly be of immense value. Several companies and research groups are therefore working to make such a microdevice—with sample preparation module and sensing module, as the two primary modules of the microdevice.

The sample preparation module is required because almost 45% of human blood comprises cellular elements (solid components in blood), which needs to be discarded before conducting the test. Red blood cells form the most important cellular component, while white blood cells and platelets are the next major cellular components. The liquid component of blood (plasma) contains a plethora of biomarkers and it is therefore vital to extract the plasma from blood for conducting the tests. The conventional method of plasma separation is through centrifugation, which is tedious, time consuming, requires human interface, and involves multiple stages of sample handling. Integrating a centrifuge with a microdevice is not easy and removes the advantages of microdevices (small sample size, portable, etc.). This requirement of separating plasma from blood in a simple and cost-efficient manner and without the use of a centrifuge has led research groups to develop technologies on microfluidics platform.

An easy-to-use on-chip microdevice for extraction of high-quality plasma, developed at the Indian Institute of Technology Bombay, is shown in Fig. 1.2a, b. Note that several of the physical effects on which the separation principle is based kick-

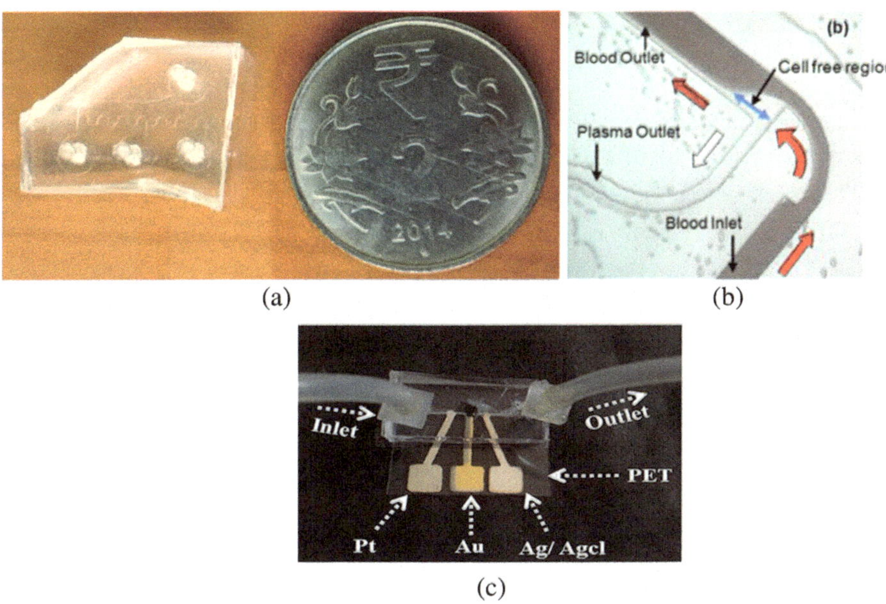

Fig. 1.2 (**a**) Microchannel for separating plasma from blood developed at the Indian Institute of Technology Bombay. A two-rupee coin has been placed next to the fabricated device for comparison. (**b**) Close-up view of flow of blood through the microdevice, with plasma passing through one branch and the remaining blood passing through the other branch [118]. (**c**) Optical image of fabricated microfluidic devices embedded with a three electrode system consisting of Platinum (Pt) counter electrode, Silver/silver chloride (Ag/AgCl) reference electrode, and Gold (Au) working electrode on a PET substrate. Working electrode Au is functionalized with C-ZNO NFs. Finally, the AntiHRP II was covalently immobilized to the nanofiber integrated sensing surface for detection of Malaria parasite [111]

in only at the microscale. Figure 1.2c shows a micro-sensor for detecting malaria, developed at the Indian Institute of Technology Hyderabad. Successful integration and field-testing of the components of the microdevices can revolutionalize the way blood tests are done in the future.

Examples 2 and 3 given above involve flow of gases through microdevices, which is the subject of this book. We will briefly comment on flow in diverging and converging passages with relevance to design of micro-nozzle and micro-thruster, and flow in other complex passages with relevance to microdevice for breath analyser, in Sect. 2.6.

1.3 Classification of Flow Regimes

The length scale of micro-scale systems is in micro-meter (or μm). Due to such small length scale, it is possible that the mean free path of the gas (λ) becomes comparable to the length scale (L). Formally, this possibility is quantified by defining a parameter Knudsen number as

$$Kn = \frac{\lambda}{L}. \tag{1.1}$$

In situations (such as external flows) where definition of L is not obvious, one can employ spatial gradient of a relevant parameter (ϕ, such as density) for defining the length scale as

$$L^{-1} = \frac{d\phi/dx}{\phi}. \tag{1.2}$$

As an example, consider flow of air in a centimeter size passage. The Knudsen number for this case is 7×10^{-6} since the mean free path of air under standard condition is about 70 nm. Therefore, $Kn \to 0$ at the conventional scale because of relatively large value of L, whereas it can be easily verified that Kn becomes non-negligible ($> 10^{-3}$) for micro-scale flows. The presence of Knudsen number as an additional parameter leads to the interesting possibility of finding new flow physics in such microscale flow systems, as discussed further in this book.

The Knudsen number can be used to demarcate the flow into four regimes: continuum ($Kn < 10^{-3}$), slip ($10^{-3} < Kn < 10^{-1}$), transition ($10^{-1} < Kn < 10^{1}$), and free-molecular ($Kn > 10$) (Table 1.1). The definition of the various regimes is therefore based on order of magnitude of the Knudsen number and should only be taken as indicative. Note that slip and transition regimes are sandwiched between two major regimes of continuum and free-molecular. Both these intermediate regimes span approximately two decades each. Some references refer to Kn between 0.01 and 0.1 as the slip regime. However since slip effect starts to show up for $Kn > 0.001$, $Kn = 0.001$ is taken here as the beginning of the slip regime, as also done in several other sources.

Table 1.1 Classification of flow regimes and models

Knudsen number	Regime	Fluid model
$Kn < 10^{-3}$	Continuum	Navier–Stokes equations with no slip boundary conditions
$10^{-3} < Kn < 10^{-1}$	Slip flow	Navier–Stokes equations with velocity slip and temperature jump boundary conditions
$10^{-1} < Kn < 10$	Transition	Higher order continuum transport equations (e.g., Burnett, Grad, OBurnett, O13, etc.)
$Kn > 10$	Free molecular flow	Collisionless Boltzmann equation

In this book, we are primarily concerned with the slip and transition regimes.

1.4 Characteristics of Microscale Flows

At the microscale, the interaction between the fluid and solid wall is different than that at the macroscale. Fundamental studies have shown that the continuum hypothesis may not be valid at microscale and some specific effects may be present that can alter the fluid flow and heat transfer characteristics significantly. At the microscale, surface area to volume ratio is much larger, which results in increased surface forces, which may produce large pressure drops, compressibility effects, and viscous dissipation; decreased inertial forces, which allows diffusion and conduction processes to become relatively more significant; and increased heat transfer, which may lead to variable fluid properties and creep flow. Therefore, these effects need to be considered to predict the fluid flow and heat transfer characteristics. A detailed discussion of these effects is presented in the following sections.

1.4.1 Rarefaction

Rarefaction refers to the reduction of molecular density in the fluid flow medium and usually, it is encountered in micro and nano-scale flows. In such flows, the molecular mean free path may be of the same order as that of the characteristic dimension of the system. Microscale flows are characterized by Knudsen number (Eq. (1.1)). A larger Knudsen number arises either due to longer mean free path (rarefied gas flows) or small characteristic dimension of the system (microdevices). The Knudsen number is used to demarcate the fluid flow into four regimes: continuum regime

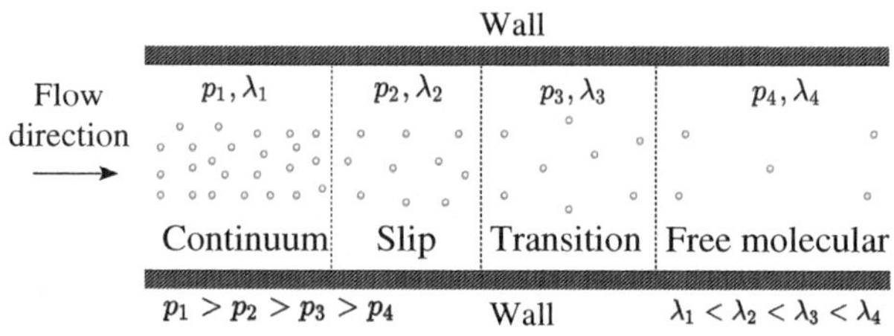

Fig. 1.3 With mean free path increasing due to reduction in pressure, the flow transits through various regimes. (Here p denotes pressure and λ denotes mean free path and hollow circles represent the molecules in random motion)

$Kn < 10^{-3}$, slip regime $10^{-3} < Kn < 10^{-1}$, transition regime $10^{-1} < Kn < 10$, and free molecular regime $Kn > 10$. These regimes can occur in isolation, or as shown in Fig. 1.3 can even co-exist in a given passage.

For very low Knudsen number, the number of collisions between the molecules is large compared to the number of collisions between the molecules and the wall. In such a case, the usual continuum concept is applicable and the Navier–Stokes equations and the Fourier heat conduction law are valid. The continuum flow is characterized by the Reynolds number and the Mach number only; the Knudsen number will not enter the problem explicitly since it is already considered to be very small.

For sufficiently large Knudsen number, the continuum concept needs to be modified to predict the fluid flow and heat transfer characteristics. In such a case, first- (or second-) order slip boundary condition needs to be considered in the slip and early transition regimes. The Navier–Stokes equations with first-order slip boundary condition are relatively accurate up to $Kn \approx 0.1$, beyond which second-order slip boundary conditions are more appropriate. At still higher Knudsen numbers, higher-order continuum transport equations need to be considered to analyze the fluid flow and heat transfer behavior.

Rarefaction affects both fluid flow and heat transfer characteristics significantly. It is responsible for a decrease in friction factor and Nusselt number. With the increase in Kn, the Nusselt number reduces due to the increased temperature jump at the wall.

1.4.2 Compressibility

Compressibility effect leads to density becoming a variable in the flow (Fig. 1.4), and therefore addition of another variable occurs in the problem. It leads to the linking of the momentum equation with the energy equation. Further, certain terms (such as pressure work) may have to be retained in the energy equation in this case. Therefore, the complexity of the problem increases substantially.

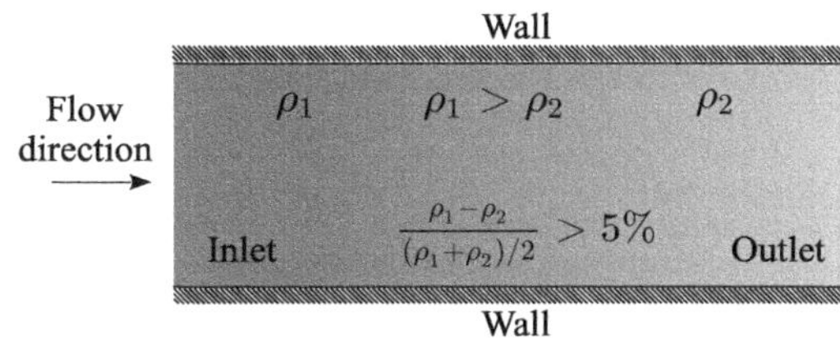

Fig. 1.4 Shading represents the gradient of density; when density difference is large, compressibility effects become important

Usually, the effect of compressibility is considered when the Mach number exceeds the value of 0.3, otherwise the flow is treated as incompressible. However, $Ma > 0.3$ criterion is only a sufficient condition for flow to be considered compressible. In case of flow through microdevices, the pressure drops rapidly along the flow direction because of viscous effect, even though the velocity may not be high enough for the Mach number to exceed the threshold of 0.3. The gas density would therefore change and these variations alter both the velocity and temperature profiles. Hence, the fluid flow and heat transfer behavior are affected significantly by compressibility effect.

The Mach number is related to the Knudsen number and Reynolds number through the relation

$$Kn = \sqrt{\frac{\pi\gamma}{2}}\frac{Ma}{Re} \tag{1.3}$$

obtained from the classical kinetic theory of gases where γ denotes the ratio of specific heats. That is, out of these three relevant non-dimensional numbers, only two of them can be independently specified in any problem.

1.4.3 Thermal Creep

In case of microflows, the slip velocity is related to the velocity gradient at the wall. Additionally, if a temperature gradient exists in the flow, the slip velocity will be altered as depicted in Fig. 1.5. The slip velocity in terms of the above effects is given by the following expression

$$u_s - u_w = \frac{2-\sigma_v}{\sigma_v}\lambda\left.\frac{\partial u}{\partial y}\right|_{wall} + \frac{3}{4}\frac{\mu}{\rho T}\left.\frac{\partial T}{dx}\right|_{wall} \tag{1.4}$$

where the last term represents the thermal creep or transpiration effect. In the above equation, u_s is velocity of gas at the wall, u_w is wall velocity, σ_v is tangential momentum accommodation coefficient, λ is mean free path of the gas, x and y are directions parallel and normal to the wall, μ is viscosity, ρ is density, and T is absolute temperature.

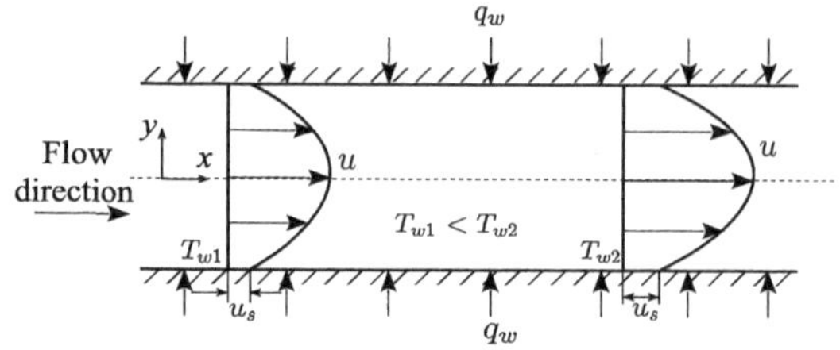

Fig. 1.5 A schematic diagram showing thermal creep phenomenon; since $T_{w1} < T_{w2}$, a macroscopic motion of gas molecules is observed from points 1 to 2 resulting in increased slip velocity at point 2

The thermal creep is defined as the macroscopic movement of rarefied gas molecules induced by a temperature gradient from lower to higher temperature zone. Note the difference in direction of gradient of the two terms. Usually, creep flow is negligible for large scale flows (since the term scales as Kn), fully developed and constant wall temperature flows, while it becomes significant for constant wall heat flux flows in the slip regime. The creep flow can alter the pressure losses and convective heat transfer rates significantly as compared to the cases where the effect of creep is negligible. The term also leads to a coupling of the momentum and energy equations.

1.4.4 Viscous Dissipation

Due to the large velocity gradient in the microchannel (brought about by small length scale), viscous dissipation (proportional to square of velocity gradient) becomes substantial (Fig. 1.6). The ratio of heat generation because of viscous dissipation and heat exchange between fluid and the wall by molecular conduction is quantified by a dimensionless parameter, known as Brinkman number. Mathematically,

$$Br = \frac{\mu u_m^2}{k(T_w - T_m)} \text{ or } \frac{\mu u_m^2}{q_w D_h} \text{ or } EcPr \tag{1.5}$$

where μ is the dynamic viscosity, u_m is the flow velocity, k is the thermal conductivity, $(T_w - T_m)$ is the temperature difference at a particular axial location, T_m is the bulk fluid temperature; T_w is the wall temperature, q_w is the wall heat flux, Pr is the Prandtl number, and Ec is the Eckert number. Note that the definition of Brinkman number varies with the type of boundary condition at the wall. A larger value of Brinkman number corresponds to larger viscous dissipation compared to the rate of heat conduction, and hence larger is the local temperature rise. The positive value of Brinkman number indicates heat transfer from wall to the fluid (i.e. heating), while the opposite is true for negative value of Brinkman number (i.e. cooling).

Although Br is usually neglected in low-speed and low-viscosity flows through conventionally sized channels of short lengths, it may become important for flows

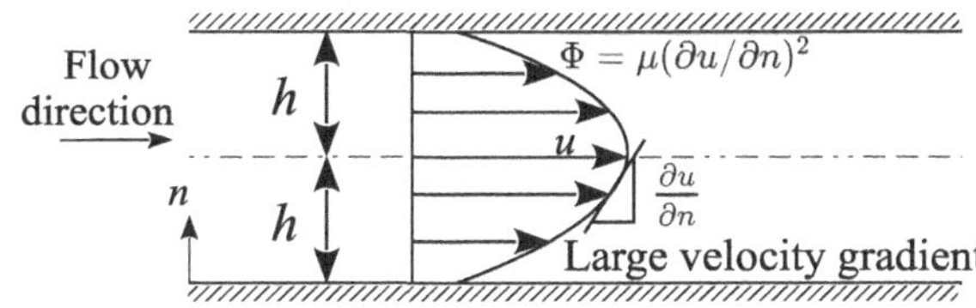

Fig. 1.6 Schematic showing the significance of viscous dissipation

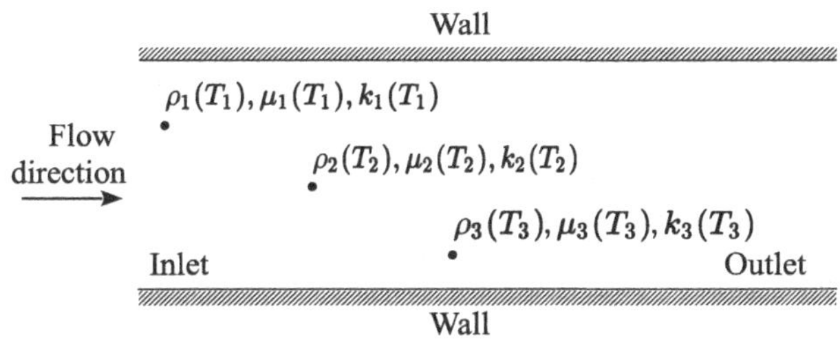

Fig. 1.7 With density (ρ), viscosity (μ), and thermal conductivity (k) being different in different regions of the flow, due to steep variation in temperature, the property variation in fluid can become substantial

through conventionally sized long passages. For flows in microchannel, the length-to-diameter ratio is much larger as compared to flows through conventionally sized long passages; thus, Br may become important in microchannels. For macroscale flows, the effect of viscous dissipation is significant either for higher viscous flows or flows with higher velocity, while for microscale flows, the effect of viscous dissipation is significant even for low velocity flow because of smaller dimensions. Similar to thermal creep, viscous dissipation causes additional coupling between the momentum and energy equations.

1.4.5 Property Variation

Since microscale flows are characterized by large temperature gradient, and thermophysical properties of fluid are a function of temperature, it can be expected that the thermophysical properties can vary in the flow as depicted in Fig. 1.7. Both the flow and heat transfer behavior can differ substantially in presence of property variation as compared to the constant property case. The property variation can affect both the developing and fully developed regions of the flow. The amount of difference is a function of Knudsen number, aspect ratio, and temperature difference between the channel inlet and walls.

Property variation is therefore relevant for both macro- and microscale flows. This effect needs to be suitably accounted for through appropriate equations for temperature-varying property over the temperature range of interest. Of course, this introduces additional variables and equations in the problem, thereby increasing the complexity.

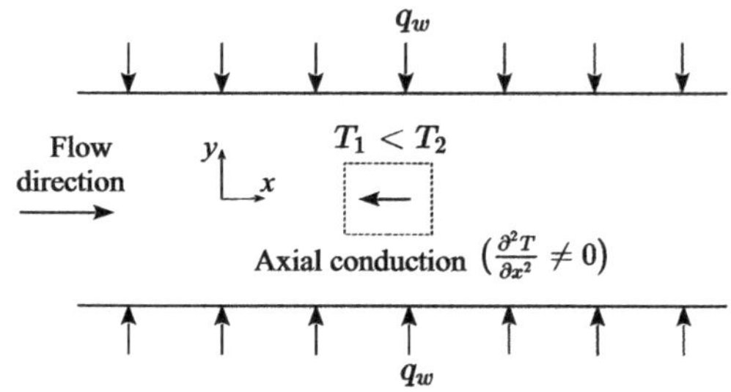

Fig. 1.8 The second derivative of temperature with respect to axial direction (representing axial conduction) is usually non-negligible at the microscale

1.4.6 Axial Conduction

In the analysis of flow at the conventional scale, it is customary to assume that axial conduction can be neglected. This simplifies the analysis by removing the $\frac{\partial^2 T}{\partial x^2}$ term (where x is the streamwise coordinate) from the equation. The effect of axial conduction is quantified by Peclet number defined as

$$Pe = \frac{\rho c_p u_m D_h}{k} = \frac{\rho u_m D_h}{\mu} \frac{\mu c_p}{k} = RePr \tag{1.6}$$

Thus, Peclet number signifies the strength of convection to the strength of conduction. In the case of macro-sized flow, because of relatively large value of Reynolds number, the Peclet number is usually large ($Pe > 150$) and the effect of axial conduction may be neglected. On the contrary, in case of microscale flow with smaller channel dimensions and Reynolds number, the effect of axial conduction becomes much more significant (Fig. 1.8). Therefore, such an assumption may not be justifiable at the microscale.

1.4.7 Conjugate Heat Transfer

Normally it is assumed that heat transfer to the fluid is not affected by heat transfer inside the solid walls surrounding the fluid. That is, the heat supplied to the walls is faithfully transferred to the fluid and any heat transfer in the walls is insignificant and therefore need not be accounted for. However, if the wall thickness becomes comparable to the flow passage length scale (a situation applicable to microscale flows) as shown in Fig. 1.9, this assumption ceases to apply. The implication of the relaxation of this assumption can be quite severe in that the boundary condition applied on the outer wall of the passage is not transmitted faithfully to the inner wall of the passage (the difference being due to heat conduction in the wall). In other words, the boundary condition seen by the fluid is not the boundary condition that is applied in the problem. Since the boundary condition has a strong impact on

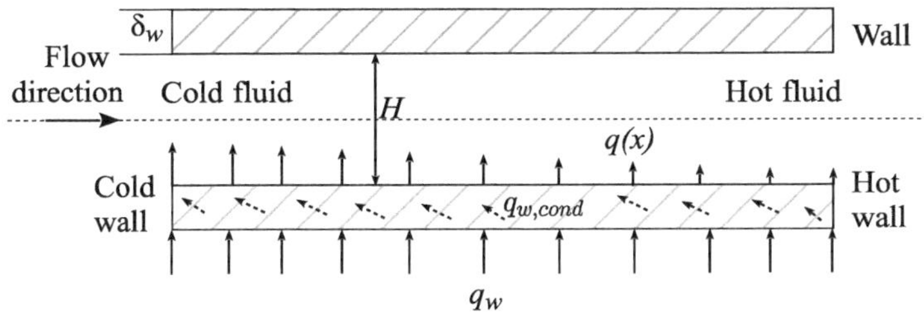

Fig. 1.9 A schematic showing the conjugate effect at microscale where the wall thickness δ_w is comparable to the channel height H. The constant heat flux applied at the outer wall, q_w and the heat flux at the inside the wall, $q(x)$ are different because of the heat conduction taking place in the wall thickness region

the obtained solution, this conjugate effect needs to be suitably accounted for in the analysis.

Conjugate effects are more pronounced at microscale as compared to macroscale because the hydraulic dimension is comparable to the channel thickness in the former case. Therefore, the standard correlations for idealized constant wall temperature or uniform wall heat flux boundary conditions have to be applied with more care. The conjugate effect can be quantified by a parameter defined as

$$M = \frac{\psi_{cond}}{\psi_{conv}} \tag{1.7}$$

where ψ_{cond} is a heat flux characterizing axial heat transfer in the wall, ψ_{conv} is the total convective heat flux. The quantity M is therefore the ratio of conduction in the wall to the heat transferred as convection from the surface. For $M > 0.01$, the conjugate effect cannot be neglected. Problems involving substantial conjugate effect therefore requires simultaneous solution in both the solid and fluid domains. The coupling of the solid and fluid domains obviously increases the complexity.

Other than the aforementioned generic characteristics of microscale flows, some specific characteristics of microscale flows are discussed in Chaps. 2 and 3.

1.5 Mathematical Modelling of Microscale Flows

After examining the characteristics of microscale flows, we discuss ways to mathematically model microscale flows. We first provide a brief introduction to the Navier–Stokes equations, which are used for modelling conventional scale flows, followed by two examples illustrating the need for altering the boundary conditions and even the governing equations for modelling high Knudsen number flows.

1.5.1 The Navier–Stokes Equations

The Navier–Stokes equations are used to model the flow of fluids at the conventional scale. The Navier–Stokes equations are mathematical representation of the physical principle on conservation of momentum, as applied to a differential element in the flow. The left-hand side of the equation represents the rate of change of momentum of fluid in the control volume, while the right-hand side represents the external forces acting on the fluid. (The mathematical form of these equations is provided in the next chapter.) The external forces are due to gravity or electric/magnetic field (body forces), pressure or viscous effects (surface forces), surface tension (line force), etc. While the body and surface forces are part of the governing equation, the line force is generally accounted for through the boundary condition. Note that the body forces scale as L^3 (where L is the characteristic dimension), while the surface and line forces scale as L^2 and L^1, respectively. Since $L^3 \ll L^2$ for small values of L, the body force (gravity, etc.) typically has only a minor role to play at the microscale. That is, the orientation of the microdevice is not expected to significantly alter the performance of the microdevice.

Being differential equations, the Navier–Stokes equations require specification of initial and boundary conditions for their complete solution. The most important boundary condition of concern here is that at the fluid–wall interface. The no-slip boundary condition, whereby it is assumed that the tangential velocity of the fluid at the wall is equal to the velocity of the wall, is normally imposed in conventional scale flows. The fluid is therefore not allowed to slip on the wall (similarly, temperature of the fluid at the wall is assumed equal to the wall temperature, in case this is required during solution of the energy equation). Further, the wall-normal velocity component is zero for a non-porous wall.

Although the Navier–Stokes equations refer to the momentum equations, in this book we have used the term "Navier–Stokes equations" to refer to the conventional mass, momentum, and energy equations taken together.

1.5.2 Limitations of Conventional Equations and Boundary Conditions

In this section, we show that the solution for flow in the slip and transition regimes differs substantially from its continuum regime counterpart. Towards this, we examine two problems for which obtaining solution is relatively easy. We then give an overview of approaches available to model the flow in the slip and transition regimes.

Upon imposing a uniform shear to the flow (for example, through the movement of one large plate with respect to another large stationary plate; the sides of the plates being large with respect to the normal distance between the plates), a linear velocity profile is expected to be set up in the flow (Fig. 1.10a). This expectation is

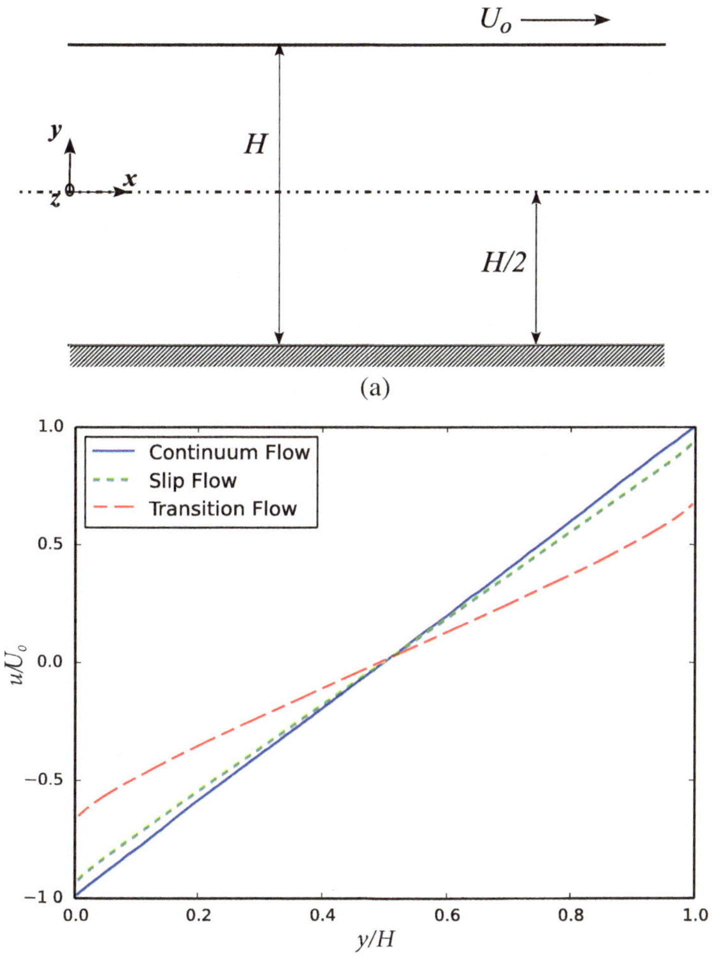

(b)

Fig. 1.10 (**a**) Schematic of flow between two parallel plates, upper plate is moving at speed U_o in the positive x-direction while lower plate is stationary. (**b**) Velocity profile in different flow regimes obtained using DSMC (We are grateful to Mr. Abhimanyu Gavasane for these computations)

in line with the observation at the conventional scale (distance between the plates is much larger than the mean free path of the gas in the gap, or $Kn \approx 0$). However, upon reducing the distance between the plates or increasing the mean free path of the gas (for example, by reducing the pressure), the velocity starts to deviate from the classical profile (Fig. 1.10b). In particular, a finite velocity slip is observed at the plates at $Kn \approx 10^{-3}$. Further, the velocity profile becomes non-linear in the near plate region when the distance between the plates becomes comparable to the mean free path of the gas or $Kn \approx 10^{-1}$.

The above observation points to modelling two important aspects of the flow present at finite Knudsen numbers ($Kn > 10^{-3}$): velocity slip at the wall and non-linearity in the velocity profile in the near plate region. Presence of wall-slip suggests that the no-slip boundary condition has to be relaxed at finite Knudsen numbers; we will examine several slip models in Sect. 2.3. It turns out that in most situations, we can model flow in a large part of the slip regime through the use of the Navier–Stokes equations but with modification of the boundary condition from no-slip to slip. The non-linear nature of the velocity profile at still higher Knudsen numbers, however, cannot be modelled within the framework of the Navier–Stokes equations. This is true for the present problem even after accounting for property variation, for example, due to variation in temperature if any.

A similar observation is noted for flow between two large parallel plates, with the flow being driven by a pressure gradient between the inlet and outlet (Fig. 1.11a). In this situation, the velocity profile is expected to be parabolic with a constant pressure along the wall-normal direction; these are supported by observations. However, with an increase in the Knudsen number, the flow starts to slip at the wall in the slip regime (Fig. 1.11b). Besides the wall-slip, a non-linear pressure variation is observed along the wall-normal direction in the transition regime (Fig. 1.11c). As before, the first observation can be modelled through appropriate wall-slip models, while the latter observation is beyond the modelling capability of the Navier–Stokes equations.

1.5.3 Approach to Modelling Microscale Flows

The two simple flow problems presented in the previous section bring us to the interesting question: How do we model the flow at high Knudsen number? These examples have exposed the limitation of the linear constitutive relation (relation between stress and strain rate) involved in the classical Navier–Stokes equations. A non-linear constitutive relation applies to non-Newtonian fluids; however, the behavior of non-Newtonian fluids is not dependent on the value of the Knudsen number, implying that non-Newtonian fluid models are not applicable in the present scenario. Other modifications to the Navier–Stokes equations have also not been successful as discussed in Chap. 4. We have to rather consider equations containing terms of higher order in Knudsen number, termed as "*higher order continuum transport equations*." The considered equations should obviously reduce to the Navier–Stokes equations in the limit of low Knudsen number. It can be anticipated that the *higher order continuum transport equations* will contain additional terms as compared to the Navier–Stokes equations, with the additional terms being of the order Knudsen number squared (or higher). Also the non-linear nature of the constitutive equation will make these models much more involved to work with.

It is however not obvious how to obtain higher-order models from the conventional method of deriving the transport equations. Although there are several distinct approaches in the conventional method to arrive at the transport equations,

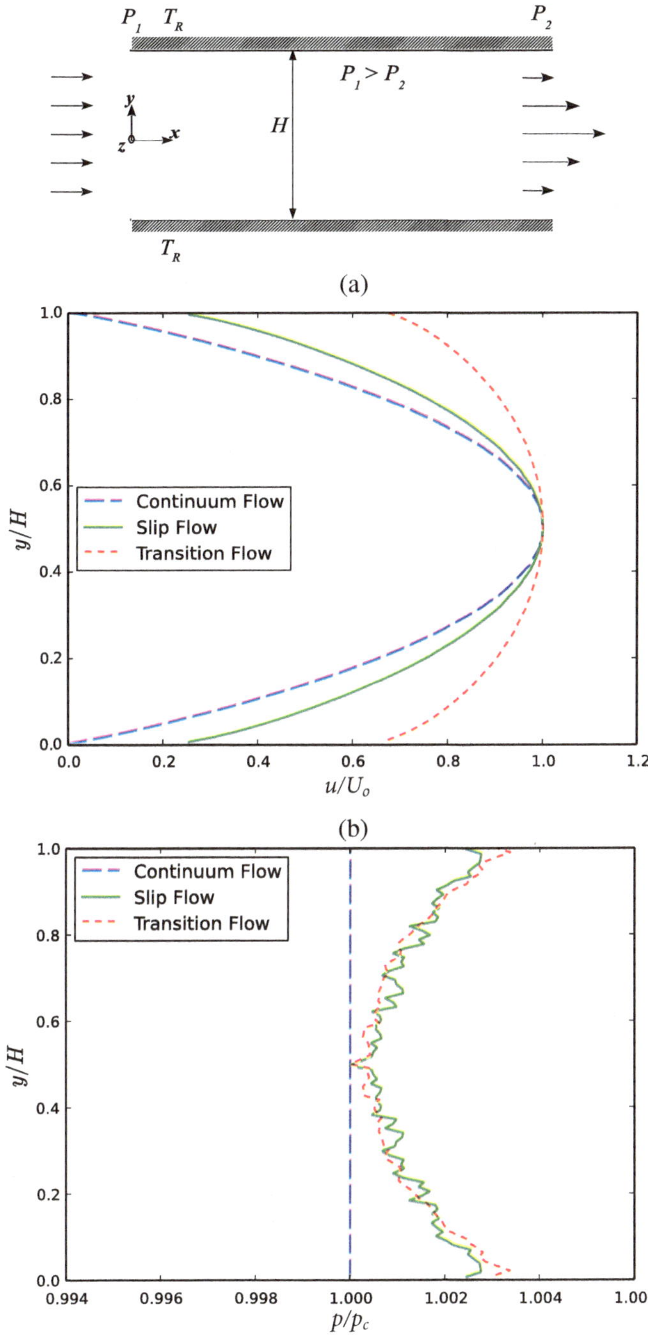

Fig. 1.11 (**a**) Schematic of flow between two parallel plates, the flow being driven by a pressure gradient. (**b**) Velocity and (**c**) pressure profiles across the channel height in different flow regimes obtained using DSMC, profiles are non-dimensionalized using the centerline values. (We are grateful to Mr. Abhimanyu Gavasane for these computations)

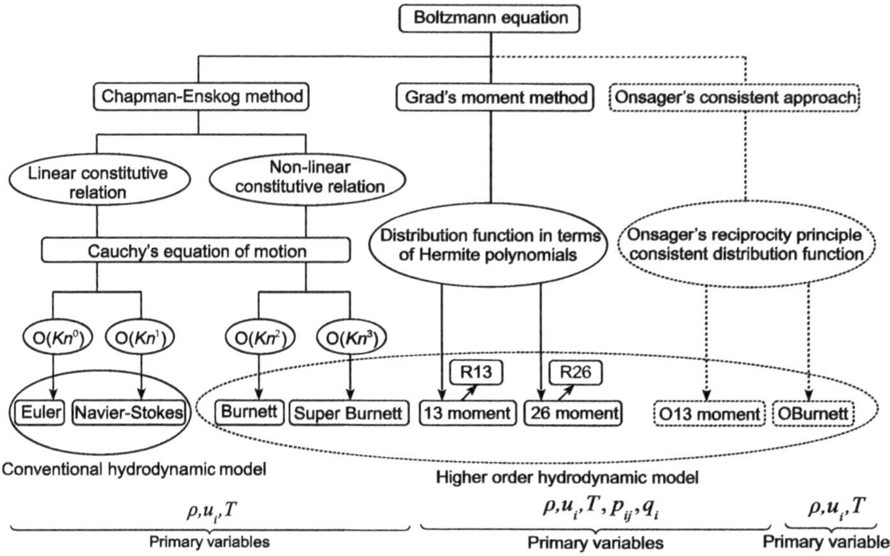

Fig. 1.12 Different schemes for modelling of fluid flow

no extra term that could have been retained is seen in any of these derivation procedures. Therefore, these conventional approaches cannot be extended to obtain higher-order models. The alternative then is to change the starting point and resort to the Boltzmann equation. The conservation equations for mass, momentum, and energy can be obtained easily from the Boltzmann equation. Interestingly, as will be shown later in Sect. 5.2.4, the exact form of the distribution function (solution of the Boltzmann equation) will not even be needed to arrive at the conservation equations. However, the problem of closure shows up at this stage. This problem of closure is same as that seen with the conventional method. An empirically obtained relation between stress and strain rate (through Newton's law of viscosity), and heat flux and temperature gradient (through Fourier's law of heat conduction) are supplied in the conventional method to close the system of equations. This way of closure is however not followed in the alternative method which makes this Boltzmann equation based approach distinct from the conventional method.

There are several approaches of deriving transport equations starting from the Boltzmann equation. An essential requirement of these approaches is that they should faithfully reproduce the known equations (the Navier–Stokes equations) at the lower-order, before the higher-order equations can be trusted. The two successful methods for solving the Boltzmann equation (and considered here) are the Chapman-Enskog expansion and the Grad moment method (Fig. 1.12). Both these methods provide a way to evaluate the distribution function, which can then be used to close the equations.

In the Chapman-Enskog method, the distribution function is expanded in a perturbation series around equilibrium (i.e., Maxwell-Boltzmann distribution function corresponding to gas at rest). The method leads to both linear and non-linear constitutive relations depending on where the expansion is terminated. The linear constitutive relations are of Knudsen number order one, and substitution of the constitutive model in the conservation equation leads to the Navier–Stokes and energy equations. The non-linear terms are of Knudsen number order two (or higher); such equations lead to higher-order models, which are known as the Burnett (or super Burnett) equations. The inviscid Euler equations can also be obtained using this approach. A detailed derivation of the Burnett equations is given in Chap. 5.

Note that the number of independent variables in all the above equations is five. These variables are density (ρ), velocity vector (v_i), and temperature (T). (Note that tensor notation has been used extensively throughout the book and an introduction to tensors is included in Appendix A.) In contrast, the moment method does not seek a relation between stress and strain rate, rather treats stress (σ_{ij}) and heat flux (q_i) as independent variables. The number of independent variables therefore increases to 13 (or even 20 or 26) as separate evolution equations are written for each of these variables. The problem of closure is then transferred to the unknown terms appearing in the stress and heat flux evolution equations. Distribution function, written in terms of Hermite polynomials, is finally employed to obtain closure. The Grad 13-moment and 20-moment equations are discussed in detail in Chap. 6.

Some variants of the Burnett and Grad equations and other alternative approaches are introduced in Chap. 7. Equations obtained from a recent approach, which employs a thermodynamically consistent distribution function, with the approach being very different from the two aforementioned classical approaches, are also presented in Chap. 7. This approach, although promising, still needs to be established and is hence shown by dotted line in Fig. 1.12.

1.6 Relevance and Scope of the Book

The book considers several topics that have not been dealt with adequately in the past. For example, while some researchers know that the Navier–Stokes equations can be derived from the Boltzmann equation, how exactly can it be derived is not so well understood. Here, we provide the derivation in a relatively simple way. Such an understanding may also help better appreciate numerical methods such as the Lattice Boltzmann method, which is based on the Boltzmann equation and Chapman-Enskog expansion; the Lattice Boltzmann method has become very popular and is being widely used for solving flow problems.

Similarly, while the need for looking beyond the Navier–Stokes equations (especially for cases where the continuum hypothesis is violated) has already been felt, the alternate set of equations to be solved is not known. Microscale flows put the need for having better higher-order transport equations in perspective. The subject of higher-order continuum transport equations is still under development, with research groups coming up with their own versions of the equations. It is generally difficult to follow all the physical and mathematical details behind these equations, due to the inherent complexity involved in the process of derivation. For people not so well versed with this line of investigation, it becomes further difficult to appreciate these new developments. The book attempts to fill this gap by exposing the required details behind the idea and derivation in relatively simple terms. The hope is that the reader will be better prepared to follow the newly proposed equations, and even contribute towards their derivation.

Further, we realize that the various available equations need to be rigorously tested for their accuracy by numerous people by applying them to different problems. Obviously, this can happen only after a certain level of comfort has been achieved with the available equations.

We also compile and discuss the few known exact solutions of the Navier–Stokes and higher-order continuum transport equations for problems in the slip regime. To the best of the authors' knowledge, such a complete compilation (especially for higher-order continuum transport equations) is not previously available.

The book is therefore organized in two parts, with the first part dealing with physical aspects of microscale flows and heat transfer. The second part deals with the mathematical modelling of microscale (and other high Knudsen number) flows, and application of obtained equations to a few simple problems.

In Chap. 2, we discuss the various fluid flow solutions within the framework of the Navier–Stokes equations and slip boundary condition. The slip model is derived followed by a survey of the measured values of the slip coefficients. Analytical solution for a few useful cases is derived in reasonable detail, and a discussion of the obtained solution provides pertinent insights into the fluid flow behavior. A compilation of some interesting known observations in straight as well as complex passages is also provided. This is followed by a compilation of some useful correlations.

In Chap. 3, we consider the energy equation along with temperature jump condition at the wall. We consider analytical solution for problems involving temperature jump and viscous dissipation, while the effect of other complicating parameters is only briefly commented upon. The objective therefore is not to provide all known solutions of microscale flows, rather discuss some typical effects at the microscale. Our approach is similar to what is followed in the literature, in that a single (or a few, but not all) complexities are dealt with in a given problem; however, since experiments and practical situations have many more complexities, we find that there is a further need for theoretical development of the subject.

In Chap. 4, we compile problems for which the Navier–Stokes equations are found to be inadequate. Since such a compilation is being attempted for the first time, it is likely to be incomplete. We consider some known extensions of the Navier–Stokes equations; also a popular non-Fourier model is discussed. We argue that these extensions are still inadequate for studying problems in the transition regime, and we have to rather resort to an entirely different approach.

Accordingly, the Burnett equations are introduced in Chap. 5. We understand that the awareness and appreciation of these equations is rather limited. The derivation is therefore explained in detail, and further summarized through a flow chart. An order of magnitude analysis and stability analysis of the Burnett equations are also presented. A novel iterative approach to solve these equations, which can further be applied to other partial differential equations, and other recent solutions of the Burnett equations are also provided, and results are compared against DSMC data. The Burnett equations in cylindrical coordinates are also provided for ready reference.

The Grad equations are next introduced in Chap. 6. The derivation is again stepwise and aided with a flow chart, with most of the derivation being sufficiently original. The required boundary conditions are obtained next. The linearized Grad equations are then solved for an impulsively started plate problem, and a comparison of the solution with both the Navier–Stokes and kinetic equations provides useful insights. The problem of cylindrical Couette flow is also solved using the linearized Grad equations. There are only a handful of examples where analytical solution of the Grad equations are available, making this compilation original and useful.

The strengths and limitations of the Burnett and Grad equations are discussed in Chap. 7. This helps appreciate the need for having variants of these equations. Some other promising recent works on higher-order continuum transport equations are also covered in this chapter. A comparison of the available variants for two problems is also provided. Several other examples where variants are compared are available in the literature, which can be further referred to.

It is recognized that the analytical approach adopted in the earlier chapters is not always sufficient to understand flows at the microscale. Therefore, a brief introduction to the relevant numerical and experimental techniques is provided in Chap. 8. Some broad conclusions and scope for further work are documented in Chap. 9.

The book therefore aims to expound on the influence of additional physical effects present at the microscale on the fluid flow and heat transfer aspects of gas flow in microchannel. The book also exposes the derivation and application of higher-order models, making them accessible to a wide readership.

Chapter 2
Microscale Flows

In this chapter, fundamentals of fluid flow and flow modelling aspects in the slip regime have been presented. This includes a brief introduction to the Navier–Stokes equations and the slip boundary condition. A few analytical solutions for flow in the slip regime are then obtained. Finally, we briefly examine the flow in some complex passages and present some useful empirical correlations. These solutions and insights help appreciate the flow physics in the slip regime better.

2.1 Introduction

At the microscale, the interaction between the fluid and solid wall is different than that at the macro-scale, due to the relatively large value of Knudsen number. It is readily apparent from Eq. (1.1) that a larger Knudsen number can arise either due to long mean free path (rarefied gas flows) or for a small characteristic dimension of the system (micro-devices) or both. In this chapter, we present the concept of velocity slip along with the factors affecting slip and the value of the slip coefficient. Further, the solutions of the Navier–Stokes equations in the slip regime and implications of slip on the flow behavior are also presented. We also briefly discuss some experimental results for flows in relatively complex passages.

While numerous studies are available on the fundamental properties of fluid flow at the macroscale, relatively few resources are available for the microscale. Such resources are important since microscale flows occur in a variety of practical applications operating in the slip regime. These applications include micro-motors, micro-actuators, micro-turbines, micro-sensors, micro-reactors, micro-valves, micro-pumps to name a few. Such resources are also relevant for rarefied gas flows and particularly low-pressure applications such as high-altitude aircraft flight and vacuum devices.

A. Agrawal et al., *Microscale Flow and Heat Transfer*, Mechanical Engineering Series, https://doi.org/10.1007/978-3-030-10662-1_2

2.2 Governing Equations for Fluid Flow

Governing equations for fluid flow can be derived from *first principle*, based on the physical principles for conservation of mass, momentum, and energy [102]. The equations can be cast in either integral or differential form, corresponding to finite- and infinitesimal-size control volume, respectively. Only the differential form of mass (continuity), momentum (Navier–Stokes), and energy equations are given here.

2.2.1 *Tensorial Form*

The continuity equation is

$$\frac{\partial \rho}{\partial t} + \frac{\partial}{\partial x_i}(\rho u_i) = 0 \tag{2.1}$$

The equation of motion for a Newtonian fluid is obtained by substituting the following constitutive equation,

$$\tau_{ij} = -p\delta_{ij} + \mu\left(\frac{\partial u_i}{\partial x_j} + \frac{\partial u_j}{\partial x_i}\right) + \mu_2 \frac{\partial u_k}{\partial x_k}\delta_{ij} \tag{2.2}$$

into the Cauchy's equation of motion,

$$\rho\left(\frac{\partial u_j}{\partial t} + u_i \frac{\partial u_j}{\partial x_i}\right) = \frac{\partial \tau_{ij}}{\partial x_i} + \rho g_j \tag{2.3}$$

to obtain

$$\rho\left(\frac{\partial u_j}{\partial t} + u_i \frac{\partial u_j}{\partial x_i}\right) = -\frac{\partial p}{\partial x_j} + \frac{\partial}{\partial x_i}\left[\mu\left(\frac{\partial u_j}{\partial x_i} + \frac{\partial u_i}{\partial x_j}\right) + \mu_2 \frac{\partial u_k}{\partial x_k}\delta_{ij}\right] + \rho g_j \tag{2.4}$$

Note that $(\partial p/\partial x_i)\delta_{ij} = \partial p/\partial x_j$ has been employed in the above derivation g_j is external body force per unit mass.

The second coefficient of viscosity, μ_2 is related to dynamic viscosity μ and the coefficient of bulk viscosity μ_v through $\mu_2 = (\mu_v - \frac{2}{3}\mu)$. As per Stokes' hypothesis, the coefficient of bulk viscosity is zero giving $\mu_2 = -(2/3)\mu$; this hypothesis can be shown to apply well to monatomic gases (see Sect. 5.4.2 and Eq. (5.129)). In general, the viscosities, μ and μ_v, can depend on the thermodynamic state. Indeed, μ for most fluids displays a rather strong dependence on temperature, decreasing with temperature for liquids and increasing more rapidly than $T^{1/2}$ (where T is the absolute temperature) for gases. As shown in the next section, the constitutive equations for normal and shear stresses (Eq. (2.2)) can be obtained as an analogy to generalized Hooke's law in solid mechanics.

Using $\mu_2 = -2\mu/3$ as per Stokes' hypothesis, Eq. (2.4) can be written as

$$\rho\left(\frac{\partial u_j}{\partial t}+u_i\frac{\partial u_j}{\partial x_i}\right)=-\frac{\partial p}{\partial x_j}+\frac{\partial}{\partial x_i}\left[\mu\left(\frac{\partial u_j}{\partial x_i}+\frac{\partial u_i}{\partial x_j}\right)-\frac{2}{3}\mu\frac{\partial u_k}{\partial x_k}\delta_{ij}\right]+\rho g_j \tag{2.5}$$

The thermal energy equation in tensorial notation is expressed as

$$\rho\frac{D\epsilon}{Dt}=-\frac{\partial q_k}{\partial x_k}-p\frac{\partial u_k}{\partial x_k}+\Phi \tag{2.6}$$

where ϵ is the specific internal energy and Φ is the viscous dissipation given as

$$\begin{aligned}\Phi &= 2\mu e_{ij}e_{ij}+\mu_2\left(\frac{\partial u_k}{\partial x_k}\right)^2\\ &= 2\mu e_{ij}e_{ij}-\frac{2}{3}\mu\left(\frac{\partial u_k}{\partial x_k}\right)^2\\ &= 2\mu\left[e_{ij}-\frac{1}{3}\frac{\partial u_k}{\partial x_k}\delta_{ij}\right]^2\end{aligned} \tag{2.7}$$

where $e_{ij} = \frac{1}{2}\left(\frac{\partial u_i}{\partial x_j}+\frac{\partial u_j}{\partial x_i}\right)$ is the strain rate tensor (symmetric part of velocity gradient tensor). Note that Stokes' hypothesis, $\mu_2 = -2\mu/3$ is employed in the second step. In the last step, with some algebra, complete square can be obtained which signifies that viscous dissipation, Φ is *always* positive. As evident in Eq. (2.7), viscous dissipation is proportional to dynamic viscosity and represents the irreversible conversion of kinetic energy into internal energy due to viscous effects.

Equation (2.6) states that the internal energy can increase because of heat coming into the differential element (first term on RHS is negative of divergence of heat flux vector), volume compression and heating due to viscous dissipation. Expressing heat flux vector in terms of temperature gradient ($q_i = -k\partial T/\partial x_i$) using Fourier's law of heat conduction, Eq. (2.6) becomes

$$\rho\frac{D\epsilon}{Dt}=\frac{\partial}{\partial x_k}\left(k\frac{\partial T}{\partial x_k}\right)-p\frac{\partial u_k}{\partial x_k}+\Phi \tag{2.8}$$

The system of equations (Eqs. (2.1), (2.5), and (2.8)) form an unclosed set of five equations with seven variables namely, u_i, p, ρ, T, and ϵ. Assuming the gas to be a perfect gas, we have equation of state as the sixth equation,

$$p = \rho RT \tag{2.9}$$

where R is the specific gas constant. The seventh equation can be obtained using a thermodynamic relation for a calorically perfect gas (constant specific heats) as

$$\epsilon = c_v T \tag{2.10}$$

Substituting for ϵ (Eq. (2.10)) and Φ (Eq. (2.7)) and using $\frac{D}{Dt} = \frac{\partial}{\partial t} + u_k \frac{\partial}{\partial x_k}$ to replace the material derivative term, the final form of thermal energy equation can be written as

$$\rho c_v \left(\frac{\partial T}{\partial t} + u_k \frac{\partial T}{\partial x_k} \right) = \frac{\partial}{\partial x_k} \left(k \frac{\partial T}{\partial x_k} \right) - p \frac{\partial u_k}{\partial x_k}$$
$$+ 2\mu \left[\frac{1}{2} \left(\frac{\partial u_i}{\partial x_j} + \frac{\partial u_j}{\partial x_i} \right) - \frac{1}{3} \frac{\partial u_k}{\partial x_k} \delta_{ij} \right]^2 \tag{2.11}$$

Using Eq. (2.9), one can replace ρ with p, and obtain a closed set of five equations with u_i, p, and T as the dependent field variables. To obtain solutions for these variables, second-order coupled non-linear partial differential equations (2.1), (2.5), and (2.11) need to be solved along with specification of appropriate number of boundary conditions. The specification of boundary conditions at a solid wall-gas interface is discussed in Sect. 2.3.

2.2.2 Constitutive Relations

In this section, we derive the constitutive relation for Newtonian fluids. While the shear stress is more straightforward, the normal stresses are obtained in a manner analogous to that for solids [126].

As per Hooke's law, a normal stress τ_{xx} will produce a strain ε_{xx} where $\varepsilon_{xx} = \tau_{xx}/E$, with E being the modulus of elasticity. The normal stress also produces lateral strains given by $\varepsilon_{yy} = \varepsilon_{zz} = -\nu\tau_{xx}/E$ where ν is the Poisson's ratio. Therefore, in general, when τ_{xx}, τ_{yy}, and τ_{zz} are all present, ε_{xx} will be given by (since both τ_{yy} and τ_{zz} will also produce a lateral strain)

$$\begin{aligned} \varepsilon_{xx} &= \frac{\tau_{xx}}{E} - \nu \left(\frac{\tau_{yy}}{E} + \frac{\tau_{zz}}{E} \right) \\ &= (1+\nu)\frac{\tau_{xx}}{E} - \nu \left(\frac{\tau_{xx}}{E} + \frac{\tau_{yy}}{E} + \frac{\tau_{zz}}{E} \right) \\ &= (1+\nu)\frac{\tau_{xx}}{E} - 3\nu\frac{\bar{\tau}}{E} \end{aligned} \tag{2.12}$$

where $\bar{\tau} = (\tau_{xx} + \tau_{yy} + \tau_{zz})/3$. Equation (2.12) can be rewritten as

$$\tau_{xx} = \frac{E}{1+\nu}\varepsilon_{xx} + \frac{3\nu}{1+\nu}\bar{\tau}. \tag{2.13}$$

Since shear modulus (G) is related to the modulus of elasticity and the Poisson's ratio, as $G = \frac{E}{2(1+\nu)}$ we can write Eq. (2.13) (with the last term split into two terms) as

$$\tau_{xx} = 2G\varepsilon_{xx} + \bar{\tau} + \frac{(2\nu - 1)}{1 + \nu}\bar{\tau}. \tag{2.14}$$

Similar to equation for ε_{xx} (Eq. (2.12)), we can write equations for ε_{yy} and ε_{zz}. Summing up the three equations for the normal strains, we obtain

$$\varepsilon_{xx} + \varepsilon_{yy} + \varepsilon_{zz} = \frac{1 - 2\nu}{E}(\tau_{xx} + \tau_{yy} + \tau_{zz})$$

$$\text{or,} \quad \frac{E}{3(1 - 2\nu)}(\varepsilon_{xx} + \varepsilon_{yy} + \varepsilon_{zz}) = \bar{\tau}. \tag{2.15}$$

Substituting Eq. (2.15) for the last term in Eq. (2.14):

$$\tau_{xx} = \bar{\tau} + 2G\varepsilon_{xx} - \frac{2}{3}G(\varepsilon_{xx} + \varepsilon_{yy} + \varepsilon_{zz}) \tag{2.16}$$

which is the required equation for solids.

For applying the above relation to fluids, we replace G with μ and the strain with the strain-rate (denoted by a dot over the relevant symbol). We further identify $\dot{e}_{xx}$, $\dot{e}_{yy}$, and $\dot{e}_{zz}$ as $\partial u/\partial x$, $\partial v/\partial y$, and $\partial w/\partial z$, respectively (u, v, w are the velocity components). Finally, we put $\bar{\tau} = -p$, to obtain

$$\tau_{xx} = -p + 2\mu\frac{\partial u}{\partial x} - \frac{2}{3}\mu\left(\frac{\partial u}{\partial x} + \frac{\partial v}{\partial y} + \frac{\partial w}{\partial z}\right) \tag{2.17}$$

which completes our derivation for normal stress in fluids. Note that the above equation is same as Eq. (2.2) (for $i = 1$ and $j = 1$) with $\mu_2 = -(2/3)\mu$.

The shear stress is directly proportional to the rate of shear strain for Newtonian fluids, with viscosity as the constant of proportionality. Therefore,

$$\begin{aligned}\tau_{xy} &= \mu\dot{e}_{xy} \\ &= \mu\left(\frac{\partial u}{\partial y} + \frac{\partial v}{\partial x}\right).\end{aligned} \tag{2.18}$$

The other normal and shear stress components can be obtained by a suitable change of variables as given in Table 2.1.

Table 2.1 Base equation and change of variables to be followed while evaluating other components of stress tensor

Variable	Base equation	Change of variables
τ_{yy}	τ_{xx}	$u \to v$, $x \to y$, $v \to u$, and $y \to x$
τ_{zz}	τ_{xx}	$u \to w$, $x \to z$, $w \to u$, and $z \to x$
τ_{yz}	τ_{xy}	$u \to v$, $x \to y$, $v \to w$, $y \to z$, $w \to u$, and $z \to x$
τ_{zx}	τ_{xy}	$u \to w$, $x \to z$, $v \to u$, $y \to x$, $w \to v$, and $z \to y$

2.2.3 Compressible Navier–Stokes Equations

Strictly speaking, the Navier–Stokes equation is only a statement of the momentum balance given in Eq. (2.4). The expanded form of the equations in the compressible form is given in this section. Three separate equations can be written from Eq. (2.5) by successively taking j (the free index) as 1, 2, 3. On the other hand, i is the repeated index in Eq. (2.4) and, therefore, we need to sum over i (with i going from 1 to 3). For example, from Eq. (2.4) we can write the x-momentum ($j = 1$) equation as

x-momentum:

$$\rho\left(\frac{\partial u}{\partial t}+u\frac{\partial u}{\partial x}+v\frac{\partial u}{\partial y}+w\frac{\partial u}{\partial z}\right)$$
$$=-\frac{\partial p}{\partial x}+\frac{\partial}{\partial x}\left\{\mu\left[2\frac{\partial u}{\partial x}-\frac{2}{3}\left(\frac{\partial u}{\partial x}+\frac{\partial v}{\partial y}+\frac{\partial w}{\partial z}\right)\right]\right\}$$
$$+\frac{\partial}{\partial y}\left[\mu\left(\frac{\partial u}{\partial y}+\frac{\partial v}{\partial x}\right)\right]+\frac{\partial}{\partial z}\left[\mu\left(\frac{\partial u}{\partial z}+\frac{\partial w}{\partial x}\right)\right]+\rho g_x \tag{2.19}$$

Note that in the above equation, we replaced (u_1, u_2, u_3) by (u, v, w), and (x_1, x_2, x_3) by (x, y, z). Further, recall that $\delta_{ij} = 1$ for $i = j$, and zero otherwise.

Similarly, the y- and z-momentum equations can be written by taking $j = 2$ and 3, respectively.

y-momentum:

$$\rho\left(\frac{\partial v}{\partial t}+u\frac{\partial v}{\partial x}+v\frac{\partial v}{\partial y}+w\frac{\partial v}{\partial z}\right)$$
$$=-\frac{\partial p}{\partial y}+\frac{\partial}{\partial y}\left\{\mu\left[2\frac{\partial v}{\partial y}-\frac{2}{3}\left(\frac{\partial u}{\partial x}+\frac{\partial v}{\partial y}+\frac{\partial w}{\partial z}\right)\right]\right\}$$
$$+\frac{\partial}{\partial x}\left[\mu\left(\frac{\partial u}{\partial y}+\frac{\partial v}{\partial x}\right)\right]+\frac{\partial}{\partial z}\left[\mu\left(\frac{\partial v}{\partial z}+\frac{\partial w}{\partial y}\right)\right]+\rho g_y \tag{2.20}$$

z-momentum:

$$\rho\left(\frac{\partial w}{\partial t}+u\frac{\partial w}{\partial x}+v\frac{\partial w}{\partial y}+w\frac{\partial w}{\partial z}\right)$$

$$= -\frac{\partial p}{\partial z} + \frac{\partial}{\partial z}\left\{\mu\left[2\frac{\partial w}{\partial z} - \frac{2}{3}\left(\frac{\partial u}{\partial x} + \frac{\partial v}{\partial y} + \frac{\partial w}{\partial z}\right)\right]\right\}$$
$$+ \frac{\partial}{\partial x}\left[\mu\left(\frac{\partial w}{\partial x} + \frac{\partial u}{\partial z}\right)\right] + \frac{\partial}{\partial y}\left[\mu\left(\frac{\partial v}{\partial z} + \frac{\partial w}{\partial y}\right)\right] + \rho g_z \tag{2.21}$$

The continuity equation is a scalar equation (no free index) and summation over i yields

$$\frac{\partial \rho}{\partial t} + \frac{\partial(\rho u)}{\partial x} + \frac{\partial(\rho v)}{\partial y} + \frac{\partial(\rho w)}{\partial z} = 0 \tag{2.22}$$

The thermal energy equation (Eq. (2.11)) in expanded form is

$$\rho c_v\left(\frac{\partial T}{\partial t} + u\frac{\partial T}{\partial x} + v\frac{\partial T}{\partial y} + w\frac{\partial T}{\partial z}\right) = \frac{\partial}{\partial x}\left(k\frac{\partial T}{\partial x}\right) + \frac{\partial}{\partial y}\left(k\frac{\partial T}{\partial y}\right) + \frac{\partial}{\partial z}\left(k\frac{\partial T}{\partial z}\right)$$
$$- p\left(\frac{\partial u}{\partial x} + \frac{\partial v}{\partial y} + \frac{\partial w}{\partial z}\right) + \Phi \tag{2.23}$$

where the viscous dissipation term Φ in expanded form is

$$\Phi = 2\mu\left[\left(\frac{\partial u}{\partial x}\right)^2 + \left(\frac{\partial v}{\partial y}\right)^2 + \left(\frac{\partial w}{\partial z}\right)^2\right] + \mu\left(\frac{\partial u}{\partial y} + \frac{\partial v}{\partial x}\right)^2$$
$$+ \mu\left(\frac{\partial v}{\partial z} + \frac{\partial w}{\partial y}\right)^2 + \mu\left(\frac{\partial w}{\partial x} + \frac{\partial u}{\partial z}\right)^2 - \frac{2}{3}\mu\left(\frac{\partial u}{\partial x} + \frac{\partial v}{\partial y} + \frac{\partial w}{\partial z}\right)^2 \tag{2.24}$$

Compressibility is usually associated with high speed flows ($Ma > 0.3$, where Ma is Mach number). However, for low Mach number, the flow can still be compressible. For example, strong wall heating or cooling can cause a sufficient change in density and the incompressible flow criterion breaks down even at low speeds. Less known is the situation encountered in some micro-devices where the pressure may strongly change due to viscous effects even though the speeds may not be high enough for the Mach number to go above the traditionally applied threshold of 0.3 as also discussed in Sect. 1.4.2. Corresponding to the changes in pressure would be changes in density that must be taken into account when writing the equations of motion. The compressibility effects therefore become significant at high temperature or at high speed and in micro-devices involving gas flow.

2.2.4 Incompressible Navier–Stokes Equations

The incompressible assumption applies when the fluid density can be assumed to be a constant over the entire flow domain. This assumption amounts to taking

$\frac{D\rho}{Dt} = \frac{\partial \rho}{\partial t} + u_i \frac{\partial \rho}{\partial x_i} \approx 0$, which, in turn, implies that the spatial and temporal gradients are necessarily small enough to be neglected. Thus, along with questioning the nature of the density field, one also needs to question the nature of the density gradients. For example, even in low speed microscale flows (with $Ma < 0.3$), the spatial gradient of density becomes large enough to invalidate the incompressibility assumption. Similarly, a cavity with a diaphragm vibrating at a high frequency causes sufficiently large density fluctuations in the vicinity of the diaphragm. This density fluctuation disallows the flow to be treated as incompressible even when the flow speed is enforced to be small.

Upon using the incompressible flow assumption (i.e. $\frac{D\rho}{Dt} = \frac{\partial \rho}{\partial t} + u_i \frac{\partial \rho}{\partial x_i} \approx 0$), the continuity equation (2.1) simplifies to

$$\frac{\partial u}{\partial x} + \frac{\partial v}{\partial y} + \frac{\partial w}{\partial z} = 0 \tag{2.25}$$

Using continuity equation $\partial u_k / \partial x_k$, the momentum equation (2.5) in tensorial form simplifies as

$$\rho \left(\frac{\partial u_j}{\partial t} + u_i \frac{\partial u_j}{\partial x_i} \right) = -\frac{\partial p}{\partial x_j} + \frac{\partial}{\partial x_i} \left[\mu \left(\frac{\partial u_j}{\partial x_i} + \frac{\partial u_i}{\partial x_j} \right) \right] + \rho g_j \tag{2.26}$$

When dynamic viscosity is taken as a constant, the momentum equation in tensorial form simplifies as

$$\begin{aligned} \rho \left(\frac{\partial u_j}{\partial t} + u_i \frac{\partial u_j}{\partial x_i} \right) &= -\frac{\partial p}{\partial x_j} + \mu \frac{\partial}{\partial x_i} \left(\frac{\partial u_j}{\partial x_i} + \frac{\partial u_i}{\partial x_j} \right) + \rho g_j \\ &= -\frac{\partial p}{\partial x_j} + \mu \frac{\partial^2 u_j}{\partial x_i \partial x_i} + \mu \frac{\partial}{\partial x_j} \left(\frac{\partial u_i}{\partial x_i} \right) + \rho g_j \\ \therefore \rho \left(\frac{\partial u_j}{\partial t} + u_i \frac{\partial u_j}{\partial x_i} \right) &= -\frac{\partial p}{\partial x_j} + \mu \frac{\partial^2 u_j}{\partial x_i \partial x_i} + \rho g_j \end{aligned} \tag{2.27}$$

where we have used continuity equation ($\partial u_i / \partial x_i = 0$) in the last step. In expanded form, these equations can be written as

x-momentum:

$$\rho \left(\frac{\partial u}{\partial t} + u \frac{\partial u}{\partial x} + v \frac{\partial u}{\partial y} + w \frac{\partial u}{\partial z} \right) = -\frac{\partial p}{\partial x} + \mu \left(\frac{\partial^2 u}{\partial x^2} + \frac{\partial^2 u}{\partial y^2} + \frac{\partial^2 u}{\partial z^2} \right) + \rho g_x \tag{2.28}$$

y-momentum:

$$\rho\left(\frac{\partial v}{\partial t}+u\frac{\partial v}{\partial x}+v\frac{\partial v}{\partial y}+w\frac{\partial v}{\partial z}\right)=-\frac{\partial p}{\partial y}+\mu\left(\frac{\partial^2 v}{\partial x^2}+\frac{\partial^2 v}{\partial y^2}+\frac{\partial^2 v}{\partial z^2}\right)+\rho g_y \tag{2.29}$$

z-momentum:

$$\rho\left(\frac{\partial w}{\partial t}+u\frac{\partial w}{\partial x}+v\frac{\partial w}{\partial y}+w\frac{\partial w}{\partial z}\right)=-\frac{\partial p}{\partial z}+\mu\left(\frac{\partial^2 w}{\partial x^2}+\frac{\partial^2 w}{\partial y^2}+\frac{\partial^2 w}{\partial z^2}\right)+\rho g_z \tag{2.30}$$

Energy equation:

$$\rho c_p\left(\frac{\partial T}{\partial t}+u\frac{\partial T}{\partial x}+v\frac{\partial T}{\partial y}+w\frac{\partial T}{\partial z}\right)=\frac{\partial}{\partial x}\left(k\frac{\partial T}{\partial x}\right)+\frac{\partial}{\partial y}\left(k\frac{\partial T}{\partial y}\right)+\frac{\partial}{\partial z}\left(k\frac{\partial T}{\partial z}\right)+\Phi \tag{2.31}$$

where the viscous dissipation term Φ is

$$\Phi=2\mu\left[\left(\frac{\partial u}{\partial x}\right)^2+\left(\frac{\partial v}{\partial y}\right)^2+\left(\frac{\partial w}{\partial z}\right)^2\right]+\mu\left(\frac{\partial u}{\partial y}+\frac{\partial v}{\partial x}\right)^2+\mu\left(\frac{\partial v}{\partial z}+\frac{\partial w}{\partial y}\right)^2+\mu\left(\frac{\partial w}{\partial x}+\frac{\partial u}{\partial z}\right)^2 \tag{2.32}$$

The incompressible flows are not affected by the second coefficient of viscosity μ_2 because the second-last term in Eq. (2.4) drops off due to $\partial u_k/\partial x_k = 0$ (for incompressible flows, from continuity equation (2.25)). Note the appearance of c_p instead of c_v in the energy equation (Eq. (2.31)). The energy equation for incompressible flows is actually the convection of enthalpy and not internal energy that is balanced by heat conduction and viscous dissipation [57].

Although density is no longer a variable, the number of variables (u_i, p, and T) does not reduce. Note that the equation of state (2.9) is not applicable to incompressible flows. To illustrate this point, consider any isothermal incompressible flow. Here, T is constant (because of isothermal assumption), ρ is constant (because of incompressible flow assumption), yet, in general, pressure varies in the flow, clearly showing that $p = \rho RT$ is not applicable to this (and all incompressible flow) situation.

2.3 Boundary Conditions

The no-slip boundary condition is used to analyze the fluid flow at macroscale. As per this condition, the velocity of the fluid at the wall is equal to the velocity of the wall. This boundary condition has been shown to apply well to a wide range of fluid flow problems at the conventional scale.

However, the no-slip condition does not apply to micro-scale flows, where, rather, slip occurs at the boundary. Initially, the concept of slip boundary condition was presented by Navier. According to his concept, the magnitude of the slip velocity is proportional to the magnitude of shear rate at the wall. This idea was found tenable by Kundt and Warburg who performed experiments on rarefied gas flow. They found that slipping of gas is indeed proportional to the velocity gradient close to the wall, i.e.,

$$u_s = \zeta \left.\frac{du}{dy}\right|_{wall} \tag{2.33}$$

where ζ is a constant, called the coefficient of slip and represents a length.

2.3.1 *Maxwell's Slip Theory*

Maxwell explored the concept of slip further and quantified the slip length near the wall. During gas–wall interaction, the gas molecules are classified into two streams relative to the wall—approaching and receding (Fig. 2.1a). Maxwell assumed that all

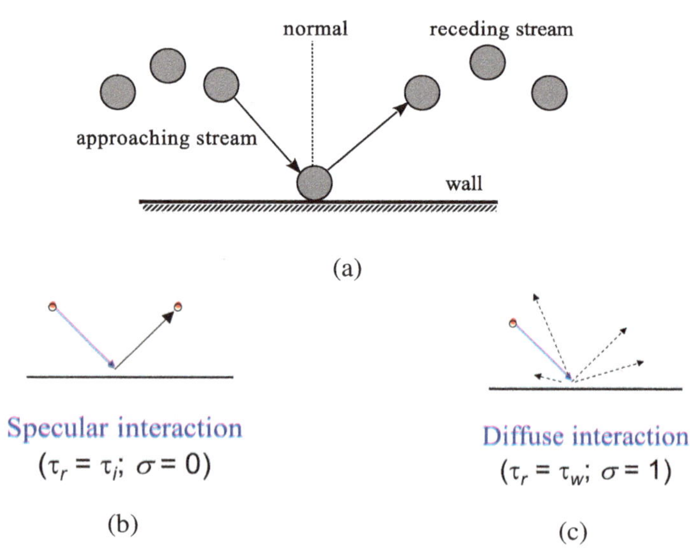

Fig. 2.1 (**a**) Gas–wall interactions, (**b**) specular interactions, and (**c**) diffuse interactions

interactions can be classified as either specular or diffuse and defined a parameter, the tangential momentum accommodation coefficient (TMAC) to quantify the interaction as

$$\sigma_v = \frac{\tau_i - \tau_r}{\tau_i - \tau_w}. \tag{2.34}$$

In the above equation, τ_i and τ_r are the tangential momentum of the incoming and reflected molecules, respectively, and τ_w is the tangential momentum of the wall. Notice that $\tau_w = 0$ for a stationary wall; τ_w is also the tangential momentum of the re-emitted molecules that may have been adsorbed to the surface. The subscript v in σ is to emphasize that we are referring to accommodation of velocity (later in Chap. 3, we will introduce thermal accommodation coefficient).

During specular interaction (which is equivalent to reflection of light from a surface; Fig. 2.1b), the tangential momentum of the incoming and reflected molecules is equal (while the normal component of their momentum is reversed). That is, $\tau_i = \tau_r$ and from Eq. (2.34) $\sigma_v = 0$ for specular interaction. On the other hand, during diffuse interaction, since the molecule can move in any random direction after interaction with the wall, on an *average* its momentum is zero for a stationary wall (and equal to τ_w otherwise). That is, $\tau_r = \tau_w$ for diffuse interaction and $\sigma_v = 1$. The specular and diffuse interactions therefore nicely define the limits of σ_v; all practical values are expected to lie within these two limits. (Note that σ_v can become greater than unity, if a molecule is reflected back in the direction from where it came; a situation called back scattering. However, this situation does not arise frequently.) In any flow, if f is the fraction of diffuse interaction out of the total interactions, then the effective value of TMAC is

$$\sigma_v = f \times \sigma_{v,\text{diffuse}} + (1 - f) \times \sigma_{v,\text{specular}} = f \times 1 + (1 - f) \times 0 = f. \tag{2.35}$$

Therefore, σ_v also denotes the fraction of diffuse interactions.

During these interactions, the wall experiences a viscous drag due to the difference between the tangential momentum of the approaching stream and that carried away by the receding stream. Maxwell assumed that the approaching stream exhibits a velocity gradient extending uniformly up to some value u_0, while u_s is the slip velocity at the wall as shown in Fig. 2.2. The tangential momentum brought by the approaching stream to the wall (per unit time per unit area) can then be analyzed

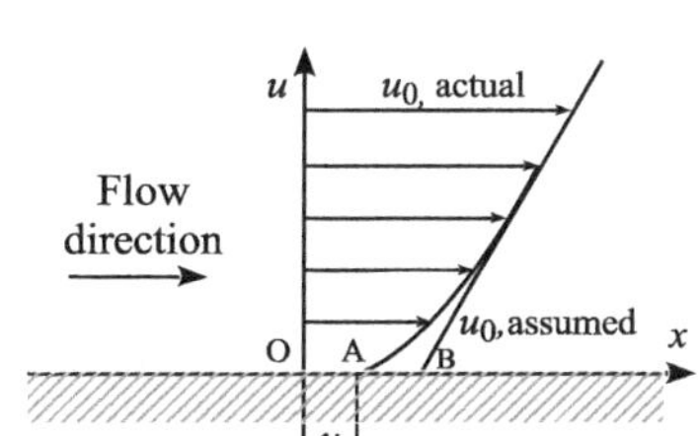

Fig. 2.2 Actual velocity profile extended linearly up to the wall

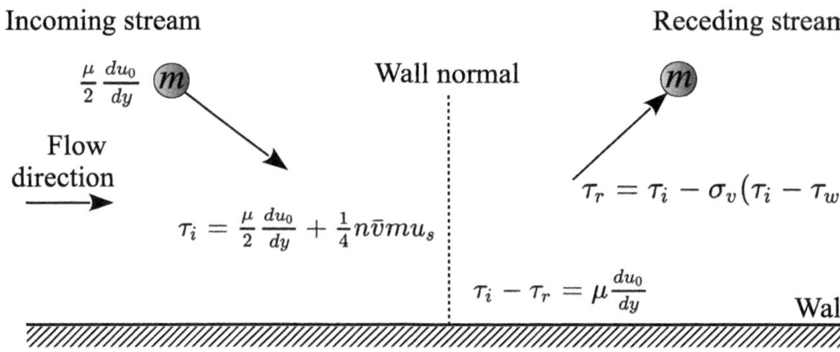

Fig. 2.3 Momentum of the incoming and receding molecular stream

as [82]: The momentum that is brought to the wall is equal in magnitude to that which is transmitted elsewhere in the gas by the corresponding molecular stream. Therefore, the incoming momentum is $\frac{\mu}{2}\frac{du_0}{dy}$ per unit area per unit time, where μ is the coefficient of viscosity, and y is normal to the wall. From kinetic theory of gases, the number of molecules incident per unit wall area is $\frac{1}{4}n\bar{v}$ for a gas with n molecules per unit volume having a mean speed of $\bar{v}$. These molecules will carry a momentum of mu_s where m is the mass of the molecule. Therefore, the momentum carried away by the gas per unit area per unit time due to the slip velocity u_s in the direction of the flow is $\frac{1}{4}nm\bar{v}$ times u_s. The incoming momenta is therefore the sum of these two components. The above argument only applies to diffusively reflected molecules, while specularly reflected molecules do not contribute towards drag as apparent from Fig. 2.1(b). Since drag on the wall being difference of incoming and outgoing momenta, we obtain

$$\sigma_v\left(\frac{1}{2}\mu\frac{du_0}{dy} + \frac{1}{4}\rho u_s\bar{v}\right) = \mu\frac{du_0}{dy} \tag{2.36}$$

where we have utilized $\rho = mn$ the density. Hence, the slip velocity u_s can be written as

$$u_s = 2\left(\frac{2-\sigma_v}{\sigma_v}\right)\frac{\mu}{\rho\bar{v}}\frac{du_0}{dy}. \tag{2.37}$$

Upon inserting $\bar{v} = 2(2RT/\pi)^{1/2}$, a standard result from kinetic theory, in the above equation, we obtain

$$u_s = \left(\frac{2-\sigma_v}{\sigma_v}\right)\lambda\left.\frac{du}{dy}\right|_w \tag{2.38}$$

where we have additionally utilized expressions for mean free path $\lambda = (\mu/p)\sqrt{\pi RT/2}$ and the equation of state in arriving at Eq. (2.38).

Note that only one slip coefficient σ_v (or $C_1 = \frac{2-\sigma_v}{\sigma_v}$) along with the first derivative of velocity is involved in the above equation. The above model is therefore known as "first-order" slip boundary condition, as opposed to higher-order slip models introduced in the next two sections. The accuracy of the above model has been verified by different researchers in various ways, including a direct comparison with DSMC data by Gavasane et al. [59].

2.3.2 Derivation of Higher-Order Slip Boundary Condition

In this section, we derive a second-order velocity slip model. Although the derivation is somewhat heuristic, it gives important clues about how the model derived above can be extended.

Considering momentum crossing a control surface located at a distance of $\lambda/2$ away from the wall, where λ is the mean free path of the gas (Fig. 2.4). One-half of the momentum crossing this control surface is brought by gas at a distance of λ from the surface (where the gas velocity is u_λ), while the other half is brought by gas after interaction from the wall. The magnitude of this latter momentum comprises components due to both specular and diffuse reflections at the wall. The diffuse component contributes $\sigma_v u_w$ while the specular contributes $(1-\sigma_v)u_\lambda$. Therefore, we obtain

$$u_{\lambda/2} = \frac{u_\lambda}{2} + \frac{\sigma_v u_w + (1-\sigma_v)u_\lambda}{2}. \tag{2.39}$$

If we expand u_λ in Taylor series about $\lambda/2$

$$u_\lambda = u_{\lambda/2} + \frac{\lambda}{2}\frac{\partial u}{\partial y}\bigg|_{\lambda/2} + \frac{\lambda^2}{8}\frac{\partial^2 u}{\partial y^2}\bigg|_{\lambda/2} + \cdots \tag{2.40}$$

and substitute the above series into Eq. (2.39), we can easily obtain

$$(u_{\lambda/2} - u_w) = \frac{\lambda}{2}\frac{2-\sigma_v}{\sigma_v}\frac{\partial u}{\partial y}\bigg|_{\lambda/2} + \frac{\lambda^2}{8}\frac{2-\sigma_v}{\sigma_v}\frac{\partial^2 u}{\partial y^2}\bigg|_{\lambda/2} + \cdots \tag{2.41}$$

If we now move the control surface from a distance of $\lambda/2$ to close to the wall, we will obtain

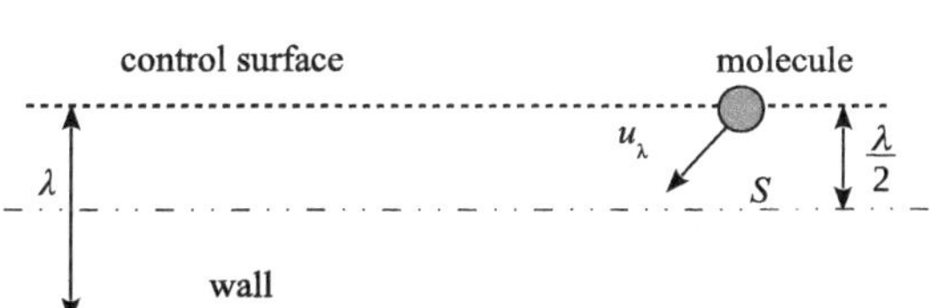

Fig. 2.4 Second-order slip boundary condition

$$(u_{\lambda/2} - u_w) = \lambda \frac{2-\sigma_v}{\sigma_v} \frac{\partial u}{\partial y}\bigg|_{\lambda/2} + \lambda^2 \frac{2-\sigma_v}{\sigma_v} \frac{\partial^2 u}{\partial y^2}\bigg|_{\lambda/2} + \cdots \tag{2.42}$$

Replacing $u_{\lambda/2}$ by u_g and defining $u_g - u_w = u_s$, we obtain (upon neglecting higher order terms)

$$u_s = \frac{2-\sigma_v}{\sigma_v} \left(\lambda \frac{\partial u}{\partial y}\bigg|_{\lambda/2} + \lambda^2 \frac{\partial^2 u}{\partial y^2}\bigg|_{\lambda/2} \right) \tag{2.43}$$

which is the general second-order slip model. Note that the above derivation is not rigorous; more accurate derivation can be done in a manner discussed in Sect. 6.5.

2.3.3 *Alternate Slip Models*

In the past, several researchers have extended the basic slip model of Maxwell. These extensions have been presented here for easy reference.

Diessler [40] realized that the Maxwell's model is applicable only when velocity and temperature profiles are linear over a mean free path. However, the model needs to be generalized in case this requirement is violated. He undertook detailed kinetic theory based calculations to arrive at the following generalized second order slip model:

$$u_s = \pm \frac{2-\sigma_v}{\sigma_v} \lambda \frac{\partial u}{\partial y}\bigg|_w - \frac{9}{16}\lambda^2 \left[2\frac{\partial^2 u}{\partial y^2}\bigg|_w + \frac{\partial^2 u}{\partial x^2}\bigg|_w + \frac{\partial^2 u}{\partial z^2}\bigg|_w \right] \tag{2.44}$$

where x, y, and z refer to the stream wise co-ordinate, transverse co-ordinate along the height and transverse co-ordinate along the width, respectively and subscript w signifies that the partial derivatives are evaluated at the wall. Neglecting the tangential derivative of velocity, the above model reduces to

$$u_s = \pm \frac{2-\sigma_v}{\sigma_v} \lambda \frac{\partial u}{\partial y}\bigg|_w - \frac{9}{8}\lambda^2 \frac{\partial^2 u}{\partial y^2}\bigg|_w \tag{2.45}$$

The above model was generalized by Sreekanth [137] as

$$u_s = -C_1 \lambda \frac{\partial u}{\partial y}\bigg|_w - C_2 \lambda^2 \frac{\partial^2 u}{\partial y^2}\bigg|_w \tag{2.46}$$

This is a widely employed model and we refer to it as *Sreekanth's slip model.*

Since these early improvements by Deissler [40] and Sreekanth [137], new slip models have been proposed. For example, Beskok and Karniadakis [81] modified the second-order slip model as

$$u_s = \pm \frac{2-\sigma_v}{\sigma_v} \frac{Kn}{1-bKn} \frac{\partial u}{\partial \bar{y}}\bigg|_w \tag{2.47}$$

to aid in the numerical implementation of the boundary condition. In the above equation, $\bar{y}$ denotes dimensionless wall-normal coordinate, b is an empirical parameter which depends on the geometric configuration. The value of b can be determined as $b = u_0'/2u_0''$ with u_0' and u_0'' being first- and second-order derivative of velocity normal to the surface for no-slip level of approximation. Xue and Fan [55] expressed the second-order boundary condition in the following form:

$$u_s = \pm \frac{2-\sigma_v}{\sigma_v} \tanh(Kn) \frac{\partial u}{\partial \bar{y}}\bigg|_w \tag{2.48}$$

Jie et al. [80] reported a hybrid boundary condition, given as

$$u_s = \pm \frac{2-\sigma_v}{\sigma_v} \left[Kn \frac{\partial u}{\partial \bar{y}}\bigg|_w + Re \frac{Kn}{2} \frac{\partial p}{\partial \bar{x}}\bigg|_w \right] \tag{2.49}$$

where $\bar{x}$ and $\bar{y}$ denote dimensionless streamwise and normal coordinates, respectively, and Re denotes the local Reynolds number. Lockerby et al. [89] proposed the linearized Maxwell–Burnett boundary condition as given below.

$$\begin{aligned} u_s = & \pm \frac{2-\sigma_v}{\sigma_v} \lambda \left(\frac{\partial u}{\partial y} + \frac{\partial u}{\partial x} \right)\bigg|_w + \frac{3}{4} \frac{\mu}{\rho T} \frac{\partial T}{\partial x}\bigg|_w \\ & + \frac{2-\sigma_v}{\sigma_v} \lambda \left(2 \frac{\mu}{\rho^2} \frac{\partial^2 \rho}{\partial x \partial y} - \frac{\mu}{\rho T} \frac{\partial^2 T}{\partial x \partial y} \right)\bigg|_w \\ & + \frac{3}{16\pi} \frac{\gamma - 1}{\gamma} Pr \lambda^2 \left[(45\gamma - 61) \frac{\partial^2 u}{\partial x^2} + (45\gamma - 49) \frac{\partial^2 v}{\partial x \partial y} - 12 \frac{\partial^2 u}{\partial y^2} \right]\bigg|_w \end{aligned} \tag{2.50}$$

where u and v are the velocity components tangential and normal to the wall. The second order slip models are derived from the kinetic theory by considering the momentum transfer rate of gas molecules impinging on the wall [168]. Chen and Tian [33] proposed

$$u_s = (1-\alpha) u_g + \alpha u_w; \quad \text{where } \alpha = \frac{1}{1 + 4\omega Kn/p} \tag{2.51}$$

u_g is the velocity of the gas at the wall, and the role of ω in the above equation is similar to that of TMAC. This last model accounts for a finite residence time of the molecule at the wall before getting reflected from the wall.

In this book, we adopt the model of Sreekanth (Eq. (2.46)), which is usually the case, given its general nature and ease of use. The model is similar to Eq. (2.43) derived in the previous section, in that Sreekanth's model is of second-order in terms of Knudsen number and involves the first- and second derivatives of velocity. There are, however, some subtle differences between the two models: note that the derivatives are calculated at the control surface (introduced in the derivation) in Eq. (2.43), whereas the derivatives are calculated at the wall in Sreekanth's model, making the latter model more convenient to apply. This change of location where the gradients are calculated leads to the appearance of two independent slip coefficients as opposed to a single coefficient in Eq. (2.43). We examine the value of the slip coefficients proposed in the literature in the next section.

2.3.4 Value of Slip Coefficients

In this section, we compile the values of slip coefficients and tangential momentum accommodation coefficient in Tables 2.2 and 2.3, respectively, as reported by different researchers. The information will be useful while using the first-order and second-order slip models (Eqs. (2.38) and (2.46)) introduced earlier. The value of slip coefficient C_1 is usually related to TMAC through the following equation

$$C_1 = \frac{2 - \sigma_v}{\sigma_v}. \tag{2.52}$$

Most of the values provided in this section were compiled by Agrawal and Prabhu [5], who documented these values along with the parameter range applicable and the associated uncertainty in the measurements. The values have been suitably updated by the present authors. Compiling such information allows the trend in the values with respect to Knudsen number to be better gleaned from the data (Fig. 2.5). As evident from Fig. 2.5a, the value of TMAC for monatomic gases varies over a small range even with six-decade change in Knudsen number. Further, the variation in the value of TMAC for the five monatomic gases is small (Table 2.4). Therefore, Agrawal and Prabhu [4] suggested to assign a constant value of 0.926 for all monatomic gases. However, the data for non-monatomic gases in Fig. 2.5b seems to show a decreasing trend with Knudsen number. The following empirical correlation was proposed by Agrawal and Prabhu [4] based on data in the figure:

$$\sigma_v = 1 - \log(1 + Kn^{0.7}) \tag{2.53}$$

The value of TMAC being close to unity suggests that the interaction of the gas molecules with the wall is mostly diffuse in nature, with less than 8% interaction being specular for monatomic gases (see Eq. (2.35)) irrespective of the amount of rarefaction of the gas. Note that the value of 0.926 for monatomic gases applies to most surface materials and temperatures above the room temperature. The overly diffuse nature of the interaction suggests that the momentum of the reflected molecule is almost independent of the momentum of the incident molecule;

Table 2.2 The values of TMAC in the slip and transition regimes as reported by different authors (updated from Agrawal and Prabhu [5])

Method	Source	Gas	Kn range	σ_v
Flow through capillary or microchannel	Sreekanth [137]	Nitrogen	0.007–0.03	1
			0.03–0.13	0.9317
			0.13–0.237	0.9317
	Colin et al. [35]	Helium	0.029–0.22	0.93
		Nitrogen	0.002–0.03	1
			0.005–0.09	0.93
	Ewart et al. [54]	Helium	0.009–0.309	0.914 ± 0.009
		Argon	0.003–0.302	0.871 ± 0.017
		Nitrogen	0.003–0.291	0.908 ± 0.041
	Maurer et al. [93]	Helium	0.06–0.8	0.91 ± 0.03
		Nitrogen	0.002–0.59	0.87 ± 0.03
	Ewart et al. [53]	Helium	0.03–0.3	0.910 ± 0.004
	Arkilic et al. [12]	Argon	0.1–0.41	0.8 ± 0.1
		Nitrogen	0.1–0.34	0.83 ± 0.05
		CO_2	0.1–0.44	0.88 ± 0.06
	Jang and Wereley [79]	Air	0.0017	0.85
	Yamaguchi et al. [171]	Argon	0.05–0.3	0.869 ± 0.03
		Nitrogen	0.05–0.3	0.851 ± 0.02
		Oxygen	0.05–0.3	0.851 ± 0.02
	Vadiraj et al. [72]	Argon	0.07–1.2	0.957 ± 0.04
		Nitrogen	0.072–0.11	0.955 ± 0.03
		Oxygen	0.072–1.13	0.866 ± 0.04
Spinning rotor gauge	Tekasakul et al. [145] and Bentz et al. [19]	Helium	4.64×10^{-3}–5.83×10^{-1}	0.9410 ± 0.0328
		Argon	1.67×10^{-3}–2.10×10^{-1}	0.8968 ± 0.0354
		Krypton	1.30×10^{-3}–1.63×10^{-1}	0.8706 ± 0.0578
		Nitrogen	1.63×10^{-3}–2.58×10^{-2}	0.86 ± 0.027
		Methane	1.30×10^{-3}–2.15×10^{-2}	1.04 ± 0.022
Rotating cylinder	Agrawal and Prabhu [4] (data of Kuhlthau 1949)	Dry air	0.1–8.3	0.74
			0.04–0.08	0.94
Unsteady method	Suetin et al. [143]	Helium	0.001–2	0.895 ± 0.004
			2.8–886	0.935 ± 0.004
		Neon	0.001–2	0.865 ± 0.004
			2.8–886	0.929 ± 0.003
		Argon	0.001–2	0.927 ± 0.028
			2.8–886	0.975 ± 0.006
	Porodnov et al. [117]	Krypton	4.9×10^{-4}–9.6×10^{-3}	0.995 ± 0.026
		Xenon	3.6×10^{-4}–7.0×10^{-3}	1.010 ± 0.040
		Hydrogen	1.1×10^{-3}–2.2×10^{-2}	0.957 ± 0.015
		Deuterium	1.1×10^{-3}–2.2×10^{-2}	0.934 ± 0.006
		Nitrogen	6.0×10^{-4}–1.2×10^{-2}	0.925 ± 0.014
		CO_2	4.0×10^{-4}–7.8×10^{-3}	0.993 ± 0.009

Table 2.3 The values of slip coefficients as reported by different authors (updated from Agrawal and Prabhu [5])

Source	Gas	Kn range	C_1	C_2
Maxwell [94]	–	–	1	0
Schamberg [125]	–	–	1	$5\pi/12$
Albertoni et al. [9]	–	–	1.1466	0
Deissler [40]	–	–	1	1.6875
Cercignani [29]	–	–	1.1466	0.9756
Chapman and Cowling [31]	–	–	≈ 1	≈ 0.5
Hsia and Domoto [75]	–	–	1	0.5
Mitsuya [97]	–	–	1	2/9
Sreekanth [137]	Nitrogen	0.007–0.03	1	0
		0.03–0.13	1.1466	0
		0.13–0.237	1.1466	0.14
Ewart et al. [54]	Helium	0.009–0.309	1.052 ± 0.020	0.148 ± 0.014
	Argon	0.003–0.302	1.147 ± 0.042	0.294 ± 0.029
	Nitrogen	0.003–0.291	1.066 ± 0.088	0.231 ± 0.057
Maurer et al. [93]	Helium	0.06–0.8	1.2 ± 0.05	0.23 ± 0.1
	Nitrogen	0.002–0.59	1.3 ± 0.05	0.26 ± 0.1
Ewart et al. [53]	Helium	0.03–0.7	1.26 ± 0.022	0.17 ± 0.02
Yamaguchi et al. [171]	Argon	0.05–0.3	1.30 ± 0.09	0.063 ± 0.018
	Nitrogen	0.05–0.3	1.35 ± 0.06	0.031 ± 0.005
	Oxygen	0.05–0.3	1.35 ± 0.06	0.028 ± 0.004
Vadiraj et al. [72]	Argon	0.07–1.2	1.089 ± 0.08	0.140 ± 0.007
	Nitrogen	0.072–0.11	1.093 ± 0.07	0.151 ± 0.003
	Oxygen	0.072–1.13	1.301 ± 0.09	0.155 ± 0.004

The first eight entries in the table are theoretical values, while the remaining are experimental

suggesting that gas molecules have sufficient time to interact with the wall. This can arise if the molecule which has been directed to the surface and the molecule coming out of the surface are not the same, and therefore momentum of the incoming molecule and that of the outgoing molecule are not correlated. Alternatively, the incident molecule may undergo multiple reflections at the surface, owing to the roughness of the wall; therefore, the momentum of the molecule effectively undergoes a random change in magnitude and direction.

Based on available experimental data, Agrawal and Prabhu [5] suggested the following value of slip coefficients: $C_1 = 1.1465$ and $C_2 = 0.164$. It is suggested to use these values for all (monatomic and non-monatomic) gases in the range $0.002 \leq Kn \leq 0.8$.

A dependence of slip coefficient on the nature of the gas suggests that the flow behavior will additionally depend on the gas that we are working with. So does it mean that, for fully developed flow in a tube, the friction factor depends on additional parameters (specific to the gas) besides the Reynolds and Knudsen numbers? Of course, such a dependence will substantially complicate the analysis

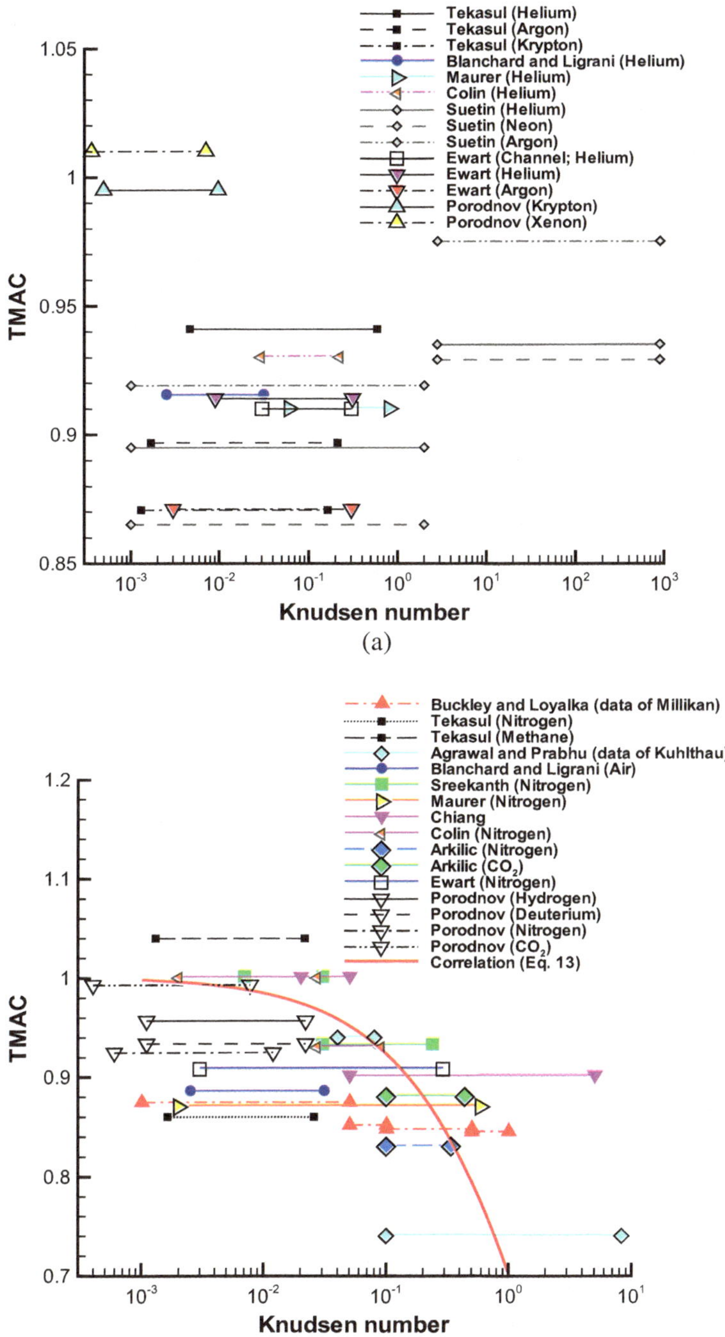

Fig. 2.5 TMAC versus Knudsen number as reported by Agrawal and Prabhu [5]. (**a**) Monatomic gases. (**b**) Non-monatomic gases

Table 2.4 Mean TMAC from all reported values for monatomic gases (n_{obs} indicates the number of observations on which the mean value is based)

Gas	Helium	Neon	Argon	Krypton	Xenon
σ_v	0.923	0.912	0.896	0.954	0.950
n_{obs}	14	6	16	8	5

These values appear valid for commonly employed surface materials and the entire range of Knudsen number (from Agrawal and Prabhu [5], with the value of Argon updated)

of slip flows. This question warrants a clear answer than what can be gleaned from the available literature. On the other hand, it also opens the exciting possibility of making some simple measurements and determining the nature of the gas involved; something which can be useful to the sensors community, interested, for example, in detecting minute amounts of explosives or harmful bio-gases.

2.4 Some Exact Solutions

In this section, we will consider some problems from the slip flow regime for which analytical solution is possible. A discussion of the results can provide insights into the slip flow behavior. These cases also serve to be useful for initial benchmarking of numerical codes and experimental setups. The theory is also useful (and has been employed) for obtaining the value of TMAC reported in the previous section.

2.4.1 Couette Flow

Couette flow is perhaps the simplest of the flow problems. Here, the upper plate is moving with velocity U while the lower plate is stationary; the distance between the plates being H (Fig. 2.6). We will assume that the flow between the plates is laminar, incompressible, steady, and fully developed, with no variation along the z-direction. The pressure is assumed to be constant along the streamwise direction ($dp/dx = 0$). The only difference with respect to the standard problem is that there is slip at the walls [110].

After applying the aforesaid assumptions, the x-momentum equation (2.13) reduces to

$$\frac{d^2u}{dy^2} = 0 \tag{2.54}$$

On integrating this equation, we have

$$u = A_1 y + A_2 \tag{2.55}$$

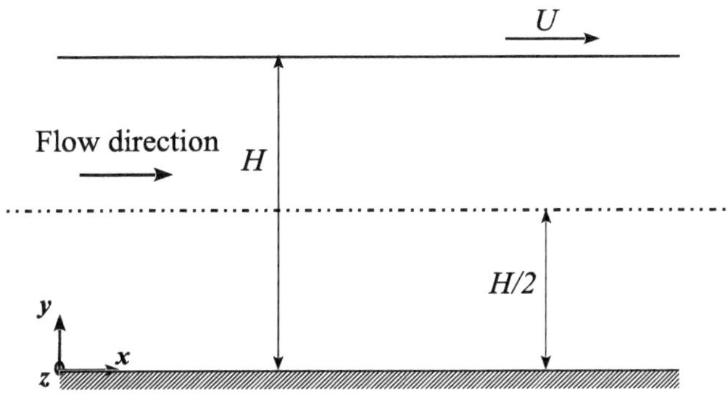

Fig. 2.6 Schematic diagram of Couette flow with upper plate moving with velocity U in x-direction and lower plate is stationary

where A_1 and A_2 are the two constants of integration. First-order slip (Eq. (2.38)) gives the following boundary conditions:

$$u_s - u_w = \frac{2-\sigma_v}{\sigma_v}\lambda \frac{\partial u}{\partial y}\bigg|_{y=0 \text{ and } y=H} \tag{2.56}$$

With $u_w = 0$ on the lower wall ($y = 0$), $u_w = U$ on the upper wall ($y = H$), and keeping in mind that $\frac{\partial u}{\partial y}\big|_{y=0} = -\frac{\partial u}{\partial y}\big|_{y=H}$, we obtain

$$A_1 = \frac{U}{H + 2\lambda\left(\dfrac{2-\sigma_v}{\sigma_v}\right)} \quad \text{and} \quad A_2 = \frac{\left(\dfrac{2-\sigma_v}{\sigma_v}\right)\lambda U}{H + 2\lambda\left(\dfrac{2-\sigma_v}{\sigma_v}\right)}.$$

Upon substituting the values of A_1 and A_2 in Eq. (2.55), we obtain

$$u = \frac{Uy + \left(\dfrac{2-\sigma_v}{\sigma_v}\right)\lambda U}{H + 2\lambda\left(\dfrac{2-\sigma_v}{\sigma_v}\right)} \tag{2.57}$$

as the required velocity profile. We can express the above velocity profile in dimensionless form as

$$\frac{u}{U} = \frac{\dfrac{y}{H} + Kn\left(\dfrac{2-\sigma_v}{\sigma_v}\right)}{1 + 2Kn\left(\dfrac{2-\sigma_v}{\sigma_v}\right)} \tag{2.58}$$

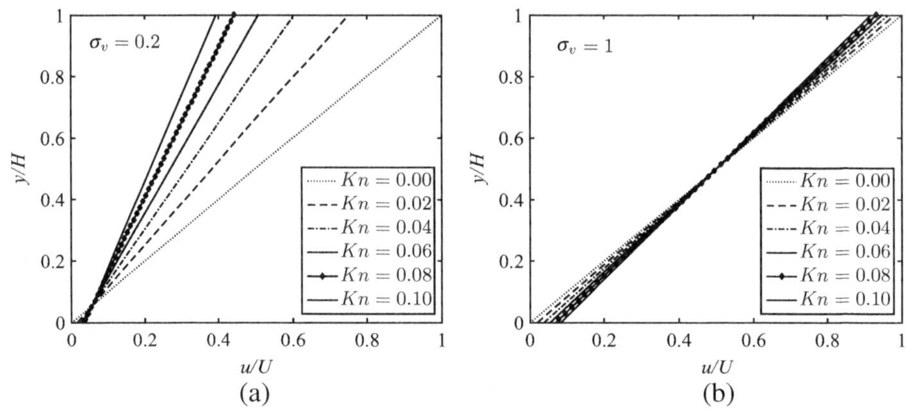

Fig. 2.7 Velocity distribution across the channel height for different values of Knudsen number for (**a**) $\sigma_v = 0.2$ (**b**) $\sigma_v = 1$

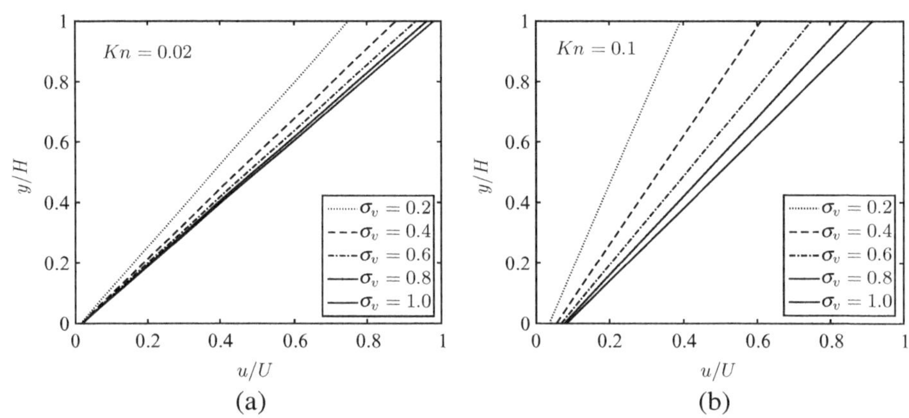

Fig. 2.8 Velocity distribution across the channel height for different values of TMAC for (**a**) $Kn = 0.02$ (**b**) $Kn = 0.1$

where $Kn = \lambda/H$. The solution given by Eq. (2.58) indicates not only a velocity slip at the wall but also a correction to the slope of profile for the Couette flow in presence of slip at the wall. The velocity profile is a function of Knudsen number and TMAC as evident from Eq. (2.58). Notice that there is an increase in the slip velocity with an increase in the Knudsen number for a fixed value of tangential momentum accommodation coefficient (Fig. 2.7). Similarly, the slip velocity increases with a decrease in the value of tangential momentum accommodation coefficient (or the gas-wall interaction becoming more specular) as shown in Fig. 2.8.

The volume flow rate can be calculated using Eq. (2.57) as

$$\dot{Q} = \int_0^H u dy \tag{2.59}$$

which yields

$$\frac{\dot{Q}}{UH} = \frac{1}{2}. \tag{2.60}$$

Notice that the volume flow rate depends only on the speed of the plate and the distance between the plates, and is independent of both the Knudsen number and TMAC. The slip flow solution therefore gives the same volume flow rate as that of the no-slip case. This occurs because the increase in velocity owing to slip in the lower-half is exactly compensated by a reduction in velocity in the upper-half; as evident from Fig. 2.7b.

We next compute the skin friction coefficient from

$$C_f = \frac{\tau_w}{(1/2)\rho U^2} \tag{2.61}$$

where τ_w $(= \mu du/dy)$ is the wall shear stress. For the case of slip at the wall we obtain

$$C_{f,slip} = \frac{2}{Re}\left(\frac{1}{1 + 2\left(\frac{2-\sigma_v}{\sigma_v}\right)Kn}\right) \tag{2.62}$$

where $Re = \rho UH/\mu$. For the case of no-slip at the wall $(Kn \rightarrow 0)$, the above equation reduces to

$$C_{f,no\text{-}slip} = \frac{2}{Re} \tag{2.63}$$

Hence,

$$\frac{C_{f,slip}}{C_{f,no\text{-}slip}} = \frac{1}{1 + 2\left(\frac{2-\sigma_v}{\sigma_v}\right)Kn} \tag{2.64}$$

Equation (2.62) can alternatively be written as

$$C_{f,slip}Ma = \frac{2}{\left(\frac{Re}{Ma}\right)}\left(\frac{1}{1 + \sqrt{2\pi\gamma}\left(\frac{2-\sigma_v}{\sigma_v}\right)\frac{Ma}{Re}}\right) \tag{2.65}$$

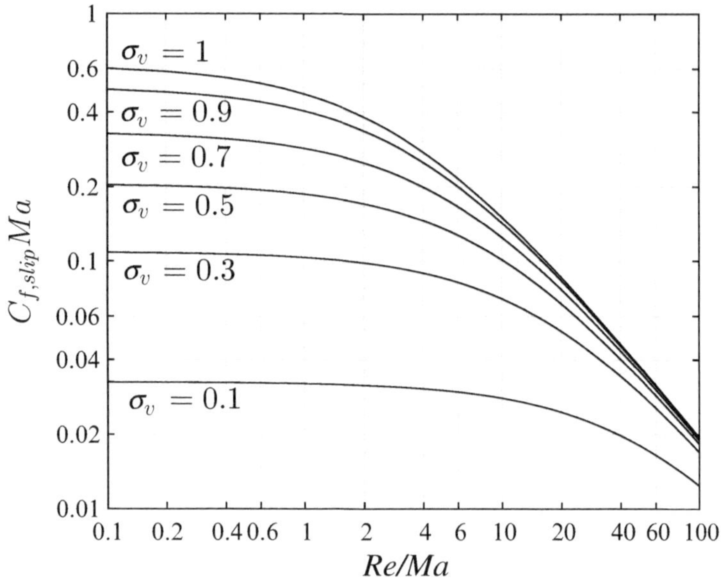

Fig. 2.9 Variation of $C_{f,slip}Ma$ with respect to Re/Ma

where we have additionally used the relation:

$$Kn = \sqrt{\frac{\pi\gamma}{2}}\frac{Ma}{Re} \tag{2.66}$$

In Fig. 2.9, $C_{f,slip}Ma$ has been plotted versus Re/Ma with σ_v as the parameter. It is clear that the quantity $C_{f,slip}Ma$ decreases with the increase in parameter Re/Ma. Equation (2.65) can therefore be used for determining the value of TMAC from measurement of skin friction coefficient and knowing the values of Reynolds and Mach numbers.

2.4.2 Flow in a Microchannel

Several researchers have attempted to obtain analytical solution of gaseous slip flow in a microchannel. The problem is much more complicated than flow in a conventional channel because of both compressibility effect and slip at the wall. The two most important approaches to solve the problem are provided in this and the next subsection. The first approach, due to Arkilic et al. [11], involves solution of the Navier–Stokes equations, whereas the integral form of the momentum equation is solved in the other approach. A comparison of the two approaches is provided at the end of the next subsection.

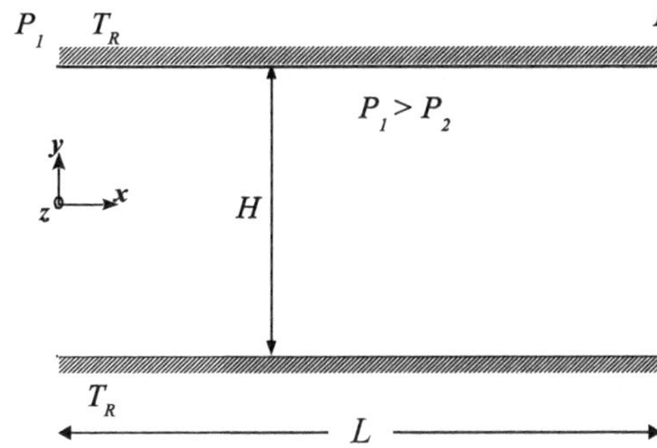

Fig. 2.10 Schematic of flow between two parallel plates, driven by a pressure gradient

2.4.2.1 Perturbation of Navier–Stokes Equations

Following Arkilic et al. [11], we consider steady isothermal flow through a microchannel of large depth such that the two-dimensional flow assumption applies (Fig. 2.10). This assumption is valid for microchannels of high aspect ratio, and allows the w component of velocity and any variation along the z-direction to be neglected. The isothermal flow assumption simplifies the analysis, as it obviates the need for considering the energy equation. Isothermal condition applies if effects such as heat generated due to viscous dissipation and cooling due to the expansion of the gas are properly mitigated, for example, through the use of a good heat conducting material of the microchannel.

Under the aforementioned assumptions, the continuity (2.22) and Navier–Stokes equations (2.27), (2.28) in the absence of body force terms reduce to

$$\frac{\partial(\rho u)}{\partial x} + \frac{\partial(\rho v)}{\partial y} = 0, \tag{2.67}$$

$$\rho\left(u\frac{\partial u}{\partial x} + v\frac{\partial u}{\partial y}\right) = -\frac{\partial p}{\partial x} + \mu\left[\frac{\partial^2 u}{\partial x^2} + \frac{\partial^2 u}{\partial y^2} + \frac{1}{3}\left(\frac{\partial^2 u}{\partial x^2} + \frac{\partial^2 v}{\partial x \partial y}\right)\right], \tag{2.68}$$

$$\rho\left(u\frac{\partial v}{\partial x} + v\frac{\partial v}{\partial y}\right) = -\frac{\partial p}{\partial y} + \mu\left[\frac{\partial^2 v}{\partial x^2} + \frac{\partial^2 v}{\partial y^2} + \frac{1}{3}\left(\frac{\partial^2 v}{\partial y^2} + \frac{\partial^2 u}{\partial x \partial y}\right)\right]. \tag{2.69}$$

Here, u and v are the streamwise and wall-normal components of velocity, i.e. along the x- and y-directions, respectively. We non-dimensionalize the above equations using the following scales (the non-dimensional variables are represented by a ˜ symbol),

$$\tilde{x} = \frac{x}{L};\ \tilde{y} = \frac{y}{H};\ \tilde{u} = \frac{u}{\bar{u}};\ \tilde{v} = \frac{v}{\bar{u}};\ \tilde{p} = \frac{p}{\bar{p}};\ \tilde{\rho} = \frac{\rho}{\bar{\rho}} \tag{2.70}$$

where L and H are the microchannel length and height, respectively, $\bar{u}$ is the velocity at the centerline and at the exit of the microchannel, $\overline{p}$ and $\overline{\rho}$ are the exit pressure and density values, respectively.

Invoking isothermal assumption and introducing equation of state for $\tilde{\rho}$ as $\tilde{p} = \tilde{\rho} RT$, we get the above equations in the non-dimensional form as

$$\left(\frac{H}{L}\right)\frac{\partial(\tilde{\rho}\tilde{u})}{\partial \tilde{x}} + \frac{\partial(\tilde{\rho}\tilde{v})}{\partial \tilde{y}} = 0,$$

$$\left(\frac{\bar{\rho}\bar{u}H}{\mu}\right)\tilde{\rho}\left[\left(\frac{H}{L}\right)\tilde{u}\frac{\partial \tilde{u}}{\partial \tilde{x}} + \tilde{v}\frac{\partial \tilde{u}}{\partial \tilde{y}}\right] = -\left(\frac{\bar{p}H^2}{\bar{u}L\mu}\right)\frac{\partial \tilde{p}}{\partial \tilde{x}} + \left(\frac{H}{L}\right)^2\frac{\partial^2 \tilde{u}}{\partial \tilde{x}^2} + \frac{\partial^2 \tilde{u}}{\partial \tilde{y}^2} + \frac{1}{3}\left[\left(\frac{H}{L}\right)^2\frac{\partial^2 \tilde{u}}{\partial \tilde{x}^2} + \left(\frac{H}{L}\right)\frac{\partial^2 \tilde{v}}{\partial \tilde{x}\partial \tilde{y}}\right],$$

$$\left(\frac{\bar{\rho}\bar{u}H}{\mu}\right)\tilde{\rho}\left[\left(\frac{H}{L}\right)\tilde{u}\frac{\partial \tilde{v}}{\partial \tilde{x}} + \tilde{v}\frac{\partial \tilde{v}}{\partial \tilde{y}}\right] = -\left(\frac{\bar{p}H}{\mu\bar{u}}\right)\frac{\partial \tilde{p}}{\partial \tilde{y}} + \left(\frac{H}{L}\right)^2\frac{\partial^2 \tilde{v}}{\partial \tilde{x}^2} + \frac{\partial^2 \tilde{v}}{\partial \tilde{y}^2} + \frac{1}{3}\left[\frac{\partial^2 \tilde{v}}{\partial \tilde{y}^2} + \left(\frac{H}{L}\right)\frac{\partial^2 \tilde{u}}{\partial \tilde{x}\partial \tilde{y}}\right].$$

Introducing the ratio of the channel height to its length as $\varepsilon = H/L$ (considered to be small), the Reynolds number, $Re = \bar{\rho}\bar{u}H/\mu$ and the exit Mach number, $Ma = \bar{u}/c$ $(c = \sqrt{\gamma RT})$,

$$\varepsilon\frac{\partial(\tilde{\rho}\tilde{u})}{\partial \tilde{x}} + \frac{\partial(\tilde{\rho}\tilde{v})}{\partial \tilde{y}} = 0, \tag{2.71}$$

$$Re\ \tilde{\rho}\left(\varepsilon\tilde{u}\frac{\partial \tilde{u}}{\partial \tilde{x}} + \tilde{v}\frac{\partial \tilde{u}}{\partial \tilde{y}}\right) = -\frac{\varepsilon Re}{\gamma Ma^2}\frac{\partial \tilde{p}}{\partial \tilde{x}} + \varepsilon^2\frac{\partial^2 \tilde{u}}{\partial \tilde{x}^2} + \frac{\partial^2 \tilde{u}}{\partial \tilde{y}^2} + \frac{1}{3}\left(\varepsilon^2\frac{\partial^2 \tilde{u}}{\partial \tilde{x}^2} + \varepsilon\frac{\partial^2 \tilde{v}}{\partial \tilde{x}\partial \tilde{y}}\right), \tag{2.72}$$

$$Re\ \tilde{\rho}\left(\varepsilon\tilde{u}\frac{\partial \tilde{v}}{\partial \tilde{x}} + \tilde{v}\frac{\partial \tilde{v}}{\partial \tilde{y}}\right) = -\frac{Re}{\gamma Ma^2}\frac{\partial \tilde{p}}{\partial \tilde{y}} + \varepsilon^2\frac{\partial^2 \tilde{v}}{\partial \tilde{x}^2} + \frac{\partial^2 \tilde{v}}{\partial \tilde{y}^2} + \frac{1}{3}\left(\frac{\partial^2 \tilde{v}}{\partial \tilde{y}^2} + \varepsilon\frac{\partial^2 \tilde{u}}{\partial \tilde{x}\partial \tilde{y}}\right). \tag{2.73}$$

Note that the numerator of the Reynolds number is equal to the mass flow rate per unit depth through the microchannel, and the viscosity in the denominator is constant because of the isothermal assumption; therefore, the Reynolds number does not vary with position in the microchannel.

The boundary conditions are such that the lateral velocity $\tilde{v}$ vanishes at the solid walls, and that the streamwise velocity $\tilde{u}$ is equal to the slip velocity. Therefore, Eq. (2.38) becomes:

$$\tilde{u}|_{wall} = C_1 Kn\left.\frac{\partial \tilde{u}}{\partial \tilde{y}}\right|_{wall}. \tag{2.74}$$

Additional boundary conditions required for the complete solution will be introduced later.

For flow in conventional channels, it is assumed that $\partial()/\partial x = 0$ (other than for pressure), which allows simplification of the equations. However, this cannot be done here, as the velocity can increase with x given the compressible nature of the flow. This behavior has led to a complex set of governing equations (2.71)–(2.73). In order to simplify the governing equations, we apply the perturbation technique and expand $\tilde{u}$, $\tilde{v}$, and $\tilde{p}$ in powers of ε (which is already assumed to be small):

$$\tilde{u} = \tilde{u}_o + \varepsilon\tilde{u}_1 + \varepsilon^2\tilde{u}_2 + \cdots \tag{2.75}$$

$$\tilde{v} = \varepsilon\tilde{v}_1 + \varepsilon^2\tilde{v}_2 + \cdots \tag{2.76}$$

$$\tilde{p} = \tilde{p}_o + \varepsilon\tilde{p}_1 + \varepsilon^2\tilde{p}_2 + \cdots \tag{2.77}$$

Note that in order to satisfy the law of conservation of mass, Eq. (2.71) dictates that the highest order terms in $\tilde{u}$ and $\tilde{v}$ be of the same order in ε, i.e. $\tilde{v}$ must be of $O(\varepsilon)$. Thus, the zeroth-order wall-normal velocity is not included in Eq. (2.76). Upon substituting the above series into the governing equations leads to a series of equations in power of ε, which are relatively easier to solve than the original governing equations.

Notice that the coefficients of certain terms in Eqs. (2.71)–(2.73) depend on the values of Re and Ma, which can independently be $O(\varepsilon)$, $O(1)$, or $O(1/\varepsilon)$, respectively, leading to nine different cases. Since Kn is related to Ma and Re through Eq. (2.66), its order gets fixed upon choosing the orders of Re and Ma. Here, we will consider $Ma = O(\varepsilon)$ and $Re = O(\varepsilon)$, which implies that $Kn = O(1)$; this case can represent typical microscale flows well. Thus, Re/Ma^2 involved in the momentum equations is $O(1/\varepsilon)$.

It is apparent that the pressure term in Eq. (2.73) being of $O(1/\varepsilon)$ is much larger than all the other terms in the equation. Therefore, Eq. (2.73) simplifies to

$$\frac{\partial \tilde{p}}{\partial \tilde{y}} = 0 \tag{2.78}$$

which leads to $\tilde{p} = \tilde{p}_o(\tilde{x})$. Using this result, the $\tilde{x}$-momentum equation (2.72), at $O(1)$, can be simplified as

$$\frac{\varepsilon Re}{\gamma Ma^2}\frac{d\tilde{p}_o}{d\tilde{x}} = \frac{\partial^2 \tilde{u}_o}{\partial \tilde{y}^2} \tag{2.79}$$

Notice that Eqs. (2.78), (2.79) are similar to what one would obtain for a conventional channel flow. In particular, the absence of inertial terms in Eq. (2.79) is underscored. However, in the present case, the streamwise velocity is dependent upon both the wall-normal ($\tilde{y}$) and streamwise ($\tilde{x}$) directions. This essential difference is due to the fact that the streamwise velocity in a microchannel keeps increasing, leading to $\partial \tilde{u}/\partial \tilde{x} > 0$. The implication of this continuous flow acceleration is that the pressure variation is no longer linear (although it is still only a function of streamwise coordinate) and the lateral velocity component is no longer zero (although it is a few orders of magnitude smaller than the streamwise velocity component; see Eq. (2.76)).

The $\tilde{x}$-momentum equation (Eq. (2.79)) can now be integrated twice with respect to $\tilde{y}$. The two constants of integration can be obtained by utilizing the symmetry condition ($\partial \tilde{u}_o/\partial \tilde{y} = 0$ at the microchannel centerline) and the slip-flow boundary condition (Eq. (2.74)), to obtain

$$\tilde{u}_o(\tilde{x}, \tilde{y}) = -\frac{\varepsilon Re}{8\gamma Ma^2}\frac{d\tilde{p}_o}{d\tilde{x}}\left(1 - 4\tilde{y}^2 + 4C_1\frac{Kn}{\tilde{p}_o}\right). \tag{2.80}$$

The last term in the above solution goes to zero as $Kn \to 0$; clearly, slip at the wall affects the velocity profile. The velocity profile in dimensional form is

$$u_o(x, y) = -\frac{1}{8}\frac{H^2}{\mu}\frac{dp_o(x)}{dx}\left(1 - 4\frac{y^2}{H^2} + 4C_1\frac{\lambda(x)}{H}\frac{\overline{p}}{p_o(x)}\right) \tag{2.81}$$

which shows the dependence of various terms involved in the above equation on either x or y. It is apparent that although the velocity profile is parabolic, the coefficients are a function of the streamwise coordinate.

The solution for $\tilde{u}_o$ can now be substituted into the continuity equation (2.71) to obtain $\tilde{v}_1$ as

$$\tilde{v}_1(\tilde{x}, \tilde{y}) = \frac{\varepsilon Re}{8\gamma Ma^2}\frac{1}{\tilde{p}_o}\left[\frac{1}{2}\frac{d^2(\tilde{p}_o)^2}{d\tilde{x}^2}\left(\tilde{y} - \frac{4}{3}\tilde{y}^3\right) + 4C_1 Kn\tilde{y}\frac{d^2\tilde{p}_o}{d\tilde{x}^2}\right]. \tag{2.82}$$

Since the wall-normal velocity must vanish at the wall ($\tilde{y} = \pm 1/2$), the term in the bracket should go to zero, leading (finally!) to the following equation for pressure:

$$\frac{d^2(\tilde{p}_o)^2}{d\tilde{x}^2} + 12C_1 Kn\frac{d^2\tilde{p}_o}{d\tilde{x}^2} = 0. \tag{2.83}$$

The solution of Eq. (2.83) can be found by integrating the equation twice. The boundary conditions are $\tilde{p}_o = 1$ at the outlet (since pressure has been normalized by the outlet pressure) and the pressure ratio across the microchannel is specified as, say, $\mathbb{P}$ (i.e. $\tilde{p} = \mathbb{P}$ at $x = 0$). The quadratic equation obtained upon integration can be easily solved to arrive at the following expression for the zeroth-order pressure distribution:

$$\tilde{p}_o(\tilde{x}) = -6C_1 Kn + \sqrt{(6C_1 Kn)^2 + (1 + 12C_1 Kn)\tilde{x} + (\mathbb{P}^2 + 12C_1 Kn\mathbb{P})(1 - \tilde{x})}. \tag{2.84}$$

It is apparent from the above solution that the variation of pressure in a microchannel is non-linear, as already suspected.

Note that only the first-order term in the perturbation series (Eqs. (2.80)–(2.83)) has been explicitly obtained here. However, as will become apparent, this suffices for our purpose. Having obtained the complete solution, the derived quantities like mass flow rate and friction factor can be obtained. The dimensional mass flow rate through the microchannel is given by

$$\dot{m}_{slip} = \frac{H^3 w P_o^2}{24\mu L R T}\left[\mathbb{P}^2 - 1 + 12C_1 Kn(\mathbb{P} - 1)\right] \tag{2.85}$$

where P_o is the (dimensional) outlet pressure, and w is the micro channel width. Notice the particularly strong dependence of the mass flow rate on the height of the microchannel H, which is qualitatively similar to that observed in conventional channels. To see the effect of slip on mass flow rate, we take the ratio of the above obtained mass flow rate with mass flow rate ($\dot{m}_{no\text{-}slip}$) obtained by putting $Kn = 0$ in the above expression. This yields

$$\frac{\dot{m}_{slip}}{\dot{m}_{no\text{-}slip}} = 1 + \frac{12C_1 Kn}{\mathbb{P} + 1}. \tag{2.86}$$

It is apparent from the above equation that the mass flow rate through the microchannel is enhanced due to wall slip. The amount of enhancement is larger at higher Knudsen number, larger value of slip coefficient (or more specular reflection at the wall), and smaller value of pressure ratio.

2.4.2.2 Integral Approach

In this section, we follow Dongari et al. [44] and obtain an analytical solution of gaseous slip flow in a long channel, by solving the integral form of the momentum equation. This methodology results in a more generalized solution due to the following reasons: a second-order slip boundary condition is applied thereby making the solution valid for a larger range of Knudsen number; inertial terms are retained to reveal a dependence on the Reynolds number.

Just like the previous section, the assumptions of steady isothermal fully developed flow are applicable in the present case as well. Moreover, entry and exit effects along with viscous compressive stresses are assumed to be negligible. These assumptions allow us to safely assume a parabolic velocity profile which takes the form

$$u(x, y) = A(x)y^2 + B(x)y + C(x) \tag{2.87}$$

Note that only the shape of the velocity profile is assumed and that the velocity field is still two-dimensional; also, the density (or pressure) field is a function of the streamwise coordinate only, as is demonstrated further.

For a microchannel of height H, it is worthwhile to note the following points at this stage:

1. The hydraulic diameter $D_h = 2H$ is chosen as the characteristic length scale. Accordingly, the Knudsen and Reynolds numbers are $Kn = \frac{\lambda}{2H}$ and $Re = \frac{\rho\bar{u}(2H)}{\mu}$ where λ is the mean free path length and $\bar{u} = \bar{u}(x)$ is the cross-section averaged streamwise velocity.
2. The Knudsen number is directly proportional to the rarefaction or inversely proportional to the pressure (p), i.e. $Kn \propto (1/p) \implies p \times Kn = \text{constant}$.
3. The ideal gas equation is employed, i.e. $p = \rho RT$, where R is the gas constant.

For a fully developed flow, the maximum velocity for any streamwise location is evaluated to be the centerline velocity. Therefore, using the condition that $\frac{\partial u}{\partial y}|_{y=\frac{H}{2}} = 0$, we can evaluate $B(x) = -AH$. Further, at $y = H$, using Eq. (2.46) and noting that $Kn = \frac{\lambda}{2H}$, we get $C(x) = (-AH^2)\big(2C_1Kn + 8C_2Kn^2\big)$. Lastly, keeping in mind that the flow fields are unit-depth quantities, we obtain $\bar{u} = \frac{\dot{Q}}{A} = \frac{1}{H}\int\limits_0^H u dy = (-AH^2)\big(\frac{1}{6} + 2C_1Kn + 8C_2Kn^2\big)$. Thus, making use of $B(x)$, $C(x)$ and $\bar{u}$, we can obtain the velocity profile as

$$u(x, y) = \bar{u}(x)\frac{y/H - (y/H)^2 + 2C_1Kn + 8C_2Kn^2}{1/6 + 2C_1Kn + 8C_2Kn^2} \tag{2.88}$$

Recalling that Re is constant in the streamwise direction ($Re = \dot{m}_{flow}/\mu$, $\dot{m}_{flow}$ is constant because of the steady state assumption and μ is constant due to the isothermal assumption) and using the equation of state, we can write $\bar{u}$ as

$$\bar{u}(x) = \frac{Re\mu RT}{p(x)D_h} \tag{2.89}$$

Therefore Eq. (2.88) can be written as

$$u(x, y) = \frac{Re\mu RT}{p(x)D_h}\left(\frac{y/H - (y/H)^2 + 2C_1Kn(x) + 8C_2Kn^2(x)}{1/6 + 2C_1Kn(x) + 8C_2Kn^2(x)}\right). \tag{2.90}$$

Knowing the velocity profile, the shear stress at the wall can be obtained from Eq. (2.88) as

$$\tau_w = -\mu\frac{\partial u}{\partial y}\bigg|_{y=H} = \frac{3Re\mu^2}{\rho H^2(1 + 12C_1Kn + 48C_2Kn^2)}. \tag{2.91}$$

In order to evaluate the pressure field, we make use of the integral formulation of the momentum balance in the streamwise direction. Consider a control volume of axial length δx located at a distance x from the leading edge of the microchannel. The forces acting on this elemental control volume are shown in Fig. 2.11. The momentum equation for this finite elemental volume spanning the cross section of the microchannel can be obtained using the integral momentum balance as

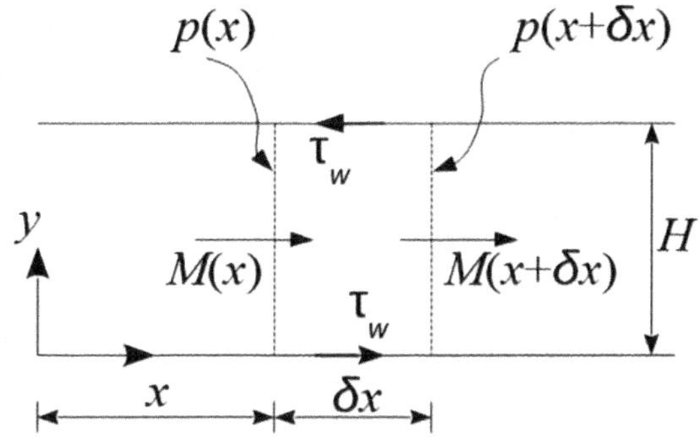

Fig. 2.11 Control volume for the integral momentum balance. $M(x) = \int_A \rho u^2 dA$, $M(x)$ is the momentum flux, A is the cross-sectional area into the plane of the page, $p(x)$ is pressure and τ_w is wall shear stress

$$
\begin{aligned}
&A\big[p(x) - p(x+\delta x)\big] + (-2\tau_w \times A_{lateral}) = M(x+\delta x) - M(x) \\
&\therefore A\left\{p(x) - \left[p(x) + \delta x \frac{dp}{dx} + \cdots\right]\right\} - 2\tau_w\left(\frac{A}{H}\delta x\right) \\
&= \left\{\left[M(x) + \delta x \frac{dM}{dx} + \cdots\right] - M(x)\right\} \\
&\therefore -A\delta x \frac{dp}{dx} - 2\frac{A}{H}\tau_w \delta x = \frac{dM(x)}{dx}\delta x + O(\delta x^2) \\
&\therefore -Adp - 2\frac{A}{H}\tau_w dx = d\left[\int_A \rho u^2 dA\right] + O(\delta x)
\end{aligned}
\tag{2.92}
$$

In arriving at Eq. (2.92), we have used Taylor series expansion in the second step and finally cast the equation in a variable separable form. To proceed further, we make the approximation that since δx is small, we can neglect the $O(\delta x)$ terms. Thus, the integral formulation of the momentum equation for a rectangular microchannel is

$$
-Adp - \frac{2A\tau_w}{H}dx = d\left[\int_A \rho u^2 dA\right] \tag{2.93}
$$

To solve for p, we first make use of Eq. (2.88) to simplify the integral appearing in Eq. (2.93) as

$$
\mathbb{I} = \int_A \rho u^2 dA = \rho \int_0^H u^2 dy = \rho \bar{u}^2 H \chi \tag{2.94}
$$

where

$$
\chi = \frac{\frac{1}{30} + \frac{2}{3}C_1 Kn + 4C_1^2 Kn^2 + \frac{8}{3}C_2 Kn^2 + 64C_2^2 Kn^4 + 32C_1C_2Kn^3}{(1/6 + 2C_1 Kn + 8C_2 Kn^2)^2} \tag{2.95}
$$

Substituting τ_w from Eq. (2.91) and $\mathbb{I}$ from Eq. (2.94) into Eq. (2.93), we get

$$-Hdp - 2\left[\frac{3Re\mu^2}{\rho H^2(1 + 12C_1Kn + 48C_2Kn^2)}\right]dx = d\left[\rho\bar{u}^2H\chi\right] \tag{2.96}$$

where we have made use of the fact that $A = H$.

Using $Kn\ p$ = constant, we get $Kn = (Kn_0 \times p_0)/p$ where Kn_0 and p_0 are known at some reference position x_0. Further, we have already established in Eq. (2.89) that $\bar{u} = \bar{u}(x)$. Thus, Eq. (2.96) can be expressed in a variable separable form to obtain the pressure field $p(x)$ as

$$\begin{aligned}\left(\frac{p}{p_0}\right)^2 &- 1 + 24C_1Kn_0\left(\frac{p}{p_0} - 1\right) + 96C_2Kn_0^2\log\frac{p}{p_0} \\ &+ 2Re^2\beta\chi\left[12C_1Kn_0\left(\frac{p_0}{p} - 1\right)\right. \\ &\left.+ 24C_2Kn_0^2\left(\left(\frac{p_0}{p}\right)^2 - 1\right) - \log\frac{p}{p_0}\right] = -96Re\beta\frac{x - x_0}{D_h}\end{aligned} \tag{2.97}$$

where χ is given in Eq. (2.95) and $\beta = (\mu^2RT)/(p_0^2D_h^2)$. Note the dependence of pressure on both Reynolds and Knudsen numbers from the above expression, with a more dominant effect of the latter non-dimensional number.

The lateral component of velocity (v) can now be obtained by substituting the expressions for $\rho(x)$ and $u(x, y)$ into the continuity equation, which yields

$$\begin{aligned}v(x, y) = &\left(\frac{12C_1Kn(x) + 96C_2Kn^2(x)}{1 + 12C_1Kn(x) + 48C_2Kn^2(x)}\right)\frac{1}{p^2(x)}\frac{Re\mu RT}{p(x)D_h}\frac{dp}{dx} \\ &\times y\left[1 - \frac{3y/H - 2(y/H)^2 + 12C_1Kn(x) + 48C_2Kn^2(x)}{1 + 12C_1Kn(x) + 48C_2Kn^2(x)}\right]\end{aligned} \tag{2.98}$$

The mass flow rate per unit depth through the microchannel can be obtained from the Reynolds number as

$$\dot{m} = \frac{Re\mu}{2}. \tag{2.99}$$

However, if the pressure ratio across the microchannel ($\mathbb{P}$) is specified for a given gas and microchannel, the mass flow rate per unit depth flowing through the microchannel can be determined from

$$\dot{m} = \frac{\mu}{2}\left[\frac{-a_2 + \sqrt{a_2^2 - 4a_1a_3}}{2a_1}\right] \tag{2.100}$$

where a_1, a_2, and a_3 are given as

$$a_1 = 2\beta\chi\left[12C_1Kn_0(\mathbb{P}-1) + 24C_2Kn_0^2(\mathbb{P}^2-1) + \log\mathbb{P}\right],$$

$$a_2 = \frac{48\beta L}{H}$$

$$a_3 = (1/\mathbb{P}^2 - 1) + 24C_1Kn_0(1/\mathbb{P} - 1) - 96C_2Kn_0^2\log\mathbb{P}$$

The Reynolds number can subsequently be calculated from Eq. (2.99).

Note on Use of the Equations

A note on the use of the above solution has been added here for convenience. Given is a microchannel of known dimensions (verify that it is long and two-dimensional) and a known gas flows through it at a given temperature (assumed to be constant). The interest is in one of the two cases: (1) finding out the mass flow rate for given inlet and outlet pressures (pressure driven system), or (2) finding out the pressure drop for a given mass flow rate with either inlet or outlet pressure specified (mass flow driven system). Of further interest would be finding the complete velocity and pressure fields in the microchannel.

First we pick values of slip coefficients: $C_1 = 1.1465$ and $C_2 = 0.164$ (Sect. 2.3.4). The solution proceeds by solving the equations in the following order:

Case 1 (Pressure Driven System) Calculate outlet Kn (since outlet pressure is known) and χ from Eq. (2.98). Find the mass flow rate from Eq. (2.100) (since $\mathbb{P}$ is known). Now calculate Re from Eq. (2.99). Take different values of pressure (between inlet and outlet) and find the corresponding x from Eq. (2.97) to obtain the streamwise variation of pressure in the microchannel. Finally, find $u(x, y)$ and $v(x, y)$ from Eqs. (2.90) and (2.98), respectively, since all the required variables are known by this stage.

Case 2 (Mass Flow Driven System with Specified Outlet Pressure) Calculate outlet Kn (since outlet pressure is known) and χ from Eq. (2.98). Calculate Re from Eq. (2.99) since $\dot{m}$ is known in this case. Guess different values of pressure (greater than the outlet pressure) and find the corresponding x from Eq. (2.97). Verify that the obtained value of x lies within the bounds of the microchannel. This gives the streamwise variation of pressure in the microchannel. Finally, find $u(x, y)$ and $v(x, y)$ from Eqs. (2.90) and (2.98), respectively, since all the required variables are known by this stage.

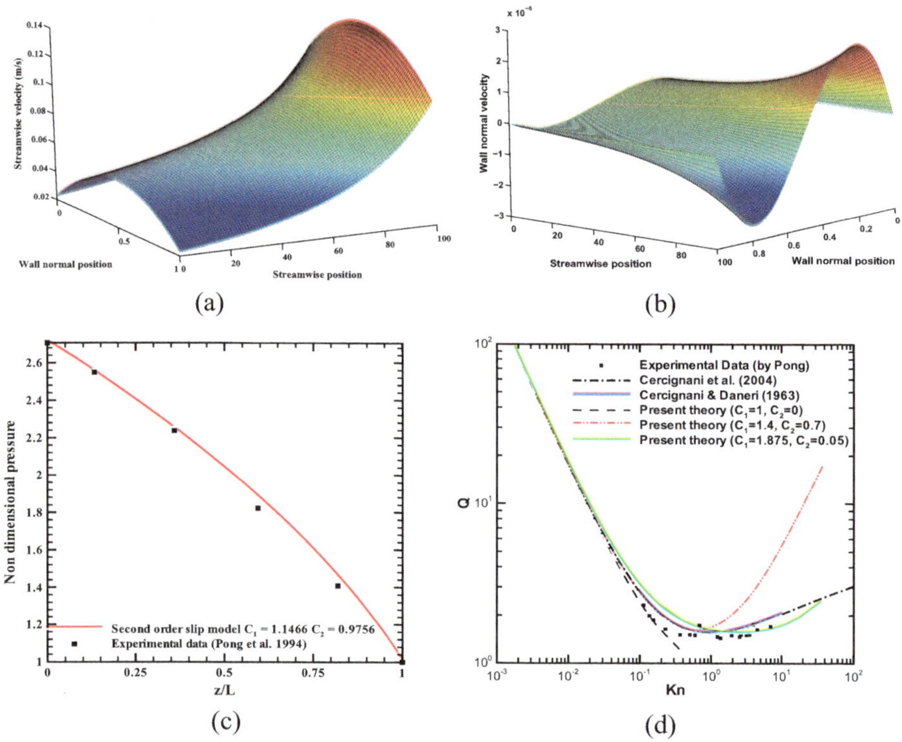

Fig. 2.12 (**a**) Variation of streamwise velocity as predicted by the theory. (**b**) Variation of cross stream velocity as predicted by the theory. (**c**) Variation of pressure as predicted by the theory, and comparison against the experimental data of Pong et al. [115], for outlet Knudsen number of 0.059. (**d**) Variation of normalized volume flux versus Knudsen number, and comparison against the theoretical results of Cercignani et al. [30]. All figures taken from Dongari et al. [44]

2.4.2.3 Comments

The variations of streamwise velocity, cross stream velocity, and pressure with streamwise coordinate are plotted in Fig. 2.12. The streamwise velocity is maximum at the centerline and increases with streamwise coordinate owing to the acceleration of the flow (Fig. 2.12a). Notice the slip at the wall, with the magnitude of slip velocity increasing with streamwise coordinate. The cross stream velocity is zero at the centerline (due to symmetry, the flow cannot move towards or away from the centerline), and also zero at the walls (due to non-porous nature of the walls) (Fig. 2.12b). For the case considered, the magnitude of cross stream velocity is about four orders of magnitude smaller than the streamwise velocity. The small difference in the amount of flow acceleration with lateral position leads to a finite (but small) amount of cross stream velocity.

The pressure drops non-linearly with streamwise coordinate (Fig. 2.12c) because of additional contribution from the change in momentum of the gas (Eq. (2.131)). Pressure is easier to measure than velocities in a microchannel, although measurement of pressure at intermediate locations is much more involved than measuring pressures at the two ends of the microchannel. A comparison of the theoretical pressure variation with the experimental data of Pong et al. [115] is also shown in the figure.

Figure 2.12d shows the variation of non-dimensional volume flow rate versus Knudsen number, and the theoretical results are seen to compare well against some available experimental and linearized Boltzmann equation based theoretical results. Note the appearance of a minima in the curve—known as the "Knudsen minima." Since the expressions are given in terms of slip coefficients, one can change the value of the slip coefficients and probe the magnitude and location of the minima. Interestingly, it is seen that the minima does not appear upon choosing $C_2 = 0$, i.e., for first-order slip model, while the minima appears for all cases tested with $C_2 \neq 0$. The two sets of values for slip coefficients used in Fig. 2.12d, although hypothetical, help underscore that the Navier–Stokes equation based solution can be extended well into the transition regime with the help of a second-order slip model, as the proposed solution matches the linearized Boltzmann equation based theoretical results till $Kn \approx 1$ and even higher depending on the value of C_1 and C_2. However picking the right values of slip coefficients remains an issue in this approach.

We now show that the present solution is more general than the one obtained in Sect. 2.4.2.1. On differentiating Eq. (2.97) twice, with respect to x and substituting $C_2 = 0$, we obtain:

$$\frac{d^2}{dx^2}\left(\frac{p}{p_o}\right)^2 + \frac{24C_1Kn_o}{p_o}\frac{d^2p}{dx^2}$$
$$+ 2Re^2\beta\chi\left(\frac{24C_1Kn+1}{p^2}\left[\frac{dp}{dx}\right]^2 - \frac{12C_1Kn+1}{p}\frac{d^2p}{dx^2}\right) = 0 \qquad (2.101)$$

which can be compared to Eq. (2.83). The coefficient of d^2p/dx^2 in Eq. (2.101) is different because of the use of hydraulic diameter as the length scale instead of H, which changed the definition of Kn by a factor of 2. The extra terms in Eq. (2.101) are because the inertial terms are retained in the integral analysis; the magnitude of the terms is proportional to the square of the Reynolds number. Since in Sect. 2.4.2.1 we assumed that Reynolds number is of order ϵ, the extra terms do not appear there, thereby confirming the more general nature of the solution in Sect. 2.4.2.2.

2.4.3 Flow in a Microtube

The solution of gas flow in microtube essentially follows the same steps as for gas flow in a microchannel discussed in Sect. 2.4.2.2. The assumptions of the flow being two-dimensional, locally fully developed and isothermal with negligible viscous compressive stresses applies to the present case as well.

For a second-order slip boundary condition, we can obtain the velocity distribution as

$$\frac{u}{\bar{u}} = \frac{1 + 4C_1 Kn + 8C_2 Kn^2 - r^2/R^2}{1/2 + 4C_1 Kn + 8C_2 Kn^2} \tag{2.102}$$

Knowing the velocity profile, the shear stress at the wall can be obtained as

$$\bar{\tau}_w = \frac{8Re\mu^2}{\rho D^2(1 + 8C_1 Kn + 16C_2 Kn^2)} \tag{2.103}$$

where $\bar{u}$ = mean velocity, $Re = \rho \bar{u} D/\mu$, $Kn = \lambda/D$, and D is the microtube diameter.

As with Eq. (2.93), we can express the momentum balance between two cross sections of the microtube as

$$-\frac{\pi D^2}{4} dp - \bar{\tau}_w \pi D dz = d\left[\int_0^{\frac{D}{2}} \rho u^2 (2\pi r) dr\right] \tag{2.104}$$

where z is the coordinate along the axis of the tube.

Integrating Eq. (2.104) with the help of Eqs. (2.102) and (2.103), we obtain the following expression for pressure:

$$\left(\frac{p}{p_o}\right)^2 - 1 + 16C_1 Kn_o\left(\frac{p}{p_o} - 1\right) + 32C_2 Kn_o^2 \log\frac{p}{p_o} + 2Re^2\beta\chi$$
$$\left\{8C_1 Kn_o\left(\frac{p_o}{p} - 1\right) + 8C_2 Kn_o^2\left[\left(\frac{p_o}{p}\right)^2 - 1\right] - \log\frac{p}{p_o}\right\} = -64Re\beta\frac{z - z_o}{D} \tag{2.105}$$

where p_0 is the pressure at some reference position z_0, $\beta = \frac{\mu^2 RT}{p_o^2 D^2}$, and

$$\chi = \frac{1/3 + 4C_1 Kn + 16C_1^2 Kn^2 + 8C_2 Kn^2 + 64C_2^2 Kn^4 + 64C_1 C_2 Kn^3}{(1/2 + 4C_1 Kn + 8C_2 Kn^2)^2}.$$

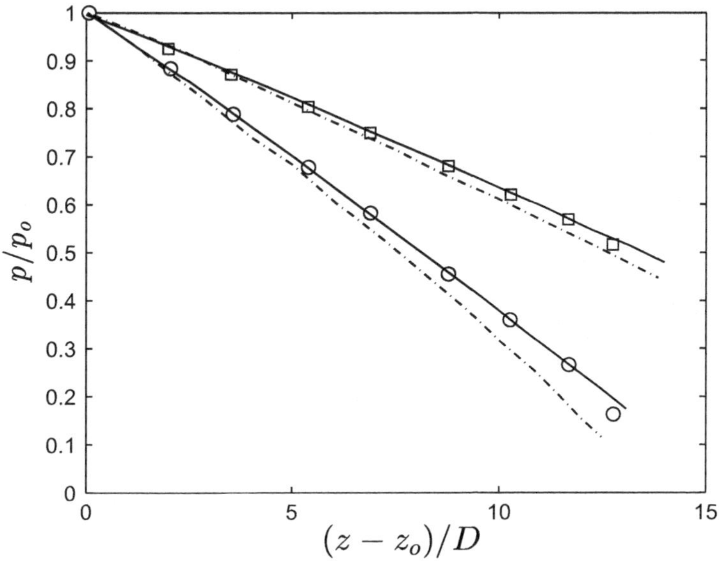

Fig. 2.13 Pressure drop according to slip theory compared with the experimental results [137], solid line represents second order slip theory results ($C_1 = 1.1466, C_2 = 0.14$), dashed line represents first order theory results ($C_1 = 1.1466$), circles correspond to experimental results for $Kn = 0.265$, $Re = 0.147$, squares correspond to experimental results for $Kn = 0.254$, $Re = 0.093$, [137]

The variation of pressure as a function of position in the microtube is plotted in Fig. 2.13, from where it is clear that the nature of the graph is qualitatively similar to the pressure variation in a microchannel (Fig. 2.12c). The theoretical results are also compared with the experimental data of Sreekanth [137] at reasonably high values of Kn and a good match is seen, especially when second-order slip model is employed.

The solution for microtube obtained in this section can be used for a pressure driven or mass flow driven system in a similar manner as described for microchannels.

2.4.4 Flow in an Arbitrary Cross Section Microchannel

In the above sections we have considered flow in passages which have a regular cross section. Since high aspect ratios could be realized for these cross sections, we consider the flow to be two-dimensional. This assumption however does apply, for example, where the side walls are not very far away from the centerline of the passage (i.e., the width of the rectangular cross section passage is comparable to its height). Fabricating circular or rectangular cross section microchannels also turns out to be difficult, whereas trapezoidal (or triangular) cross section microchannels are more common in practice. This motivates us to examine flow in microchannel of alternate cross sections.

Gas flow in an arbitrary cross section microchannel was solved by Duan [46] under the conditions that the flow is locally fully developed and isothermal. Any change in momentum of the flow was neglected; therefore, the pressure force only helps in overcoming the friction force but does not accelerate the flow.

Duan assumed the slip model to be of the form:

$$w|_{y=a,b} = -\frac{2-\sigma_v}{\sigma_v}\lambda \left.\frac{\partial w}{\partial y}\right|_{y=a,b} - C_2\lambda^2\left(\frac{2-\sigma_v}{\sigma_v}\right)^2\left(\left.\frac{\partial^2 w}{\partial y^2}\right|_{y=a,b} + \frac{1}{2}\left.\frac{\partial^2 w}{\partial x^2}\right|_{y=a,b}\right) \tag{2.106}$$

where w is the streamwise velocity component, x and y are the coordinates along the lateral directions, and a and b are the base width and height of the microchannel. The velocity distribution and average velocity $\bar{w}$ was obtained as

$$w(x,y) = -\frac{b^2}{\mu}\frac{dp}{dz}\left[\sum_{i,j=1}^{\infty} A_{ij}\cos\left(\varphi_i \frac{x}{a\varepsilon}\right)\cos\left(\psi_j \frac{y}{b}\right) + 2C_2\frac{4}{(1+\varepsilon)^2}Kn^{*2}\right] \tag{2.107}$$

$$\bar{w}(x,y) = -\frac{b^2}{\mu}\frac{dp}{dz}\left[\sum_{i,j=1}^{\infty} A_{ij}\frac{\varepsilon\sin\left(\frac{\varphi_i}{\varepsilon}\right)\sin\psi_j}{\varphi_i\psi_j} + 2C_2\frac{4}{(1+\varepsilon)^2}Kn^{*2}\right] \tag{2.108}$$

where ε is the effective aspect ratio of the microchannel (and related appropriately to a and b through the geometry of the cross section; Fig. 2.14), Kn^* is modified Knudsen number defined as

$$Kn^* = \frac{2-\sigma_v}{\sigma_v}Kn,$$

φ_i and ψ_j are eigenvalues obtained from the following equations

$$\frac{4}{(1+\varepsilon)}Kn^*\varphi_i \tan\frac{\varphi_i}{\varepsilon} + 2C_2\frac{4}{(1+\varepsilon)^2}Kn^{*2}\varphi_i^2 = 1$$

$$\frac{4}{(1+\varepsilon)}Kn^*\psi_j \tan\psi_j + 2C_2\frac{4}{(1+\varepsilon)^2}Kn^{*2}\psi_j^2 = 1$$

and

$$A_{ij}=\frac{1}{\varphi_i^2+\psi_j^2}\left[\frac{4\sin(\varphi_i/\varepsilon)\sin\psi_j}{\varphi_i\psi_j}+4C_2\frac{2}{1+\varepsilon}Kn^*\left(\frac{\sin(\varphi_i/\varepsilon)\cos\psi_j}{\varphi_i}+\frac{\cos(\varphi_i/\varepsilon)\sin\psi_j}{\psi_j}\right)\right]$$
$$\left[\left(\frac{1}{\varepsilon}+\frac{\sin(2\varphi_i/\varepsilon)}{2\varphi_i}\right)\left(1+\frac{\sin 2\psi_j}{2\psi_j}\right)+2C_2\frac{2}{1+\varepsilon}Kn^*\left(\cos^2\psi_j\left(\frac{1}{\varepsilon}+\frac{\sin(2\varphi_i/\varepsilon)}{2\varphi_i}\right)\right.\right.$$
$$\left.\left.+\cos^2\frac{\varphi}{\varepsilon}\left(1+\frac{\sin 2\psi_j}{2\psi_j}\right)\right)\right]^{-1}$$

The pressure as a function of position (z) can be related to the inlet and outlet pressures (p_i and p_o, respectively) through

$$\left[\frac{p_i^2}{p_o^2}-\frac{p^2}{p_o^2}+2\alpha_1 Kn_o^*\left(\frac{p_i}{p_o}-\frac{p}{p_o}\right)+2C_2\alpha_2 Kn_o^{*2}\ln\frac{p_i}{p}\right]$$
$$=\frac{z}{L}\left[\frac{p_i^2}{p_o^2}-1+2\alpha_1 Kn_o^*\left(\frac{p_i}{p_o}-1\right)+2C_2\alpha_2 Kn_o^{*2}\ln\frac{p_i}{p}\right] \tag{2.109}$$

where L is the length of the microchannel, and coefficients α_1 and α_2 are related to the aspect ratio of the microchannel through relations

$$\alpha_1 = 12 - 10.598\varepsilon + 8.654\varepsilon^2 - 2.231\varepsilon^3$$
$$\alpha_2 = 48 - 112.7\varepsilon + 194.2\varepsilon^2 - 155.7\varepsilon^3 + 42.66\varepsilon^4$$

The mass flow rate ($\dot{m}$) through the microchannel can be found as

$$\dot{m} = \rho\bar{w}A$$
$$=\left(\frac{A^{5/2}}{P}\right)\frac{p_0^2}{\mu RTL(fRe_{\sqrt{A}})_{ns}}\left[\frac{p_i^2}{p_o^2}-1+2\alpha_1 Kn_o^*\left(\frac{p_i}{p_o}-1\right)\right.$$
$$\left.+2C_2\alpha_2 Kn_o^{*2}\ln\frac{p_i}{p_o}\right] \tag{2.110}$$

where P is the wetted perimeter of the microchannel, square root of the cross-sectional area ($\sqrt{A}$) has been used as the length scale of the microchannel (instead of hydraulic diameter), and $(fRe_{\sqrt{A}})_{ns}$ is obtained by setting $Kn^* = 0$ in Eq. (2.136).

Duan [46] claims that the above result can be used for describing slip flow in circular and non-circular cross sections such as (Fig. 2.14): rectangular, trapezoidal, double-trapezoidal, triangular, rhombic, hexagonal, octagonal, elliptical, semi-elliptical, parabolic, circular sector, annular sector, rectangular duct with unilateral elliptical or circular end, and annular. Therefore, the analysis presented in this section can be applied to a large number of practical situations.

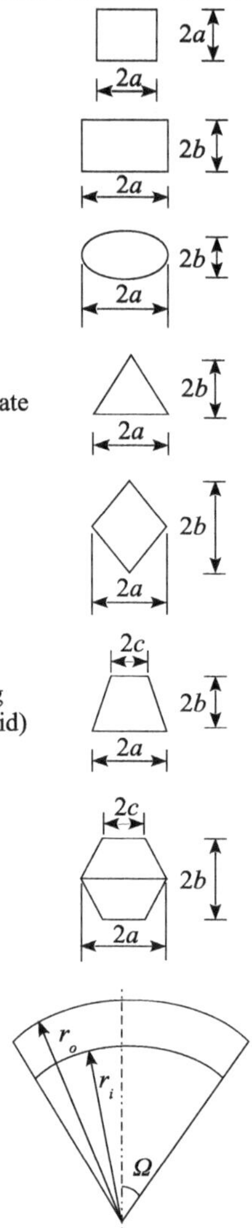

Geometry	Cross section	Effective aspect ratio	
Regular polygons		$\varepsilon = 1 \quad N \geq 4$	
Rectangle		$\varepsilon = \frac{b}{a}$	
Ellipse		$\varepsilon = \frac{b}{a}$	
Arbitrary triangle (including approximate triangle)		$\varepsilon = \left[\left(\frac{2b}{a}\right)^n + \left(\frac{2a}{b}\right)^n\right]^{-1/n}$	$n = 0.53$
Rhombus		$\varepsilon = \left[\left(\frac{2b}{a}\right)^n + \left(\frac{2a}{b}\right)^n\right]^{-1/n}$	$n = 0.68$
Trapezoid (including approximate trapezoid)		$\varepsilon = \left[\left(\frac{2b}{a+c}\right)^n + \left(\frac{a+c}{2b}\right)^n\right]^{-1/n}$	$n = 1.7$
Double trapezoid		$\varepsilon = \left[\left(\frac{a+c}{2b}\right)^n + \left(\frac{1+\frac{c}{a}}{2a/b}\right)^n\right]^{-1/n}$	$n = 2.8$
Annular sector		$\varepsilon = \frac{1-r}{\Omega(1+r)}$ where, $r = \frac{r_i}{r_o}$	

Fig. 2.14 Effective aspect ratio for different arbitrary cross sections [46]

2.4.5 Flow in the Annulus of Rotating Sphere and Cylinder

In this section we will briefly examine the theory of the spinning rotor gauge in the slip regime. As seen in Sect. 2.3.4, spinning rotor gauge has been employed for measuring the value of TMAC, and therefore its study is relevant. The spinning rotor gauge comprises a small sphere magnetically levitated within a cylindrical tube

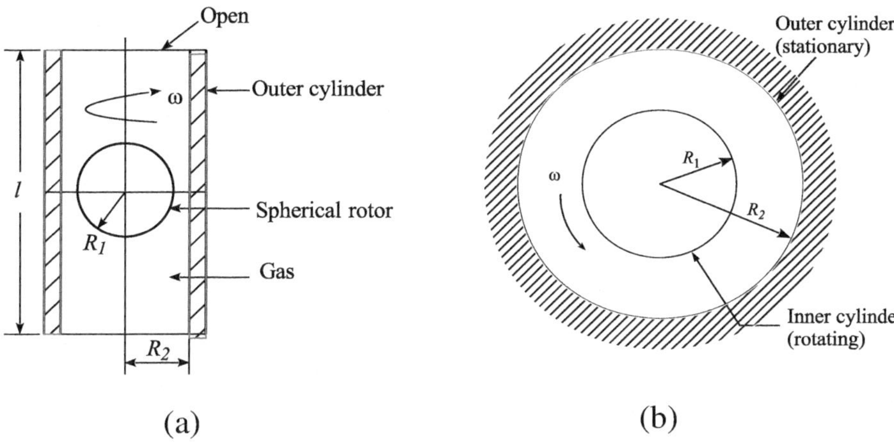

Fig. 2.15 Schematic of (**a**) spinning rotor gauge (**b**) two concentric cylinders

(Fig. 2.15). The sphere is levitated so as to avoid any physical contact between the sphere and the tube. The sphere is first rotated electromagnetically through the field generated by coils surrounding the tube. The driving force is then removed; owing to friction between the sphere and the surrounding gas, the sphere starts to decelerate. From measurement of deceleration of the sphere, obtained through variations in the field induced on the coils, the resulting torque on the sphere can be obtained, from which the value of TMAC can be deduced. The interest therefore is in obtaining an expression for torque in terms of TMAC.

The solution of the spinning rotor gauge was obtained by Loyalka [90]. He neglected the inertial terms in the Navier–Stokes equations due to the low Reynolds number of the flow, and cast the resulting Stokes equations in spherical coordinates. The slip condition was applied at the surfaces of both the inner sphere and the outer cylinder. After applying Jeffery's transformation and some involved algebra, the following expression was obtained:

$$\frac{T}{\mu\omega} = 8\pi R_1^3 C_o(R_1/R_2)\left(1 + \frac{3C_1\lambda}{R_1} + \frac{C_1\lambda}{R_2}\right) \tag{2.111}$$

where T is the torque acting on the sphere, ω is the angular speed of the sphere, R_1 is the radius of the sphere, R_2 is the radius of the tube in which the sphere is suspended, C_1 is the slip coefficient $(= (2-\sigma_v)/\sigma_v)$, and coefficient C_o is given by

$$C_o(R_1/R_2) = \left\{1 - 0.79682417\left(\frac{R_1}{R_2}\right)^3 - 0.060047040\left(\frac{R_1}{R_2}\right)^{10}\right\}^{-1}.$$

Interested readers can refer to the original paper for further details about the analysis.

Two concentric cylinders, with the inner cylinder rotating while the outer cylinder is stationary is another configuration used to obtain the value of TMAC, and therefore a brief description of this case is included here. Agrawal and Prabhu [4] obtained an expression for torque acting on the cylinder in slip regime. The cylindrical form of the Navier–Stokes equations for the problem simplifies to

$$-\frac{u_\theta^2}{r} = -\frac{1}{\rho}\frac{dp}{dr} \tag{2.112}$$

$$\mu\frac{d}{dr}\left(\frac{1}{r}\frac{d}{dr}(ru_\theta)\right) = 0 \tag{2.113}$$

where u_θ is the tangential velocity and r is the radial coordinate. The above equations are momentum equations in radial and tangential directions, respectively. The slip condition at the walls gives

$$u_\theta = \omega R_1 + C_1\lambda\left(\frac{du_\theta}{dr} - \frac{u_\theta}{r}\right) \quad \text{at } r = R_1 \tag{2.114}$$

$$u_\theta = -C_1\lambda\left(\frac{du_\theta}{dr} - \frac{u_\theta}{r}\right) \quad \text{at } r = R_2 \tag{2.115}$$

where R_1 and R_2 are the radius of the inner and outer cylinder, respectively, and ω is the angular speed of the inner cylinder. Note that the slip velocity is related to the shear stress at the wall which leads to the extra term in cylindrical coordinate.

The solution for velocity can be obtained from Eq. (2.113) in a manner similar to conventional flows (albeit with slip boundary condition) as

$$u_\theta = \frac{\omega}{A - B}\left(Ar - \frac{1}{r}\right) \tag{2.116}$$

where

$$A = \frac{1}{R_2^2}\left(1 - C_1\frac{2\lambda}{R_2}\right),$$

$$B = \frac{1}{R_1^2}\left(1 + C_1\frac{2\lambda}{R_1}\right)$$

from where the torque (T) on the inner cylinder can be found as

$$T = \frac{4\pi\mu\omega L}{B - A} \tag{2.117}$$

where L is the length of the cylinders.

2.5 Observations on Flow in Straight Passages

Some interesting observations in microchannels such as Knudsen minima is reviewed in this section. We also comment on two other practical aspects: effect of transients and roughness on the flow.

2.5.1 *Appearance of Knudsen Minima*

An interesting phenomenon exhibited by gas flow in microchannel is the appearance of a minimum in normalized flow rate plotted against rarefaction factor. This was first noted in the measurements of Knudsen in 1909 while investigating flow of carbon dioxide in small capillaries. Subsequent measurements have confirmed the presence of this "Knudsen minima" at Kn close to unity. For example, Fig. 2.16 (taken from Hemadri et al. [70]) shows measurements from two different research groups, with three different gases showing this minima. Steckelmacher [138] reviewed the numerous developments that immediately followed the seminal work of Knudsen. The paper presents formula for flow resistance in various cross section tubes, in the free molecular regime, as given by Knudsen and corrected by Smoluchowski. He also mentions that the appearance of Knudsen minima occurs only for relatively long tubes.

Some researchers view the flow of gas under an imposed pressure gradient as a superimposition of flow due to convection and molecular diffusion [114, 161]. The contribution of convection dominates at higher pressure; however, its contribution

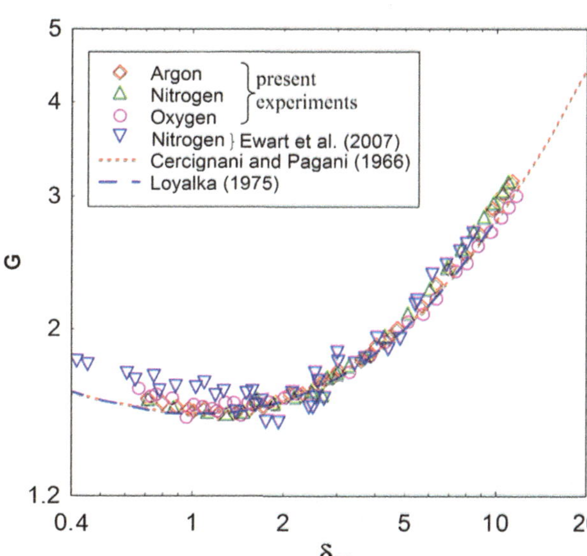

Fig. 2.16 Comparison of non-dimensional mass flow rate G with respect to mean rarefaction parameter δ_m (δ_m is inversely proportional to Knudsen number and given as $\delta_m = \sqrt{\pi}/(2Kn)$), Hemadri et al. [70]

to the overall flow reduces with a reduction in pressure. On the other hand, the contribution of molecular diffusion is insignificant at high pressure, but the contribution increases with a reduction in pressure. Therefore, the overall flow rate first reduces before increasing, and leads to the appearance of a minima.

Alternatively, Gu et al. [69] viewed the phenomenon in terms of centerline velocity and slip velocity. Recall, that for a given flow rate, in presence of a slip, the centerline velocity will be smaller as compared to its no-slip counterpart. Therefore, any change in flow rate in the slip regime will be accompanied by changes in both the centerline velocity and the slip velocity. It was observed that initially, the rate of decrease in centerline velocity is greater than the rate of increase in slip velocity, which leads to a reduction in the flow rate. Subsequently, the trends reverse and the flow rate starts increasing, leading to the appearance of the minima. It needs to be underscored that the dimensional flow rate decreases monotonically with increase in rarefaction; the minimum is observed only upon normalization.

The Knudsen minima is seen to occur in passages of both uniform cross section and non-uniform cross section. The latter has been shown only recently through the experiments of Hemadri et al. [70].

2.5.2 *Flow in Rough Microchannel*

Surface roughness (ε) is expected to play a significant role at the microscale and its role has been actively investigated. This is because the value of roughness in a relative sense (i.e., ε/D_h) increases when the length scale (D_h) decreases. A schematic diagram of cross section of rough microtube is shown in Fig. 2.17. Because of the random nature of the surface roughness, it is extremely difficult to accurately predict the local radius r. Instead, mean radius a which is the average value of the local radius, r, is used. As per our conventional understanding, surface roughness is not important in the laminar regime and relevant only in the turbulent regime (one can refer, for example, to Moodys chart in this context). Since, flows in the microscale are mostly laminar (because of the low value of Reynolds number), will the relatively large value of ε/D_h still affect the fluid flow and heat transfer, or can our expectation from conventional flows be simply extended to microscale, therefore becomes a relevant question.

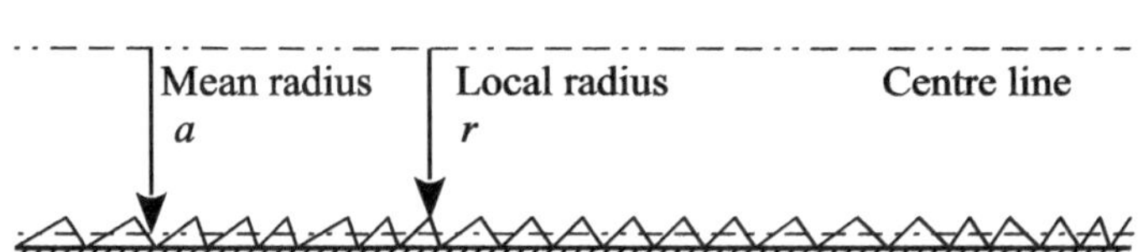

Fig. 2.17 Schematic diagram of cross section of rough microtube

Techniques used to enhance roughness are adherence of sand particles, microwire electrical discharge machining, laser drilling, electrochemical etching, and chemical etching. Acid based etching of the microchannels has also been suggested for producing rough microchannels, with the grade of roughness being varied by the etching time and acid concentration. Several instruments are available to characterize the features on a surface (such as white light interferometry, digital microscope, and stylus-based profilometer) that can help determine the value of surface roughness. Cleanliness of the surface plays a crucial role, since presence of impurities can change the surface roughness characteristics. Another way to characterize surface roughness qualitatively is by measuring the dynamic advancing and receding contact angles of the surface. The contact angle increases with increasing surface roughness due to enhanced wetting characteristics.

Measurements on rough microchannel are problematic because of the difficulty in controlling and characterizing the amount of roughness. Further, roughening of the microchannel tends to change its hydraulic diameter; although this change is small, the effect gets amplified due to the high sensitivity of pressure drop on hydraulic diameter. However, the change in hydraulic diameter is much more difficult to characterize, primarily because of the difficulty in finding a reference from where the measurements of height and width should be undertaken. As the standard deviation of the hydraulic diameter measurements increases, the uncertainty in friction factor also increases. Turner et al. [151] did not find any significant effect of surface roughness in slip regime, because of high experimental uncertainty masking any differences observed due to surface roughness. Some authors found friction factor to enhance at microscale [1, 144, 162, 169], while some found friction factor to reduce [34, 113, 174]. It is ambiguous that this increment or decrement owes to compressibility, rarefaction, or roughness effect.

On the other hand, simulations become difficult because of the requirement to resolve the region close to the roughness element well, which increases the requirement for the total number of grid points in the domain. Further, performing simulations with random roughness is difficult. Due to these reasons, many simulations have considered much larger roughness elements arranged periodically on a surface. This shows that there is no consensus at present in the literature about the effect of surface roughness on friction factor in gaseous flows at microscale. Even no claim can be made, on the effect of roughness shape geometries, height, width, spacing, density distribution, and angle of roughness inclination on friction factor in microchannel flows. The scales of microfluidic systems would continue to decrease in future, which would increase the relative significance of exploring the influence of surface roughness.

2.5.3 Transient Flow in a Capillary

A long capillary connected to a reservoir at each of its two ends with the pressure of the reservoirs varying with time has also been considered. Interest is in finding the evolution of the pressure in the two end-reservoirs as a function of the rarefaction parameter. Practically, the problem is relevant for finding out how rapidly a rarefied flow will respond to a sudden change to an externally imposed pressure gradient coming, for example, from a vacuum pump or a valve.

Lihnaropoulos and Valougeorgis [87] solved the above problem by employing linearized unsteady BGK equation, along with the Maxwell's boundary conditions (assuming purely diffuse interaction). The gas starts moving due to the suddenly applied pressure gradient across the ends of the tube, and the variation of velocity as a function of radial position and time can be calculated. With passage of time, the gas reaches the steady-state flow field in an asymptotic manner. The time required for the flow to reach its fully developed characteristics is minimum close to Knudsen minimum of unity. Sharipov and Graur [127] found that the time for attaining the equilibrium pressure is maximum in the transition regime (i.e., close of the value of Knudsen minimum) and decreases rapidly in the continuum regime.

2.6 Observations on Flow in Complex Passages

A microchannel is unlikely to be straight in most practical applications; it would rather involve bends, sudden or gradual change in cross-sectional area, bifurcation, junction, and other complex features. The interest is therefore in understanding flows in "complex" microchannels comprising one of these features. A clear understanding of the flow behavior through such sections is unarguably of both practical and fundamental interest.

2.6.1 Flow in Sudden Expansion/Contraction Microchannel

A sudden change in the cross-sectional area can have a strong effect on the local flow dynamics, and the disturbance created by the junction to the flow can travel several diameters upstream and downstream of the junction. The strength of the disturbance depends on the area ratio of expansion/contraction, the geometry of the cross section, the Reynolds number, and the Knudsen number of the flow. Appearance of a corner eddy at the sudden expansion/contraction section is the most noticeable feature of the flow in the continuum regime (Fig. 2.18a). Will the corner vortex still form if the Knudsen number is changed from zero to a finite value, and how far upstream and downstream will the disturbance propagate, are therefore natural questions to wonder about.

The measurements of Varade et al. [156] in a scaled-up facility comprising two tubes of different cross-sectional areas connected to form a junction, surprisingly showed that corner eddies are not formed in the slip flow regime. The maximum Reynolds number covered in these measurements is 837 and area ratio is 64. Far from the junction, the pressure was found to drop non-linearly in each of the two tubes and the pressure variation compares well with the theoretical results

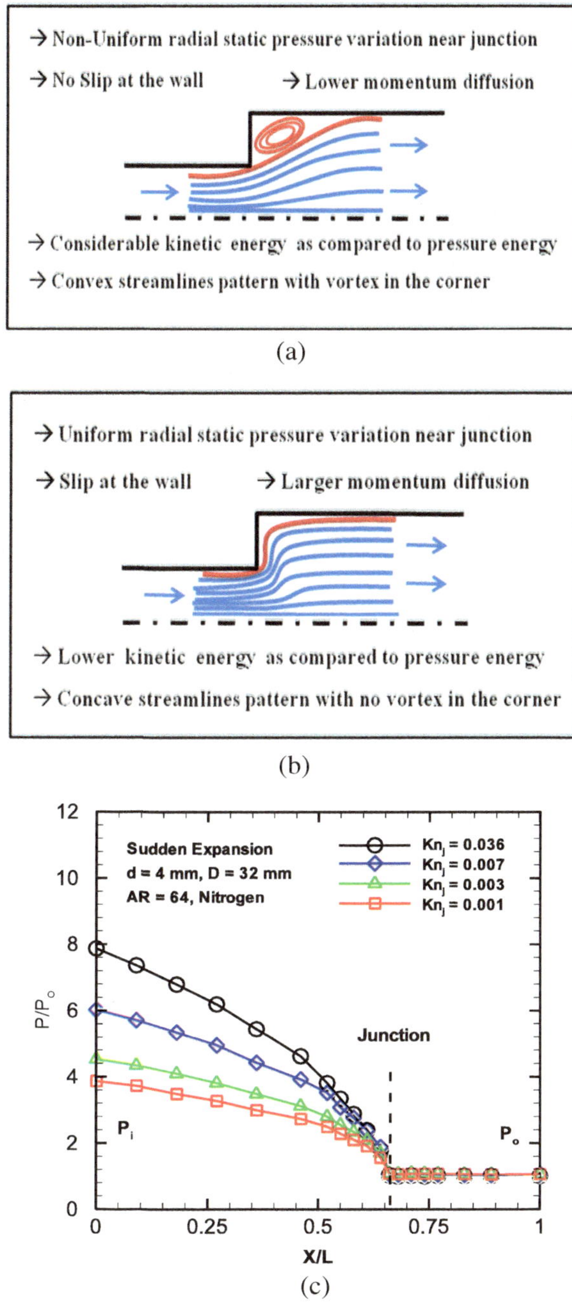

Fig. 2.18 (**a**) Schematic streamlines of laminar, incompressible, separated flow near the junction of a tube with a sudden change in cross-sectional area. (**b**) Schematic streamlines of rarefied gas flow near the junction. (**c**) Experimentally measured static pressure variation (normalized with outlet pressure) along the wall for flow of low pressure gas in a tube with a sudden change in cross-sectional area. All figures taken from Varade et al. [156]

of Sect. 2.4.3. However, the pressure was found to drop more rapidly than the theoretical pressure close to the junction; the pressure profiles also exhibited a discontinuity in slope at the junction (Fig. 2.18c). The variation of pressure is in contrast to observation in the continuum regime, where no such discontinuity is seen. In fact, in the continuum regime, the pressure curve shows an adverse pressure gradient corresponding to the flow separation region, a feature which was not seen in the experiments of Varade et al. The authors explained the observations through an increased momentum owing to slip at the wall, and enhanced radial diffusion of flow downstream of the junction, due to enhanced diffusivity brought about by the reduction in density (or pressure) in their system (Fig. 2.18b). However, these measurements do not rule out the possibility of finding very small corner eddies, of size smaller than the resolution in their measurement.

The following correlation was proposed for calculating the additional pressure drop (p_{add}) in sudden expansion case [156]:

$$K_{add,SE} = \frac{16.36(AR)^{0.63}}{(Kn_j)^{0.45}(Re_s)^{1.2}} \tag{2.118}$$

where AR is the area ratio, Kn_j is the Knudsen number at the junction, Re_s is the Reynolds number in smaller tube. Knowing K_{add}, the additional pressure drop can be calculated as $p_{add} = (1/2)K_{add}\rho_j U_j^2$, where ρ_j is the density of gas at the junction and U_j is the average velocity at the junction in the smaller tube just upstream of the junction. The above empirical correlation is valid for $3.74 \leq AR \leq 64; 0.0002 \leq Kn_j \leq 0.075; 0.32 \leq Re_s \leq 837$. Similarly, the following correlation for sudden contraction case was proposed by Varade et al. [157]:

$$K_{add,SC} = \frac{(AR)^{2.7}(Kn_j)^{0.50}}{3(Re_s)^{1.2}}. \tag{2.119}$$

The above empirical correlation is valid for $12.43 \leq AR \leq 64$; $0.0003 \leq Kn_j \leq 0.02$; $0.56 \leq Re_s \leq 837$. It is interesting to see from Eqs. (2.118) and (2.119) the same dependence of Reynolds number on causing an additional pressure drop. Further notice that the Knudsen number also has approximately the same value of exponent but with different roles in the two cases.

2.6.2 *Flow in Diverging/Converging Microchannel*

At the conventional scale, flow in diverging or converging microchannel is referred to as Jeffery-Hamel flow, where the walls are replaced by a row of mass source or sink. The Navier–Stokes equations reduce to non-linear ordinary differential equation, which can subsequently be solved. The above approach however does not seem to have been extended to the slip regime. Graur et al. [65] obtained an *approximate* analytical solution by solving the Stokes equation, along with application of slip boundary condition.

Diverging/converging microchannels were employed by Stemme and Stemme [139] to control the flow direction, instead of using micro-valves. Their construction therefore reduces the number of moving parts, thereby making it attractive. The flow rate in a diverging microchannel is different in a converging microchannel, for the same pressure difference acting across the length of the microchannel. This disparity in flow rate is utilized to cause a flow rectification, and can be quantified in terms of a parameter, diodicity D defined as [65]:

$$D = \frac{\dot{m}^{conv}/((p_1^{conv})^2 - (p_2^{conv})^2)}{\dot{m}^{div}/((p_1^{div})^2 - (p_2^{div})^2)} \tag{2.120}$$

where $\dot{m}$ refers to mass flow rate, p is pressure, superscripts "conv" and "div" refer to converging and diverging directions, respectively, and subscripts "1" and "2" refer to inlet and outlet of the microchannel, respectively. For a given flow rate, pressure drop in the converging microchannel is greater than its diverging counterpart [70, 71]. A large absolute value of diodicity leads to a better efficiency of valve-less micro-pump, with the effect vanishing at $D = 1$. The experiments of Duryodhan et al. [49] and numerical simulation of Graur et al. [65] suggest a maximum value of 1.2 for diodicity.

A fundamental question to address while employing diverging/converging microchannel is definition of characteristic length, required, for example, in calculation of Reynolds number. The issue is not straightforward because of the continuous change in cross-sectional area and therefore the hydraulic diameter. Using the hydraulic diameter at the entrance or at the mid-length of the passage is not scientifically compelling. This issue was addressed by Duryodhan et al. [49] with liquid flow and the location where the hydraulic diameter should be calculated was expressed as a fraction of microchannel length. The location was determined by equating the pressure drop in the diverging (or converging) microchannel to an equivalent straight microchannel. Based on this equivalence, the hydraulic diameter at $L/3$ (where L is the length of the microchannel) from the narrower end was proposed as the "equivalent hydraulic diameter" (Fig. 2.19a). Interestingly, the proposed location is reasonably invariant of the divergence angle, length and cross-sectional dimensions of the microchannel, and also the flow rate (or Reynolds number) but shows a dependence on the Knudsen number [70]. The location changes to $L/3.6$ from the narrower end in case of a converging microchannel, for the case of liquid flow (Fig. 2.19b). These length scales have also been employed in the calculation of Nusselt number.

Note that the difference in location between the diverging and converging microchannels gives rise to diodicity. With an increase in Knudsen number, the location of the equivalent hydraulic diameter moves to the narrower end, for both diverging and converging microchannels [70]. This suggests that the diodic effect vanishes (i.e., D becomes unity) with increasing rarefaction of the flow, and the design of Stemme and Stemme [139] cannot be used to pump low-pressure gases. In Sect. 3.7, we will discuss a Knudsen pump which is used to pump low pressure gases based on temperature difference at the two ends of a microchannel.

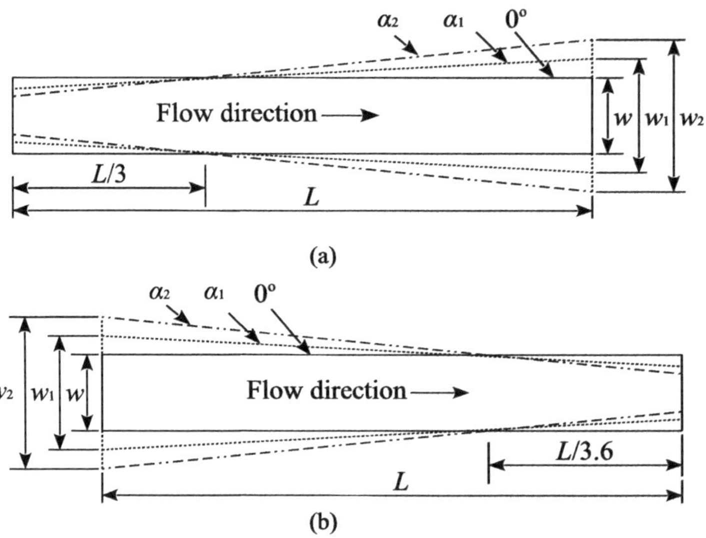

Fig. 2.19 A schematic illustrating the concept of equivalent hydraulic diameter for (**a**) diverging and (**b**) converging microchannel. The pressure drop and heat transfer in diverging/converging microchannel with liquid flow is same as a straight microchannel, provided that the straight microchannel has the dimension at the location marked in the figure. Taken from Duryodhan et al. [51]

2.6.3 Flow in a Bend Microchannel

Similar to flow in a sudden expansion/contraction passage, an eddy is expected to form at the corner of a bend microchannel. Does the formation of this corner eddy get delayed in presence of slip as in the case of a sudden expansion/contraction microchannel? Towards this, Varade et al. [158] conducted measurements in a scaled-up facility and obtained local wall pressure distribution along the inner and outer walls of a 90° bend tube. It was surprisingly found that an adverse pressure gradient occurs near the bend, indicating flow separation, at much lower value of Reynolds number as compared to conventional flow. The numerically obtained data of Varade et al. [158], Agrawal et al. [6], and White et al. [165] also confirm this observation. The numerically obtained velocity field near the bend plane clearly indicates secondary flows at Reynolds number as low as unity. The flow acceleration and the presence of secondary flows near the bend causes a larger pressure drop as compared with a straight tube.

Based on the experimental data, the following correlation was proposed for $f\,Re$ for a bend tube in presence of slip:

$$f\,Re = \frac{60}{1 - 0.64\exp(-607Kn_m)} \tag{2.121}$$

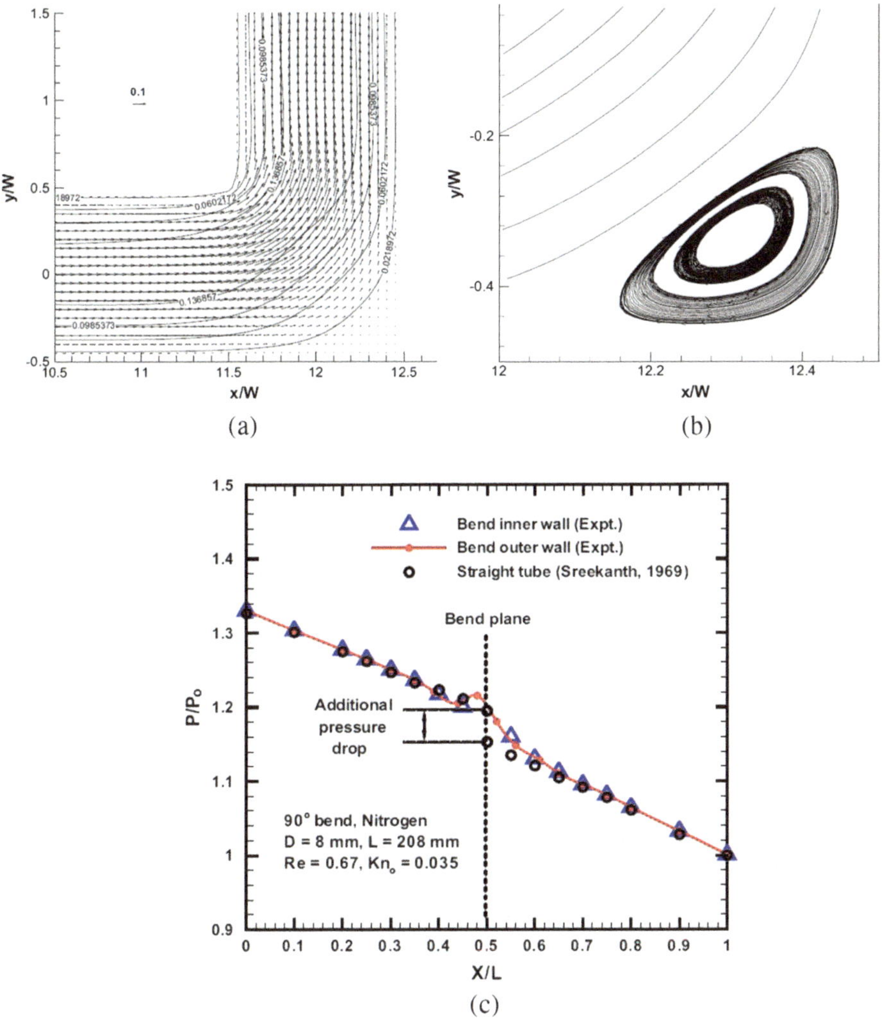

Fig. 2.20 (**a**) Velocity vectors near the bend for $Kn = 0.202$, numbers denote the local Mach number (Agrawal et al. [6]). (**b**) Streamlines in the zoomed view of the bend region showing clearly the presence of the corner vortex (Agrawal et al. [6]). (**c**) Experimentally measured static pressure variation for bend microchannel compared with straight tube for $Re = 0.67$, $Kn = 0.035$ (Varade et al. [158])

Here, Kn_m is the mean Knudsen number defined as the average of the inlet (Kn_i) and outlet (Kn_o) Knudsen numbers (i.e., $Kn_m = (Kn_i + Kn_o)/2$). The above correlation applies to $0.27 \leq Re \leq 418.5$; $0.0007 \leq Kn \leq 0.0359$. The correlation suggests that $f Re$ decreases with an increase in the Knudsen number; further, $f Re$ becomes independent of rarefaction for $Kn_o > 0.005$. An empirical correlation for computing the additional pressure loss coefficient was also presented.

The contrasting behavior of flow in sudden expansion/contraction versus bend microtube in the slip regime is interesting and worth further study and confirmation.

2.7 Useful Empirical Correlations

In this section, we provide some useful empirical correlations for calculating the developing length and friction factor in microchannel and microtubes.

2.7.1 *Developing Length in Microtube and Microchannel*

When the flow enters a passage, it takes a certain length to develop, which is called the developing length. Once the flow becomes fully developed, the shape of the velocity profile does not change. The developing region is more difficult to analyze than the fully developed region as there is an additional change in inertia of the fluid in this region. Appropriate correlations have therefore been proposed to estimate the developing length in the literature. Information about the developing length is useful, for example, for deciding the location of the first pressure tap while making measurements in a long microchannel.

For laminar flow in a pipe in the continuum regime, one of the following correlations can be employed to estimate the developing length (L_d) normalized by the pipe diameter (D):

$$\frac{L_d}{D} = [(0.619)^{1.6} + (0.0567Re)^{1.6}]^{(1/1.6)}. \tag{2.122}$$

The above correlation is by Durst et al. [48] while the following correlation is by Dombrowski [43].

$$\frac{L_d}{D} = 0.260 + 0.055Re + 0.379\exp(-0.148Re) \tag{2.123}$$

Chen [32] proposed:

$$\frac{L_d}{D} = \frac{0.6}{0.035Re + 1} + 0.056Re. \tag{2.124}$$

All the above three correlations give reasonably close values of developing length for laminar continuum flow in a tube (i.e., within 6% of each other; except around $Re = 10$ where the difference increases to 12%) over the Reynolds number range of 0–3000.

In presence of slip at the wall, Sreekanth [137] suggested to calculate the developing length as

$$\frac{L_d}{D} = \frac{kRe}{4} \tag{2.125}$$

where k takes a value between 0.15 and 0.2 depending on the Reynolds number. Duan and Muzychka [47] modified the Chen's formula given above to include the effect of slip at the wall as

$$\frac{L_d}{D} = \frac{0.60}{0.035Re + 1} + 0.0396Re\left(1 + 3.7\frac{2-\sigma_v}{\sigma_v}Kn - 15\left(\frac{2-\sigma_v}{\sigma_v}Kn\right)^2\right) \tag{2.126}$$

where σv is tangential momentum accommodation coefficient discussed in Sect. 2.3. Note that the form of the correlation, in both continuum and slip regimes, has to be chosen such that the developing length is not suggested to be zero when the Reynolds number is zero.

For laminar flow in the continuum regime in a two-dimensional microchannel, one of the following correlations can be employed:

$$\frac{L_d}{D_h} = [(0.631)^{1.6} + (0.0442Re)^{1.6}]^{(1/1.6)} \tag{2.127}$$

suggested by Durst et al. [48] or

$$\frac{L_d}{D_h} = \frac{0.315}{0.0175Re + 1} + 0.011Re \tag{2.128}$$

given by Chen [32]. In presence of slip, Duan and Muzychka [47] modified the Chen's formula as

$$\frac{L_d}{D_h} = \frac{0.315}{0.0175Re + 1} + 0.0112Re\left(1 + 6.7\frac{2-\sigma_v}{\sigma_v}Kn - 37\left(\frac{2-\sigma_v}{\sigma_v}Kn\right)^2\right) \tag{2.129}$$

Barber and Emerson [18] suggested

$$\frac{L_d}{D_h} = \frac{0.332}{0.0271Re + 1} + 0.011Re\left(\frac{1 + 14.78\left(\frac{2-\sigma_v}{\sigma_v}\right)Kn}{1 + 9.78\left(\frac{2-\sigma_v}{\sigma_v}\right)Kn}\right) \tag{2.130}$$

Since the Reynolds number of the flow is small at the microscale, primarily because of the small length scale, the developing length is small in most cases; and the influence of the developing region on the overall pressure drop in a long microchannel can be neglected.

2.7.2 *Friction Factor for Microchannel of Various Cross Sections*

Measurement of pressure drop as a function of flow rate is among the first exercise undertaken in practice to test a freshly fabricated microchannel. The pressure drop is normalized into friction factor and flow rate into Reynolds number, using the values of the geometric parameters of the passage and the fluid properties. The value of friction factor (or Poiseuille number, which is the product of friction factor and Reynolds number) is then compared against available data or empirically obtained correlations. A good comparison would point towards the experimental setup and measurement procedure being satisfactory. Similarly, numerically obtained value of friction factor is compared with its experimentally determined value. This shows the importance of having high-quality empirical correlations for friction factor. Each correlation has a validity over a certain parameter range, and the comparison should be limited to this range.

The friction factor (f) for flow in a microchannel can be calculated from the expression given by Wu and Little

$$\frac{dp}{dx} = -\frac{f}{D_h}\frac{\rho\bar{u}^2}{2} + \rho\bar{u}\frac{d\bar{u}}{dx} \tag{2.131}$$

which shows that the pressure drop depends on flow friction as well as acceleration effect caused by the change of density. Integrating the above equation across the length of the channel and using the equation of state, we obtain [44]:

$$f = \frac{\mathbb{P}^2 - 1}{\mathbb{P}^2}\left[\frac{\pi}{2Re^2Kn^2} + \log\mathbb{P}\right]\frac{2D_h}{L}. \tag{2.132}$$

where $\mathbb{P}$ is pressure ratio.

For flow in microtube involving density variation and rarefaction, Verma et al. proposed the following correlation based on compilation of a large amount of data from the literature:

$$f\,Re = \frac{64}{1 + 14.9Kn} \tag{2.133}$$

Demsis et al. [42] proposed the following correlation based on measurements for flow in a smooth square tube:

$$f\,Re = \frac{57}{1 + 34.5Kn} \tag{2.134}$$

The above correlation is valid in the range $0.0022 < Kn < 0.024$ and $0.54 < Re < 13.2$ range. Both the above correlations give the correct value of $f\,Re$ in the continuum regime (i.e., as $Kn \to 0$).

Niazmand et al. [105] reported correlations to evaluate friction factor through rectangular and trapezoidal channels, given by

$$(f Re)_{fd} = \left[13.9\left(\frac{90}{\phi}\right)^{-0.07} + 10.4\exp\left(-3.25\alpha\left(\frac{90}{\phi}\right)^{0.23}\right)\right]G_1 \qquad (2.135)$$

where $G_1 = 1 - 2.48\,Kn^{0.64}(1 - 0.2\tanh(3\alpha))$.

The friction factor for an arbitrary cross section microchannel can be obtained from [46]

$$\begin{aligned} f Re_{\sqrt{A}} &= \frac{2(-A/P)(dp/dz)\sqrt{A}}{\mu\bar{w}} \\ &= \frac{4}{\sqrt{\epsilon}(1+\epsilon)\left[\sum_{i,j=1}^{\infty} A_{ij}\dfrac{\epsilon\sin(\varphi_i/\epsilon)\sin\psi_j}{\varphi_i\psi_j} + 2C_2\dfrac{4Kn^{*2}}{(1+\epsilon)^2}\right]} \end{aligned} \qquad (2.136)$$

where the symbols have been defined in Sect. 2.4.4. Clearly, the Poiseuille number (product of friction factor and Reynolds number) depends on the geometry of the cross section and rarefaction. Additionally, correlations for microchannel of various useful shapes have been documented in a tabular form in Morini et al. [101].

All the above correlations suggest that the friction factor reduces with an increase in rarefaction or Knudsen number.

2.8 Summary

In this chapter, we have shown that slip at the wall (present in gaseous microscale flows) can bring about substantial difference in the obtained solution with respect to the no-slip case (conventional scale flows). Towards this, we reviewed the relevant slip models and the empirically obtained values of the slip coefficients. The available data for slip coefficients suggests some interesting trends—such as dependence of the slip coefficients on the atomicity of the gas, which needs further verification.

The process of obtaining an analytical solution in presence of both slip at the wall and compressibility indeed becomes complicated, and therefore exact solutions are available for only a handful of cases, as reviewed here. A successful comparison of these theoretical solutions against experimental and DSMC data shows the applicability of the Navier–Stokes equations in the slip regime, but with a modification of the boundary condition to first-order slip, and with a further increase in Knudsen number to second-order slip. However, the validity of the Navier–Stokes equations in the transition regime even with second-order slip boundary condition becomes questionable.

Information of compressible slip flow in alternate cross section straight microchannels and complex passages of uniform cross section is rather limited. These are interesting cases to study from both practical and fundamental viewpoints. Available information suggests that flow in complex microchannels can be substantially different from its counterpart in the conventional range of parameters, and further study can help uncover other unexpected flow behavior in such complex scenarios.

Chapter 3
Microscale Heat Transfer

In this chapter, we present the fundamental aspects of microscale heat transfer for gas flow through different geometries. Such a study is motivated by interest in cooling of electronic components, energy conversion devices, and other MEMS and bio-medical applications. The heat transfer at microscale is different than that of macroscale primarily due to the presence of velocity slip and temperature jump at the wall. The physics pertinent to microscale heat transfer is reasonably complex and solutions of various simplified models are available in the literature. Here we confine our presentation to the slip flow regime for flow through three configurations: parallel plates, microtube, and micro-annulus. We also briefly discuss the effect of other complicating factors and comment on comparison with experiments. A discussion on Knudsen pump and useful empirical correlations are also provided.

3.1 Introduction

This chapter deals with the fundamentals of heat transfer of gas flow at microscale with special attention to its basic concepts including slip velocity, temperature jump, rarefaction, and viscous dissipation. The presence of slip velocity at the wall can lead to an enhancement in Nusselt number, while the temperature jump tends to reduce the value of Nusselt number, as the flow now 'feels' a smaller temperature difference. A careful analysis of heat transfer in the slip regime therefore becomes particularly interesting.

Further complications arise since the effect of axial conduction can also become significant due to the low Peclet number ($Pe = RePr$) brought about by the low value of Reynolds number. Variation in thermo-physical properties of the fluid, due to variation in temperature, is another concern especially when the temperature difference between fluid and wall is large. Several theoretical studies also assume a relatively large effect of viscous dissipation on heat transfer. The compressibility

A. Agrawal et al., *Microscale Flow and Heat Transfer*, Mechanical Engineering Series, https://doi.org/10.1007/978-3-030-10662-1_3

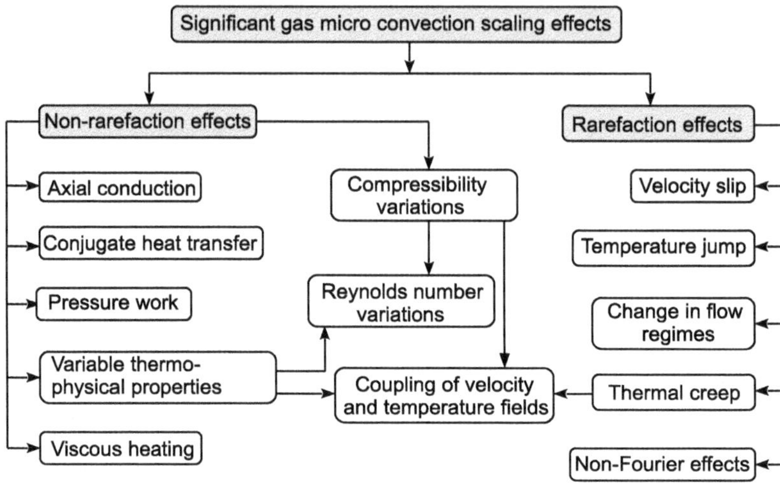

Fig. 3.1 Non-continuum effects in gas micro-convection (adapted with modifications from Prabhu and Mahulikar [119])

effect already discussed in Chap. 2 is also relevant here. Conduction in the wall (conjugate effect) can also become important because the thickness of wall is comparable to the lateral dimensions of the flow passage. These various effects can be classified as *rarefaction* and *non-rarefaction effects* (Fig. 3.1), where rarefaction effects are due to the relatively large value of Knudsen number whereas non-rarefaction effects refer to other microscale effects.

The study of heat transfer in the slip regime is therefore quite involved, as should be further evident from our discussion in this chapter.

3.2 Governing Equations

The thermal energy equation (Eq. (2.11)), which is the primary governing equation for heat transfer problems is provided in this section in both tensorial and expanded forms. In tensorial notation,

$$\rho c_v \left(\frac{\partial T}{\partial t} + u_k \frac{\partial T}{\partial x_k} \right) = \frac{\partial}{\partial x_k} \left(k \frac{\partial T}{\partial x_k} \right) - p \frac{\partial u_k}{\partial x_k} + 2\mu \left[\frac{1}{2} \left(\frac{\partial u_i}{\partial x_j} + \frac{\partial u_j}{\partial x_i} \right) - \frac{1}{3} \frac{\partial u_k}{\partial x_k} \delta_{ij} \right]^2 \tag{3.1}$$

and in expanded form,

$$\rho c_v\left(\frac{\partial T}{\partial t}+u\frac{\partial T}{\partial x}+v\frac{\partial T}{\partial y}+w\frac{\partial T}{\partial z}\right)=\frac{\partial}{\partial x}\left(k\frac{\partial T}{\partial x}\right)+\frac{\partial}{\partial y}\left(k\frac{\partial T}{\partial y}\right)+\frac{\partial}{\partial z}\left(k\frac{\partial T}{\partial z}\right)$$
$$-p\left(\frac{\partial u}{\partial x}+\frac{\partial v}{\partial y}+\frac{\partial w}{\partial z}\right)+\Phi \tag{3.2}$$

where viscous dissipation term Φ in expanded form is

$$\Phi=2\mu\left[\left(\frac{\partial u}{\partial x}\right)^2+\left(\frac{\partial v}{\partial y}\right)^2+\left(\frac{\partial w}{\partial z}\right)^2\right]+\mu\left(\frac{\partial u}{\partial y}+\frac{\partial v}{\partial x}\right)^2$$
$$+\mu\left(\frac{\partial v}{\partial z}+\frac{\partial w}{\partial y}\right)^2+\mu\left(\frac{\partial w}{\partial x}+\frac{\partial u}{\partial z}\right)^2-\frac{2}{3}\mu\left(\frac{\partial u}{\partial x}+\frac{\partial v}{\partial y}+\frac{\partial w}{\partial z}\right)^2 \tag{3.3}$$

3.3 Boundary Conditions

Usually, no-temperature jump boundary condition is considered in the study of heat transfer phenomena at the macroscale. However this boundary condition cannot be used at microscale, rather a temperature jump boundary condition is applicable here (Fig. 3.2).

Similar to TMAC defined in Sect. 2.3, here we define thermal accommodation coefficient,

$$\sigma_T=\frac{E_i-E_r}{E_i-E_w} \tag{3.4}$$

where E_i and E_r are the energy fluxes of incoming and reflected molecules per unit time, and E_w is the energy flux of re-emitted molecules corresponding to the

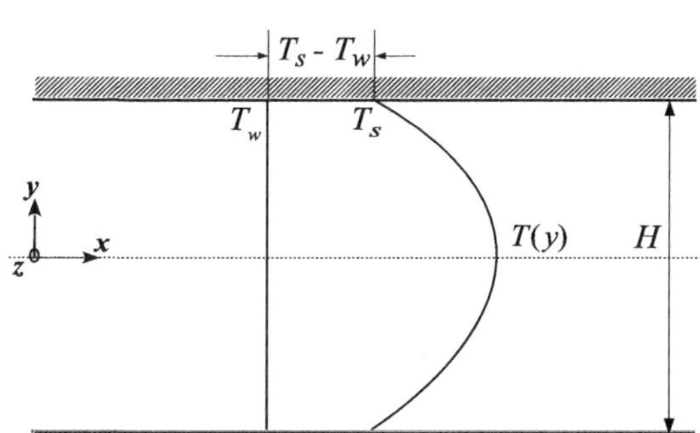

Fig. 3.2 Schematic diagram showing temperature jump at the wall in a parallel plate channel; the walls are maintained at temperature T_w while the gas layer adjacent to the wall experiences a temperature T_s showing a jump in temperature $(T_s - T_w)$ at the wall

surface temperature T_w. Note that for the case of perfect energy exchange between the gas and wall, $E_r = E_w$, and we obtain $\sigma_T = 1$. On the other hand, for no energy exchange between the gas and wall, the outgoing molecules carry the same energy that they came in with. Therefore, in this case, $E_r = E_i$ and $\sigma_T = 0$. So we see that σ_T is a parameter that nicely describes the amount of "accommodation" that the outgoing molecules have with the wall. The actual value of thermal accommodation coefficient depends on the wall temperature, gas pressure, surface finish, and surface roughness of the wall.

3.3.1 Derivation of First Order Temperature Jump Condition

As done earlier with momentum, we consider the energy of incoming and outgoing streams separately [82] as shown in Fig. 3.3. The incoming stream, say, coming from a higher temperature region away from the wall, brings in additional energy with it. If $(\partial T/\partial y)|_w$ is the temperature gradient, then $(k/2)(\partial T/\partial y)|_w$ (where k is the thermal conductivity of the gas) is the excess energy brought in by the incoming stream. Here, we have assumed that half of the excess energy is brought to the wall, while the other half is transmitted elsewhere in the gas. The incoming gas also brings the internal and kinetic energies with it. Recall that from kinetic theory $\frac{1}{4}n\bar{v}$ is the number of molecules incident per unit area in a gas, where n is the molecules per unit volume and $\bar{v}$ is their mean speed. These molecules will carry an energy of $m(c_v + (1/2)R)T_s$ where m is the mass of the molecule, c_v is the specific heat at constant volume, R is the gas constant, and T_s is the temperature of the gas slightly away from the wall (Fig. 3.2). Therefore, we obtain $E_i = (k/2)(\partial T/\partial y)\big|_w + \frac{1}{4}n\bar{v}m(c_v + (1/2)R)T_s$. The outgoing gas will carry away their internal and kinetic energies with them. The energy term (E_w) can be written as $\frac{1}{4}n\bar{v}m(c_v + (1/2)R)T_w$ where T_w is the temperature of the wall.

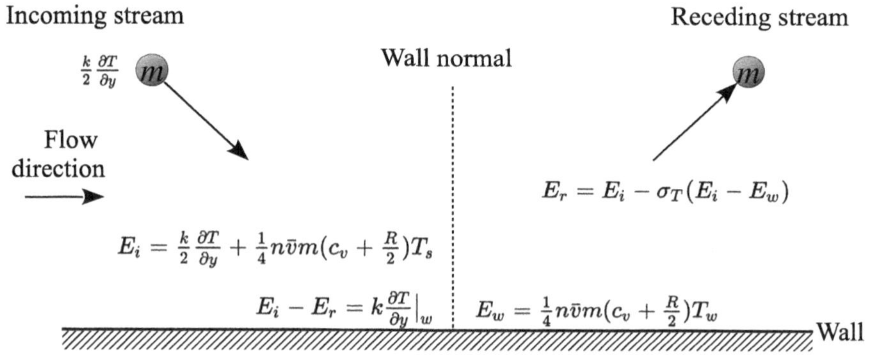

Fig. 3.3 Energy of incoming and outgoing stream of molecules close to the wall

Therefore

$$E_i - E_w = \frac{k}{2}\frac{\partial T}{\partial y}\bigg|_w + \frac{1}{4}n\bar{v}m\left(c_v + \frac{1}{2}R\right)(T_s - T_w) \tag{3.5}$$

Using $\rho = mn$, $c_v + R/2 = (\gamma + 1)c_v/2$, $\bar{v} = 2(2RT/\pi)^{1/2}$, we can write the above equation as,

$$E_i - E_w = \frac{k}{2}\frac{\partial T}{\partial y}\bigg|_w + \frac{\gamma + 1}{2}\frac{c_v(T_s - T_w)p}{\sqrt{2\pi RT}} \tag{3.6}$$

The $E_i - E_r$ term (energy delivered to the wall) is

$$E_i - E_r = k\frac{\partial T}{\partial y}\bigg|_w \tag{3.7}$$

where E_r denotes the energy carried away by the reflected stream.

Upon substituting the above equations into Eq. (3.4),

$$k\frac{\partial T}{\partial y}\bigg|_w = \sigma_T\left[\frac{k}{2}\frac{\partial T}{\partial y}\bigg|_w + \frac{\gamma + 1}{2}\frac{c_v p(T_s - T_w)}{\sqrt{2\pi RT}}\right] \tag{3.8}$$

On re-arranging the above equation along with using $\lambda = \frac{\mu}{p}\sqrt{\frac{\pi RT}{2}}$, $\gamma = \frac{c_p}{c_v}$, $Pr = \frac{\mu c_p}{k}$, we obtain the desired form of the temperature jump condition,

$$T_s - T_w = \frac{2 - \sigma_T}{\sigma_T}\frac{2\gamma}{Pr(\gamma + 1)}\lambda\frac{\partial T}{\partial y}\bigg|_w \tag{3.9}$$

Equation (3.9) can also be written

$$T_s - T_w = \zeta_T\frac{\partial T}{\partial y}\bigg|_w \tag{3.10}$$

where ζ_T denotes the temperature jump length,

$$\zeta_T = \frac{2 - \sigma_T}{\sigma_T}\frac{2\gamma}{Pr(\gamma + 1)}\lambda \tag{3.11}$$

3.4 Some Exact Solutions

As summarized in Fig. 3.1, there are several effects that can arise simultaneously in heat transfer problems; because of these effects many more terms (as compared to the corresponding problem at the conventional scale) have to be retained in the governing equations. This makes obtaining analytical solution of the governing equations difficult. Here we consider only two such effects—rarefaction and viscous dissipation where obtaining analytical solution is relatively easy. A discussion of the other effects on the solution is provided briefly in a subsequent section.

Specifically we present exact solutions for heat transfer of gas flow through microchannel, microtube, and microannulus. The flow is assumed to be steady, two-dimensional, laminar, incompressible, and fully developed both hydrodynamically and thermally in all the cases. Thermophysical properties of the Newtonian fluid involved are assumed to be constant. The effect of axial conduction is neglected both in the fluid and through the wall.

3.4.1 Heat Transfer in Microchannel

We present closed form expressions for temperature distribution and Nusselt number for microchannel gas flow, first derived by Aydin and Avci [15]. Consider flow between two stationary parallel plates at a distance W apart (Fig. 3.4). The flow is driven by a constant pressure gradient along the x-axis. Two cases can arise: Uniform heat flux is supplied to the walls of microchannel due to which there is a transfer of heat from wall to the flowing gas, or the walls are maintained at a temperature which is different from that of the flowing gas.

For the present situation, the solution of the momentum equation is not coupled to the energy equation, and therefore the mass and momentum equations are solved before the energy equation. Further due to the fully developed flow assumption, the velocity u varies only in the y direction. The velocity profile, along with slip at the wall can be derived as in Chap. 2 to obtain

$$\frac{u}{u_m} = \frac{3}{2}\left(\frac{1 + 4Kn - (y/w)^2}{1 + 6Kn}\right) \tag{3.12}$$

where u_m denotes the mean velocity and Kn ($=\lambda/2w$) is the Knudsen number. Note that the first order slip model with complete accommodation (i.e., $\sigma_v = 1$) has been utilized in writing the above solution.

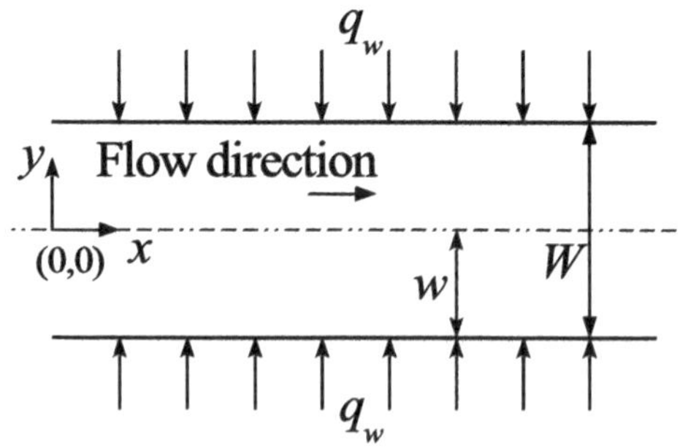

Fig. 3.4 Schematic of microchannel gas flow with uniform wall heat flux boundary condition

The energy equation considering viscous dissipation is

$$u\frac{\partial T}{\partial x} = \alpha\frac{\partial^2 T}{\partial y^2} + \frac{\nu}{c_p}\left(\frac{\partial u}{\partial y}\right)^2 \tag{3.13}$$

where the last term denotes the effect of viscous dissipation (obtained through simplification of Eq. (3.3)) and ν is the kinematic viscosity ($= \mu/\rho$). The thermal boundary condition at the center of microchannel (due to symmetry) is

$$\left.\frac{\partial T}{\partial y}\right|_{y=0} = 0. \tag{3.14}$$

3.4.1.1 Uniform Heat Flux

We will first consider the case of uniform heat flux condition. Application of constant wall heat flux boundary condition at wall is

$$q_w = k\left.\frac{\partial T}{\partial y}\right|_{y=w} \tag{3.15}$$

where q_w is positive when its direction is to the fluid (heated wall), otherwise it is negative (cooled wall). As in the conventional case,

$$\frac{\partial T}{\partial x} = \frac{dT_s}{dx} = \frac{dT_m}{dx} \tag{3.16}$$

where T is the absolute temperature, T_s is the temperature of gas at the wall, and T_m is the bulk mean temperature (defined below through Eq. (3.25)).

Define the non-dimensional variables,

$$Y = \frac{y}{w},\quad \theta = \frac{T_s - T}{T_s - T_c},\quad \theta_q = \frac{T - T_s}{q_w w/k},\quad Br = \frac{\mu u_m^2}{k(T_s - T_c)},\quad Br_q = \frac{\mu u_m^2}{w q_w} \tag{3.17}$$

where T_c is the centerline temperature.

In the fully developed region the axial temperature gradient is obtained by applying energy balance to an elemental area,

$$\frac{dT_m}{dx} = \frac{q_w P + \int \mu \Phi dA}{\dot{m} c_p} \tag{3.18}$$

where P is heated perimeter. Using Eq. (3.18) the axial temperature gradient is evaluated,

$$\frac{dT_m}{dx} = \frac{q_w}{\rho u_m c_p w}\left[1 + \frac{3Br_q}{(1+6Kn)^2}\right] \tag{3.19}$$

where P is heated perimeter. Using Eqs. (3.12), (3.15), (3.16), (3.17), and (3.19) the energy equation (3.13) in non-dimensional form is

$$\frac{d^2\theta_q}{dY^2} = \frac{3}{2}\left(\frac{1}{1+6Kn} + \frac{3Br_q}{(1+6Kn)^3}\right)(1+4Kn-Y^2) - \frac{9Br_qY^2}{(1+6Kn)^2} \tag{3.20}$$

The non-dimensional boundary conditions are

$$\frac{\partial\theta_q}{\partial Y} = 1 \text{ at } Y = 1$$
$$\theta_q = 0 \text{ at } Y = 1 \tag{3.21}$$

The non-dimensional energy equation (3.20) can be integrated twice with the thermal boundary condition equation (3.21), gives the temperature distribution in terms of modified Brinkman number,

$$\theta_q = \frac{T-T_s}{q_w w/k} = \left[\left(\frac{1}{1+6Kn} + \frac{3Br_q}{(1+6Kn)^3}\right)\left(-\frac{5}{8}+\frac{3}{4}Y^2-\frac{1}{8}Y^4 + 3Kn(Y^2-1)\right)\right] - \frac{3Br_q}{4(1+6Kn)^2}(Y^4-1) \tag{3.22}$$

The above solution is however in terms of T_s; since the temperature of the gas at the wall is not explicitly known, we express the above equation in terms of wall temperature T_w. For this, we utilize the temperature jump condition given in Eq. (3.9), which expressed in non-dimensional form becomes

$$\frac{T_s - T_w}{q_w w/k} = -\frac{4Kn\gamma}{Pr(\gamma+1)} \tag{3.23}$$

where we have assumed $\sigma_T = 1$ in writing the above equation.

The non-dimensional temperature distribution can therefore be written as

$$\theta_q^* = \frac{T-T_w}{q_w w/k} = \left(\frac{1}{1+6Kn} + \frac{3Br_q}{(1+6Kn)^3}\right)\left(-\frac{5}{8}+\frac{3}{4}Y^2-\frac{1}{8}Y^4 + 3Kn(Y^2-1)\right) - \frac{3Br_q}{4(1+6Kn)^2}(Y^4-1) - \frac{4Kn\gamma}{Pr(\gamma+1)} \tag{3.24}$$

where a "*" has been added to the symbol to emphasize that the solution incorporates the temperature jump condition, while subscript "q" emphasizes the use of modified Brinkman number.

The bulk mean temperature of the fluid is usually estimated on the basis of thermal energy transported by the fluid stream that passes through a given cross-sectional area A. T_m can be calculated using the temperature distribution,

$$T_m = \frac{\int_A \rho c_p u T dA}{\int_A \rho c_p u dA} \tag{3.25}$$

The non-dimensional bulk mean temperature finally becomes

$$\theta_{q,m}^* = \frac{T_m - T_w}{q_w w/k} = -\frac{1}{3}\left(1 + \frac{12Kn\gamma}{Pr(\gamma+1)}\right) - \frac{2Br_q}{35(1+6Kn)^4} - \frac{11Br_q}{35(1+6Kn)^3}$$
$$- \frac{2(1+21Br_q)}{105(1+6Kn)^2} - \frac{2}{15(1+6Kn)} \tag{3.26}$$

The Nusselt number is defined,

$$Nu_q = \frac{h(2w)}{k} = \frac{q_w(2w)}{k(T_w - T_m)} = -\frac{2}{\theta_{q,m}^*} \tag{3.27}$$

where h is the convective heat transfer coefficient.

The Nusselt number expression in terms of modified Brinkman number is

$$Nu_q = 2\left[\frac{1}{3}\left(1 + \frac{12Kn\gamma}{Pr(\gamma+1)}\right) + \frac{2Br_q}{35(1+6Kn)^4}\right.$$
$$\left. + \frac{11Br_q}{35(1+6Kn)^3} + \frac{2(1+21Br_q)}{105(1+6Kn)^2} + \frac{2}{15(1+6Kn)}\right]^{-1} \tag{3.28}$$

3.4.1.2 Constant Wall Temperature Case

The constant wall temperature case suggests that $dT_s/dx = 0$, and as in the conventional case,

$$\frac{\partial T}{\partial x} = \left(\frac{T - T_s}{T_c - T_s}\right)\frac{dT_c}{dx} \tag{3.29}$$

Using Eqs. (3.13), (3.17), and (3.29) we write

$$\frac{d^2\theta}{dY^2} = \frac{3a}{2}\left(\frac{1+4Kn-Y^2}{1+6Kn}\right)\theta + \frac{9BrY^2}{(1+6Kn)^2} \tag{3.30}$$

where $a = -[u_m w^2/(\alpha(T_s - T_c))](dT_c/dx)$. A closed form solution of the above equation is however not possible.

3.4.1.3 Parametric Variation

We examine the effect of viscous dissipation and rarefaction on the heat transfer characteristics for the uniform heat flux case. Prandtl number is taken as 0.71. The Knudsen number values are considered between 0 and 0.10, while the modified Brinkman number is varied from −0.65 to 0.65.

The non-dimensional velocity profiles are plotted by using Eq. (3.12) for different values of Knudsen number as shown in Fig. 3.5. The velocity profile for no slip condition is shown by $Kn = 0$, while for slip condition the profile is denoted by non-zero Knudsen number. It is clear from Eq. (3.12) that the velocity varies in y direction only, which is due to the fully developed assumption. Also, it is interesting to note that the slip velocity increases with increasing Knudsen number, while the maximum velocity decreases with the increase in Knudsen number, therefore, the velocity profile becomes relatively flat.

The variation of non-dimensional temperature distribution given by Eq. (3.24) at various modified Brinkman numbers is shown in Fig. 3.6. Notice that the temperature profile is significantly distorted for finite viscous dissipation and Knudsen number. However, the effect of modified Brinkman number on the temperature profile is more profound as commented below.

Figure 3.7 depicts the variation of Nusselt number with the Knudsen number for different values of the modified Brinkman number. The Nusselt number decreases monotonically with an increase in Knudsen number. The Nusselt number also shows a strong dependence on modified Brinkman number, with an increase (decrease) in Nusselt number with negative (positive) value of modified Brinkman number. This can be understood as follows: the viscous dissipation term in the equation acts as a heat source in the flow. An increased dissipation increases the bulk temperature of the fluid due to internal heating. For the case of wall heating, this increase in the fluid temperature decreases the temperature difference between the wall and bulk fluid, which results in a decrease in the Nusselt number. While for the case of cooling, the

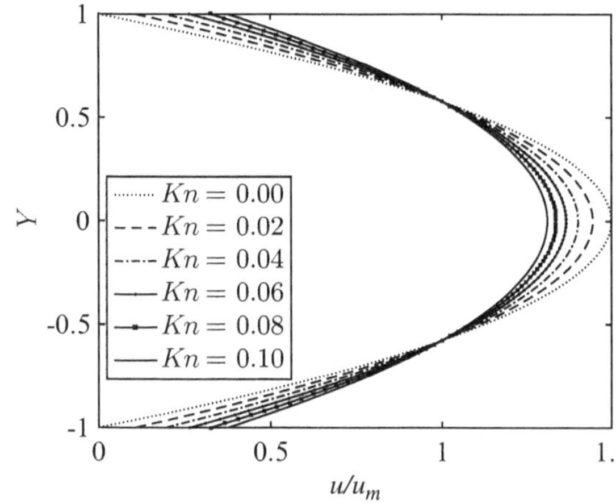

Fig. 3.5 Non-dimensional velocity profile at different Knudsen number

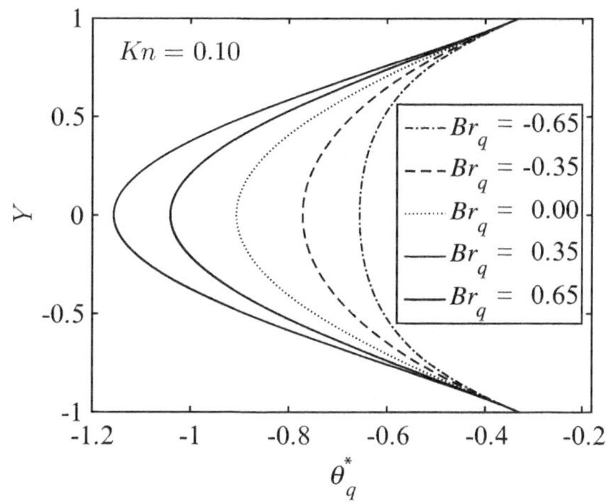

Fig. 3.6 Non-dimensional temperature profile for various modified Brinkman number

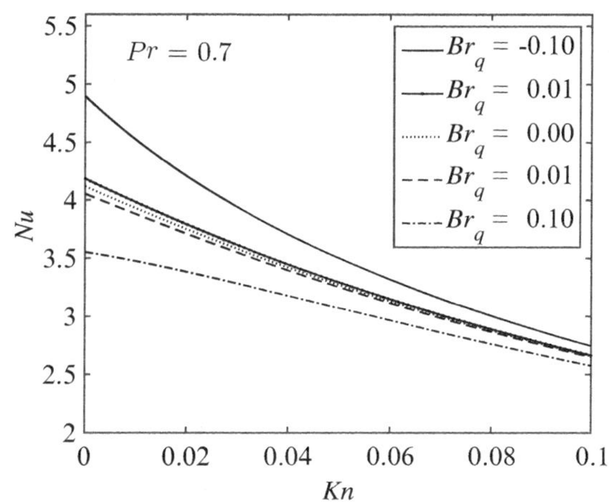

Fig. 3.7 Variation of Nusselt number with Knudsen number at different modified Brinkman number

reverse applies: the temperature difference is increased with an increase in modified Brinkman number, which leads to an increase in the Nusselt number.

Figure 3.8 shows the variation of Nusselt number with modified Brinkman number for different values of the Knudsen number. The figure shows a singularity and even negative values of Nusselt number, with the location of singularity being dependent on the Knudsen number. These observations may appear to be counter-intuitive but can be explained as follows: as already seen, the temperature distribution is significantly distorted due to viscous dissipation, which alters the Nusselt number as well. The difference in mean fluid temperature and wall temperature changes its sign across the singular point, while the difference is zero at the singular point.

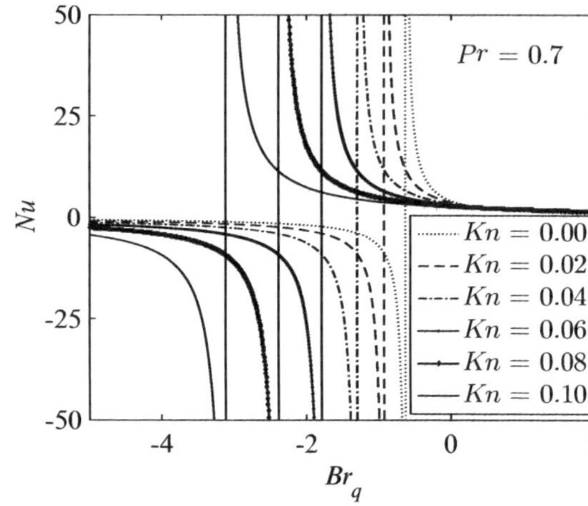

Fig. 3.8 Variation of Nusselt number with modified Brinkman number at different Knudsen number

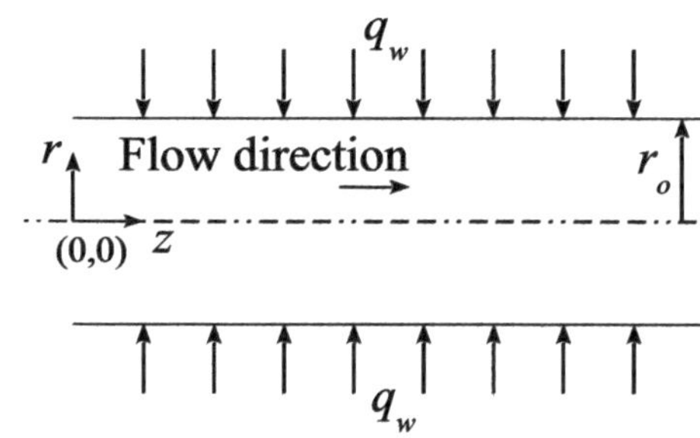

Fig. 3.9 Schematic diagram of gas flow through a micropipe with uniform wall heat flux boundary condition

3.4.2 Heat Transfer Analysis Through a Micropipe

We now consider a fully developed laminar gas flow through a micropipe; such an analysis was carried out earlier by Aydin and Avci [14]. The direction of fluid flow is along z axis, with r axis perpendicular to it. Note that the fluid flow is independent of z, the velocity has only component u in the r direction (Fig. 3.9).

The fully developed velocity profile is obtained by solving the z-momentum equation by employing the first-order slip condition with complete accommodation at the wall. The velocity normalized by the mean velocity is

$$\frac{u}{u_m} = 2\left(\frac{1 + 4Kn - (r/r_o)^2}{1 + 8Kn}\right) \tag{3.31}$$

The energy equation including the effect of viscous dissipation is

$$u\frac{\partial T}{\partial z} = \frac{\alpha}{r}\frac{\partial}{\partial r}\left(r\frac{\partial T}{\partial r}\right) + \frac{\nu}{c_p}\left(\frac{\partial u}{\partial r}\right)^2 \tag{3.32}$$

where the second term in the right-hand side quantifies the effect of viscous dissipation. The thermal boundary condition at the center ($r = 0$) is

$$\left.\frac{\partial T}{\partial r}\right|_{r=0} = 0 \tag{3.33}$$

The constant heat flux boundary condition at the wall is

$$q_w = k\left.\frac{\partial T}{\partial r}\right|_{r=r_o} \tag{3.34}$$

where q_w is taken to be positive when its direction is to the fluid (the hot wall), otherwise it is negative (the cold wall).

The non-dimensional variables are defined,

$$R = \frac{r}{r_o},\ \theta = \frac{T_s - T}{T_s - T_c},\ \theta_q = \frac{T - T_s}{q_w r_o/k},\ Br = \frac{\mu u_m^2}{k(T_s - T_c)},\ Br_q = \frac{\mu u_m^2}{D q_w} \tag{3.35}$$

Now using Eq. (3.18) the axial temperature gradient in the fully developed region is evaluated,

$$\frac{dT_m}{dz} = \frac{2q_w}{\rho u_m c_p R}\left[1 + \frac{8Br_q}{(1+8Kn)^2}\right] \tag{3.36}$$

Using Eqs. (3.16), (3.31), (3.34), (3.35), and (3.36), the energy equation (3.32) in non-dimensional form is

$$\frac{d}{dR}\left(R\frac{d\theta_q}{dR}\right) = \left(\frac{4}{1+8Kn} + \frac{32Br_q}{(1+8Kn)^3}\right)(R + 4KnR - R^3) - \frac{32Br_q R^3}{(1+8Kn)^2} \tag{3.37}$$

The non-dimensional boundary conditions are

$$\begin{aligned}\frac{\partial \theta_q}{\partial R} &= 1 \text{ at } R = 1\\ \theta_q &= 0 \text{ at } R = 1\end{aligned} \tag{3.38}$$

The solution of the non-dimensional energy equation (3.37), using the thermal boundary conditions given by Eq. (3.38) gives the temperature distribution in terms of modified Brinkman number,

$$\theta_q = \frac{T - T_s}{q_w r_o/k} = \left(\frac{4}{1+8Kn} + \frac{32Br_q}{(1+8Kn)^3}\right)\left(-\frac{3}{16} + \frac{R^2}{4} - \frac{R^4}{16} + Kn(R^2-1)\right) - \frac{2Br_q}{(1+8Kn)^2}(R^4-1) \quad (3.39)$$

Utilizing Eq. (3.9), the temperature jump at the wall is

$$\frac{T_s - T_w}{q_w r_o/k} = -\frac{4Kn\gamma}{Pr(\gamma+1)} \quad (3.40)$$

Using Eqs. (3.39) and (3.40) the non-dimensional temperature distribution with temperature jump in terms of modified Brinkman number is

$$\theta_q^* = \frac{T - T_w}{q_w r_o/k} = \left(\frac{4}{1+8Kn} + \frac{32Br_q}{(1+8Kn)^3}\right)\left(-\frac{3}{16} + \frac{R^2}{4} - \frac{R^4}{16} + Kn(R^2-1)\right) - \frac{2Br_q}{(1+8Kn)^2}(R^4-1) - \frac{4Kn\gamma}{Pr(\gamma+1)} \quad (3.41)$$

Using Eq. (3.25) the bulk mean temperature in terms of modified Brinkman number is

$$\theta_{q,m}^* = \frac{T_m - T_w}{q_w r_o/k} = -\frac{1}{4}\left(1 + \frac{16Kn\gamma}{Pr(\gamma+1)}\right) - \frac{Br_q}{3(1+8Kn)^4} - \frac{Br_q}{(1+8Kn)^3} - \frac{1+16Br_q}{24(1+8Kn)^2} - \frac{1}{6(1+8Kn)} \quad (3.42)$$

The Nusselt number is defined,

$$Nu_q = \frac{hD}{k} = -\frac{2}{\theta_{q,m}^*} \quad (3.43)$$

Using Eqs. (3.42) and (3.43), the Nusselt number expression in terms of modified Brinkman number is

$$Nu_q = 2\left[\frac{1}{4}\left(1 + \frac{16Kn\gamma}{Pr(\gamma+1)}\right) + \frac{Br_q}{3(1+8Kn)^4} + \frac{Br_q}{(1+8Kn)^3} + \frac{1+16Br_q}{24(1+8Kn)^2} + \frac{1}{6(1+8Kn)}\right]^{-1} \quad (3.44)$$

3.4.2.1 Constant Wall Temperature Case

The constant wall temperature case suggests that $dT_s/dz = 0$, and as in the conventional case,

$$\frac{\partial T}{\partial z} = \left(\frac{T - T_s}{T_c - T_s}\right)\frac{dT_c}{dz} \tag{3.45}$$

Using Eqs. (3.31), (3.32), (3.35), and (3.45) we write

$$\frac{d}{dR}\left(R\frac{d\theta}{dR}\right) = 2a\left(\frac{R + 4KnR - R^3}{1 + 8Kn}\right)\theta + \frac{16BrR^3}{(1 + 8Kn)^2} \tag{3.46}$$

where $a = -[u_m r_o^2/(\alpha(T_s - T_c))](dT_c/dz)$. A closed form solution of the above equation is however not possible.

3.4.2.2 Parametric Variation

We presented the effects of modified Brinkman number and Knudsen number on Nusselt number for both hydrodynamically and thermally fully developed flow. Note that $Kn = 0$ represents the macroscale case, while $Kn > 0$ shows the microscale case, and $Br_q = 0$ represents the case of no-viscous dissipation. We employed a constant heat flux boundary condition at the wall.

The velocity profiles are scaled with u_m from Eq. (3.50) and plotted for different values of the Knudsen number, i.e. (Kn = 0.00, 0.02, 0.04, 0.06, 0.08, and 0.1) as shown in Fig. 3.10. Here, velocity profile drawn for $Kn = 0$ represents the no-slip condition at wall, while the velocity profiles drawn for non-zero Knudsen numbers represent the slip condition at wall. Note that the velocity varies in radial direction only, which is due to the fully developed assumption. As explained earlier, the slip velocity increases and maximum velocity decreases with the increase in Knudsen number, which makes the velocity profile flat.

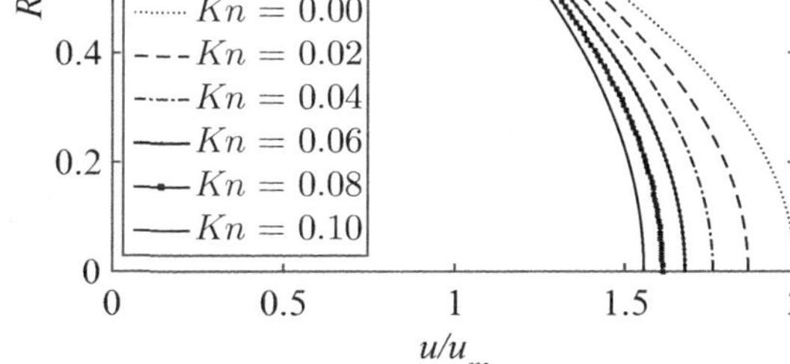

Fig. 3.10 Velocity distribution across the non-dimensional radial coordinate in micropipe at different Knudsen numbers

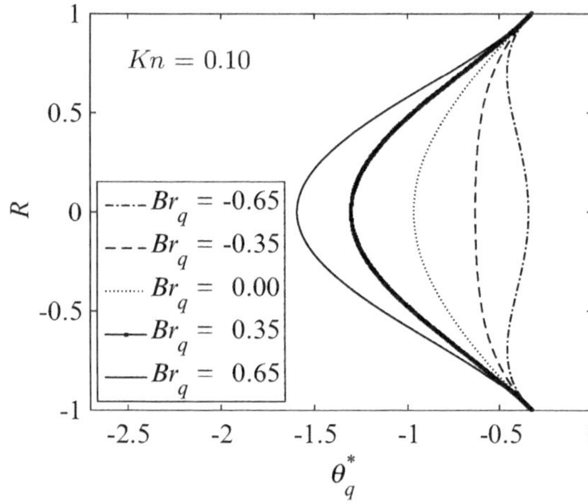

Fig. 3.11 Temperature distribution across the non-dimensional radial coordinate at different modified Brinkman number

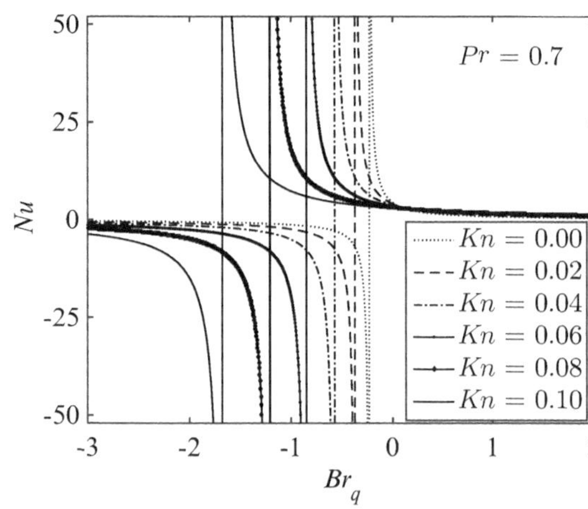

Fig. 3.12 Variation of Nusselt number with modified Brinkman number at different Knudsen number

Figure 3.11 depicts the non-dimensional temperature profiles obtained from Eq. (3.41) using different values of modified Brinkman number ($Br_q = -0.65$, -0.35, 0.00, 0.35, and 0.65) and a fixed value of Knudsen number. The positive value of modified Brinkman number is used to represent the wall heating case, negative values to the wall cooling case, and $Br_q = 0$ represents the case of no viscous dissipation. The viscous dissipation acts as an energy source and increases the fluid temperature, especially near the wall due to higher shear rate. Here, the effect of viscous dissipation on the non-dimensional temperature profile is clearly visible from Fig. 3.11.

The variation of Nusselt number with the modified Brinkman number for different Knudsen number ($Kn = 0.00$ to 0.10) is shown in Fig. 3.12. As explained earlier, we observed the singularities (shown by vertical lines) in Nusselt number for a particular Knudsen number, which discloses the fact that the heat supplied by the wall to fluid is balanced with the heat generated due to the effect of viscous dissipation.

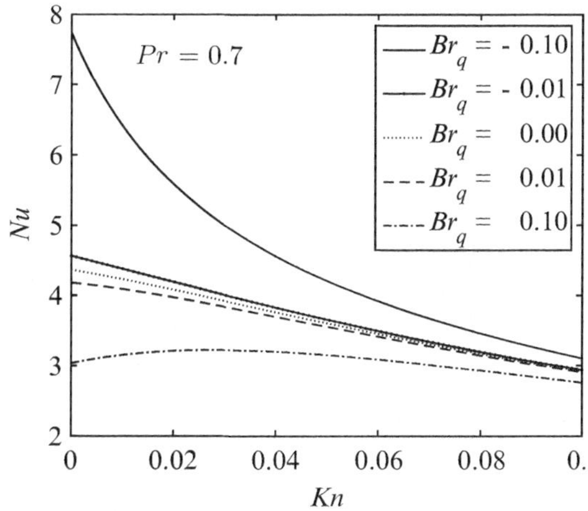

Fig. 3.13 Variation of Nusselt number with Knudsen number at different modified Brinkman number

Figure 3.13 depicts the variation of Nusselt number with the Knudsen number at different modified Brinkman number, i.e. ($Br_q = -0.10, -0.01, 0.00, 0.01$, and 0.10). The Nusselt number is found to decrease with the increase in Knudsen number. Also, for lower values of Knudsen number the effect of viscous dissipation is more pronounced as compared to the higher Knudsen number. For the case of no-viscous dissipation, i.e. $Br_q = 0$, the Nusselt number decreases with the increase in Knudsen number. For positive values of modified Brinkman number the fluid is considered to be heated ($T_w > T_m$) and opposite is true for the negative values of Br_q. For the case of wall heating, viscous dissipation decreases the temperature difference between the wall and bulk fluid. On the contrary, for the cooling case it increases the temperature difference between both the wall and bulk fluid by increasing the fluid temperature more. It is interesting to note that the Nusselt number is higher for negative modified Brinkman number as compared to the positive values. The Nusselt number decreases with the increase in Knudsen number. However, for $Br_q = 0.10$, the Nusselt number increases up to a certain value of $Kn(\approx 0.01)$, beyond it the trend is found to be similar again.

3.4.3 Heat Transfer Through a Micro-Annulus

We presented the heat transfer characteristics of a gas flow though a micro-annulus reported by Aydin and Avci [13]. Two types of thermal boundary conditions are considered, namely, outer wall at constant heat flux and inner wall insulated (case A, refer Fig. 3.14a); and outer wall insulated and inner wall at constant heat flux (case B, refer Fig. 3.14b).

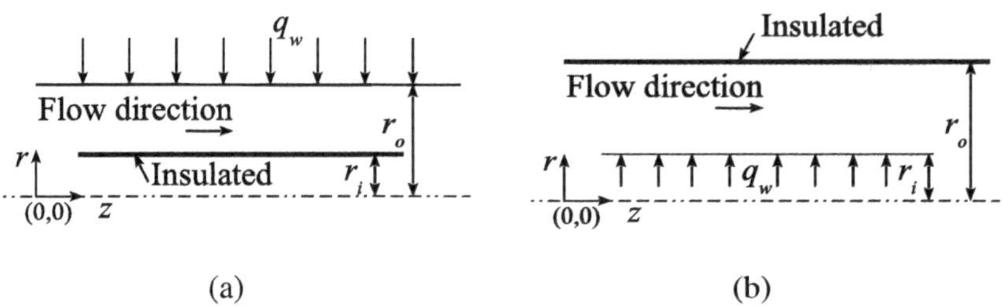

Fig. 3.14 Schematic diagram of gas flow through a micro-annulus (**a**) Outer wall kept at constant heat flux and inner wall insulated (case A) (**b**) Outer wall insulated and inner wall kept at constant heat flux (case B)

The momentum equation in z-direction is first solved:

$$\frac{1}{r}\frac{\partial}{\partial r}\left(r\frac{\partial u}{\partial r}\right) = \frac{1}{\mu}\frac{dp}{dz} = \text{constant} \tag{3.47}$$

The velocity slip boundary conditions at the wall are

$$u = u_s = \frac{2-\sigma_v}{\sigma_v}\lambda\frac{\partial u}{\partial r}\bigg|_{r=r_i}$$
$$u = u_s = -\frac{2-\sigma_v}{\sigma_v}\lambda\frac{\partial u}{\partial r}\bigg|_{r=r_o} \tag{3.48}$$

where σ_v denotes tangential momentum accommodation coefficient and taken as unity for most of the engineering problems.

The non-dimensional variables are

$$R = \frac{r}{r_o},\ r^* = \frac{r_i}{r_o},\ U = \frac{u}{u_m} \tag{3.49}$$

The fully developed velocity profile is derived from the momentum equation by employing the first order velocity slip condition at the wall. By using Eqs. (3.48) and (3.49) the velocity distribution is determined as

$$U = 2\left[\frac{1 - R^2 + 2r_m^{*2}\ln R + A}{B}\right] \tag{3.50}$$

where

$$A = 4Kn(1 - r^*)(1 - r_m^{*2}),$$
$$B = 1 - r^{*2} - 4r_m^{*2}\left(\frac{1}{2} + \frac{r^{*2}}{1 - r^{*2}}\ln r^*\right)$$
$$+8Kn(1 - r^*)(1 - r_m^{*2}),$$

Kn is the Knudsen number $\left(Kn = \frac{\lambda}{D_h}\right)$, and $r_m^* (= \frac{r_m}{r_o})$ is the dimensionless mean radius. This result will be used while solving the energy equation later on.
The point of maximum velocity can be obtained by setting $\frac{\partial u}{\partial r} = 0$. The non-dimensional radius is

$$r_m^* = \left[\frac{(1 - r^{*2})(1 + 4Kn)}{2 \ln \frac{1}{r^*} - 4Kn \left(\frac{r^{*2} - 1}{r^{*2}} \right)} \right]^{1/2} \tag{3.51}$$

Further, for the limiting condition $Kn = 0$, the above equation reduces to

$$r_m^* = \left[\frac{1 - r^{*2}}{2 \ln(1/r^*)} \right]^{1/2} \tag{3.52}$$

which is a well-established result reported in the literature (Rohsenow et al. [121]).

The energy equation including the effect of viscous dissipation is

$$u \frac{\partial T}{\partial z} = \frac{\alpha}{r} \frac{\partial}{\partial r} \left(r \frac{\partial T}{\partial r} \right) + \frac{\nu}{c_p} \left(\frac{\partial u}{\partial r} \right)^2 \tag{3.53}$$

The constant heat flux boundary condition at the wall is

$$q_w = k \left. \frac{\partial T}{\partial r} \right|_{r=r_o} \tag{3.54}$$

The non-dimensional variables are

$$\theta = \frac{T - T_s}{q_w r_o / k}, \quad \theta_m^* = \frac{T_m - T_w}{q_w r_o / k}, \quad Br = \frac{\mu u_m^2}{q_w r_o} \tag{3.55}$$

Using Eqs. (3.16), (3.50), (3.54), and (3.55), the energy equation (3.53) in non-dimensional form is

$$\frac{1}{R} \frac{d}{dR} \left(R \frac{d\theta}{dR} \right) = aU - Br \left(\frac{dU}{dR} \right)^2 \tag{3.56}$$

where $a = \frac{u_m}{\alpha} \frac{k r_o}{q_w} \frac{dT_s}{dz}$.

For case A, the non-dimensional boundary conditions are

$$\theta = 0, \quad \left. \frac{\partial \theta}{\partial R} \right|_{R=1} = 1$$
$$\left. \frac{\partial \theta}{\partial R} \right|_{R=r^*} = 0 \tag{3.57}$$

The solution of non-dimensional energy equation (3.56) is obtained by integrating it twice with the thermal boundary conditions given by Eq. (3.57),

$$\theta(R) = \frac{T - T_s}{q_w r_o/k} = \frac{a}{2B}\left(-\frac{3}{4} - A + 2r_m^{*2} + R^2\left(1 + A - 2r_m^{*2} - \frac{R^2}{4}\right) - \ln R \left(1 + 2A - 2r_m^{*2}(1 + R^2)\right)\right) + \frac{Br}{B^2}\left((1 - R^2)(1 + R^2 - 8r_m^{*2}) + 4\ln R(1 - 4r_m^{*2}) - 8r_m^{*4}(\ln R)^2\right) + \ln R \tag{3.58}$$

where

$$a = \left[\frac{-2B^2 + 8Br(r^{*2} - 1)(1 + r^{*2} - 4r_m^{*2}) + 32Br\, r_m^{*4}\ln r^*}{B(1 + 2A - 2r_m^{*2} - r^{*2})(r^{*2} - 1) + 4B\, r_m^{*2} r^{*2}\ln r^*}\right]$$

For case B, the non-dimensional boundary conditions are

$$\theta = 0, \quad \left.\frac{\partial\theta}{\partial R}\right|_{R=r^*} = -1$$
$$\left.\frac{\partial\theta}{\partial R}\right|_{R=1} = 0 \tag{3.59}$$

Similarly, the solution of non-dimensional energy equation (3.56) is obtained by integrating it twice with the thermal boundary conditions given by Eq. (3.59),

$$\theta(R) = \frac{T - T_s}{q_w r_o/k} = \frac{a}{2B}\left((R^2 - r^{*2})(1 + A - 2r_m^{*2}) - ((R^4 - r^{*4})/4) - (\ln R - \ln r^*)(1 + 2A - 2r_m^{*2}) + 2r_m^{*2}(R^2\ln R - r^{*2}\ln r^*)\right) + \frac{Br}{B^2}\left((R^2 - r^{*2})(8r_m^{*2} - (R^2 + r^{*2})) + 4(\ln R - \ln r^*)(1 - 4r_m^{*2}) - 8r_m^{*4}((\ln R)^2 - (\ln r^*)^2)\right) \tag{3.60}$$

where

$$a = \left[\frac{-2B^2 r^* + 8Br(r^{*2} - 1)(1 + r^{*2} - 4r_m^{*2}) + 32Br\, r_m^{*4}\ln r^*}{B(1 + 2A - 2r_m^{*2} - r^{*2})(r^{*2} - 1) + 4B\, r_m^{*2} r^{*2}\ln r^*}\right]$$

Using Eq. (3.9) the temperature jump at the wall can be written

$$\frac{T_s - T_w}{q_w r_o/k} = -\frac{4Kn\gamma}{Pr(\gamma + 1)}(1 - r^*) \tag{3.61}$$

Using Eqs. (3.58) and (3.61), the non-dimensional temperature distribution with temperature jump is

$$\theta^*(R) = \frac{T - T_w}{q_w r_o/k} = \frac{a}{2B}\Bigg(-\frac{3}{4} - A + 2r_m^{*2} + R^2\Big(1 + A - 2r_m^{*2} - \frac{R^2}{4}\Big)$$
$$- \ln R\Big(1 + 2A - 2r_m^{*2}(1 + R^2)\Big)\Bigg) + \frac{Br}{B^2}\Big((1 - R^2)(1 + R^2 - 8r_m^{*2})$$
$$+ 4\ln R(1 - 4r_m^{*2}) - 8r_m^{*4}(\ln R)^2\Big) + \ln R - \frac{4Kn\gamma}{Pr(\gamma + 1)}(1 - r^*) \tag{3.62}$$

Similarly, using Eqs. (3.60) and (3.61), the non-dimensional temperature distribution is

$$\theta^*(R) = \frac{T - T_w}{q_w r_o/k} = \frac{a}{2B}\Big((R^2 - r^{*2})(1 + A - 2r_m^{*2}) - ((R^4 - r^{*4})/4)$$
$$- (\ln R - \ln r^*)(1 + 2A - 2r_m^{*2}) + 2r_m^{*2}(R^2 \ln R - r^{*2}\ln r^*)\Big)$$
$$+ \frac{Br}{B^2}\Big((R^2 - r^{*2})(8r_m^{*2} - (R^2 + r^{*2})) + 4(\ln R - \ln r^*)(1 - 4r_m^{*2})$$
$$- 8r_m^{*4}((\ln R)^2 - (\ln r^*)^2)\Big) - \frac{4Kn\gamma}{Pr(\gamma + 1)}(1 - r^*) \tag{3.63}$$

Using Eq. (3.25) the non-dimensional bulk mean temperature for case A is

$$\theta_m^* = \frac{4(p_1 + p_2 + p_3)}{1 + 2A - 2r_m^{*2}} - \frac{4Kn\gamma}{Pr(\gamma + 1)}(1 - r^*) \tag{3.64}$$

where

$$p_1 = \frac{a}{2B}\Bigg(\frac{(1+2A-2r_m^{*2})(-3-4A+8r_m^{*2})}{16} + \frac{(3+6A-6r_m^{*2})(2+6A-3r_m^{*2})}{72}$$
$$- \frac{(3+12A-4r_m^{*2})}{72} - \frac{(1+2A)(-3-4A+8r_m^{*2})}{16} + \frac{r_m^{*2}(-32-45A+81r_m^{*2})}{72}\Bigg)$$

$$p_2 = \frac{Br}{B^2}\Bigg(\frac{15 + 24A - 32r_m^{*2} - 24r_m^{*2}(4 + 6A - 9r_m^{*2})}{72}$$
$$+ \frac{(1 - 4r_m^{*2})(-3 - 4A + 8r_m^{*2})}{4} - \frac{r_m^{*4}(7 + 8A - 24r_m^{*2})}{4}\Bigg)$$

and

$$p_3 = \frac{-5 - 9A + 18r_m^{*2}}{36}.$$

Similarly the non-dimensional bulk mean temperature for case B is

$$\theta_m^* = \frac{4(p_4 + p_5)}{1 + 2A - 2r_m^{*2}} - \frac{4Kn\gamma}{Pr(\gamma + 1)}(1 - r^*) \tag{3.65}$$

where

$$\begin{aligned} p_4 = \frac{a}{2B}\Bigg[&(1 + A - 2r_m^{*2})\left(\frac{2 + 6A - 3r_m^{*2}}{24} - r^{*2}\frac{1 + 2A - 2r_m^{*2}}{4}\right) \\ &- \frac{1}{4}\left(\frac{3 + 12A - 4r_m^{*2}}{72} - r^{*4}\frac{1 + 2A - 2r_m^{*2}}{4}\right) \\ &- (1 + 2A - 2r_m^{*2})\left(\frac{-3 - 4A + 8r_m^{*2}}{16} - \ln r^*\frac{1 + 2A - 2r_m^{*2}}{4}\right) \\ &+ \frac{r_m^{*2}}{2}\left(\frac{(-5 - 9A + 9r_m^{*2})}{36} - r^{*2}\ln r^*(1 + 2A - 2r_m^{*2})\right)\Bigg] \end{aligned}$$

and

$$\begin{aligned} p_5 = \frac{Br}{B^2}\Bigg[&\left(\frac{52r_m^{*2} - 72r_m^{*4} - 12A(1 - 12r_m^{*2}) - 3}{72} - r^{*2}\frac{(8r_m^{*2} - r^{*2})(1 + 2A - 2r_m^{*2})}{4}\right) \\ &+ 4(1 - 4r_m^{*2})\left(\frac{-3 - 4A + 8r_m^{*2}}{16} - \ln r^*\frac{1 + 2A - 2r_m^{*2}}{4}\right) \\ &- 8r_m^{*4}\left(\frac{7 + 8A - 24r_m^{*2}}{32} - (\ln r^*)^2\frac{1 + 2A - 2r_m^{*2}}{4}\right)\Bigg] \end{aligned} \tag{3.66}$$

Using expressions derived above, the Nusselt number can be obtained from,

$$Nu = \frac{q_w D_h}{k(T_w - T_m)} = -\frac{2}{\theta_m^*}(1 - r^*) \tag{3.67}$$

3.4.3.1 Parametric Variation

We presented the results of Avci and Aydin [13] for gas flow through a concentric cylindrical annulus by employing two different thermal boundary conditions, such as: uniform heat flux at the outer wall and adiabatic inner wall (case A) and uniform heat flux at the inner wall and adiabatic outer wall (case B). The effect of rarefaction and viscous dissipation on temperature distribution and Nusselt number is discussed.

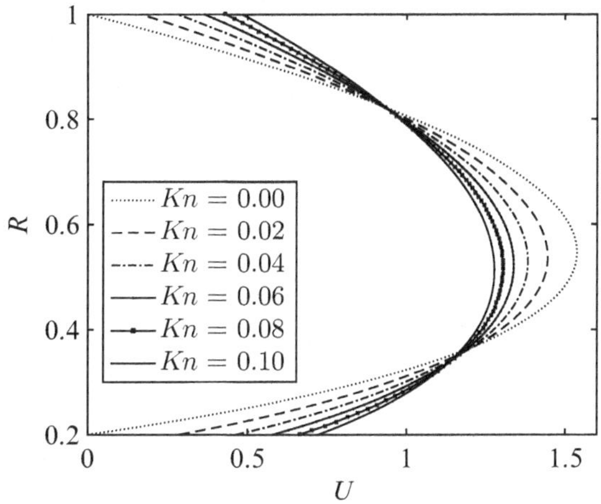

Fig. 3.15 Variation of non-dimensional velocity with non-dimensional radial co-ordinate at different Knudsen number

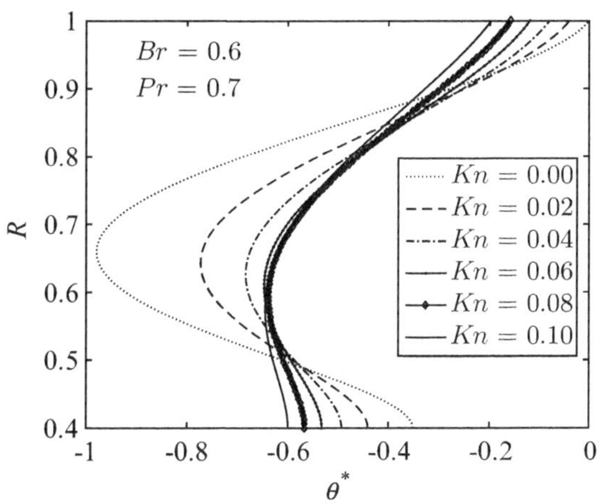

Fig. 3.16 Variation of non-dimensional temperature with non-dimensional radial co-ordinate at different Knudsen number (case A)

Figure 3.15 illustrates the non-dimensional velocity profiles obtained for different values of the Knudsen number at $r^* = 0.2$. The velocity values at $R = 0.2$ and $R = 1$ (denotes the inner and outer walls of the annulus) represent the non-dimensional values of the velocity slip. The velocity distribution and the slip velocities are not symmetrical at the walls. A point having maximum value of the velocity occurs near to the inner wall and can be obtained by setting $\frac{\partial u}{\partial r} = 0$. Note that with the increase in Knudsen number, the slip velocity increases at the inner wall and outer wall of the micro annulus, while the maximum value of the velocity decreases with the increase in Knudsen number, which makes the velocity profile flat, as shown in Fig. 3.15.

For both cases A and B, the variation of non-dimensional temperature with non-dimensional radial co-ordinate for different values of Knudsen numbers at $Br = 0.6$ is shown in Figs. 3.16 and 3.17, respectively. Here, $Kn = 0$ represents the temperature profile without rarefaction, while non-zero Kn represent the

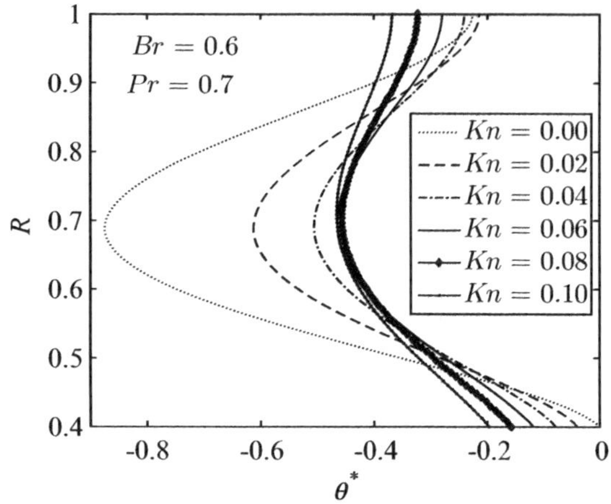

Fig. 3.17 Variation of non-dimensional temperature with non-dimensional radial co-ordinate at different Knudsen number (case B)

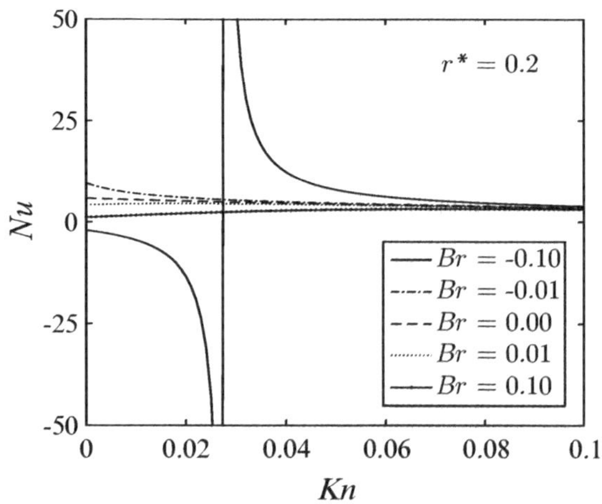

Fig. 3.18 Variation of Nusselt number with Knudsen number at different Brinkman number (case A)

temperature profile with the effect of rarefaction. The rarefaction is found to affect the temperature distribution in the presence of viscous dissipation. It is noticed that the increase in Knudsen number flattens the temperature distribution significantly.

For case A (outer wall kept at constant heat flux and inner wall insulated), the variation of Nusselt number with Knudsen number for various values of Brinkman number at $r^* = 0.2$ is shown in Fig. 3.18. For $Br = 0$, the effect of viscous dissipation is absent and fully developed Nusselt number decreases with the increase in Knudsen number due to the increased rarefaction effect. As Knudsen increases the temperature jump at wall increases, which leads to a decrease in heat transfer. It is interesting to note that the effect of viscous dissipation is more pronounced at lower Knudsen number, while for higher values of Knudsen number this effect is insignificant. In addition, viscous dissipation causes singularity in the Nusselt number at a particular Knudsen number. For negative Brinkman number, the Nusselt

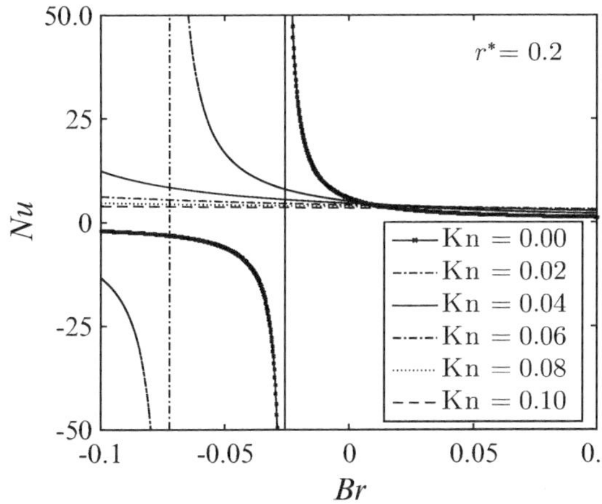

Fig. 3.19 Variation of Nusselt number with Brinkman number at different Knudsen number (case A)

number is found to be more as compared to the case of positive Brinkman number. This is because near the wall, fluid temperature increases due to viscous dissipation, therefore, the temperature difference between wall and fluid increases, which results in an enhanced heat transfer as compared to the case of no viscous dissipation. Note that the heat transfer behavior is affected by the combined effect of rarefaction and viscous dissipation as shown in Fig. 3.18.

Similarly, for case A, the variation of Nusselt number with Brinkman number for various values of the Knudsen number at $r^* = 0.2$ is shown in Fig. 3.19. Here, the vertical lines show the singularity in Nusselt number at a particular Knudsen number, that is because the mean temperature reaches the wall temperature. While, the negative Nusselt number shows that the mean temperature is higher as compared to the wall temperature. In addition, it is observed that the effect of viscous dissipation, quantified by Brinkman number, is effective for lower values of Knudsen number, as shown in Fig. 3.19.

For case B (outer wall insulated and inner wall kept at constant heat flux) Fig. 3.20 describes the variation of Nusselt number with Knudsen number for different values of the Brinkman number at $r^* = 0.2$. The effect of rarefaction, quantified by Knudsen number, is to reduce heat transfer due to the increased temperature jump, which lowers the fluid temperature due to large temperature gradients close to the wall.

For case B (outer wall insulated and inner wall kept at constant heat flux) the variation of Nusselt number with the Brinkman number for different values of the Knudsen number is shown in Fig. 3.21. The vertical lines show the points of singularity in Nusselt number, however there is no physical significance of the singular points; these points indicate that $T_m = T_w$. It is observed that the effect of viscous dissipation is more pronounced at lower values of Knudsen number, while for high Kn value, it is insignificant as demonstrated in Fig. 3.21.

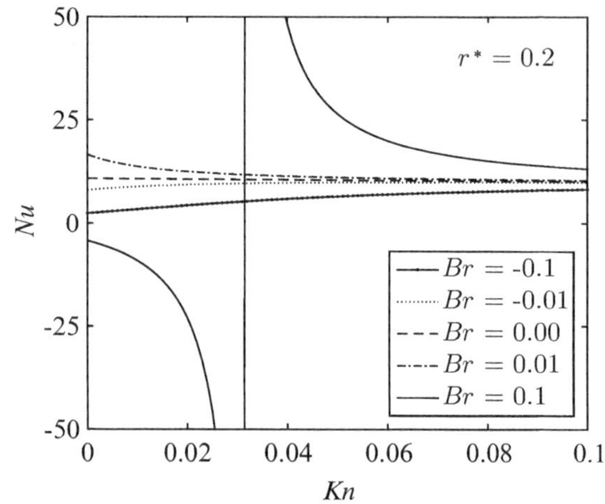

Fig. 3.20 Variation of Nusselt number with Knudsen number at different Brinkman number (case B)

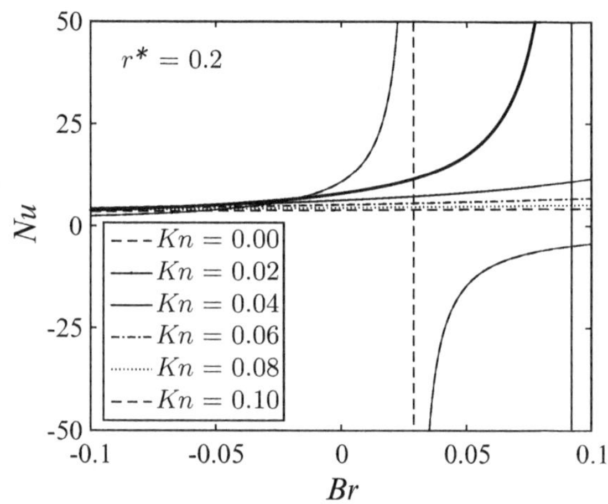

Fig. 3.21 Variation of Nusselt number with Brinkman number at different Knudsen number (case B)

3.5 Observations on Other Effects

In the above section, we have obtained exact solution of the governing equation for a few relatively simple cases. However, in presence of additional effects—such as axial conduction, compressibility, wall conduction, property variation although governing equations can still be written—the analytical solution of the governing equations becomes difficult and one rather solves the equations numerically.

3.5.1 Variation in Thermophysical Properties

Among the relevant thermophysical properties, viscosity, density, and thermal conductivity show a stronger dependence on temperature as compared to heat capacity. Therefore, the variation of these properties with temperature has to be suitably accounted for in understanding the variation of velocity and temperature profiles and the Nusselt number. The influence of density and thermal conductivity variations due to temperature gradient in the flow is measured in terms of entropy generation. The rate of entropy generation per unit volume in the fluid $\dot{S}$ is directly proportional to the amount of viscous dissipation (Φ, which depends on the square of velocity gradient; Eq. (3.3)) and the square of temperature gradient; mathematically,

$$\dot{S} = \frac{\Phi}{T} + \frac{k}{T^2}(\nabla T)^2 \tag{3.68}$$

The variation of properties leads to an increase in the velocity (and temperature) gradient in certain region and a decrease in some other region, with respect to the baseline case of constant property, which alters the rate of entropy generation per unit volume in the fluid. Therefore, the integral value of $\dot{S}$ over the entire region is computed in order to access the effect of property variation better.

In the context of flow in a microtube, Prabhu and Mahulikar [119] reported that the axial and radial variations in density cause a distortion in the velocity profile. The density gradients cause flattening of axial velocity profile leading to an effect that is similar to "flow un-development." Similar to flow in the developing region, this induces a radial flow in order to satisfy the condition for continuity of mass. The presence of radial velocity also leads to a radial convection, which can significantly affect the overall convection in the microtube. The variation of thermal conductivity along the flow direction leads to axial conduction and an increase in the fluid temperature, while the radial variation of thermal conductivity causes a flattening of the temperature profile. The overall rate of entropy generation was found to increase in presence of property variation.

3.5.2 Conduction in the Substrate

Frequently, the thickness of the substrate on which the microchannels are etched is relatively large compared to the flow passage. This leads to substantial amount of conduction of heat in the substrate itself. So if a uniform heat flux is applied on the outer wall of the microchannel, which is typically the case, the heat flux on the inner wall is non-uniform (Fig. 1.9). The amount of non-uniformity is a function of various parameters including wall thermal conductivity and wall thickness.

Maranzana et al. [92] defined a parameter called the axial wall conduction number to quantify when this wall conduction will become large enough so that we need to account for it. This parameter, M, is a ratio of heat transferred as conduction in the wall (opposite to the direction of the flow) to the heat transferred as convection from the surface, and expressed mathematically,

$$M = \frac{\Phi_{cond}}{\Phi_{conv}} = \frac{k_{so} A_{so}(T_{s,o} - T_{s,i})}{L\dot{m}c_p(T_{f,o} - T_{f,i})} \tag{3.69}$$

where Φ_{cond} is a heat flux characterizing axial heat transfer in the wall, Φ_{conv} is the total convective heat flux, k_{so}, A_{so}, and L are, respectively, the thermal conductivity, cross-sectional area and length of the microchannel solid substrate. Further, $T_{s,o}$ and $T_{s,i}$ are the outlet and inlet temperature of substrate, while $T_{f,o}$ and $T_{f,i}$ are the outlet and inlet temperature of fluid. The threshold value of M beyond which the wall conduction effect should be considered is 0.01. A detailed analysis of this conjugate effect in the context of a microchannel has been made by Mohrana and Khandekar [100]. Further, the conjugate effect has been utilized to obtain a uniform wall temperature on the inner wall of a diverging microchannel, while supplying a uniform heat flux on the outer wall, in Duryodhan et al. [50]. The above studies are however with liquid flow.

We note that although wall thickness at the microscale is typically smaller than that at the conventional scale in absolute sense, the relative thickness (ratio of thickness of wall to thickness of flow passage) is much larger at the microscale, making conjugate effect a very relevant issue in microchannel heat transfer problems. That is, at the conventional scale although wall conduction does occur, its relative importance with respect to convection in affecting the heat flux distribution is insignificant and therefore the conjugate effects have conventionally been neglected.

3.5.3 Axial Conduction

The importance of axial conduction can be accessed from the value of Peclet number, Pe. Since $Pe = \frac{\rho c_p u_m D_h}{k}$, upon multiplying the numerator and denominator of this equation by the axial fluid bulk temperature gradient (dT_m/dx), we can write this expression in terms of mass flow rate $(\dot{m})$ through the microchannel,

$$Pe = \frac{\dot{m} c_p D_h (dT_m/dx)}{k A_o (dT_m/dx)} \tag{3.70}$$

Thus, Peclet number signifies the ratio of rates of advection to diffusion or it represents the relative magnitude of the thermal energy transported to the fluid (as fluid enthalpy change) to the thermal energy axially conducted within the fluid. The Peclet number is also the product of the Reynolds and Prandtl numbers. Due to a reduction in Reynolds number, the Peclet number reduces, and the convection term no longer dominates over conduction of heat in the fluid. That is, the longitudinal (or axial) conduction of heat can no longer be neglected in comparison to lateral conduction or convection.

Note that in the presence of axial heat conduction both Nusselt number and thermal entry length increases. Also, the effect of axial conduction on Nusselt number increases with a decrease in Knudsen number. The thermal entrance region increases with an increase in Peclet number.

It is apparent from the theoretical analysis of Satapathy [124] that temperature starts to change much before the gas enters the heating zone for $Pe < 1$. Further, a change in the value of Knudsen number does not seem to change the variation of temperature in pre-heating zone much, although the effect is evident in the post heating zone. This analysis is however for uniform fluid velocity. A similar observation can be made from experimental data as discussed in Sect. 3.6.

3.5.4 Flow Work and Shear Work

The finite velocity at the wall (due to slip) combined with the shear stress produces a shear work at the boundary (Fig. 3.22), while this effect is absent at the macroscale due to the no-slip condition. This shear work at the solid boundaries can make a significant contribution in microscale gas flows, especially when the slip effect is substantial.

The steady energy equation considering the effects of viscous dissipation and flow work can *roughly* be written

$$\rho c_p u \frac{\partial T}{\partial x} = k \frac{\partial^2 T}{\partial y^2} + u \frac{dp}{dx} + \tau_{xy} \frac{du}{dy} \tag{3.71}$$

where $u\frac{dp}{dx}$ and $\tau_{xy}\frac{du}{dy}$ terms denote the flow work and heat generated due to viscous heating, respectively. Note that the two terms have opposite sign: The flow work is negative because pressure usually decreases with x, while viscous heating ($\mu(du/dy)^2$) is positive, and the shear work ($\tau_{xy} \times u_s = \mu u_s \frac{du}{dy}$) due to slipping of the gas at boundary balances the contribution of viscous dissipation and flow work. The contribution of shear work at macroscale is zero due to the assumption of no-velocity slip boundary condition. At the microscale shear work plays a determining role in the convective heat transfer problems under the assumption of (locally) fully developed and constant heat flux boundary condition. Therefore, it should be considered in the calculation of total heat exchange with the walls, however, it has no direct effect on the temperature field as it happens at the boundary only [58].

The term $u\frac{dp}{dx}$ acts as a distributed heat sink, with the majority of thermal energy absorbed near the center of flow, due to the larger velocity magnitude in this region. On the contrary, shear work acts as a heat source due to the thermal energy generated by the flow. In general, the two opposite effects may not cancel each other, and therefore the implications of these works on the overall flow and heat transfer behavior have to be carefully analyzed.

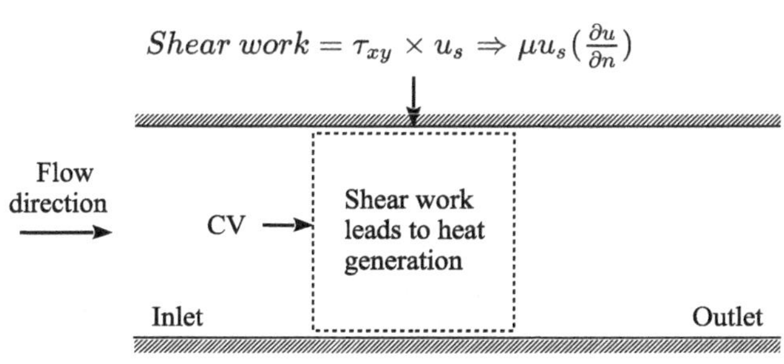

Fig. 3.22 Contribution of shear work at wall due to slipping of the gas

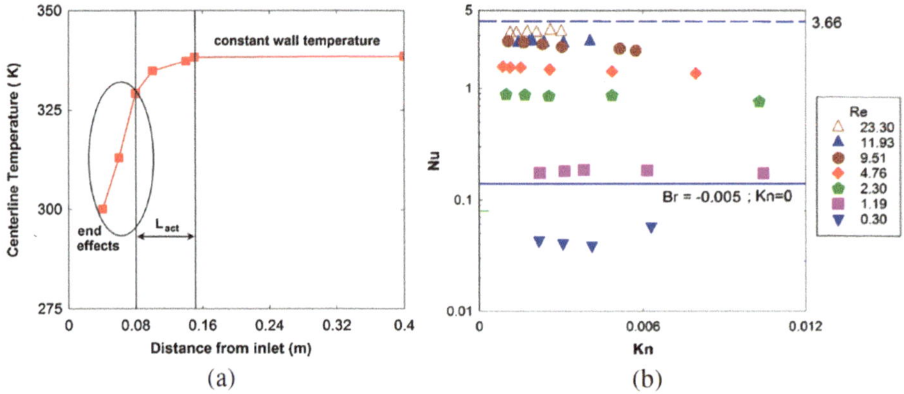

Fig. 3.23 (**a**) The axial variation of centerline fluid temperature along the length of tube (**b**) Variation of Nusselt number with Knudsen number at different Reynolds number; Figures taken from [73]

3.6 Observations from Experiments

The amount of experimental data for flow in the slip regime in any configuration is rather limited. Hemadri et al. [73] employed a counter-flow tube-in-tube heat exchanger for measurement of the overall heat transfer coefficient in the slip regime (of the type shown in Fig. 8.4c). Towards this, low-pressure nitrogen flowed through the conventional size inner tube of the heat exchanger and picked up heat from the hot water flowing in the outer jacket. Knowing the mass flow rate of nitrogen and water, and the respective inlet and outlet temperatures, allows the overall heat transfer coefficient to be determined. Further, local measurement of fluid temperature, with the help of thermocouples mounted at different axial locations at the centerline of the tube, allowed the local heat transfer coefficient to be determined.

These measurements showed a large amount of heating of the incoming flow, owing to axial back conduction of heat in the gas (Fig. 3.23a). As evident from the figure, the temperature of the gas reached the wall temperature over a relatively short distance from the entry of the tube. The Nusselt number value was found to be much lower than expected from available theoretical analysis (Fig. 3.23b). Further, the value of Nusselt number showed a stronger dependence on Reynolds number rather than Knudsen number, as apparent from the figure.

3.7 Application to Knudsen Pump

Thermal creep is an interesting flow phenomenon based on which a pump for pumping gases can be constructed. Consider two reservoirs containing same gas maintained at different temperatures, connected by a microchannel. The width of microchannel is comparable to the mean free path of the gas. Then a flow of gas occurs from the *cold* reservoir to the *hot* reservoir, and this phenomenon is called thermal creep (or thermal transpiration).

As per kinetic theory, the mass of gas molecules crossing unit area per unit time towards one side is given by

$$\Gamma_m = \frac{1}{4}\rho\bar{v} = \rho\sqrt{\frac{RT}{2\pi}} = \frac{p}{\sqrt{2\pi RT}} \tag{3.72}$$

where $\bar{v}$ is the mean molecular speed ($=\sqrt{8RT/\pi}$). Therefore, the mass flow of molecules crossing from the two sides of the reservoir is proportional to $p/\sqrt{T}$. So if the reservoirs are maintained at the same pressure but different temperatures, the flux of molecules crossing from hot to cold side, and the flux of molecules crossing from cold to hot side will be different; with the latter flux being larger than the former flux. This leads to a net movement of gas from the cold to the hot reservoir. The cross-over of molecules to the other reservoir will continue until the pressures become equal on the two sides or

$$\frac{p_1}{p_2} = \sqrt{\frac{T_1}{T_2}}. \tag{3.73}$$

Based on this idea, McNamara and Gianchandani [96] constructed a pump (called the Knudsen pump). A schematic of their micro-pump is shown in Fig. 3.24. The pressure generated by the pump can be increased by having multiple-stages of the pump.

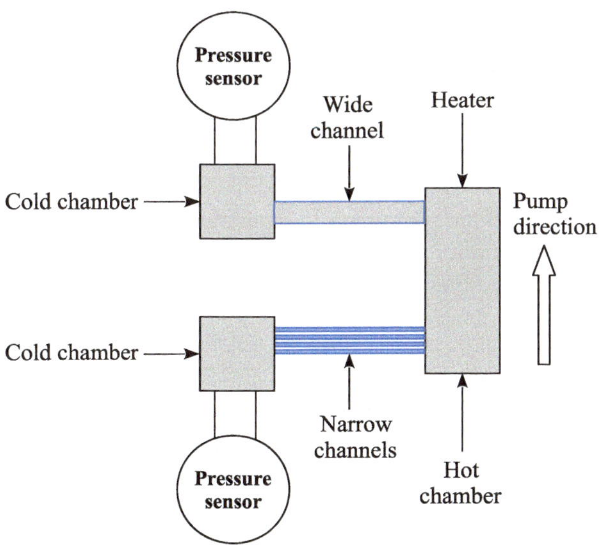

Fig. 3.24 Layout of a Knudsen pump

Note that the analysis presented here involves several simplifying assumptions. The flow in the microchannel is also required to be in the free-molecular or transition regime. However given this interesting application of gaseous flow in small passages, we have covered it in this section.

3.8 Useful Empirical Correlations

Some useful empirical correlations are compiled in this section for ready reference. Tunc and Bayazitoglu [150] proposed the following expression for local Nusselt number (Nu_x) for flow in a tube involving velocity slip, temperature jump, and viscous dissipation,

$$Nu_x = \frac{2}{\theta_s + \dfrac{4Kn\gamma}{Pr(\gamma+1)} - \theta_m} \tag{3.74}$$

where θ_s is the non-dimensional temperature of the fluid at wall and θ_m is the mean bulk temperature. The above equation is for uniform wall heat flux condition. The above correlation was obtained analytically by solving the governing equation using the integral transform method.

Similarly, Nusselt number expression for constant wall temperature condition was proposed as,

$$Nu_x = \frac{h_x D}{k} = \frac{-2\left.\dfrac{\partial\theta}{\partial\eta}\right|_{\eta=1}}{\theta_m - \dfrac{4Kn\gamma}{Pr(\gamma+1)}\left.\dfrac{\partial\theta}{\partial\eta}\right|_{\eta=1}} \tag{3.75}$$

where θ is the non-dimensional temperature and η is the non-dimensional radial co-ordinate.

Inman [77] proposed an expression to evaluate heat transfer characteristics of laminar gas flow in a parallel plate channel or circular tube by employing constant heat flux boundary condition in the slip flow regime,

$$Nu = \left[\frac{\xi}{4} + \frac{17 + 84\zeta + 105\zeta^2}{140(1+3\zeta)^2}\right]^{-1} \tag{3.76}$$

where $\zeta = 4Kn\dfrac{2-\sigma_v}{\sigma_v}$ and $\xi = \dfrac{2-\sigma_T}{\sigma_T}\dfrac{8Kn\gamma}{Pr(\gamma+1)}$.

Miyamoto et al. [98] considered argon flow in a small parallel plate channel to study heat transfer characteristics by employing the constant heat flux boundary condition at the wall. The effect of velocity slip, temperature jump, and viscous dissipation is included. The expression for Nusselt number is

$$Nu = \left[\frac{9Br}{(1+6Kn)^2} \left(\frac{3+42Kn+140Kn^2}{35(1+6Kn)} + \frac{8Kn^2\gamma}{Pr(\gamma+1)} + \frac{17+168Kn+420Kn^2}{140(1+6Kn)^2} + \frac{Kn\gamma}{Pr(\gamma+1)} \right) \right]^{-1} \quad (3.77)$$

Hooman [74] proposed a correlation to estimate heat transfer characteristics of a gas flow between parallel plate microchannels by using superposition approach,

$$Nu = \left[\frac{2-\sigma_T}{\sigma_T} \frac{Kn\gamma}{Pr(\gamma+1)} + \frac{17\sigma_v + 84Kn\sigma_v + 0.5Kn(14+70Kn)(24-12\sigma_v)}{140(1+6Kn)(\sigma_v + 0.5Kn(24-12\sigma_v))} \right]^{-1} \quad (3.78)$$

Niazmand et al. [105] reported correlations to evaluate Nusselt number for slip flow through rectangular and trapezoidal channels,

$$(Nu)_{fd} = \left[2.87 \left(\frac{90}{\phi} \right)^{-0.26} + 4.8 \exp\left(-3.9\alpha \left(\frac{90}{\phi} \right)^{0.21} \right) \right] G_2 G_3 \quad (3.79)$$

where α denotes the aspect ratio, ϕ is the acute side angle, and

$$G_2 = 1 - 1.75\, Kn^{0.64}(1 - 0.72 \tanh(2\alpha)),$$

$$G_3 = 1 + 0.075(1+\alpha)\exp(-0.45RePr).$$

3.9 Summary

In this chapter we have discussed about the microscale heat transfer and factors affecting heat transfer characteristics, such as: velocity slip, temperature jump, and viscous dissipation. Some exact solution for these cases were obtained. The strong effect of viscous dissipation on heat transfer is particularly noteworthy; the Nusselt number can even become negative when the temperature difference between the wall and mean temperature changes its sign. However, these solutions neglected factors such as axial conduction, pressure work, and compressibility effects, which can be important as evident from the experimental data. Property variation and conjugate effects can also play a substantial role and affect the velocity and/or temperature profiles. It is therefore worthwhile to innovatively design and conduct experiments in the slip regime where the effect of individual parameters on Nusselt number can be isolated, and the influence of microscale parameters can be better quantified.

Chapter 4
Need for Looking Beyond the Navier–Stokes Equations

The need for looking beyond the Navier–Stokes equations is addressed in this chapter through specific examples where these equations fail. We also examine some extensions of the Navier–Stokes equations, which have been recently proposed in the literature. Similarly, attempts to modify the Fourier law to account for non-Fourier effects are reviewed. An example of shock wave where these alternative forms of Navier–Stokes equations have been applied is also included.

4.1 Introduction

In the two previous chapters, we examined the solutions for gaseous slip flow and heat transfer under various conditions. The obtained solutions are primarily applicable in the slip regime; the extension in the range of Knudsen number over which the solution is applicable is only modest. In order to model flows at still higher Knudsen number range (particularly in the transition regime), the Navier–Stokes equations are no longer applicable, and there is a need to look beyond these conventional equations.

However, before examining extension of the flow model to higher Knudsen number (which is undertaken in the next three chapters), we examine some relatively simple extensions of the Navier–Stokes and Fourier-order energy equations in the present chapter. An important reason why such extension is being sought is that numerical methods for solving the Navier–Stokes equations (or the field of computational fluid dynamics) are already well developed [38, 128] and its extension to any additional terms in the conventional Navier–Stokes equations is deemed to be not that difficult.

A. Agrawal et al., *Microscale Flow and Heat Transfer*, Mechanical Engineering Series, https://doi.org/10.1007/978-3-030-10662-1_4

4.2 Examples Needing a Look Beyond the Navier–Stokes Equations

We start by citing few problems whose treatment within the Navier–Stokes formulation is known to be problematic.

1. *Couette flow in the later slip and early transition regimes*: As discussed in Sect. 1.5.2, DSMC simulations suggest a non-linear velocity profile in the later slip and early transition regimes (Fig. 1.10b). The Navier–Stokes equations predict a linear profile (Eq. (2.58)), suggesting that the non-linearity cannot be modelled within the framework of these equations. As an aside, we note that the DSMC data suggests that the velocity profiles become linear again at still higher Knudsen numbers [133].
2. *Poiseuille flow in the transition regime*: As discussed in Sect. 1.5.2, for flow in a microchannel, the Navier–Stokes equations predict a uniform pressure in the lateral direction (Eq. (2.78)). However, DSMC data shows that pressure varies along the lateral direction (Fig. 1.11c). This non-uniform value of pressure along the lateral direction cannot be modelled within the framework of the Navier–Stokes equations.
3. *Flow near the leading edge of a plate*: For uniform flow encountering a stationary flat plate, the Navier–Stokes equations predict an infinite coefficient of friction and an infinite Nusselt number at the leading edge of the plate. While the Navier–Stokes equations based solution is well applicable to regions slightly downstream of the leading edge, the singularity at the leading edge needs to be modelled by the Grad equation (or another higher-order continuum transport equation). This example can be extended to include flows at the entrance of a passage where such singularity appears.
4. *Impulsively started plate in a flow*: The impulsively started flow problem is analogous to the problem of flow near the leading edge of a plate. The Navier–Stokes equations are not applicable to time which is just after the start of the plate motion, although the equations are applicable to slightly later times. This point is discussed at length in Sect. 6.7 where solutions from various approaches are compared.
5. *Flow in sudden contraction*: The Navier–Stokes equations do not capture the gross features of flow in a sudden contraction in the slip regime well. In contrast, numerical results (essentially solving the Navier–Stokes equations) for the case of sudden expansion (i.e., with flow direction reversed) at otherwise comparable values of governing parameters compare reasonably well with experimental data for both global and local flow features. (This point was noted by Vijay V. Varade but the numerical evidence for it is still unpublished.) Further, the absence of flow separation in sudden expansion while early onset of flow separation in bend passages are interesting observations, and it is not clear if these observations can be quantitatively described within the framework of the Navier–Stokes equations.

6. *Shock wave*: The flow properties change rapidly across a shock with the gradients being confined in a region of the order of the mean free path of the gas on either side of the shock. The flow in the vicinity of the shock layer is far from local thermodynamic equilibrium, and the Navier–Stokes equations do not perform well in this region. Predicting the shock structure correctly therefore forms an important test case for any higher order transport equation.
7. *Propagation of waves in atmosphere of Mars*: The atmosphere of Mars is considerably more rarefied than that of other terrestrial environments. Therefore, as recognized in recent works (for example, see reference [8, 112]), the treatment of infrasonic and ultrasonic waves in the atmosphere of Mars cannot be based on the Navier–Stokes equations. These authors suggested employing higher orders of the Boltzmann equation for capturing the high Knudsen number regimes likely to be encountered in the transmission of these waves.
8. *Heat transfer in microchannels and cavities*: DSMC calculation of heat transfer in microchannel [16] and cavities [99] has shown several non-intuitive effects. For example, the direction of rotation of vortices in a cavity with Navier–Stokes equations was found to be opposite to that predicted by DSMC and some higher-order transport equations, even at moderate values of Knudsen number. Further, anti-Fourier heat transfer, i.e. transfer of heat from cold-to-hot regions of the flow field was noted from the DSMC simulations for heat transfer in a microchannel. The experimentally measured value of Nusselt number for flow of rarefied gas in a tube is much smaller than the predicted values, as seen in Sect. 3.6. Clearly, these effects are not within the ambit of the conventional equations.

The two main extensions of the Navier–Stokes equations are mentioned in the next two sections. This is followed by a discussion on a popular non-Fourier model.

4.3 Extended Navier–Stokes Equations

Brenner [25] proposed an extension to the Navier–Stokes equation, with the intention to explain the thermophoretic motion of particles. Brenner's hypothesis is that the velocity involved in the constitutive equation (Eq. (2.2)) is not same as velocity which brings in mass in to a control volume (Eq. (2.1)). He argued that the former is volume velocity ($\mathbf{u}_v$) whereas the latter is mass velocity ($\mathbf{u}_m$). Here, the boldface symbol denotes a vector quantity. Dye added to the flow will show volume velocity, whereas the mean velocity of the molecules comprising the fluid element is mass velocity. The two velocities are however related through the following relation:

$$\mathbf{u}_v = \mathbf{u}_m + \mathbf{j}_v \tag{4.1}$$

where $\mathbf{j}_v = \alpha \nabla(\ln \rho)$ represents the diffusive flux of volume, and α is thermal diffusivity. Note that $\mathbf{j}_v = 0$ for incompressible flows, and we obtain $\mathbf{u}_v = \mathbf{u}_m$ for this special case. At the wall, the mass velocity is supposed to go to zero.

This modification, if applicable, changes the constitutive relation to

$$\sigma = \sigma^{NS} + 2\mu\overline{\nabla \mathbf{j}_v} \tag{4.2}$$

where σ is the stress tensor and the overbar denotes the symmetric and traceless part of the tensor (equivalent to angular bracket in our tensor notation). Therefore, both the Navier–Stokes and energy equations change to:

$$\rho\left(\frac{\partial \mathbf{u}_m}{\partial t} + \mathbf{u}_m \cdot \nabla \mathbf{u}_m\right) = -\nabla p + \nabla \cdot (2\mu\overline{\nabla \mathbf{u}_v}) + \nabla(\mu_2 \nabla \cdot \mathbf{u}_v) \tag{4.3}$$

$$\rho C_p\left(\frac{\partial T}{\partial t} + \mathbf{u}_m \cdot \nabla T\right) = \nabla \cdot (k\nabla T) + \nabla \cdot \left[p(\mathbf{u}_v - \mathbf{u}_m)\right] - \left(\frac{\partial \ln \rho}{\partial \ln T}\right)_p \frac{D_m p}{Dt}$$
$$+ \mu_2(\nabla \cdot \mathbf{u}_v)(\nabla \cdot \mathbf{u}_m) + 2\mu\overline{\nabla \mathbf{u}_v} : \overline{\nabla \mathbf{u}_m} \tag{4.4}$$

where μ_2 is the second coefficient of viscosity. The mass velocity satisfies the continuity equation; that is, Eq. (2.1) with $\mathbf{u} = \mathbf{u}_m$ remains unchanged. The volume velocity satisfies the kinematic constraint that $\nabla \cdot \mathbf{u}_v = 0$ irrespective of whether the flow is compressible or incompressible.

The argument that the new constitutive relation is acceptable is advanced by comparing the additional term in the constitutive equation (4.2) or

$$\sigma^+ = 2\mu\overline{\nabla j_v} \tag{4.5}$$

with two Burnett order terms

$$\sigma^+ = -\frac{\mu^2}{\rho T}(K_1\overline{\nabla\nabla T} + K_2 T^{-1}\overline{\nabla T \nabla T}) \tag{4.6}$$

(the Burnett equations are introduced formally in Chap. 5); these terms were argued to be same. Therefore, there is some justification for augmenting the Navier–Stokes equation in the manner suggested by Brenner.

The above modification however leads to several difficulties. For example:

1. The question about the order of Kn for this additional term (Eq. (4.5)) arises. As shown explicitly in Sect. 5.6, the terms in Navier–Stokes and energy equations are of order Knudsen number, whereas Burnett order stress and heat flux terms are of order Knudsen squared. Therefore, an incompatibility in the order of Kn for the various terms in the equation arises.
2. Addition of term σ^+ (Eq. (4.6)) to the constitutive relation (Eq. (4.2)) implies that mechanical stress can be induced by a purely density/thermal gradient (even in the continuum framework—and not just at Burnett order).
3. Notice that Eq. (4.6) involves second derivative of temperature, which becomes of third-order term upon substituting it in the Cauchy's equation of motion

(Eq. (2.3)). This term is known to create problem of stability and the requirement of extra boundary condition in the Burnett equation. The issue of having to prescribe extra boundary condition and the problem of stability would thereby creep into the Navier–Stokes equation as well!

4. Further, notice that the diffusion velocity is in the direction of the density gradient, whereas customary it is in the opposite direction to the density and temperature gradients.
5. The last term in Eq. (4.4) (the viscous dissipation term) is no longer a perfect square. The implication of this will have to be carefully analyzed, in that, can it lead to viscous dissipation becoming negative.

In essence, in this approach, selected terms from a higher order equation (Burnett equations) are added to a lower order equation (Navier–Stokes equations) with the aim of fixing certain deficiencies in the lower order equation. Such attempts have been made to fix the Burnett equation by adding selected terms from the super Burnett equations, as discussed in Sect. 7.3.1. In general, this approach does not lead to a better framework of equations because of the selective nature of terms being added to the base equation. Going all out to the higher order equation is rather a superior approach.

4.4 Modified Navier–Stokes Equations

Another interesting attempt to extend the Navier–Stokes equation has been made in Sambasivam [123]. Observations on flow through microchannels and shock wave, which are not well predicted by the Navier–Stokes equations provided the motivation for this work. The hypothesis is that strong gradients of density and temperature affect the flow in these situations, and therefore modification to the Navier–Stokes equations is needed to cover these cases.

A diffusive mass flux term $(-\partial \dot{m}^D/\partial x_i)$ was added to the right-hand side of the continuity equation. This term was obtained as

$$\dot{m}^D = -\mu \left(\frac{1}{p}\frac{\partial p}{\partial x_i} - \frac{1}{2T}\frac{\partial T}{\partial x_i} \right) \tag{4.7}$$

suggesting that mass diffusion depends on both pressure and temperature gradients. Alternatively, the mass diffusion term can be written in terms of gradients of density and temperature, using the ideal gas law. The constitutive relations are also affected because of similar modification (as mass diffusion) to stress and heat flux terms. The modified equations are

$$\frac{\partial \rho}{\partial t} + \frac{\partial (\rho u_i)}{\partial x_i} = -\frac{\partial \dot{m}_i^D}{\partial x_i} \tag{4.8}$$

$$\frac{\partial(\rho u_j)}{\partial t} + \frac{\partial(\rho u_i u_j)}{\partial x_i} = -\frac{\partial p}{\partial x_j} + \frac{\partial \sigma_{ij}^T}{\partial x_i} + \rho g_j \tag{4.9}$$

$$\frac{\partial(\rho \epsilon)}{\partial t} + \frac{\partial(\rho u_i \epsilon)}{\partial x_i} = -\frac{\partial q_i^T}{\partial x_i} - p\frac{\partial u_i}{\partial x_i} - \sigma_{ij}^T \frac{\partial u_j}{\partial x_i} \tag{4.10}$$

where the constitutive relations get modified to:

$$\sigma_{ij}^T = \mu \left(\frac{\partial u_j}{\partial x_i} + \frac{\partial u_i}{\partial x_j} \right) - \frac{2}{3}\mu \delta_{ij} \frac{\partial u_k}{\partial x_k} - \dot{m}_i^D u_j - \dot{m}_j^D u_i + \frac{2}{3}\delta_{ij} \dot{m}_k^D u_k \tag{4.11}$$

$$q_i^T = -k\frac{\partial T}{\partial x_i} + \dot{m}_i^D C_p T. \tag{4.12}$$

Good agreement with the experimental data of Maurer for flow in a microchannel was demonstrated through these equations. However, the results from the extended Navier–Stokes equation for complex microchannel case are only in moderate agreement with the DSMC data.

Note that the viscous dissipation term (last term in energy equation (4.10)) is no longer a perfect square—same problem as Brenner's equation. The implication of this will have to be carefully analyzed, in that, can it lead to viscous dissipation becoming negative, and thereby violating thermodynamics.

The continuity equation remains unaffected even in the case of higher order continuum transport equations—the Burnett and Grad equations and its variants, as discussed in Chaps. 5 and 6. Therefore, the added mass diffusion term is not of higher order, it is rather a correction to the transport equations, in general. Note that no need for correction to the Navier–Stokes equations has been felt otherwise in case of high pressure and temperature gradient situations. For example, this equation suggests onset of flow in the presence of a horizontal temperature gradient, whereas in such situations, buoyancy would set up a flow while no flow would occur in absence of gravity.

4.5 Non-Fourier Heat Conduction

The Fourier law suggests that the heat flux is instantaneously setup (or vanishes) upon application (or removal) of a temperature gradient. Since no process is instantaneous, correction is clearly required to the Fourier's law. Cattaneo and Vernotte suggested a simple modification, whereby a delay ($= \tau$) was introduced to Fourier's relation, i.e.,

$$q(x, t + \tau) = -k\nabla T(x, t) \tag{4.13}$$

Upon expanding the above equation in a Taylor series, and retaining only first-order terms, we obtain

$$\tau \frac{\partial q}{\partial t}(x, t) + q(x, t) = -k \nabla T(x, t) \tag{4.14}$$

The above equation is known as Cattaneo's equation, and is one of the simplest non-Fourier models available in the literature. Insertion of the above heat flux relation yields the following hyperbolic heat conduction equation:

$$\frac{1}{c^2}\frac{\partial^2 T}{\partial t^2} + \frac{1}{\alpha}\frac{\partial T}{\partial t} = \nabla^2 T + \frac{1}{k}\left(g + \frac{\alpha}{c^2}\frac{\partial g}{\partial t}\right) \tag{4.15}$$

where viscous dissipation and several other terms have been neglected for simplicity, g is volumetric heat generation, α is thermal diffusivity, and c is wave speed. Notice that the above equation suggests a wave-like nature of heat flow, instead of a diffusive nature embodied in Fourier's law. As argued in Sect. 6.10, τ is expected to be of order 10^{-10} s for air under standard conditions.

It is known that the Cattaneo's equation becomes relevant in cases involving very short (femto second) laser pulses [152]. Another application where the relaxation time effect cannot be ignored [109] is exothermal catalytic reaction in crystals, where significant temperature rise can happen within a duration of the order of 10^{-13} s. The classical theory also breaks down at very low temperatures (i.e., near absolute zero) [109]. Again, in situations involving small length scales ($[(1/T)(dT/dx)]^{-1} << $ free path of the medium), non-Fourier effects cannot be ignored [173].

Besides change in the nature of the equation from parabolic to hyperbolic, note the additional requirement of initial condition with Cattaneo's equation, when compared to the Fourier's law. Of course, the final solution for temperature (or heat flux) not only depends on how good the employed model (or constitutive relation) is, but also on how accurately this additional initial condition has been prescribed. Typically, the temporal gradient of temperature at initial time is specified as the additional initial condition for solution of various problems in the literature.

An additional non-dimensional number becomes relevant when solving the Cattaneo's equation. The additional parameter (delay time, τ) can be non-dimensionalized with characteristic length scale (L) and, in absence of a characteristic time scale, by thermal diffusivity (α). The resulting non-dimensional parameter is known as Vernotte number [60], given as $Ve = \sqrt{\alpha\tau}/L$.

While the above equation has a single relaxation time, Tzou suggested the following equation with two different relaxation times (τ_q and τ_T):

$$q(x, t + \tau_q) = -k \nabla T(x, t + \tau_T) \tag{4.16}$$

Again, expanding the above relation in Taylor's series and retaining till first-order terms yield the following constitutive relation between heat flux and temperature:

$$\tau_q \frac{\partial q}{\partial t}(x,t) + q(x,t) = -k\nabla T(x,t) + \tau_T \frac{\partial}{\partial t}[\nabla T(x,t)] \tag{4.17}$$

The above equation is clearly a generation of Cattaneo's equation. However, application of Tzou's equation is more difficult than Cattaneo's equation due to presence of an additional relaxation time in Tzou's equation. Just like Cattaneo's equation, the above equation can only be deemed as a model, till it can be derived from first-principles.

4.6 Shock Wave Flow Problem

The steady plane shock wave flow problem is a standard benchmark problem against which the results of different hydrodynamic theories can be compared. As there are no solid boundaries to deal with, the problem of prescribing boundary conditions that can plague certain hydrodynamic models is also avoided.

For this problem, the conservation equations can be readily integrated to obtain [153]:

$$\begin{aligned} \rho(x)u(x) &= A_1 \\ \rho(x)u^2(x) + P_{xx} &= A_2 \\ \left(\frac{3}{2}RT(x) + \frac{u^2(x)}{2}\right)\rho(x)u(x) + P_{xx}u(x) + q_x &= A_3 \end{aligned} \tag{4.18}$$

where x is the streamwise direction, A_i are constants, P_{xx} is a normal stress component, and q_x is heat flux component. The form of these equations is such that different models can be easily tested by substituting the corresponding expressions for P_{xx} and q_x in the above equations. For example, with Navier–Stokes equations, we obtain the following expressions for velocity and temperature

$$\begin{aligned} \frac{d}{dx}u(x) &= \frac{3}{4}\frac{A_1(u(x))^2 + A_1RT(x) - A_2u(x)}{\mu(T(x))u(x)} \\ \frac{d}{dx}T(x) &= \frac{1}{2}\frac{3A_1RT(x) - A_1(u(x))^2 + 2A_2u(x) - 2A_3}{k(T(x))} \end{aligned} \tag{4.19}$$

Uribe [153] compared the solution of Eq. (4.19) with the solution of Eq. (4.18) using Brenner's equations. This comparison is presented in Fig. 4.1. Note that the figure involves reduced variables, defined as

$$u^*(s) = \frac{u(x)}{u_o}; \quad s = x\frac{3\rho_o u_o}{4\mu_o} \tag{4.20}$$

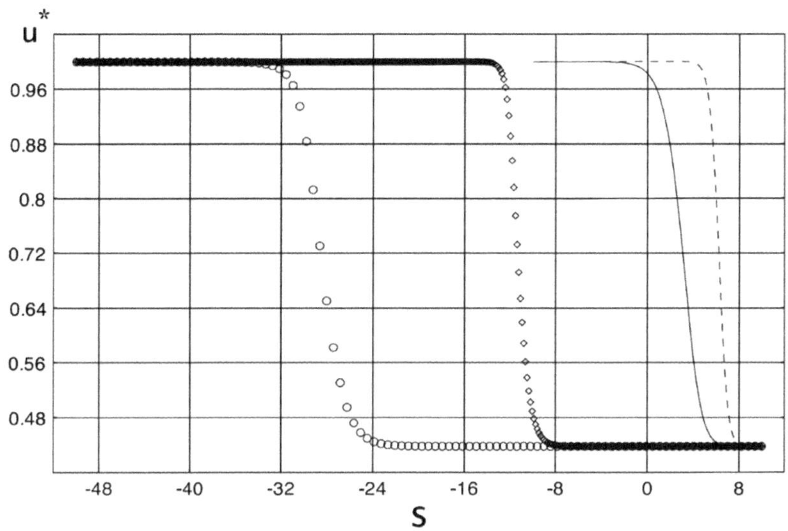

Fig. 4.1 Velocity profile in a shock wave flow problem at Ma = 2; dashed line and solid line represent results of Navier–Stokes equations for Maxwell and rigid sphere model, respectively, hollow circles and hollow diamonds represent results of Brenner's theory for Maxwell and rigid sphere model, respectively, Uribe [153]

where the subscript 'o' refers to conditions far upstream of the shock. The results are for two different models of particles (Maxwell and rigid sphere). However, the calculations of [66] for shock structure of Argon gas at various Mach numbers did not yield converged results with Brenner's equations.

4.7 Summary

Some problems needing a look beyond the Navier–Stokes equations is first presented. Certain extensions of the Navier–Stokes equations are then considered in this chapter. The extension is primarily through modification of the constitutive relations for stress and heat flux. Several shortcomings of these extensions are identified which provides the motivation for examining alternate models for describing flows in the high Knudsen number regimes.

Chapter 5
Burnett Equations: Derivation and Analysis

The Burnett equations is a super-set of the Navier–Stokes equations, and one of the most important higher-order continuum transport equations. The derivation and nature of the equation in both Cartesian and cylindrical coordinates is presented in this chapter. The equations involve large number of terms, which makes them quite formidable to solve, both analytically and numerically. Nonetheless there has been some recent success in obtaining analytical solution of the Burnett equations, as discussed here. An order of magnitude analysis of the various terms reveals that the Burnett-order terms in the equations are of order Knudsen number square. A stability analysis of the equations for one-dimensional wave reveals the unstable nature of the equations.

5.1 Introduction

The Burnett equations derived in this chapter and the Grad equations derived in the next chapter are both based on the Boltzmann equation. The Boltzmann equation is therefore briefly introduced in the next section. The solution of the Boltzmann equation is a single particle distribution function, which is a microscopic parameter with limited interest. Obtaining the exact form of the distribution function is virtually impossible and therefore, one has to work with an approximate form of the distribution function. But how far can one go without the exact form of the distribution function?

Macroscopic parameters of interest such as density, velocity, etc. are obtained by taking moments of the distribution function. The basic laws of conservation of mass, momentum, and energy can also be derived starting from the Boltzmann equation, but additional unknowns—stress tensor and heat flux vector appear in the

A. Agrawal et al., *Microscale Flow and Heat Transfer*, Mechanical Engineering Series, https://doi.org/10.1007/978-3-030-10662-1_5

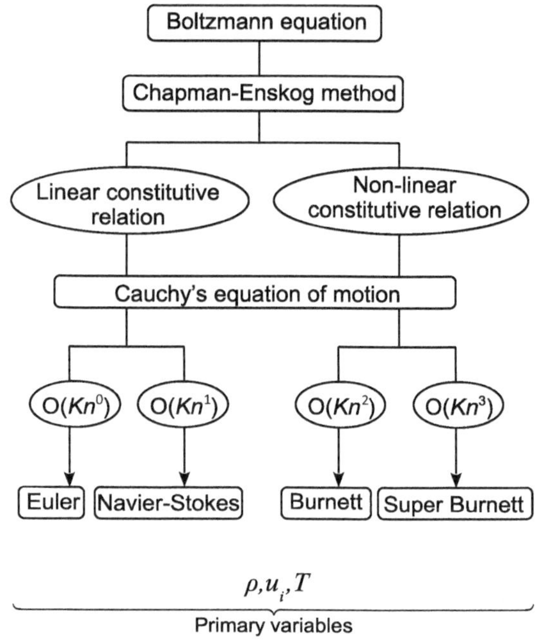

Fig. 5.1 Flowchart depicting the derivation of Euler, Navier–Stokes, Burnett, and super-Burnett equations from the Boltzmann equation using Chapman–Enskog method

equations. The exact nature of the distribution function is *not* needed till this stage. A distribution function is, however, *required for obtaining closure relations* for the stress tensor and heat flux vector.

In the present chapter, the distribution function is obtained using the Chapman–Enskog expansion method, whereby the distribution function is expressed as a perturbation series around the Maxwell–Boltzmann distribution corresponding to the gas at rest. It is shown through detailed derivation that the Chapman–Enskog expansion gives pressure at zeroth order, linear stress strain-rate relation at first order, and non-linear stress strain-rate relation at second and higher order. The stress tensor thus obtained, when substituted into the conservation equations, gives the Euler, Navier–Stokes and Burnett equations, at the zeroth, first and second order, respectively. The procedure outlined in Fig. 5.1 therefore provides an alternative way to obtain the Navier–Stokes equations in a systematic way. The detailed derivation of the Burnett equations is presented that should help in the appreciation of these (and other) higher-order continuum transport equations.

5.2 Boltzmann Equation

The Boltzmann equation is the most fundamental equation in the kinetic theory of gases. The solution of the Boltzmann equation is the single particle distribution function $f(\mathbf{x}, \mathbf{c}, t)$ (where $\mathbf{x}$ represents the position, $\mathbf{c}$ the molecular velocity, and t the time) which expresses the probability of finding a molecule in a physical volume $d\mathbf{x}(= dx_1dx_2dx_3)$ having velocities in the range $\mathbf{c}$ to $\mathbf{c} + d\mathbf{c}$ (boldface symbol denotes a vector quantity). The Boltzmann equation is represented mathematically as

$$\underbrace{\frac{\partial f}{\partial t}}_{\text{change with time}} + \underbrace{c_k\frac{\partial f}{\partial x_k}}_{\text{flux through all six faces}} + \underbrace{F_k\frac{\partial f}{\partial c_k}}_{\text{acceleration due to external forces}} = \underbrace{J(f, f_1)}_{\text{change due to collisions}} \tag{5.1}$$

where $J(f, f_1)$ represents the collision integral and $\mathbf{F}$ is the external force per unit mass acting on the molecules.

The first term on the left-hand side of Eq. (5.1) represents the change in the number of molecules in the phase space ($d\mathbf{x}d\mathbf{c}$) with respect to time. The second term represents the net rate of increase in the number of molecules in the physical space volume $d\mathbf{x}$ ($= dx_1dx_2dx_3$) caused by molecular flux through all the six faces. The third term represents the net gain of molecules in the velocity space volume $d\mathbf{c}$ ($= dc_1dc_2dc_3$) because of the external forces acting on the molecules. The right-hand side denotes the net gain of molecules entering the velocity space volume $d\mathbf{c}$ ($= dc_1dc_2dc_3$) due to the effect of the collisions between the molecules.

The detailed derivation of the Boltzmann equation with explicit form of the collision integral is well known in the literature [27, 31]. The high dimensionality (f is a function of seven independent variables) and complicated non-linear nature of the collision integral makes it particularly difficult to obtain solutions of the Boltzmann equation. The exact calculation of the distribution function is possible only for the simplest of the cases, such as a gas at rest (no gradients of density, temperature, etc.). The solution for this case is known as the Maxwell–Boltzmann distribution, f_M (defined later through Eq. (5.90)).

The macroscopic quantities such as mass density (ρ), bulk velocity ($\mathbf{u}$), and absolute temperature (T), which are experimentally measurable, are related to the distribution function $f(\mathbf{x}, \mathbf{c}, t)$ through its moments and are given as

$$\rho(\mathbf{x}, t) = m\int f(\mathbf{x}, \mathbf{c}, t)d\mathbf{c} \tag{5.2}$$

$$u_i(\mathbf{x}, t) = \frac{1}{n}\int c_i f(\mathbf{x}, \mathbf{c}, t)d\mathbf{c} \tag{5.3}$$

$$T(\mathbf{x}, t) = \frac{1}{3}\frac{m^2}{\rho k_B}\int C^2 f(\mathbf{x}, \mathbf{c}, t)d\mathbf{c} \tag{5.4}$$

Higher order moments namely, pressure tensor[1] P_{ij} and heat flux vector q_i can also be defined in terms of distribution function as

[1]Pressure tensor, P_{ij} and stress tensor, τ_{ij} as defined in Chap. 2 have opposite sign, i.e. $P_{ij} = -\tau_{ij}$.

$$P_{ij}(\mathbf{x}, t) = m \int C_i C_j f(\mathbf{x}, \mathbf{c}, t) d\mathbf{c}, \tag{5.5}$$

$$q_i(\mathbf{x}, t) = \frac{m}{2} \int C_i C^2 f(\mathbf{x}, \mathbf{c}, t) d\mathbf{c}. \tag{5.6}$$

where $\mathbf{c}$ is the molecular velocity and $\mathbf{C}$ is the peculiar velocity defined as $\mathbf{C}(\mathbf{x}, \mathbf{c}, t) = \mathbf{c} - \mathbf{u}(\mathbf{x}, t)$.

5.2.1 Macroscopic Quantities from Distribution Function

The quantity $f(\mathbf{x}, \mathbf{c}, t) d\mathbf{x} d\mathbf{c}$ represents the number of particles in physical volume $\{\mathbf{x}, \mathbf{x} + d\mathbf{x}\}$ having molecular velocities in interval $\{\mathbf{c}, \mathbf{c} + d\mathbf{c}\}$ at time t. Hence, the total number of particles in a system of a gas occupying a volume V can be obtained by integrating the distribution function over the physical space and the velocity space as

$$N = \iiint_V \iiint_{-\infty}^{\infty} f(\mathbf{x}, \mathbf{c}, t) d\mathbf{c} d\mathbf{x} \tag{5.7}$$

When integration is performed only over velocity space and divided by the physical space cell volume $d\mathbf{x}$, we obtain number density (or number of particles per unit volume) as

$$n = \iiint_{-\infty}^{\infty} f(\mathbf{x}, \mathbf{c}, t) d\mathbf{c} \tag{5.8}$$

For better readability, we simplify the notation as

$$n = \int f(\mathbf{x}, \mathbf{c}, t) d\mathbf{c} \tag{5.9}$$

We know that mass density is given as

$$\rho = \frac{M}{V} = \frac{mN}{V} = mn = m \int f(\mathbf{x}, \mathbf{c}, t) d\mathbf{c} \tag{5.10}$$

where M is the mass of the gas occupied in volume V and m is the mass of each gas molecule.

The macroscopic velocity of the gas which we directly perceive of the random molecular motion of molecules is given as

$$u_i = \frac{1}{n} \int c_i f(\mathbf{x}, \mathbf{c}, t) d\mathbf{c} \tag{5.11}$$

When macroscopic velocity is multiplied by mass density, we have momentum density (or mass flow in the i-th direction) as

$$\rho u_i = \frac{\rho}{n}\int c_i f(\mathbf{x}, \mathbf{c}, t)d\mathbf{c} = m\int c_i f(\mathbf{x}, \mathbf{c}, t)d\mathbf{c} \tag{5.12}$$

More often, it is convenient to work in a frame of reference moving with gas at local velocity u_i. As such, we define peculiar velocity as

$$\begin{aligned} \mathbf{C}(\mathbf{x}, \mathbf{c}, t) &= \mathbf{c} - \mathbf{u}(\mathbf{x}, t) \\ C_i(\mathbf{x}, \mathbf{c}, t) &= c_i - u_i(\mathbf{x}, t), \qquad \text{[In indicial notation]} \end{aligned} \tag{5.13}$$

The transformation from the microscopic velocity $\mathbf{c}$ to peculiar velocity $\mathbf{C}$ implies a translation in the velocity space. As such, we can write

$$d\mathbf{C} = d\mathbf{c} \tag{5.14}$$

Using the above definitions, it can easily be shown that the first moment of f over C_i is zero, i.e.,

$$m\int C_i f d\mathbf{c} = m\int (c_i - u_i) f d\mathbf{c} = \rho u_i - \rho u_i = 0 \tag{5.15}$$

Assuming the gas to be monatomic so that it only has three translational degrees of freedom, its energy density (e is specific energy) can be obtained by integrating the particle kinetic energy over the velocity space as

$$\rho e = \frac{m}{2}\int c^2 f d\mathbf{c} \tag{5.16}$$

Splitting the molecular velocity c into bulk velocity u and peculiar velocity C using Eq. (5.13),

$$\begin{aligned} \rho e &= \frac{m}{2}\int (C^2 + 2C_k u_k + u^2) f d\mathbf{c} \\ &= \frac{m}{2}\int C^2 f d\mathbf{c} + m u_k \int C_k f d\mathbf{c} + \frac{m}{2}u^2 \int f d\mathbf{c} \\ &= \frac{m}{2}\int C^2 f d\mathbf{c} + 0 + \frac{\rho}{2}u^2 \end{aligned} \tag{5.17}$$

where Eqs. (5.10) and (5.15) are used. The term $\frac{\rho}{2}u^2$ can be easily identified with macroscopic kinetic energy density, while the term $\frac{m}{2}\int C^2 f d\mathbf{c}$ is described as the internal (or thermal) energy of the gas. Thus, we introduce internal energy per unit mass as ϵ and multiplying by the mass density ρ, we get internal energy density (internal energy per unit volume) $\rho\epsilon$ as

$$\rho\epsilon = \frac{m}{2}\int C^2 f d\mathbf{c} \tag{5.18}$$

Hence, the total energy density can be decomposed into internal energy density and macroscopic kinetic energy density as

$$\rho e = \rho\epsilon + \frac{\rho}{2}u^2 \tag{5.19}$$

From kinetic theory of gases, temperature is directly proportional to the mean kinetic energy measured with respect to the macroscopic flow velocity, $\mathbf{u}$. For a monatomic gas, there are only three translational degrees of freedom and each degree of freedom contributes $\frac{1}{2}RT$ (R is specific gas constant) to the specific internal energy ϵ. Accordingly, for a monatomic gas, we can then write

$$\epsilon = \frac{3}{2}RT \tag{5.20}$$

Substituting expression for specific internal energy ϵ from Eq. (5.18), we have

$$\frac{m}{2\rho}\int C^2 f d\mathbf{c} = \frac{3}{2}RT \quad \Rightarrow \quad T = \frac{1}{3}\frac{m^2}{\rho k_B}\int C^2 f d\mathbf{c} \tag{5.21}$$

where we have used $R = k_B/m$ and k_B is Boltzmann constant. Thus, the absolute temperature is a certain second moment of the particle distribution function, f. This is the standard definition of temperature as given in kinetic theory of gases and is valid for all cases.

The pressure tensor P_{ij} as defined in Eq. (5.5) can be decomposed into two groups, the divergence-less stress tensor and the contraction tensor as

$$P_{ij} = p_{ij} + \frac{p}{3}\delta_{ij}. \tag{5.22}$$

with the property that p_{ij} is traceless, i.e., $p_{ii} = 0$. The term $p = P_{ii}/3$ is the mean of the normal pressures across the plane and is known as hydrostatic pressure at a point. Performing contraction for the pressure tensor equation (Eq. (5.5)), we get an expression for the hydrostatic pressure as

$$p = \frac{P_{ii}}{3} = \frac{m}{3}\int C^2 f d\mathbf{c} \tag{5.23}$$

Therefore, stress tensor can be written in integral form as

$$p_{ij} = m\int C_i C_j f d\mathbf{c} - \frac{m}{3}\int C^2\delta_{ij} f d\mathbf{c} = m\int C_{\langle i}C_{j\rangle} f d\mathbf{c} \tag{5.24}$$

where we have used $C_iC_j - \frac{1}{3}C^2\delta_{ij} = C_{\langle i}C_{j\rangle}$ and the angular brackets denote the symmetric and trace-free part of the tensor.

Using the definition of the absolute temperature (Eq. (5.21)), the ideal gas equation can be obtained as

$$p = \frac{\rho k_B}{m}\frac{1}{3}\frac{m^2}{\rho k_B}\int C^2 f d\mathbf{c} = \rho\frac{k_B}{m}T = \rho RT. \tag{5.25}$$

Using $R = k_B/m$ and $\rho = nm$, ideal gas equation can also be written as

$$p = nk_BT \tag{5.26}$$

5.2.2 *Physical Significance of Higher Order Moments*

The momentum and energy flow are important quantities that are needed in the theoretical formulation of higher order continuum transport equations. In this section, we try to explain how these quantities arise from higher order moments and also briefly give the physical significance of these moments.

Consider a certain quantity having density $\mathbf{G}$ ($\mathbf{G}$ can be a tensor of any order) in phase space (just like we defined momentum density as $\rho u_i = m\int c_i f(\mathbf{x}, \mathbf{c}, t)d\mathbf{c}$ where $G = f$) as shown in Fig. 5.2. The flow of this quantity through a surface dS by a class of molecules belonging to physical volume $\{\mathbf{x}, \mathbf{x} + d\mathbf{x}\}$ having molecular velocities $\{\mathbf{c}, \mathbf{c} + d\mathbf{c}\}$ can be written as c_nGdSdt where c_n is the component of $\mathbf{c}$ along the normal to the surface element dS. The total amount of this quantity passing through surface dS can be computed by performing integration over all possible velocities as $\int c_nGdSdtd\mathbf{c}$ and dividing by $dSdt$, we get flow of this quantity per unit area per unit time as $\int c_nGdSdtd\mathbf{c}/(dSdt) = \int c_nGd\mathbf{c}$.

In order to compute the momentum flow, we take $G_j = c_jf$ (since momentum is a vector quantity), and the flow of the j-th component of momentum in the i-th direction becomes

$$m\int c_iG_jd\mathbf{c} = m\int c_i(c_jf)d\mathbf{c} = m\int c_ic_jfd\mathbf{c}. \tag{5.27}$$

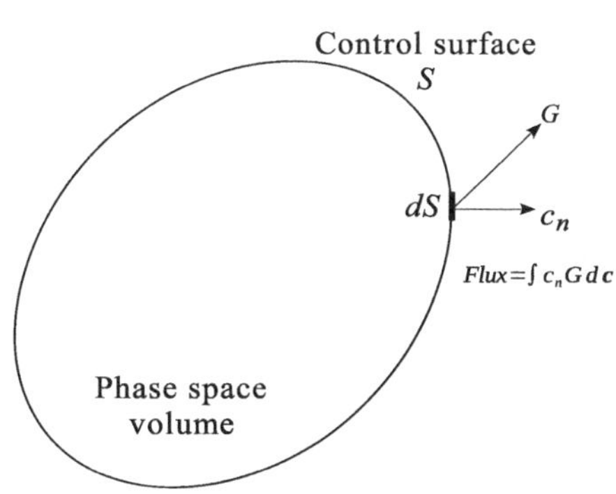

Fig. 5.2 Flux of G through control surface dS in the direction c_n

Note that the microscopic momentum flow is being represented by a symmetric second order tensor [27]. Now, consider a special case where a gas is macroscopically at rest, implying that there is no macroscopic momentum flow. However, the integral in Eq. (5.27) will be different from zero, unless for a given value of c_i, both positive and negative values of c_j are equally likely (that is, c_i and c_j are entirely uncorrelated). Since this is not true in general, the microscopically evaluated momentum flow is not entirely zero, and this microscopic momentum flow will manifest at the macroscopic scale in some form.

When the gas is in motion, we can split the microscopic velocity (**c**) into bulk velocity (**u**) and the peculiar velocity (**C**) using Eq. (5.13) as

$$
\begin{aligned}
m \int c_i c_j f d\mathbf{c} &= m \int (C_i + u_i)(C_j + u_j) f d\mathbf{c} \\
&= m \int C_i C_j f d\mathbf{c} + m u_j \int C_i f d\mathbf{c} + m u_i \int C_j f d\mathbf{c} \\
&\quad + m u_i u_j \int f d\mathbf{c} \\
&= m \int C_i C_j f d\mathbf{c} + 0 + 0 + \rho u_i u_j
\end{aligned} \tag{5.28}
$$

where Eqs. (5.10) and (5.15) are used. Thus, the microscopically evaluated momentum flow can be discerned into two parts, the *hidden* momentum flow due to the random motion of the molecules represented by the first term and the macroscopic momentum flow i.e., momentum density (ρu_i) times the velocity (u_j) represented by the second term. We still need to understand as to how the first part manifests itself in a macroscopic description.

To explain this, consider a fixed region (or control volume, CV) of gas in physical space. As per Reynolds' transport theorem, the momentum of this fixed region can change due to either external forces acting on the CV or momentum entering and leaving the CV. The macroscopic momentum flow term ($\rho u_i u_j$) in Eq. (5.28) represents the change of momentum in the CV due to matter entering and leaving the CV. The other term, $m \int C_i C_j f d\mathbf{c}$ can therefore be attributed to external force (or, more specifically the pressure tensor in the present case). This point will become more clear when we derive the basic conservation laws from the Boltzmann equation. It will be shown that the term $m \int C_i C_j f d\mathbf{c}$ appears in the conservation equation of momentum. When momentum conservation equation is derived from the macroscopic consideration (see Eq. (2.3) in Chap. 2), the term $m \int C_i C_j f d\mathbf{c}$ can only be equated with the pressure tensor P_{ij}. Hence, the pressure tensor can be written as

$$
P_{ij}(\mathbf{x}, t) = m \int C_i C_j f(\mathbf{x}, \mathbf{c}, t) d\mathbf{c} \tag{5.29}
$$

Simply put, P_{ij} is the contribution to the pressure tensor which is due to the momentum transfer resulting from the random molecular motion.

To determine the energy flow, we take $G = \frac{1}{2}c^2 f$, and the total energy flow in the i-th direction for a simple monatomic gas is given as

$$m \int c_i G d\mathbf{c} = m \int c_i \left(\frac{1}{2}c^2\right) f d\mathbf{c} = \frac{m}{2} \int c_i c^2 f d\mathbf{c} \tag{5.30}$$

Again splitting the microscopic velocity ($\mathbf{c}$) into bulk velocity ($\mathbf{u}$) and the peculiar velocity ($\mathbf{C}$) using Eq. (5.13) as was done previously

$$\begin{aligned}
\frac{m}{2} \int c_i c^2 f d\mathbf{c} &= \frac{m}{2} \int (C_i + u_i)(C_j + u_j)^2 f d\mathbf{c} \\
&= \frac{m}{2} \int (C_i + u_i)(C^2 + 2C_j u_j + u^2) f d\mathbf{c} \\
&= \frac{m}{2} \int C_i C^2 f d\mathbf{c} + m u_j \int C_i C_j f d\mathbf{c} + \frac{m}{2} u^2 \int C_i f d\mathbf{c} \\
&\quad + \frac{m}{2} u_i \int C^2 f d\mathbf{c} + m u_i u_j \int C_j f d\mathbf{c} + \frac{m}{2} u_i u^2 \int f d\mathbf{c} \\
&= \frac{m}{2} \int C_i C^2 f d\mathbf{c} + u_j P_{ij} + 0 + u_i \rho \epsilon + 0 + \frac{1}{2} u_i u^2 \rho
\end{aligned}$$

$$\therefore \quad \frac{m}{2} \int c_i c^2 f d\mathbf{c} = \frac{m}{2} \int C_i C^2 f d\mathbf{c} + u_j P_{ij} + \rho u_i \left(\epsilon + \frac{1}{2} u^2\right) \tag{5.31}$$

where Eqs. (5.5), (5.10), (5.15), and (5.18) are used. It is quite obvious that the last two terms manifest at the macroscopic level while the first term on the RHS represents as another kind of energy flow due to random molecular motion. The physical significance of each of the terms appearing on the right-hand side can be discerned as follows: The term $u_j P_{ij}$ can be interpreted as the work done by the stresses per unit time, while the term $\rho u_i \left(\epsilon + \frac{1}{2}u^2\right)$ is the energy flow due to macroscopic convection.

The first term $\frac{m}{2} \int C_i C^2 f d\mathbf{c}$ is generally called as heat flux vector and denoted by $\mathbf{q}$ as

$$q_i = \frac{m}{2} \int C_i C^2 f d\mathbf{c} \tag{5.32}$$

Just as we identified the quantity P_{ij} as the pressure tensor, the quantity $\mathbf{q}$ can be identified as heat flux vector appearing in the energy conservation equation when we derive these conservation equations from the macroscopic considerations. As Cercignani [27] pointed out, the name *heat flux vector* is somewhat misleading because we have certain situations where $q_i \neq 0$ and the temperature is practically

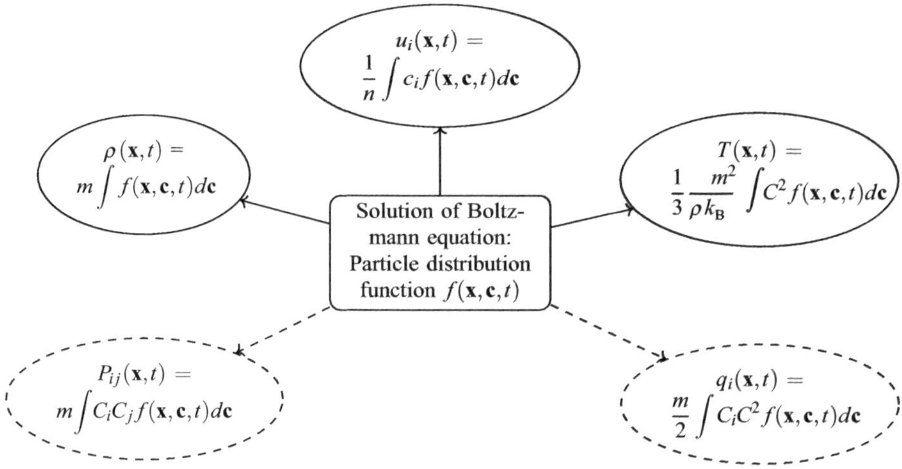

Fig. 5.3 Expressions for mass density ρ, bulk velocity u_i, temperature T, pressure tensor P_{ij}, and heat flux vector q_i in terms of distribution function f; variables in solid circles represent primary variables in the Chapman–Enskog method while that in dashed circles represent variables for which constitutive relationships are obtained in terms of primary variables and their gradients

constant everywhere. He suggested *non-convective energy flow* for quantity $\mathbf{q}$; however, the term heat flux vector remains prevalent in the literature, and is retained here.

Summarizing the above discussion, the distribution function $f(\mathbf{x}, \mathbf{c}, t)$ forms an important bridge between the detailed (in fact, too detailed!) microscopic description and the desired macroscopic description. Once $f(\mathbf{x}, \mathbf{c}, t)$ is known, we can take the moments and obtain the desirable and experimentally measurable macroscopic quantities such as mass density ρ, bulk velocity $\mathbf{u}$, absolute temperature T, pressure tensor P_{ij}, and heat flux vector $\mathbf{q}$ as shown in Fig. 5.3. In the next section, we derive a general evolution equation for moments starting from the Boltzmann equation.

5.2.3 Evolution Equations for Moments

Evolution equations for moments can be generated by multiplying the Boltzmann equation (5.1) by some function $m\xi(\mathbf{x}, \mathbf{c}, t)$ and then performing the integration over the velocity space as

$$m\int \xi \frac{\partial f}{\partial t} d\mathbf{c} + m\int \xi c_k \frac{\partial f}{\partial x_k} d\mathbf{c} + m\int \xi F_k \frac{\partial f}{\partial c_k} d\mathbf{c} = m\int \xi J(f, f_1) d\mathbf{c} \qquad (5.33)$$

We employ angular brackets $\langle\ \rangle$ notation for moments for better readability, defined as

$$\rho \langle \xi \rangle = m \int \xi f d\mathbf{c} \quad \Rightarrow \quad \langle \xi \rangle = \frac{1}{n} \int \xi f d\mathbf{c} \qquad [\text{Using } \rho = nm] \tag{5.34}$$

Using simple rules of integration and applying Leibnitz's integral rule, we evaluate the various terms of Eq. (5.33) as follows:

$$\begin{aligned}
(a) &= m \int \xi \frac{\partial f}{\partial t} d\mathbf{c} \\
&= m \int \frac{\partial (\xi f)}{\partial t} d\mathbf{c} - m \int f \frac{\partial \xi}{\partial t} d\mathbf{c} \\
&= \frac{\partial \rho \langle \xi \rangle}{\partial t} - \rho \left\langle \frac{\partial \xi}{\partial t} \right\rangle \\
(b) &= m \int \xi c_k \frac{\partial f}{\partial x_k} d\mathbf{c} \\
&= m \int \frac{\partial (\xi c_k f)}{\partial x_k} d\mathbf{c} - m \int f \frac{\partial \xi c_k}{\partial x_k} d\mathbf{c} \\
&= \frac{\partial \rho \langle \xi c_k \rangle}{\partial x_k} - \rho \left\langle c_k \frac{\partial \xi}{\partial x_k} \right\rangle \qquad [\because c_i \text{ and } x_i \text{ are independent variables}] \\
(c) &= m \int \xi F_k \frac{\partial f}{\partial c_k} d\mathbf{c} \\
&= m \int \frac{\partial (\xi F_k f)}{\partial c_k} d\mathbf{c} - m \int f \frac{\partial \xi F_k}{\partial c_k} d\mathbf{c}
\end{aligned}$$

For first term in (c), we apply Gauss divergence theorem to convert volume integral to surface integral as $\int_{-\infty}^{+\infty} \frac{\partial (m\xi F_k f)}{\partial c_k} d\mathbf{c} = \oint_{c \to \infty} m\xi F_k f n_k dA$. This surface integral needs to be performed over the control surface for $c \to \infty$. As there are no particles having infinite molecular velocities, i.e., $f(c \to \infty) = 0$, this surface integral vanishes [27, 141]. For second term in (c), we assume external forces to be independent of particle velocities, hence we take F_k out of the differentiation term.

$$\therefore \qquad (c) = 0 - F_k \rho \left\langle \frac{\partial \xi}{\partial c_k} \right\rangle$$

$$(d) = m \int \xi J(f, f_1) d\mathbf{c} = P_\xi$$

Term P_ξ signifies the rate of variation of molecular property ξ, per unit volume, because of collisions and is known as production term.

Combining all the terms, we have the following general equation of transfer:

$$\frac{\partial \rho\langle\xi\rangle}{\partial t} + \frac{\partial \rho\langle\xi c_k\rangle}{\partial x_k} = \rho\left\langle\frac{\partial \xi}{\partial t}\right\rangle + \rho\left\langle c_k\frac{\partial \xi}{\partial x_k}\right\rangle + F_k\rho\left\langle\frac{\partial \xi}{\partial c_k}\right\rangle + P_\xi. \tag{5.35}$$

5.2.4 Conservation Laws

The general equation of transfer derived in the previous section will now be utilized to obtain the conservation laws for mass, momentum, and energy by successively substituting $\xi^{(j)} = \{1, c_i, \frac{1}{2}c^2\}_j$ in Eq. (5.35). During elastic collisions between the molecules, the mass, momentum, and energy is conserved and therefore the production term P_ξ vanishes for all these cases.

5.2.4.1 Conservation of Mass

For $\xi = 1$, all the terms on the right-hand side vanish and using Eq. (5.34),

$$\langle\xi\rangle = \frac{1}{n}\int (1) f d\mathbf{c} = 1$$

$$\langle\xi c_k\rangle = \frac{1}{n}\int c_k f d\mathbf{c} = u_k$$

where we have used Eqs. (5.9) and (5.11). With these results, we obtain continuity equation as

$$\frac{\partial \rho}{\partial t} + \frac{\partial(\rho u_k)}{\partial x_k} = 0. \tag{5.36}$$

5.2.4.2 Conservation of Momentum

For $\xi = c_i$, the equation of transfer becomes

$$\frac{\partial \rho\langle c_i\rangle}{\partial t} + \frac{\partial \rho\langle c_i c_k\rangle}{\partial x_k} = \rho\left\langle\frac{\partial c_i}{\partial t}\right\rangle + \rho\left\langle c_k\frac{\partial c_i}{\partial x_k}\right\rangle + F_k\rho\left\langle\frac{\partial c_i}{\partial c_k}\right\rangle + 0 \tag{5.37}$$

Recall that we are working in seven-dimensional phase space $(x, y, z, c_x, c_y, c_z, t)$. Because of the independent nature of the variables, the first two terms on the right-hand side vanishes. With this simplification, the right-hand side takes the form

$$RHS = 0 + 0 + F_k\rho\,\langle\delta_{ik}\rangle = \rho F_i$$

Decomposing the molecular velocity into bulk velocity and peculiar velocity using Eq. (5.13), the left-hand side is evaluated as

$$\begin{aligned} LHS &= \frac{\partial \rho \langle c_i \rangle}{\partial t} + \frac{\partial \rho \langle c_i c_k \rangle}{\partial x_k} \\ &= \frac{\partial \rho u_i}{\partial t} + \frac{\partial}{\partial x_k} \left(\rho \langle u_i u_k + u_i C_k + u_k C_i + C_i C_k \rangle \right) \\ &= \frac{\partial}{\partial t}(\rho u_i) + \frac{\partial}{\partial x_k}(\rho u_i u_k + 0 + 0 + P_{ik}) \end{aligned} \tag{5.38}$$

where we have used Eqs. (5.15) and (5.5). Summarizing the results, we obtain the conservation equation for momentum as

$$\frac{\partial(\rho u_i)}{\partial t} + \frac{\partial(\rho u_i u_k + P_{ik})}{\partial x_i} = \rho F_i \tag{5.39}$$

In Lagrangian framework, following a particle, material derivative is given as

$$\frac{D}{Dt} = \frac{\partial}{\partial t} + u_j \frac{\partial}{\partial x_j}$$

and using continuity equation (Eq. (5.36)), we obtain

$$\frac{Du_i}{Dt} + \frac{1}{\rho}\frac{\partial P_{ik}}{\partial x_k} = F_i, \tag{5.40}$$

Decomposing pressure tensor into stress tensor and pressure (Eq. (5.22)), we finally obtain Cauchy's equation of motion as

$$\frac{\partial u_i}{\partial t} + u_k \frac{\partial u_i}{\partial x_k} + \frac{1}{\rho}\frac{\partial p}{\partial x_i} + \frac{1}{\rho}\frac{\partial p_{ik}}{\partial x_k} = F_i, \tag{5.41}$$

5.2.4.3 Conservation of Energy

To obtain conservation equation for energy, we substitute $\xi = \frac{1}{2}c^2$ in the transfer equation as

$$\underbrace{\frac{\partial \rho \langle \frac{1}{2}c^2 \rangle}{\partial t}}_{(a)} + \underbrace{\frac{\partial \rho \langle \frac{1}{2}c^2 c_k \rangle}{\partial x_k}}_{(b)} = \underbrace{\rho \left\langle \frac{\partial \frac{1}{2}c^2}{\partial t} \right\rangle}_{(c)} + \underbrace{\rho \left\langle c_k \frac{\partial \frac{1}{2}c^2}{\partial x_k} \right\rangle}_{(d)} + \underbrace{F_k \rho \left\langle \frac{\partial \frac{1}{2}c^2}{\partial c_k} \right\rangle}_{(e)} + 0 \tag{5.42}$$

For the terms on the right-hand side, only the term (e) survives; the remaining terms (c) and (d) vanish because of the independent nature of the variables involved. For term (e),

$$(e) = F_k\rho\left\langle\frac{\partial\frac{1}{2}c^2}{\partial c_k}\right\rangle = F_k\rho\left\langle\frac{\partial\frac{1}{2}c_ic_i}{\partial c_k}\right\rangle = F_k\rho\left\langle\frac{1}{2}2c_i\frac{\partial c_i}{\partial c_k}\right\rangle = F_k\rho\left\langle c_i\delta_{ik}\right\rangle = F_k\rho\left\langle c_k\right\rangle$$

$$\therefore \qquad (e) = F_k m\int c_k f d\mathbf{c} = \rho F_k u_k \tag{5.43}$$

For the evaluation of the left-hand side terms, we proceed as follows:

$$(a) = \frac{\partial\rho\langle\frac{1}{2}c^2\rangle}{\partial t} = \frac{\partial}{\partial t}\left(\frac{m}{2}\int c^2 f d\mathbf{c}\right) = \frac{\partial\rho e}{\partial t} = \frac{\partial(\rho\epsilon + \frac{1}{2}\rho u^2)}{\partial t} \tag{5.44}$$

where we have used Eqs. (5.16) and (5.19). For term (b), we split the microscopic velocity ($\mathbf{c}$) into bulk velocity ($\mathbf{u}$) and the peculiar velocity ($\mathbf{C}$) using Eq. (5.13) and evaluate the term $\frac{1}{2}c^2c_k$ as

$$\begin{aligned}\frac{1}{2}c^2c_k &= \frac{1}{2}c_ic_ic_k = \frac{1}{2}(C_i+u_i)(C_i+u_i)(C_k+u_k)\\ &= \frac{1}{2}C^2C_k + \frac{1}{2}C^2u_k + C_iu_iC_k + C_iu_iu_k + \frac{1}{2}u^2C_k + \frac{1}{2}u^2u_k\end{aligned} \tag{5.45}$$

$$\begin{aligned}\therefore \quad \rho\left\langle\frac{1}{2}c^2c_k\right\rangle &= m\int\frac{1}{2}C^2C_k f d\mathbf{c} + m\int\frac{1}{2}C^2u_k f d\mathbf{c} + m\int C_iu_iC_k f d\mathbf{c}\\ &\quad + m\int C_iu_iu_k f d\mathbf{c} + m\int\frac{1}{2}u^2C_k f d\mathbf{c} + m\int\frac{1}{2}u^2u_k f d\mathbf{c}\\ &= \frac{m}{2}\int C^2C_k f d\mathbf{c} + u_k\frac{m}{2}\int C^2 f d\mathbf{c} + u_i m\int C_iC_k f d\mathbf{c}\\ &\quad + u_iu_k m\int C_i f d\mathbf{c} + u^2\frac{m}{2}\int C_k f d\mathbf{c} + u^2u_k\frac{m}{2}\int f d\mathbf{c}\\ &= q_k + u_k\rho\epsilon + u_iP_{ik} + 0 + 0 + \frac{1}{2}u^2u_k\rho\end{aligned} \tag{5.46}$$

where we have used Eqs. (5.5), (5.6) (5.10), (5.13), and (5.18). Hence, the term (b) is evaluated as

$$(b) = \frac{\partial\rho\langle\frac{1}{2}c^2c_k\rangle}{\partial x_k} = \frac{\partial(\frac{1}{2}\rho u^2u_k + \rho\epsilon u_k + u_iP_{ik} + q_k)}{\partial x_k} \tag{5.47}$$

Combining the final results for terms (a), (b), and (e), we obtain the final form of the conservation equation for total energy as

$$\frac{\partial(\rho\epsilon + \frac{1}{2}\rho u^2)}{\partial t} + \frac{\partial(\frac{1}{2}\rho u^2 u_k + \rho\epsilon u_k + u_i P_{ik} + q_k)}{\partial x_k} = \rho F_k u_k, \tag{5.48}$$

This form is actually the first law of thermodynamics in the differential form having both mechanical and thermal energy terms in it. In Lagrangian framework, following a particle, $\frac{D}{Dt} = \frac{\partial}{\partial t} + u_j \frac{\partial}{\partial x_j}$ and using continuity equation (Eq. (5.36)), with some algebra, we obtain

$$\rho \frac{D(\epsilon + \frac{1}{2}u^2)}{\partial t} + \frac{\partial(u_i P_{ik} + q_k)}{\partial x_k} = \rho F_k u_k. \tag{5.49}$$

From the above equation, if we subtract ρu_i times Eq. (5.40), we obtain equation for balance of internal energy as

$$\rho \frac{D\epsilon}{Dt} + p\frac{\partial u_i}{\partial x_i} + p_{ij}\frac{\partial u_i}{\partial x_j} + \frac{\partial q_k}{\partial x_k} = 0, \tag{5.50}$$

This form states that the rate of change of internal energy plus the sum of compressive work, dissipative work, and the heat flow is equal to zero. Using $p = \frac{2}{3}\rho\epsilon$ and continuity equation (Eq. (5.36)), an alternate form of energy can be obtained as

$$\frac{\partial p}{\partial t} + \frac{\partial}{\partial x_k}(pu_k) + \frac{2}{3}p\frac{\partial u_i}{\partial x_i} + \frac{2}{3}p_{ij}\frac{\partial u_i}{\partial x_j} + \frac{2}{3}\frac{\partial q_k}{\partial x_k} = 0 \tag{5.51}$$

Substituting $\epsilon = c_v T = 3RT/2$ (for a monatomic gas, $c_v = 3R/2$) in Eq. (5.50), we get the internal energy equation in terms of temperature as

$$\frac{3}{2}\rho R\frac{\partial T}{\partial t} + \frac{3}{2}\rho R u_k\frac{\partial T}{\partial x_k} + p\frac{\partial u_i}{\partial x_i} + p_{ij}\frac{\partial u_i}{\partial x_j} + \frac{\partial q_k}{\partial x_k} = 0 \tag{5.52}$$

5.2.5 Constitutive Relations

Summarizing the conservation laws for mass, momentum, and energy derived in the previous section as

$$\frac{\partial \rho}{\partial t} + \frac{\partial(\rho u_k)}{\partial x_k} = 0 \tag{5.53}$$

$$\frac{\partial u_i}{\partial t} + u_k \frac{\partial u_i}{\partial x_k} + \frac{1}{\rho}\frac{\partial p}{\partial x_i} + \frac{1}{\rho}\frac{\partial p_{ik}}{\partial x_k} = F_i \tag{5.54}$$

$$\frac{3}{2}\rho R \frac{\partial T}{\partial t} + \frac{3}{2}\rho R u_k \frac{\partial T}{\partial x_k} + p\frac{\partial u_i}{\partial x_i} + p_{ij}\frac{\partial u_i}{\partial x_j} + \frac{\partial q_k}{\partial x_k} = 0 \tag{5.55}$$

Equations (5.53)–(5.55) form the basic conservation laws with mass density ρ, velocity u_i, and temperature T as the primary variables. This set of equations is exact and does not involve any assumption or approximation till now. However, we have eight additional unknowns, five components of stress tensor p_{ij} (since stress tensor p_{ij} is trace-free and symmetric) and three components of heat flux vector. Thus, to determine the thirteen hydrodynamic variables $(1+3+1+5+3 = 13)$, we have only five equations, rendering this system of equations insoluble. To achieve closure, we need to solve the Boltzmann equation, get the particle distribution function, and then evaluate the stress tensor and heat flux vector using Eqs. (5.24) and (5.6), respectively. The final expression for stress tensor and heat flux vector in terms of primary variables $(\rho, \mathbf{u}, T)$ and their gradients are called the constitutive relations.

In continuum theory, within the Navier–Stokes framework under phenomenological considerations, we assume that the stress tensor is proportional to rate of strain tensor (as derived in Sect. 2.2.2), and heat flux vector is proportional to temperature gradient through the Fourier's law. In this phenomenological derivation, the linear relation between the stress tensor (p_{ij}) and the strain tensor (e_{mn}) assumes the form as $p_{ij} = K_{ijmn}e_{mn}$ [85]. To determine this relationship completely, we need to introduce all the 81 components of fourth order tensor K_{ijmn}. Upon invoking assumptions of isotropic and homogeneous medium and Stokes' hypothesis, these 81 components reduce to two which we identify as absolute viscosity (μ) and second coefficient of viscosity (μ_2). The phenomenological theory does not give the value of these transport coefficients, and it becomes necessary to take recourse to experiments to determine them. On the contrary, the kinetic theory based approach permits us to obtain Navier–Stokes equations with explicit expressions for the transport coefficients.

In certain cases, the linear constitutive relationships and the Navier–Stokes equations are not able to discern the flow physics accurately (see Chap. 4). This motivated researchers to go beyond the Navier–Stokes equations using either the Chapman–Enskog method or the Grad moment method. In the present chapter, we focus on the Chapman–Enskog method by which the superset of the Navier–Stokes equations known as Burnett equations can be derived. The crux of the Burnett equations can be stated as follows: The constitutive relationships for the stress tensor and heat flux vector are obtained by appending several non-linear terms containing higher order derivatives of velocity, temperature, and density and their products to the linear relationships of the Navier–Stokes laws. These non-linear constitutive relationships are expected to capture the flow physics accurately in extreme flow conditions.

In the sections to follow, we present the derivation of the Navier–Stokes equations starting from the Boltzmann equation in sufficient detail such that it can be easily followed. Such explicit derivation does not seem to have been given earlier, although, of course, it is well known to practitioners in the field. We believe that this explicit derivation will provide an alternative way to the fluid dynamics community to arrive at their favorite Navier–Stokes equations.

5.3 The Chapman–Enskog Method

This section is organized as follows: In the first subsection, we present the non-dimensional form of the Boltzmann equation and show explicitly the role Knudsen number plays in transition from the kinetic theory to the hydrodynamics regime. The second subsection introduces the basic idea behind the Chapman–Enskog method and explicitly shows the derivation of distribution function at each successive approximation. In the final subsection, we take this functional form of the distribution function and evaluate the pressure tensor (Eq. (5.5)) and the heat flux vector (using Eq. (5.6)) at each successive approximation. Finally, we substitute these constitutive relationships in the conservation laws to achieve closure, and we obtain Euler equations, Navier–Stokes equations, and Burnett equations at first, second, and third approximation, respectively. Readers primarily interested in the derivation of these equations can skip Sect. 5.3.3 entirely without any loss.

5.3.1 Dimensionless Boltzmann Equation

In this section, we non-dimensionalize the Boltzmann equation in order to bring forward the dynamics associated with it. In the absence of external body forces, the Boltzmann equation (Eq. (5.1)) can be written as

$$\frac{\partial f}{\partial t} + c_k \frac{\partial f}{\partial x_k} = J(f, f_1) \tag{5.56}$$

where collision integral, $J(f, f_1)$ is

$$J(f, f_1) = \int \left(f_1' f' - f_1 f\right) g r \, dr \, d\alpha \, d\mathbf{c_1} \tag{5.57}$$

where g is the relative speed of two molecules before collision; r, α are polar coordinates in the plane passing through the first molecule and perpendicular to the initial trajectory of the second molecule, with r being perpendicular distance from the first molecule to relative path of second molecule before collision, and α being the polar angle. Subscript 1 refers to the second molecule which is colliding with the first molecule and symbol prime ($\prime$) denotes the quantities after collision. For more intricacies of the collision integral, readers are referred to standard texts on the Boltzmann equation [27, 31, 84]. The left-hand side of the Boltzmann equation is known as the transport operator, which predominantly governs the convection phenomena, while the collision integral dominates the diffusion phenomena.

To non-dimensionalize the Boltzmann equation, we carefully choose macroscopic length scale $\tilde{l_o}$, time scale $\tilde{t_o}$, and a reference temperature $\tilde{T_o}$. With this selection, we now have two options to define the velocity scale: (1) velocity scale based on the transport of some macroscopic portion of gas over a distance l_o in time t_o, i.e., l_o/t_o (2) velocity scale based on the thermal speed of the molecules having internal energy $3k_B T_o/2m$. As we will see, it is more convenient to define the velocity scale using the second option but with some minor modification as

$$c_o = \sqrt{\frac{5}{3}\frac{k_B T_o}{m}}$$

which is actually the speed of sound in a monatomic gas at temperature T_o.

Accordingly, we define the dimensionless variables as

$$\tilde{\mathbf{x}} = \frac{\mathbf{x}}{\tilde{l_o}}, \quad \tilde{t} = \frac{t}{\tilde{t_o}}, \quad \tilde{\mathbf{c}} = \frac{\mathbf{c}}{\tilde{c_o}}, \quad \tilde{\mathbf{g}} = \frac{\mathbf{g}}{\sqrt{2}c_o}, \quad \tilde{r} = \frac{r}{d},$$
$$\tilde{\alpha} = \frac{\alpha}{\pi}, \quad \tilde{f}(\tilde{\mathbf{x}}, \tilde{\mathbf{c}}, \tilde{t}) = \frac{l_o^3 c_o^3}{N} f(\mathbf{x}, \mathbf{c}, t) = \frac{c_o^3}{n} f(\mathbf{x}, \mathbf{c}, t) \tag{5.58}$$

where d is the particle diameter, N is the total number of molecules in a volume l_o^3, and $n = N/l_o^3$ is the particle number density of the gas. In terms of non-dimensional variables, the Boltzmann equation can be reformulated as

$$\frac{l_o}{c_o t_o}\frac{\partial \tilde{f}}{\partial \tilde{t}} + \tilde{c}_k \frac{\partial \tilde{f}}{\partial \tilde{x}_k} = l_o \sqrt{2}\pi d^2 n \int \left(\tilde{f}_1' \tilde{f}' - \tilde{f}_1 \tilde{f}\right) \tilde{g}\tilde{r}d\tilde{r}d\tilde{\alpha}d\tilde{\mathbf{c}}_1 \tag{5.59}$$

The coefficient of the first term on the left-hand side can be identified as the kinetic Strouhal number (an analog of the Strouhal number that appears in the continuum regime). Since mean free path λ is equal to $\frac{1}{\sqrt{2}\pi d^2 n}$, the coefficient of the collision integral term is the inverse of the Knudsen number as

$$\mathrm{Str} = \frac{l_o}{c_o t_o}, \qquad \mathrm{Kn} = \frac{\lambda}{l_o} = \frac{1}{l_o \sqrt{2}\pi d^2 n} \tag{5.60}$$

The final form of the dimensionless Boltzmann equation is (dropping all tildas)

$$\mathrm{Str}\frac{\partial f}{\partial t} + c_k \frac{\partial f}{\partial x_k} = \frac{1}{\mathrm{Kn}} J(f, f_1) \tag{5.61}$$

If instead of the thermal velocity scale, we choose the other velocity scale ($= l_o/t_o$) such that l_o, t_o, and T_o are consistent with the geometry of the domain, the ratio l_o/t_o corresponds to the bulk velocity u of the flow and Strouhal number reduces to Mach number, $\mathrm{Ma} = u/c_o$.

We know that the Knudsen number is a measure of the degree of rarefaction of a gas. When Kn $>>$ 1, the collision integral term in Eq. (5.61) vanishes and we have free molecular regime where gas–surface interaction becomes dominant as compared to gas–gas collisions. In the other extreme condition, when $Kn << 1$, the collision integral term predominantly determines the distribution function and state of the gas is described by the continuum regime. This implies that the Knudsen number is the deciding factor in transition from kinetic theory to hydrodynamics, as stated earlier.

5.3.2 Central Idea of the Chapman–Enskog Method

The central idea of the Chapman–Enskog method is to expand the particle distribution function in an infinite series as [31],

$$f = f^{(0)} + f^{(1)} + f^{(2)} + \cdots = \sum_{r=0}^{\infty} f^{(r)} \tag{5.62}$$

where $f^{(0)}$ describes the state of the gas at rest, $f^{(1)}$ describes the state of the gas in motion in a quasi-equilibrium state, while higher order distribution function ($f^{(2)}$, $f^{(3)}$, and so on) are further corrections to the state of the gas in motion and account for non-equilibrium effects. Each of the variables defined in Eqs. (5.2)–(5.6) in terms of the distribution function can be defined to any level of approximation. Of interest here, is definition of pressure tensor and heat flux vector (Eqs. (5.5) and (5.6)) which are also expanded in an infinite series as

$$P_{ij} = P_{ij}^{(0)} + P_{ij}^{(1)} + P_{ij}^{(2)} + P_{ij}^{(3)} \cdots = \sum_{r=0}^{\infty} P_{ij}^{(r)} \tag{5.63}$$

$$q_i = q_i^{(0)} + q_i^{(1)} + q_i^{(2)} + q_i^{(3)} + \cdots = \sum_{r=0}^{\infty} q_i^{(r)} \tag{5.64}$$

where
$$P_{ij}^{(r)} = m \int C_i C_j f^{(r)} d\mathbf{c}, \tag{5.65}$$

$$q_i^{(r)} = \frac{m}{2} \int C_i C^2 f^{(r)} d\mathbf{c}. \tag{5.66}$$

For $r > 0$, an important physical constraint is placed on higher order approximations of distribution function. This constraint states that the primary variables: mass density, momentum density, and internal energy density (ρ, ρu_i and $\rho\epsilon = 3\rho RT/2$) must remain same at any level of approximation. For this constraint to hold, the zeroth approximation $f = f^{(0)}$ should reproduce these primary variables and contribution of further higher order approximations ($f^{(1)}$, $f^{(2)}$, etc.) to these primary variables should be exactly zero. These constraints are known as compatibility conditions and mathematically represented as

$$m\int f^{(0)} d\mathbf{c} = \rho, \quad m\int c_i f^{(0)} d\mathbf{c} = \rho u_i, \quad \frac{m}{2}\int C^2 f^{(0)} d\mathbf{c} = \frac{3}{2}\rho RT.$$

$$\text{For } r > 0, \quad m\int f^{(r)} d\mathbf{c} = 0, \quad m\int c_i f^{(r)} d\mathbf{c} = 0, \quad \frac{m}{2}\int C^2 f^{(r)} d\mathbf{c} = 0. \tag{5.67}$$

In physical sense, these compatibility conditions make sure that the conservation laws are satisfied at each successive approximation. Hence, in the derivation of the higher order approximations to the distribution function, it is very critical to ensure that the higher approximations only correct the pressure tensor and heat flux vector without violating the basic conservation laws.

5.3.3 Derivation of Distribution Function

In the present section, we derive the functional form of the distribution function at each successive approximation. Particularly, the Enskog's method of successive approximation to solve the Boltzmann equation has been presented [31]. This section has been included for the sake of completeness but can be skipped in the first reading. What is of interest is to use the distribution function to derive the transport equations, which is elaborated upon in the next section.

In the most general form, the Boltzmann equation (Eq. (5.1)) is expressed as

$$\xi(f) = J(f, f_1) + \mathscr{D} f = 0, \tag{5.68}$$

$$\text{where operator,} \quad \mathscr{D} = \frac{\partial}{\partial t} + c_k\frac{\partial}{\partial x_k} + F_k\frac{\partial}{\partial c_k} \tag{5.69}$$

and $\xi(f)$ denotes the result of certain operations performed on the single particle distribution function f. Note that the quadratic terms inside the collision integral (Eq. (5.57)) are reversed, which turns the negative sign into positive. Now substitute the series (5.62) in $\xi(f)$ and *assume* that when ξ operates on this series, the final result can be expressed as a series in which the rth term involves only the first r terms of the series in (5.62). That is, we write

$$\begin{aligned}\xi(f) &= \xi(f^{(0)} + f^{(1)} + f^{(2)} + \cdots) \\ &= \xi^{(0)}(f^{(0)}) + \xi^{(1)}(f^{(0)}, f^{(1)}) + \xi^{(2)}(f^{(0)}, f^{(1)}, f^{(2)}) + \cdots\end{aligned} \tag{5.70}$$

In addition to condition that the sum (5.62) is a solution of $\xi(f) = 0$, functions $f^{(r)}$ are assumed to satisfy the following equations,

$$\xi^{(0)}(f^{(0)}) = 0, \tag{5.71}$$

$$\xi^{(1)}(f^{(0)}, f^{(1)}) = 0, \tag{5.72}$$

$$\xi^{(2)}(f^{(0)}, f^{(1)}, f^{(2)}) = 0, \tag{5.73}$$

$$\cdots\cdots\cdots\cdots\cdots\cdots\cdots\cdots$$

$$\xi^{(n)}(f^{(0)}, f^{(1)}, \dots f^{(n)}) = 0. \tag{5.74}$$

These conditions together ensure that $\xi(f) = 0$. From Eq. (5.71), $f^{(0)}$ is determined and subsequently, $f^{(1)}$, $f^{(2)}$, ... $f^{(n)}$ are determined from Eqs. (5.72), (5.73), ... (5.74), respectively; since each equation contains only one unknown when all the previous equations have been solved.

Substituting the series (5.62) in Eq. (5.68), we have

$$\begin{aligned}
\xi(f) &= J\left\{\left(\sum_{r=0}^{\infty} f^{(r)} \sum_{s=0}^{\infty} f_1^{(s)}\right)\right\} + \mathscr{D}\sum_{r=0}^{\infty} f^{(r)} \\
&= J\left\{\left(\sum_{k=0}^{\infty}\sum_{l=0}^{k} f^{(l)} f_1^{(k-l)}\right)\right\} + \mathscr{D}\sum_{r=0}^{\infty} f^{(r)} \\
&= \sum_{k=0}^{\infty}\sum_{l=0}^{k} J(f^{(l)} f_1^{(k-l)}) + \sum_{r=0}^{\infty} \mathscr{D} f^{(r)} \\
&= \sum_{r=0}^{\infty}\left(\sum_{l=0}^{r} J(f^{(l)} f_1^{(r-l)}) + \mathscr{D} f^{(r)}\right) \\
&= \sum_{r=0}^{\infty}\left(J^{(r)} + \mathscr{D} f^{(r)}\right) \qquad \text{where} \quad J^{(r)} = \sum_{l=0}^{r} J\left(f^{(l)} f_1^{(r-l)}\right)
\end{aligned} \tag{5.75}$$

Therefore, for the r-th term of ξ, we can write

$$\xi^r = J^{(r)} + \mathscr{D} f^{(r)} \tag{5.76}$$

Note that in the first step of the above equation, we applied Cauchy's rule for the product of two series (given by Eq. (5.77) below), while in the third step, we have interchanged the order of integral (involved in the collision term see Eq. 5.57) with the summation. In the last step, the dummy variable in the first term has been changed from k to r. The Cauchy's rule for the product of two series is

$$\sum_{i=0}^{\infty} a_i \sum_{j=0}^{\infty} b_j = \sum_{k=0}^{\infty}\sum_{l=0}^{k} a_l b_{k-l} \tag{5.77}$$

We now define the term $J^{(r)}$ in Eq. (5.75) as

$$J^{(r)} = \sum_{l=0}^{r} J(f^{(l)} f_1^{(r-l)}) = J(f^{(0)} f_1^{(r)}) + J(f^{(1)} f_1^{(r-1)}) + \cdots + J(f^{(r)} f_0^{(0)}) \tag{5.78}$$

For the term $\Sigma \mathscr{D} f^{(r)}$ in Eq. (5.75), Enskog suggested to divide it into a series of parts $\mathscr{D}^{(r)}$ instead of $\mathscr{D} f^{(r)}$. The expression for operator $\mathscr{D}$ (Eq. (5.69)) contains a time derivative term. Similar to f, this term is expanded in an infinite series as

$$\frac{\partial}{\partial t} = \frac{\partial_0}{\partial t} + \frac{\partial_1}{\partial t} + \frac{\partial_2}{\partial t} + \cdots = \sum_{l=0}^{\infty} \frac{\partial_l}{\partial t} \tag{5.79}$$

Substituting the series for f (Eq. (5.62)) and time derivatives (Eq. (5.79)) in the expression for operator $\mathscr{D}$ (Eq. (5.69)), and again using Cauchy's rule for product of two infinite series, we have

$$\begin{aligned}
\mathscr{D} f &= \frac{\partial f}{\partial t} + c_k \frac{\partial f}{\partial x_k} + F_k \frac{\partial f}{\partial c_k} \\
&= \left(\sum_{l=0}^{\infty} \frac{\partial_l}{\partial t} + c_k \frac{\partial}{\partial x_k} + F_k \frac{\partial}{\partial c_k} \right) \sum_{m=0}^{\infty} f^{(m)} \\
&= \sum_{r=0}^{\infty} \sum_{s=0}^{r} \frac{\partial_s f^{(r-s)}}{\partial t} + \sum_{r=0}^{\infty} \left(c_k \frac{\partial f^{(r)}}{\partial x_k} + F_k \frac{\partial f^{(r)}}{\partial c_k} \right) \\
&= \sum_{r=0}^{\infty} \left(\sum_{s=0}^{r} \frac{\partial_s f^{(r-s)}}{\partial t} + c_k \frac{\partial f^{(r)}}{\partial x_k} + F_k \frac{\partial f^{(r)}}{\partial c_k} \right) \\
&= \sum_{r=0}^{\infty} \mathscr{D}^{(r)} \qquad \text{where } \mathscr{D}^{(r)} = \sum_{s=0}^{r} \frac{\partial_s f^{(r-s)}}{\partial t} + c_k \frac{\partial f^{(r)}}{\partial x_k} + F_k \frac{\partial f^{(r)}}{\partial c_k}
\end{aligned} \tag{5.80}$$

According to Enskog, $\mathscr{D}^{(0)} = 0$, and for $r > 0$, $\mathscr{D}^{(r)}$ is dependent only on $f^{(0)}$, $f^{(1)}, \ldots, f^{(r-1)}$. As such,

$$\mathscr{D}^{(r)} \equiv \mathscr{D}^{(r)} \left(f^{(0)}, f^{(1)}, \ldots, f^{(r-1)} \right) \tag{5.81}$$

$$\begin{aligned}
\mathscr{D}^{(r)} &= \sum_{s=0}^{r-1} \frac{\partial_s f^{((r-1)-s)}}{\partial t} + c_k \frac{\partial f^{(r-1)}}{\partial x_k} + F_k \frac{\partial f^{(r-1)}}{\partial c_k} \\
&= \frac{\partial_0 f^{(r-1)}}{\partial t} + \frac{\partial_1 f^{(r-2)}}{\partial t} + \cdots + \frac{\partial_{(r-1)} f^{(0)}}{\partial t} + c_k \frac{\partial f^{(r-1)}}{\partial x_k} + F_k \frac{\partial f^{(r-1)}}{\partial c_k}
\end{aligned} \tag{5.82}$$

Each term in the above expression can be determined since $f^{(0)}, \ldots, f^{(r-1)}$ are known. Hence, at every approximation for the distribution function, the equation that needs to be solved is

$$\xi^{(r)} = J^{(r)} + \mathscr{D}^{(r)} = 0, \tag{5.83}$$

where $J^{(r)}$ is given by Eq. (5.78) and $\mathscr{D}^{(r)}$ is given by Eq. (5.82).

Substituting expressions for $J^{(r)}$ (Eq. (5.78)) and $\mathscr{D}^{(r)}$ (Eq. (5.82)) in expression for $\xi^{(r)}$ (Eq. (5.83)), we obtain

$$J(f^{(0)} f_1^{(r)}) + J(f^{(r)} f_1^{(0)}) = -\mathscr{D}^{(r)} - J(f^{(1)} f_1^{(r-1)}) - \ldots - J(f^{(r-1)} f_1^{(1)}) \tag{5.84}$$

Note that we have r terms on the right-hand side and consists of $f^{(0)}$, $f^{(1)}$, ..., $f^{(r-1)}$ which we already know are the solutions of the previous equations $\xi^{(0)} = 0, \ldots, \xi^{(r-1)} = 0$. The function to be determined $f^{(r)}$ appears linearly only on the left-hand side of the equation. So, the task at hand is to solve Eq. (5.84) which still turns out to be a complicated integral equation. The detailed procedure to solve this integral equation is given in Chapman and Cowling [31].

All the important steps in the Chapman–Enskog method that we discussed previously are outlined in a detailed flowchart as shown in Fig. 5.4 for easy reference.

5.3.3.1 Zeroth Approximation: $f^{(0)}$

For the zeroth approximation $f^{(0)}$, we substitute $r = 0$ in Eq. (5.83) and the equation to be solved becomes

$$\xi^{(0)} = J^{(0)} + \mathscr{D}^{(0)} = 0, \tag{5.85}$$

Since $\mathscr{D}^{(r)}$ is a function of $f^{(0)}, f^{(1)}, \ldots, f^{(r-1)}$, we essentially have $\mathscr{D}^{(0)} = 0$ for $r = 0$, i.e.,

$$\frac{\partial f^{(0)}}{\partial t} + c_k \frac{\partial f^{(0)}}{\partial x_k} + F_k \frac{\partial f^{(0)}}{\partial c_k} = 0 \tag{5.86}$$

This state corresponds to a uniform, steady state with no external forces. To find $J^{(0)}$, we substitute $r = 0$ in Eq. (5.78) and we get $J^{(0)} = 2J(f^{(0)} f_1^{(0)})$. Hence, the equation to be solved for $f^{(0)}$ reduces to

$$J(f^{(0)} f_1^{(0)}) = 0 \tag{5.87}$$

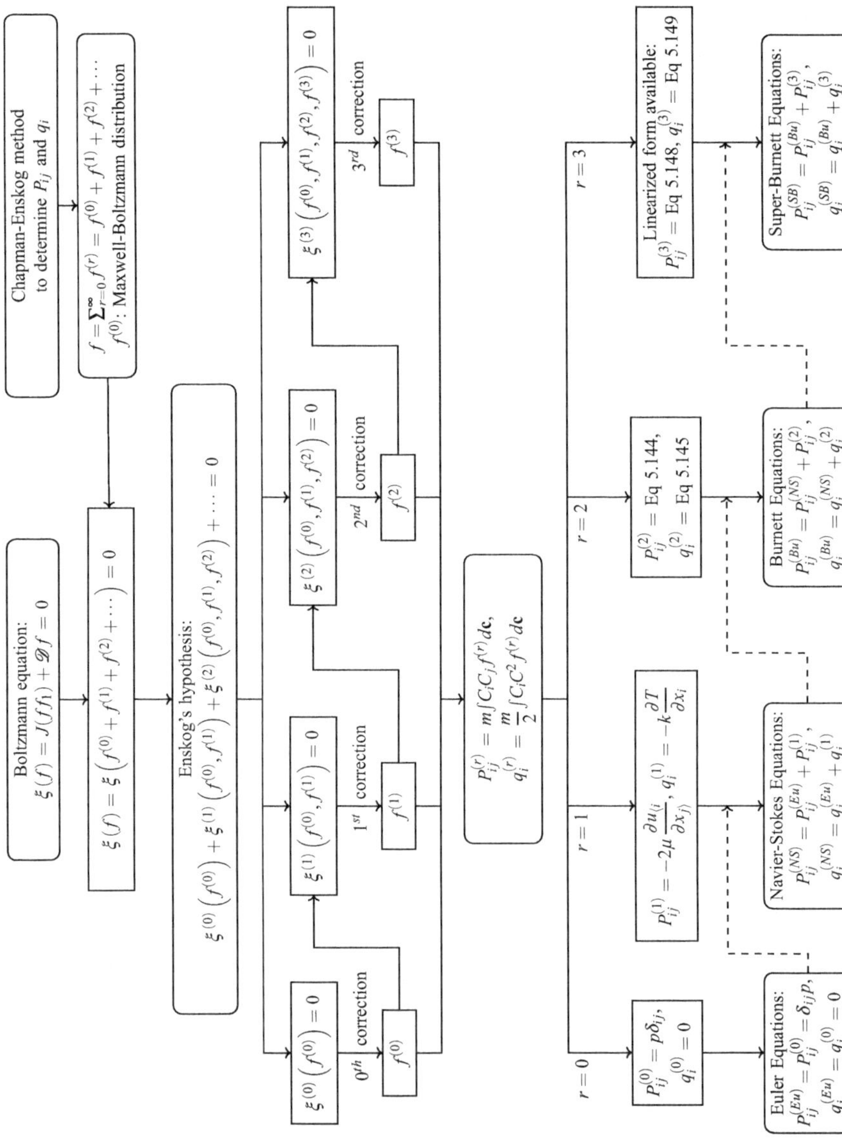

Fig. 5.4 Flowchart showing the detailed procedure of Chapman–Enskog method

This condition implies that the collision integral is zero and collisions as a whole produce no net effect as the effect of every encounter is exactly balanced by that of inverse process. Expanding the collision integral (Eq. (5.57))

$$J(f^{(0)}, f_1^{(0)}) = \int \left(f_1'^{(0)} f'^{(0)} - f_1^{(0)} f^{(0)}\right) grdrd\alpha d\mathbf{c}_1. \tag{5.88}$$

For collision integral to be zero, the only possible solution is

$$\begin{aligned} f^{(0)} f_1^{(0)} &= f^{(0)'} f_1^{(0)'} \\ \log f^{(0)} + \log f_1^{(0)} &= \log f^{(0)'} + \log f_1^{(0)'} \end{aligned} \tag{5.89}$$

Therefore, $\log f$ is a summational invariant for collisions, i.e., the quantity $\log f$ is conserved during collisions. During elastic collisions, mass, momentum, and energy are always conserved and they are the only three summational invariants. Therefore, the quantity $\log f$ should be a linear combination of the three summational invariants, $\psi^{(i)} = \left\{1, m\mathbf{c}, \frac{1}{2}mc^2\right\}_i$. After doing some algebra (see Appendix B.1), the final form of the distribution function for the zeroth approximation can be shown to be Maxwellian distribution as

$$f^{(0)} = f_M = n\left(\frac{m}{2\pi k_B T}\right)^{\frac{3}{2}} \exp\left[-\frac{mC^2}{2k_B T}\right]. \tag{5.90}$$

5.3.3.2 First Approximation: $f^{(1)}$

For the first approximation $f^{(1)}$, we substitute $r = 1$ in Eq. (5.83) and the equation to be solved becomes

$$\xi^{(1)} = J^{(1)} + \mathscr{D}^{(1)} = 0, \tag{5.91}$$

From Eq. (5.82), the differential part $\mathscr{D}^{(1)}$ depends only on the zeroth approximation $f^{(0)}$. An expression for differential part $\mathscr{D}^{(1)}$ can be obtained as (see Appendix B.2)

$$\mathscr{D}^{(1)} = f^{(0)} \left\{ \left(\frac{mC^2}{2k_B T} - \frac{5}{2}\right) C_k \frac{\partial \log T}{\partial x_k} + \frac{m}{k_B T} C_{\langle i} C_{j\rangle} \frac{\partial u_{\langle i}}{\partial x_{j\rangle}} \right\} \tag{5.92}$$

where the angular brackets denote the symmetric and trace-free part of the tensor.

The calculation of collision term $J^{(1)}$ is rather cumbersome, so we replace the non-linear collision integral by the Bhatnagar–Gross–Krook (BGK) kinetic model [20]. The mathematics involved in the evaluation of term $J^{(1)}$ with the original collision integral is complex and the essence of the Chapman–Enskog method maybe lost in the mathematical rigor. To convey the usefulness of the Chapman–Enskog method in simpler terms, the BGK kinetic model is employed and the complexity associated with the collision integral is avoided. As we will see, the derivation of the distribution function is greatly simplified, but at the same time, the BGK model retains many of the qualitative features of the original collision integral. Mathematically, the BGK kinetic model is expressed as

$$J(f, f_1) = \nu(f_M - f) \tag{5.93}$$

where f_M is the local Maxwellian and ν is the collision frequency that is assumed to be independent of the molecular velocity $\mathbf{c}$. The structure of this particular kinetic model implies that any non-equilibrium distribution function relaxes to the local Maxwellian with a relaxation time equal to the inverse of the collision frequency. The BGK kinetic model has the following minimum requirements of the Boltzmann equation: (1) The model satisfies the conservation of mass, momentum, and energy. (2) It fulfills the H theorem (or second law of thermodynamics). (3) The distribution function reduces to Maxwellian in equilibrium. However, this model fails to predict the correct value of Prandtl number which we will show in the subsequent sections.

Substituting the series of f up to first approximation ($f = f^{(0)} + f^{(1)}$) and with the knowledge of zeroth approximation being Maxwellian ($f_M = f^{(0)}$), the collision integral simplifies as

$$J(f, f_1) = \nu(f_M - f) = \nu(f_M - f^{(0)} - f^{(1)}) = -\nu f^{(1)}. \tag{5.94}$$

Substituting Eqs. (5.92) and (5.94) in Eq. (5.91), we get an expression for the first approximation of the distribution function as

$$\begin{aligned} f^{(1)} &= -\frac{f^{(0)}}{\nu}\left\{\left(\frac{mC^2}{2k_BT} - \frac{5}{2}\right)C_k\frac{\partial \log T}{\partial x_k} + \frac{m}{k_BT}C_{\langle i}C_{j\rangle}\frac{\partial u_{\langle i}}{\partial x_{j\rangle}}\right\} \\ &= -\frac{f^{(0)}}{\nu}\left\{\left(\frac{mC^2}{2k_BT} - \frac{5}{2}\right)C_k\frac{\partial \log T}{\partial x_k} + \frac{m}{k_BT}\left\{C_iC_j\frac{\partial u_i}{\partial x_j} - \frac{1}{3}C^2\delta_{ij}\frac{\partial u_i}{\partial x_j}\right\}\right\} \end{aligned} \tag{5.95}$$

where in the second step, the angular bracket notation is expanded as

$$C_{\langle i}C_{j\rangle}\frac{\partial u_{\langle i}}{\partial x_{j\rangle}} = C_iC_j\frac{\partial u_i}{\partial x_j} - \frac{1}{3}C^2\delta_{ij}\frac{\partial u_i}{\partial x_j}$$

Before proceeding to achieve closure, it is necessary to check whether the first approximation $f^{(1)}$ satisfies the compatibility conditions (Eq. (5.67)), i.e.,

$$m\int f^{(1)}d\mathbf{C} = 0, \quad m\int c_i f^{(1)}d\mathbf{C} = 0, \quad \frac{m}{2}\int C^2 f^{(1)}d\mathbf{C} = 0 \tag{5.96}$$

On evaluating these integrals, it can be shown that the first approximation $f^{(1)}$ indeed satisfies the compatibility conditions. With this functional form of the distribution function, we proceed to evaluate the first approximations to pressure tensor $P_{ij}^{(1)}$ and heat flux vector $q_i^{(1)}$ to achieve closure.

5.4 Derivation of Hydrodynamic Equations

The functional form of the distribution function that we derived previously will now be utilized to evaluate the pressure tensor and heat flux vector at each approximation level. After substituting these constitutive relationships in the basic conservation laws, it will be shown that we get Euler equations at the zeroth level, Navier–Stokes–Fourier equations at the first level, Burnett equations at the second level, and super-Burnett equations at the third level.

5.4.1 *Euler Equations*

Utilizing the Maxwellian distribution function at the zeroth level (Eq. (5.90)), we now present a step by step procedure to evaluate the zeroth approximation of the pressure tensor $P_{ij}^{(0)}$ and heat flux vector $q_i^{(0)}$. Readers interested in the mathematical details can follow this section, else this section can be skipped without loss of generality.

Putting Maxwellian distribution ($f^{(0)} = f_M$) in integral equation for pressure tensor (Eq. (5.65)) and proceeding as

$$
\begin{aligned}
P_{ij}^{(0)} &= m \int C_i C_j f_M d\mathbf{C} \\
&= mn \left(\frac{m}{2\pi k_B T}\right)^{\frac{3}{2}} \int C_i C_j \exp\left[-\frac{mC^2}{2k_B T}\right] d\mathbf{C}
\end{aligned}
\tag{5.97}
$$

Evaluation of the above triple integral is somewhat tricky, so we transform from Cartesian coordinate system $\{C_x, C_y, C_z\}$ to spherical coordinate system $\{C, \theta, \phi\}$, where θ is the azimuthal angle and ϕ is the zenith angle [141, Appendix A.3] as shown in Fig. 5.5. Also, according to Jacobian transformation, we get $dC_x dC_y dC_z = C^2 \sin\phi dC d\theta d\phi$.

Introducing the direction vector of the microscopic velocities as $\upsilon_i = C_i/C = \{\cos\theta \sin\phi, \sin\theta \sin\phi, \cos\phi\}_i$, we have

$$
P_{ij}^{(0)} = mn \left(\frac{m}{2\pi k_B T}\right)^{\frac{3}{2}} \int_{C=0}^{\infty} \int_{\phi=0}^{\pi} \int_{\theta=0}^{2\pi} C^2 \upsilon_i \upsilon_j \exp\left[-\frac{mC^2}{2k_B T}\right] C^2 \sin\phi dC d\theta d\phi
\tag{5.98}
$$

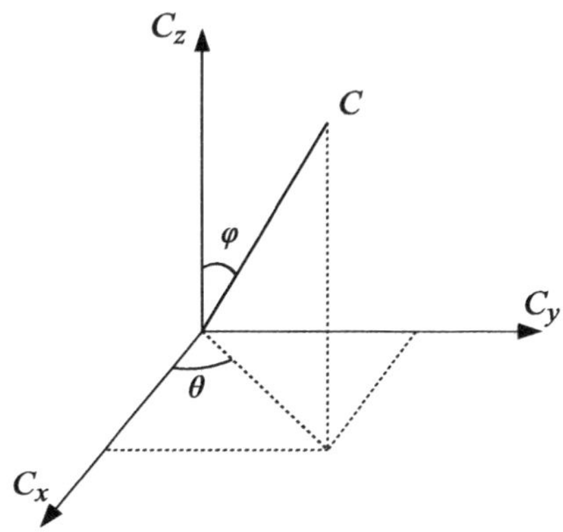

Fig. 5.5 Transformation from Cartesian $\{C_x, C_y, C_z\}$ to spherical $\{C, \theta, \phi\}$ coordinate system; $C_x = C\cos\theta\sin\phi$, $C_y = C\sin\theta\sin\phi$, $C_z = C\cos\phi$, where $C = \sqrt{C_x^2 + C_y^2 + C_z^2}$

$$\text{Let,} \qquad I_{ij} = \frac{1}{4\pi}\int_{\phi=0}^{\pi}\int_{\theta=0}^{2\pi} \upsilon_i\upsilon_j \sin\phi d\theta d\phi \tag{5.99}$$

Therefore, integral equation for the pressure tensor in the compact notation becomes

$$P_{ij}^{(0)} = mn\left(\frac{m}{2\pi k_B T}\right)^{\frac{3}{2}} 4\pi \int_0^{\infty} C^4 \exp\left[-\frac{mC^2}{2k_B T}\right] dC\, I_{ij} \tag{5.100}$$

One can evaluate the integral I_{ij} for its six distinct components. This simple symbolic integration can easily be performed with the aid of an appropriate software (such as Mathematica or Maple). This simple exercise gives all the off-diagonal terms of I_{ij} as zero and diagonal elements as $1/3$. Alternatively, one can realize that I_{ij} is symmetric and isotropic, and since the Kronecker delta δ_{ij} is the only such second-order tensor, we can express I_{ij} in terms of δ_{ij} as

$$I_{ij} = \beta\delta_{ij} \tag{5.101}$$

To find β, we multiply both sides of Eq. (5.101) by δ_{ij} as

$$I_{ij}\delta_{ij} = \beta\delta_{ij}\delta_{ij}$$

$$\therefore \quad I_{ii} = \beta\delta_{ii} \quad \Rightarrow \quad I_{ii} = 3\beta \quad [\because \delta_{ii} = 3] \tag{5.102}$$

Further, we evaluate the integral I_{ii} as

$$I_{ii} = \frac{1}{4\pi}\int_{\phi=0}^{\pi}\int_{\theta=0}^{2\pi} \upsilon_i\upsilon_i \sin\phi d\theta d\phi$$

$$\text{where} \qquad \upsilon_i\upsilon_i = (\cos\theta\sin\phi)^2 + (\sin\theta\sin\phi)^2 + (\cos\phi)^2$$

$$\therefore \quad I_{ii} = 1 \quad \Rightarrow \quad \beta = \frac{1}{3}$$

$$\therefore \quad I_{ij} = \frac{1}{3}\delta_{ij}. \tag{5.103}$$

With this result, Eq. (5.100) simplifies as

$$P_{ij}^{(0)} = \frac{1}{3}\delta_{ij}4\pi mn \left(\frac{m}{2\pi k_B T}\right)^{\frac{3}{2}} \int_0^\infty C^4 \exp\left[-\frac{mC^2}{2k_B T}\right] dC \tag{5.104}$$

The integral to be evaluated is a standard Gaussian integral defined as

$$G_n = \int_0^\infty C^n \exp[-aC^2] dC = \frac{\Gamma\left(\frac{n+1}{2}\right)}{2a^{\left(\frac{n+1}{2}\right)}} \tag{5.105}$$

where Γ is a standard gamma function often referred to as the generalized factorial function defined as

$$\Gamma(x) = \int_0^\infty t^{x-1} e^{-t} dt, \quad (x > 0)$$

In the present case, we have $a = m/2k_B T$, $n = 4$, and $\Gamma(5/2) = 3\sqrt{\pi}/4$. Substituting these expressions in Eq. (5.104), we get

$$P_{ij}^{(0)} = nk_B T\delta_{ij} = p\delta_{ij}. \tag{5.106}$$

For evaluation of heat flux vector at the zeroth approximation level, we substitute for $r = 0$, $f^{(0)} = f_M$ in Eq. (5.66) and proceed as

$$q_i^{(0)} = \frac{m}{2}\int C_i C^2 f_M d\mathbf{C} \tag{5.107}$$

Following the same procedure and expressing the integral in the compact notation as

$$q_i^{(0)} = mn\left(\frac{m}{2\pi k_B T}\right)^{\frac{3}{2}} 4\pi \int_0^\infty C^3 \exp\left[-\frac{mC^2}{2k_B T}\right] dC \; I_i$$

$$\text{where} \quad I_i = \frac{1}{4\pi}\int_{\phi=0}^{\pi}\int_{\theta=0}^{2\pi} \upsilon_i \sin\phi d\theta d\phi \tag{5.108}$$

Performing this symbolic integration for each of the three components of vector υ_i, we get

$$I_i = 0, \qquad \Rightarrow \qquad q_i^{(0)} = 0. \tag{5.109}$$

Thus, at the zeroth level, we obtain constitutive relationships for the pressure tensor and heat flux vector as

$$
\begin{aligned}
P_{ij}^{(Eu)} &= P_{ij}^{(0)} = \delta_{ij} p, \\
q_i^{(Eu)} &= q_i^{(0)} = 0
\end{aligned}
\tag{5.110}
$$

where the subscript "Eu" stands for Euler, corresponding to zeroth level of approximation. Substituting these expressions in the basic conservation equations (5.53)–(5.55) we obtain the well-known inviscid Euler equations:

$$
\begin{aligned}
\frac{\partial \rho}{\partial t} + \frac{\partial}{\partial x_i}(\rho u_i) &= 0, \\
\frac{\partial u_i}{\partial t} + u_j \frac{\partial u_i}{\partial x_j} + \frac{1}{\rho}\frac{\partial p}{\partial x_i} &= F_i, \\
\frac{3}{2}\rho R\left[\frac{\partial T}{\partial t} + u_j \frac{\partial T}{\partial x_j}\right] + p\frac{\partial u_i}{\partial x_i} &= 0.
\end{aligned}
\tag{5.111}
$$

5.4.2 Navier–Stokes Equations

An expression for the first approximation of the distribution function is obtained as (Eq. (5.95))

$$
f^{(1)} = -\frac{f^{(0)}}{\nu}\left\{\left(\frac{mC^2}{2k_B T} - \frac{5}{2}\right) C_k \frac{\partial \log T}{\partial x_k} + \frac{m}{k_B T}\left\{C_i C_j \frac{\partial u_i}{\partial x_j} - \frac{1}{3}C^2 \delta_{ij}\frac{\partial u_i}{\partial x_j}\right\}\right\}
$$

We now proceed to evaluate $P_{ij}^{(1)}$ and $q_i^{(1)}$ by utilizing $f^{(1)}$ as

$$
P_{ij}^{(1)} = m\int C_i C_j f^{(1)} d\mathbf{C}, \qquad q_i^{(1)} = \frac{m}{2}\int C_i C^2 f^{(1)} d\mathbf{C}
\tag{5.112}
$$

For first approximation of pressure tensor $P_{ij}^{(1)}$,

$$
\begin{aligned}
P_{ij}^{(1)} &= m\int C_i C_j f^{(1)} d\mathbf{C} \\
&= m\int C_i C_j \left(\frac{-f^{(0)}}{\nu}\right)\left\{\left(\frac{mC^2}{2k_B T} - \frac{5}{2}\right) C_k \frac{\partial \log T}{\partial x_k}\right. \\
&\qquad \left. + \frac{m}{k_B T}\left(C_k C_l \frac{\partial u_k}{\partial x_l} - \frac{1}{3}C^2 \delta_{kl}\frac{\partial u_k}{\partial x_l}\right)\right\} d\mathbf{C}
\end{aligned}
\tag{5.113}
$$

Note that we changed the dummy indices, i and j in the second term of the Eq. (5.95) since we already have indices i and j as the free indices.

$$P_{ij}^{(1)} = -\frac{m}{\nu}\Bigg[\frac{\partial \log T}{\partial x_k}\underbrace{\int C_i C_j C_k f^{(0)}\left(\frac{mC^2}{2k_B T}-\frac{5}{2}\right)d\mathbf{C}}_{I_1} + \frac{m}{k_B T}\frac{\partial u_k}{\partial x_l}\Bigg\{\underbrace{\int C_i C_j C_k C_l f^{(0)} d\mathbf{C}}_{I_2} - \frac{1}{3}\delta_{kl}\underbrace{\int C_i C_j C^2 f^{(0)} d\mathbf{C}}_{I_3}\Bigg\}\Bigg] \tag{5.114}$$

We encounter some common integrals in kinetic theory of gases of the type as

$$\int C_{i_1} C_{i_2}\dots C_{i_n} f_M d\mathbf{C} = n\left(\frac{m}{2\pi k_B T}\right)^{\frac{3}{2}}\int C_{i_1} C_{i_2}\dots C_{i_n}\exp\left[-\frac{mC^2}{2k_B T}\right]d\mathbf{C} \tag{5.115}$$

When we have odd number of peculiar velocities $\mathbf{C}$, i.e., for $n = 1, 3, 5, \dots$, the integral always vanishes. With this general rule, integral I_1 vanishes and integral expression for $P_{ij}^{(1)}$ reduces to

$$P_{ij}^{(1)} = -\frac{m}{\nu}\left[\frac{m}{k_B T}\frac{\partial u_k}{\partial x_l}\Bigg\{\underbrace{\int C_i C_j C_k C_l f^{(0)} d\mathbf{C}}_{I_2} - \frac{1}{3}\delta_{kl}\underbrace{\int C_i C_j C^2 f^{(0)} d\mathbf{C}}_{I_3}\Bigg\}\right] \tag{5.116}$$

As before, integrals I_2 and I_3 are evaluated by transforming from Cartesian to spherical coordinate system.

Introducing the direction vector of the microscopic velocities as $\upsilon_i = C_i/C = \{\cos\theta\sin\phi, \sin\theta\sin\phi, \cos\phi\}_i$, we evaluate integrals I_2 and I_3 separately as

$$\begin{aligned}
I_2 &= \int C_i C_j C_k C_l f^{(0)} d\mathbf{C} \\
&= n\left(\frac{m}{2\pi k_B T}\right)^{\frac{3}{2}}\int C_i C_j C_k C_l \exp\left[-\frac{mC^2}{2k_B T}\right]d\mathbf{C} \\
&= n\left(\frac{m}{2\pi k_B T}\right)^{\frac{3}{2}}\int_{C=0}^{\infty}\int_{\phi=0}^{\pi}\int_{\theta=0}^{2\pi} C^4 \upsilon_i\upsilon_j\upsilon_k\upsilon_l \exp\left[-\frac{mC^2}{2k_B T}\right]C^2\sin\phi\, dC\, d\theta\, d\phi
\end{aligned}$$

$$= n\left(\frac{m}{2\pi k_B T}\right)^{\frac{3}{2}} 4\pi \int_{C=0}^{\infty} C^6 \exp\left[-\frac{mC^2}{2k_B T}\right] dC\ I_{ijkl}$$

$$\text{where} \qquad I_{ijkl} = \frac{1}{4\pi}\int_{\phi=0}^{\pi}\int_{\theta=0}^{2\pi} \upsilon_i\upsilon_j\upsilon_k\upsilon_l \sin\phi d\theta d\phi \tag{5.117}$$

The integral I_{ijkl} is a fourth order symmetric and isotropic tensor and any fourth order isotropic tensor can be represented in terms of Kronecker delta tensor as

$$I_{ijkl} = \alpha\left[\delta_{ij}\delta_{kl} + \delta_{ik}\delta_{jl} + \delta_{il}\delta_{jk}\right] \tag{5.118}$$

To determine the value of α, we perform the contraction twice by multiplying $\delta_{ij}\delta_{kl}$,

$$I_{ijkl}\delta_{ij}\delta_{kl} = \alpha\left[\delta_{ii}\delta_{kk} + \delta_{ik}\delta_{ik} + \delta_{ik}\delta_{ik}\right] \qquad \Rightarrow \qquad I_{iikk} = \alpha[3\times 3 + \delta_{ii} + \delta_{ii}]$$

$$\therefore \qquad I_{iikk} = 15\alpha \tag{5.119}$$

and the integral I_{iikk} is evaluated as

$$I_{iikk} = \frac{1}{4\pi}\int_{\phi=0}^{\pi}\int_{\theta=0}^{2\pi} \upsilon_i\upsilon_i\upsilon_k\upsilon_k \sin\phi d\theta d\phi$$

$$\text{where} \qquad \upsilon_i\upsilon_i\upsilon_k\upsilon_k = \left[(\cos\theta\sin\phi)^2 + (\sin\theta\sin\phi)^2 + (\cos\phi)^2\right]^2$$

$$\therefore \qquad I_{iikk} = 1 \qquad \Rightarrow \qquad \alpha = \frac{1}{15}$$

$$\text{and} \qquad I_{ijkl} = \frac{1}{15}\left[\delta_{ij}\delta_{kl} + \delta_{ik}\delta_{jl} + \delta_{il}\delta_{jk}\right] \tag{5.120}$$

Substituting the final result of I_{ijkl} in integral equation for I_2,

$$I_2 = n\left(\frac{m}{2\pi k_B T}\right)^{\frac{3}{2}} 4\pi \underline{\int_{C=0}^{\infty} C^6 \exp\left[-\frac{mC^2}{2k_B T}\right] dC}\ \frac{1}{15}\left[\delta_{ij}\delta_{kl} + \delta_{ik}\delta_{jl} + \delta_{il}\delta_{jk}\right] \tag{5.121}$$

Evaluating the underlined Gaussian integral for $a = m/2k_BT, n = 6$, and $\Gamma(7/2) = \frac{15}{8}\sqrt{\pi}$,

$$\int_{C=0}^{\infty} C^6 \exp\left[-\frac{mC^2}{2k_B T}\right] dC = \frac{15\sqrt{\pi}}{16}\left(\frac{m}{2k_B T}\right)^{-\frac{7}{2}} \tag{5.122}$$

Summarizing all the results, we get expression for the integral I_2 as

$$I_2 = n\left(\frac{k_B T}{m}\right)^2 \left[\delta_{ij}\delta_{kl} + \delta_{ik}\delta_{jl} + \delta_{il}\delta_{jk}\right] \tag{5.123}$$

The evaluation of integral I_3 is relatively straightforward, and we proceed as

$$\begin{aligned}
I_3 &= \int C_i C_j C^2 f^{(0)} d\mathbf{C} \\
&= n\left(\frac{m}{2\pi k_B T}\right)^{\frac{3}{2}} \int C_i C_j C^2 \exp\left[-\frac{mC^2}{2k_B T}\right] d\mathbf{C} \\
&= n\left(\frac{m}{2\pi k_B T}\right)^{\frac{3}{2}} \int_{C=0}^{\infty}\int_{\phi=0}^{\pi}\int_{\theta=0}^{2\pi} C^2 \upsilon_i \upsilon_j C^2 \exp\left[-\frac{mC^2}{2k_B T}\right] C^2 \sin\phi dC d\theta d\phi \\
&= n\left(\frac{m}{2\pi k_B T}\right)^{\frac{3}{2}} 4\pi \int_{C=0}^{\infty} C^6 \exp\left[-\frac{mC^2}{2k_B T}\right] dC \ I_{ij} \\
&\qquad \text{where} \qquad I_{ij} = \frac{1}{4\pi}\int_{\phi=0}^{\pi}\int_{\theta=0}^{2\pi} \upsilon_i \upsilon_j \sin\phi d\theta d\phi = \frac{1}{3}\delta_{ij}
\end{aligned} \tag{5.124}$$

The Gaussian integral and the integral I_{ij} are already evaluated previously in Sect. 5.4.1, hence, the expression for integral equation I_3 is obtained as

$$I_3 = 5n\left(\frac{k_B T}{m}\right)^2 \delta_{ij}. \tag{5.125}$$

Substituting the results for integral equation I_2 (Eq. (5.123)) and I_3 (Eq. (5.125)) in expression for $P_{ij}^{(1)}$ (Eq. (5.116)), we obtain

$$\begin{aligned}
P_{ij}^{(1)} &= -\frac{m}{\nu}\left[\frac{m}{k_B T}\frac{\partial u_k}{\partial x_l}\left\{n\left(\frac{k_B T}{m}\right)^2 \left[\delta_{ij}\delta_{kl} + \delta_{ik}\delta_{jl} + \delta_{il}\delta_{jk}\right]\right.\right. \\
&\qquad \left.\left. -\frac{5}{3}\delta_{kl} n\left(\frac{k_B T}{m}\right)^2 \delta_{ij}\right\}\right] \\
&= -\frac{m}{\nu}\left[\frac{m}{k_B T} n\left(\frac{k_B T}{m}\right)^2 \frac{\partial u_k}{\partial x_l}\left\{\left[\delta_{ij}\delta_{kl} + \delta_{ik}\delta_{jl} + \delta_{il}\delta_{jk}\right] - \frac{5}{3}\delta_{kl}\delta_{ij}\right\}\right] \\
&= -\frac{m}{\nu}\frac{nk_B T}{m}\left[\frac{\partial u_k}{\partial x_l}\delta_{ij}\delta_{kl} + \frac{\partial u_k}{\partial x_l}\delta_{ik}\delta_{jl} + \frac{\partial u_k}{\partial x_l}\delta_{il}\delta_{jk} - \frac{5}{3}\frac{\partial u_k}{\partial x_l}\delta_{kl}\delta_{ij}\right] \\
&= -\frac{nk_B T}{\nu}\left[\frac{\partial u_k}{\partial x_k}\delta_{ij} + \frac{\partial u_i}{\partial x_j} + \frac{\partial u_j}{\partial x_i} - \frac{5}{3}\frac{\partial u_k}{\partial x_k}\delta_{ij}\right]
\end{aligned}$$

$$
\begin{aligned}
&= -\frac{nk_BT}{\nu}\left[\frac{\partial u_i}{\partial x_j} + \frac{\partial u_j}{\partial x_i} - \frac{2}{3}\frac{\partial u_k}{\partial x_k}\delta_{ij}\right] \\
&= -2\frac{nk_BT}{\nu}\left\{\frac{1}{2}\left(\frac{\partial u_i}{\partial x_j} + \frac{\partial u_j}{\partial x_i}\right) - \frac{1}{3}\frac{\partial u_k}{\partial x_k}\delta_{ij}\right\}
\end{aligned}
\tag{5.126}
$$

In terms of symmetric tensorial notation, and identifying the coefficient term $\frac{nk_BT}{\nu}$ as absolute viscosity (recall that, here, ν is the collision frequency and not kinematic viscosity), we obtain expression for $P_{ij}^{(1)}$ as

$$
P_{ij}^{(1)} = -2\mu\frac{\partial u_{\langle i}}{\partial x_{j\rangle}}
\tag{5.127}
$$

$$
\text{where } \mu = \frac{nk_BT}{\nu} \text{ and } \frac{\partial u_{\langle i}}{\partial x_{j\rangle}} = \frac{1}{2}\left(\frac{\partial u_i}{\partial x_j} + \frac{\partial u_j}{\partial x_i}\right) - \frac{1}{3}\frac{\partial u_k}{\partial x_k}\delta_{ij}
\tag{5.128}
$$

From the definition of decomposition of the pressure tensor (Eq. (5.22)), we can say that the first approximation to the pressure tensor is the divergence-less viscous stress tensor, i.e., $p_{ij} = P_{ij}^{(1)}$. This relation between the viscous stress tensor and velocity gradient is well-known as Navier–Stokes law.

Therefore, at the first approximation, the expression for pressure tensor is

$$
P_{ij}^{(NS)} = P_{ij}^{(Eu)} + P_{ij}^{(1)} = p\delta_{ij} - 2\mu\frac{\partial u_{\langle i}}{\partial x_{j\rangle}}.
\tag{5.129}
$$

Note that the second coefficient of viscosity and the coefficient of bulk viscosity (introduced in Sect. 2.2.1) has not appeared in Eq. (5.129). This is because the above derivation is for monatomic gases. However, polyatomic gases involve extra internal degrees of freedom and the coefficient of bulk viscosity does alter the above constitutive relation, as discussed in Kogan [83, p. 216].

To evaluate expression for $q^{(1)}$, we substitute the first approximation distribution function $f^{(1)}$ (Eq. (5.95)) in first approximation of heat flux vector (Eq. (5.66)) and proceed as

$$
\begin{aligned}
q_i^{(1)} &= \frac{m}{2}\int C_iC^2 f^{(1)} d\mathbf{C} \\
&= \frac{m}{2}\int C_iC^2\left(\frac{-f^{(0)}}{\nu}\right)\left\{\left(\frac{mC^2}{2k_BT} - \frac{5}{2}\right)C_k\frac{\partial \log T}{\partial x_k}\right. \\
&\qquad \left. + \frac{m}{k_BT}\left(C_kC_l\frac{\partial u_k}{\partial x_l} - \frac{1}{3}C^2\delta_{kl}\frac{\partial u_k}{\partial x_l}\right)\right\}d\mathbf{C}
\end{aligned}
$$

$$= -\frac{m}{2\nu}\frac{\partial \log T}{\partial x_k}\left\{\frac{m}{2k_BT}\underbrace{\int C_iC_kC^2 f^{(0)}C^2 d\mathbf{C}}_{I_1} - \frac{5}{2}\underbrace{\int C_iC_kC^2 f^{(0)}d\mathbf{C}}_{I_2}\right\}$$

$$- \frac{m}{2\nu}\frac{m}{k_BT}\frac{\partial u_k}{\partial x_l}\left\{\underbrace{\int C_iC_kC_lC^2 f^{(0)}d\mathbf{C}}_{I_3} - \frac{1}{3}\delta_{kl}\underbrace{\int C_iC^2C^2 f^{(0)}d\mathbf{C}}_{I_4}\right\}$$

For integrals I_3 and I_4, we have odd number tensors of peculiar velocities and hence, they are exactly zero. Therefore, the equation for $q^{(1)}$ reduces to

$$\therefore \quad q_i^{(1)} = -\frac{m}{2\nu}\frac{\partial \log T}{\partial x_k}\left\{\frac{m}{2k_BT}\underbrace{\int C_iC_kC^2 f^{(0)}C^2 d\mathbf{C}}_{I_1} - \frac{5}{2}\underbrace{\int C_iC_kC^2 f^{(0)}d\mathbf{C}}_{I_2}\right\} \tag{5.130}$$

Following the same procedure as outlined previously, the integrals I_1 and I_2 are evaluated as

$$\begin{aligned}
I_1 &= \int C_iC_kC^2 f^{(0)}C^2 d\mathbf{C} \\
&= n\left(\frac{m}{2\pi k_BT}\right)^{\frac{3}{2}}\int C_iC_kC^4 \exp\left[-\frac{mC^2}{2k_BT}\right]d\mathbf{C} \\
&= n\left(\frac{m}{2\pi k_BT}\right)^{\frac{3}{2}}\int_{C=0}^{\infty}\int_{\phi=0}^{\pi}\int_{\theta=0}^{2\pi} C^2 v_i v_k C^4 \exp\left[-\frac{mC^2}{2k_BT}\right]C^2\sin\phi\, dC\, d\theta\, d\phi \\
&= n\left(\frac{m}{2\pi k_BT}\right)^{\frac{3}{2}} 4\pi\int_{C=0}^{\infty} C^8 \exp\left[-\frac{mC^2}{2k_BT}\right]dC\; I_{ik}
\end{aligned} \tag{5.131}$$

We already know $I_{ik} = \delta_{ik}/3$ and evaluating the Gaussian integral for $a = m/2k_BT$, $n = 8$, and $\Gamma(9/2) = \frac{105}{16}\sqrt{\pi}$,

$$\int_{C=0}^{\infty} C^8 \exp\left[-\frac{mC^2}{2k_BT}\right]dC = \frac{105\sqrt{\pi}}{32}\left(\frac{m}{2k_BT}\right)^{-\frac{9}{2}}$$

Substituting this result, we obtain the expression for I_1 as

$$I_1 = 35n\left(\frac{k_BT}{m}\right)^3\delta_{ik} \tag{5.132}$$

The integral I_2 has already been obtained previously (Eq. (5.125)) as

$$I_2 = \int C_i C_k C^2 f^{(0)} d\mathbf{C} = 5n \left(\frac{k_B T}{m}\right)^2 \delta_{ik} \tag{5.133}$$

Finally, substituting the integrals I_1 (Eq. (5.132)) and I_2 (Eq. (5.133)) in equation for heat flux vector (Eq. (5.130)), we have

$$\begin{aligned} q_i^{(1)} &= -\frac{m}{2\nu}\frac{\partial \log T}{\partial x_k}\left\{\frac{m}{2k_B T}35n\left(\frac{k_B T}{m}\right)^3 \delta_{ik} - \frac{5}{2}5n\left(\frac{k_B T}{m}\right)^2 \delta_{ik}\right\} \\ &= -\frac{5}{2}\frac{k_B}{m}\frac{nk_B T}{\nu}\frac{\partial T}{\partial x_i} \\ \therefore \quad q_i^{(1)} &= -k\frac{\partial T}{\partial x_i}, \qquad \text{where } k = \frac{5}{2}\frac{k_B}{m}\frac{nk_B T}{\nu}. \end{aligned} \tag{5.134}$$

Therefore, at the first approximation, we obtain expression for heat flux vector as

$$q_i^{(NS)} = q_i^{(Eu)} + q_i^{(1)} = 0 - k\frac{\partial T}{\partial x_i} \tag{5.135}$$

This linear relationship between the heat flux vector and temperature gradient is famously known as Fourier's law. From the expressions for transport coefficients, μ and k, it is clear that they depend on atomic structure and only on temperature, but not on pressure, a finding well supported by the experiments.

The Prandtl number is then calculated as

$$Pr = \frac{\mu C_p}{k} = 1, \qquad [\text{where } C_p = 5R/2 = 5k_B/2m] \tag{5.136}$$

Therefore, using the BGK kinetic model, we get unity Prandtl number while the measured value of Prandtl number for a monatomic gas is equal to 2/3. This inability of the BGK model to predict the correct Prandtl number results in discrepancies in the flow description of the gas. Nonetheless, BGK kinetic model greatly simplifies the method and help us to appreciate the significance of the Chapman–Enskog method.

Summarizing the results, the constitutive relationships for the pressure tensor and heat flux vector at the first approximation are obtained as

$$P_{ij}^{(NS)} = p\delta_{ij} - 2\mu\frac{\partial u_{\langle i}}{\partial x_{j\rangle}}, \tag{5.137}$$

$$q_i^{(NS)} = -k\frac{\partial T}{\partial x_i} \tag{5.138}$$

The closure is achieved by substituting these constitutive relationships in the basic conservation laws (Eqs. (5.53)–(5.55)), and the resulting equations are the famous Navier–Stokes equations for constant thermo-physical properties.

$$\frac{\partial \rho}{\partial t} + \frac{\partial}{\partial x_i}(\rho u_i) = 0, \tag{5.139}$$

$$\frac{\partial u_i}{\partial t} + u_k \frac{\partial u_i}{\partial x_k} = -\frac{1}{\rho}\frac{\partial p}{\partial x_i} + F_i + \frac{\mu}{\rho}\left\{\frac{\partial^2 u_i}{\partial x_j \partial x_j} + \frac{1}{3}\frac{\partial}{\partial x_i}\left(\frac{\partial u_k}{\partial x_k}\right)\right\}, \tag{5.140}$$

$$\frac{3}{2}\rho R\left[\frac{\partial T}{\partial t} + u_k \frac{\partial T}{\partial x_k}\right] = -p\frac{\partial u_k}{\partial x_k} + \mu\left\{\left(\frac{\partial u_k}{\partial x_k}\right)^2 + \frac{\partial u_i}{\partial x_j}\frac{\partial u_j}{\partial x_i} - \frac{2}{3}\left(\frac{\partial u_k}{\partial x_k}\right)^2\right\}$$
$$+ k\frac{\partial^2 T}{\partial x_k \partial x_k} \tag{5.141}$$

5.4.3 Burnett Equations

The complete second order approximation to the distribution function was determined by Burnett [26] in 1935. However, extreme complexity is involved in deriving $f^{(2)}$ and therefore we do not wish to include those details in this text. As it turns out, there is no need to solve the equation for $f^{(2)}$. Based on an order of magnitude of the terms, Chapman and Cowling [31, Chapter 15] obtained the additional terms introduced by $f^{(2)}$ into the equations for pressure tensor and heat flux vector.

Here, we present only the final expressions at the third approximation which read

$$P_{ij}^{(Bu)} = P_{ij}^{(NS)} + P_{ij}^{(2)}, \tag{5.142}$$

$$q_i^{(Bu)} = q_i^{(NS)} + q_i^{(2)}, \tag{5.143}$$

where

$$P_{ij}^{(2)} = \omega_1 \frac{\mu^2}{p}\frac{\partial u_k}{\partial x_k} S_{ij} + \omega_2 \frac{\mu^2}{p}\left\{\frac{D_0 S_{ij}}{Dt} - 2\frac{\partial u_k}{\partial x_{\langle i}} S_{j\rangle k}\right\} + \omega_3 \frac{\mu^2}{\rho T}\frac{\partial^2 T}{\partial x_{\langle i}\partial x_{j\rangle}}$$
$$+ \omega_4 \frac{\mu^2}{p\rho T}\frac{\partial T}{\partial x_{\langle i}}\frac{\partial p}{\partial x_{j\rangle}} + \omega_5 \frac{\mu^2}{\rho T^2}\frac{\partial T}{\partial x_{\langle i}}\frac{\partial T}{\partial x_{j\rangle}} + \omega_6 \frac{\mu^2}{p} S_{k\langle i} S_{j\rangle k} \tag{5.144}$$

$$q_i^{(2)} = \theta_1 \frac{\mu^2}{\rho T}\frac{\partial u_k}{\partial x_k}\frac{\partial T}{\partial x_i} + \theta_2 \frac{\mu^2}{\rho T}\left\{\frac{D_0}{Dt}\left(\frac{\partial T}{\partial x_i}\right) - \frac{\partial u_k}{\partial x_i}\frac{\partial T}{\partial x_k}\right\}$$

$$+ \theta_3 \frac{\mu^2}{\rho p} S_{ik} \frac{\partial p}{\partial x_k} + \theta_4 \frac{\mu^2}{\rho} \frac{\partial S_{ik}}{\partial x_k} + 3\theta_5 \frac{\mu^2}{\rho T} S_{ik} \frac{\partial T}{\partial x_k} \tag{5.145}$$

where S_{ij} is symmetric and trace-free part of the velocity gradient tensor defined as

$$S_{ij} = \frac{\partial u_{\langle i}}{\partial x_{j\rangle}} = \frac{1}{2}\frac{\partial u_i}{\partial x_j} + \frac{1}{2}\frac{\partial u_j}{\partial x_i} - \frac{1}{3}\frac{\partial u_k}{\partial x_k}\delta_{ij}$$

The Burnett equations are obtained by substituting Eqs. (5.144)–(5.145) into Eqs. (5.142)–(5.143), respectively, and the ensuing result in the conservation laws (Eqs. (5.53)–(5.55)). This complicated set of equations was first derived by Burnett using the Chapman–Enskog method. These equations contain the material derivatives terms of S_{ij} and $\frac{\partial T}{\partial x_i}$ and this particular form of equations is known as *original* Burnett equations.

Chapman and Cowling revisited the derivation of Burnett equations using order of magnitude analysis and replaced the material derivatives terms with the spatial gradients using the inviscid Euler equations. This substitution maintained the second order accuracy of the equations and hence, was deemed reasonable. Owing to its relatively simple nature of the equations (no material derivatives terms in the equations), this set of equations as proposed by Chapman and Cowling became more prevalent in the literature and is known as *conventional* Burnett equations. With this substitution, the *conventional* Burnett equations are given as

$$P_{ij}^{(2)} = \omega_1 \frac{\mu^2}{p}\frac{\partial u_k}{\partial x_k} S_{ij} + \omega_2 \frac{\mu^2}{p}\left\{\frac{\partial}{\partial x_{\langle i}}\left(F_{j\rangle} - \frac{1}{\rho}\frac{\partial p}{\partial x_{j\rangle}}\right) - \frac{\partial u_k}{\partial x_{\langle i}}\frac{\partial u_{j\rangle}}{\partial x_k} - 2\frac{\partial u_k}{\partial x_{\langle i}} S_{j\rangle k}\right\}$$
$$+ \omega_3 \frac{\mu^2}{\rho T}\frac{\partial^2 T}{\partial x_{\langle i}\partial x_{j\rangle}} + \omega_4 \frac{\mu^2}{p\rho T}\frac{\partial T}{\partial x_{\langle i}}\frac{\partial p}{\partial x_{j\rangle}} + \omega_5 \frac{\mu^2}{\rho T^2}\frac{\partial T}{\partial x_{\langle i}}\frac{\partial T}{\partial x_{j\rangle}} + \omega_6 \frac{\mu^2}{p} S_{k\langle i} S_{j\rangle k} \tag{5.146}$$

$$q_i^{(2)} = \theta_1 \frac{\mu^2}{\rho T}\frac{\partial u_k}{\partial x_k}\frac{\partial T}{\partial x_i} + \theta_2 \frac{\mu^2}{\rho T}\left\{-\frac{2}{3}\frac{\partial}{\partial x_i}\left(T\frac{\partial u_k}{\partial x_k}\right) - 2\frac{\partial u_k}{\partial x_i}\frac{\partial T}{\partial x_k}\right\}$$
$$+ \theta_3 \frac{\mu^2}{\rho p} S_{ik} \frac{\partial p}{\partial x_k} + \theta_4 \frac{\mu^2}{\rho} \frac{\partial S_{ik}}{\partial x_k} + 3\theta_5 \frac{\mu^2}{\rho T} S_{ik} \frac{\partial T}{\partial x_k} \tag{5.147}$$

Note that the coefficients ω's and θ's involved in original as well as conventional Burnett equations are pure numbers and their values are given in Table 5.1 for Maxwell molecules (exact values) and for rigid hard spheres. The complicated structure of the equations is the reason why these equations have not been studied extensively in the past.

The numerous terms which are there in Eqs. (5.144) and (5.145) are not so apparent because of the tensor notation adopted here. The terms in expanded form are given in Sect. 5.5 albeit in cylindrical coordinates to further emphasize the complexity of these terms. The stress and heat flux terms at the Burnett order

Table 5.1 Coefficients for Burnett stress and heat flux terms [31]

Coefficients	Maxwell molecules	Hard-sphere molecules
ω_1	$4/3\,[7/2 - (T/\mu)(d\mu/dT)]$	$1.014 \times 4/3\,[7/2 - (T/\mu)(d\mu/dT)]$
ω_2	2	2.028
ω_3	3	2.418
ω_4	0	0.681
ω_5	$3(T/\mu)(d\mu/dT)$	$0.806(3T/\mu)(d\mu/dT) - 0.990$
ω_6	8	7.424
θ_1	$(15/4)\,[7/2 - (T/\mu)(d\mu/dT)]$	11.644
θ_2	45/8	5.822
θ_3	−3	−3.090
θ_4	3	2.418
θ_5	$35/4 + (T/\mu)(d\mu/dT)$	$1.627 \times 3\,[35/4 + (T/\mu)(d\mu/dT)]$

depend on various combinations of the product of gradient of velocity, temperature, and pressure—making these terms non-linear. (In contrast, the stress and heat flux terms at the Navier–Stokes order are proportional to velocity gradient, while the stress and heat flux terms at super Burnett order involve triple gradients of velocity, temperature, and pressure.) Note that the number of terms will increase further upon differentiation of the terms, as gradient of stress and heat flux terms enter the momentum and energy equation, respectively.

In the Navier–Stokes equations, non-linearity is due to the inertial terms on the left-hand side of the equations. In the Burnett equations, along with the non-linear inertial terms, the gradient of the stress and heat flux terms are also non-linear, owing to the non-linear constitutive relations as already noted above. Further, notice the presence of third-order derivative terms in the Burnett equations due to the first term of ω_2 and ω_3 terms in Eq. (5.146) and first term of θ_2 and θ_4 terms in Eq. (5.147); again referring to terms given in Sect. 5.5.2 will be helpful to see this point more clearly. The presence of third-order derivatives leads to the requirement of having to prescribe additional boundary conditions for velocity components, temperature, and pressure as compared to the Navier–Stokes equations. The Burnett equations are therefore third-order non-linear coupled partial differential equations.

Notice that the stress is non-zero even if all the velocity and pressure gradients in the flow is zero; this is due to the dependence of stress on temperature gradient through the ω_3 and ω_5 terms in Eq. (5.146). This seems strange and unphysical, as it suggests the possibility of setting up a velocity field through purely a temperature gradient. Similarly, the heat flux term depends on velocity gradient, through first of θ_2 and θ_4 terms in Eq. (5.147). At the Navier–Stokes order, the coupling between velocity and temperature fields is through the variation of fluid properties (most notably, viscosity), compressibility, or buoyancy. In contrast, the coupling between velocity and temperature fields is much more involved at the Burnett order as evident from the above discussion. (This coupling can further be through the boundary condition, for example, if the velocity slip is dependent on both normal velocity gradient and tangential temperature gradient at the wall (Eq. (1.4)).

5.4.4 Super-Burnett Equations

The third order approximation of the distribution function gives a superset of Burnett equations, known as Super-Burnett equations which are third order accurate in Knudsen number. However, because of the extreme complexity involved, full three-dimensional non-linear Super-Burnett equations are not available. Here we present the linearized form of Super-Burnett equations as obtained by Shavaliyev [129] for weakly perturbed gas flows:

$$p_{ij}^{(3)} = \frac{\mu^3}{p\rho}\left(\frac{5}{3}\frac{\partial^2}{\partial x_{\langle i}\partial x_{j\rangle}}\frac{\partial u_r}{\partial x_r} - \frac{4}{3}\frac{\partial^2}{\partial x_r \partial x_r}\frac{\partial u_{\langle i}}{\partial x_{j\rangle}}\right) \tag{5.148}$$

$$q_i^{(3)} = -\frac{\mu^3}{\rho^2}\left(\frac{157}{16}\frac{1}{T}\frac{\partial^2}{\partial x_r \partial x_r}\frac{\partial T}{\partial x_i} + \frac{5}{8}\frac{1}{\rho}\frac{\partial^2}{\partial x_r \partial x_r}\frac{\partial \rho}{\partial x_i}\right) \tag{5.149}$$

In the same work [129], Shavaliyev derived the non-linear super-Burnett equations for purely one-dimensional flows. Accordingly, the third order approximation to the stress tensor and heat flux vector is given as

$$\begin{aligned} p_{xx}^{(3)} = \frac{\mu^3}{p^2}\Bigg[& \frac{47}{3}\frac{R}{\rho}\frac{\partial T}{\partial x}\frac{\partial \rho}{\partial x}\frac{\partial u}{\partial x} \overbrace{- \frac{40}{3}\frac{RT}{\rho^2}\left(\frac{\partial \rho}{\partial x}\right)^2\frac{\partial u}{\partial x}}^{(I)_{pxx}} + \overbrace{\frac{32}{3}\frac{RT}{\rho}\frac{\partial^2 \rho}{\partial x^2}\frac{\partial u}{\partial x}}^{(II)_{pxx}} \\ & - \frac{2}{3}\frac{RT}{\rho}\frac{\partial \rho}{\partial x}\frac{\partial^2 u}{\partial x^2} - 7\frac{R}{T}\left(\frac{\partial T}{\partial x}\right)^2\frac{\partial u}{\partial x} - \frac{47}{9}R\frac{\partial T}{\partial x}\frac{\partial^2 u}{\partial x^2} \\ & - \frac{31}{9}R\frac{\partial^2 T}{\partial x^2}\frac{\partial u}{\partial x} + \frac{2}{9}RT\frac{\partial^3 u}{\partial x^3} + \frac{16}{27}\left(\frac{\partial u}{\partial x}\right)^3\Bigg] \end{aligned} \tag{5.150}$$

$$\begin{aligned} q_x^{(3)} = \frac{\mu^3}{p\rho}\Bigg[& \overbrace{-\frac{9005}{168}\frac{1}{T}\frac{\partial T}{\partial x}\left(\frac{\partial u}{\partial x}\right)^2}^{(I)_{qx}} + \overbrace{\frac{271}{21}\frac{1}{\rho}\frac{\partial \rho}{\partial x}\left(\frac{\partial u}{\partial x}\right)^2}^{(II)_{qx}} + \overbrace{\frac{421}{42}\frac{\partial u}{\partial x}\frac{\partial^2 u}{\partial x^2}}^{(III)_{qx}} \\ & + \frac{917}{8}\frac{R}{\rho T}\frac{\partial \rho}{\partial x}\left(\frac{\partial T}{\partial x}\right)^2 - \frac{1137}{16}\frac{R}{\rho^2}\frac{\partial T}{\partial x}\left(\frac{\partial \rho}{\partial x}\right)^2 + \frac{397}{16}\frac{R}{\rho}\frac{\partial \rho}{\partial x}\frac{\partial^2 T}{\partial x^2} \\ & + \frac{701}{16}\frac{R}{\rho}\frac{\partial T}{\partial x}\frac{\partial^2 \rho}{\partial x^2} - \frac{813}{16}\frac{R}{T^2}\left(\frac{\partial T}{\partial x}\right)^3 - \frac{1451}{16}\frac{R}{T}\frac{\partial T}{\partial x}\frac{\partial^2 T}{\partial x^2} \\ & - \frac{157}{16}R\frac{\partial^3 T}{\partial x^3} - \frac{41}{8}\frac{RT}{\rho^2}\frac{\partial \rho}{\partial x}\frac{\partial^2 \rho}{\partial x^2} - \frac{5}{8}\frac{RT}{\rho}\frac{\partial^3 \rho}{\partial x^3} + \frac{23}{4}\frac{RT}{\rho^3}\left(\frac{\partial \rho}{\partial x}\right)^3\Bigg] \end{aligned} \tag{5.151}$$

Table 5.2 Values of the coefficients as reported in the literature

Equation	Term	Shavaliyev [129]	Fiscko and Chapman [56]	Torrilhon and Struchtrup [148]
$p_{xx}^{(3)}$	$(I)_{p_{xx}}$	$-40/3$	$-64/9$	$-64/9$
	$(II)_{p_{xx}}$	$32/3$	$40/9$	$40/9$
$q_x^{(3)}$	$(I)_{q_x}$	$-9005/168$	$-8035/336$	$-2913/112$
	$(II)_{q_x}$	$271/21$	$166/21$	$188/21$
	$(III)_{q_x}$	$421/42$	$949/168$	$199/56$

Shavaliyev [129] believed that the solution of the problem of the structure of the shock wave can be improved further by considering these super-Burnett order terms. However, Fiscko and Chapman [56] reported that the solutions of the super-Burnett equations are actually inferior to those of the Burnett equations. A possible reason for this is to an algebraic error in the coefficient of one of the terms of these equations. It is worth noting that the coefficients of the marked terms used in Fiscko and Chapman work are different than those reported by Shavaliyev. All the remaining coefficients agree with those given by Fiscko and Chapman [56] except those for the five overlined terms.

The three-dimensional linearized super-Burnett equations [148] derived from R13 equations (which we will discuss in the next chapter) were found to be same as derived by Shavaliyev (Eqs. (5.148)–(5.149)). Once again, there was discrepancy in the values of the coefficients of the first three underlined terms of the heat flux equation (Eq. (5.151)) as reported by Torrilhon and Struchtrup [148]. Table 5.2 gives the overall summary of the values of the coefficients as reported by these researchers. Note that the coefficients differ only in magnitude and not in sign.

5.4.4.1 Summary of Chapman–Enskog Method

The entire essence of the Chapman–Enskog method is captured in a detailed flowchart as shown in Fig. 5.4. Briefly summarizing, the particle distribution function is expanded in an infinite series around Maxwellian distribution with the assumption that this series converges. This series is then substituted into the Boltzmann equation. By performing some subtle mathematical operations, Enskog hypothesized that the problem can be divided into smaller bits. The solution of the zeroth level approximation is Maxwellian distribution and with the aid of this solution, higher level approximations to the distribution function can be determined. With known functional form of the distribution function at each level, the constitutive relationships for the pressure tensor and heat flux vector are evaluated. These constitutive relationships are then substituted into Cauchy's equation of motion and the necessary closure is achieved. At zeroth, first, second and third level of approximation, we obtain Euler [$O(\mathrm{Kn}^0)$], Navier–Stokes [$O(\mathrm{Kn}^1)$], Burnett [$O(\mathrm{Kn}^2)$],and super-Burnett [$O(\mathrm{Kn}^3)$] equations, respectively.

5.5 Burnett Equations in Cylindrical Coordinates

In this section, the Burnett equations presented in Sect. 5.4.3 have been cast in cylindrical coordinates. Since the terms are in expanded form, equations in cylindrical coordinates help to appreciate the significance of the different terms and the associated complexity.

The conservation equations mass, momentum, and energy equations cast in cylindrical coordinates are

$$\frac{\partial \rho}{\partial t} + \frac{1}{r}\frac{\partial \rho u r}{\partial r} + \frac{1}{r}\frac{\partial \rho v}{\partial \theta} + \frac{\partial \rho w}{\partial z} = 0 \tag{5.152}$$

$$\begin{aligned} &\frac{\partial \rho u}{\partial t} + \frac{1}{r}\frac{\partial (\rho u u + p) r}{\partial r} + \frac{1}{r}\frac{\partial p_{rr} r}{\partial r} + \frac{1}{r}\frac{\partial \rho u v}{\partial \theta} \\ &+ \frac{1}{r}\frac{\partial p_{r\theta}}{\partial \theta} + \frac{\partial \rho w u}{\partial z} + \frac{\partial p_{rz}}{\partial z} + \frac{\rho v v + p_{\theta\theta}}{r} = 0 \end{aligned} \tag{5.153}$$

$$\begin{aligned} &\frac{\partial \rho v}{\partial t} + \frac{1}{r}\frac{\partial \rho u v r}{\partial r} + \frac{1}{r}\frac{\partial p_{\theta r} r}{\partial r} + \frac{1}{r}\frac{\partial (\rho v v + p)}{\partial \theta} \\ &+ \frac{1}{r}\frac{\partial p_{\theta\theta}}{\partial \theta} + \frac{\partial \rho w v}{\partial z} + \frac{\partial p_{\theta z}}{\partial z} + \frac{\rho u v - p_{\theta r}}{r} = 0 \end{aligned} \tag{5.154}$$

$$\frac{\partial \rho w}{\partial t} + \frac{1}{r}\frac{\partial \rho u w r}{\partial r} + \frac{1}{r}\frac{\partial p_{zr} r}{\partial r} + \frac{1}{r}\frac{\partial \rho w v}{\partial \theta} + \frac{1}{r}\frac{\partial p_{z\theta}}{\partial \theta} + \frac{\partial (\rho w w + p)}{\partial z} + \frac{\partial p_{zz}}{\partial z} = 0 \tag{5.155}$$

$$\begin{aligned} &\frac{\partial e}{\partial t} + \frac{1}{r}\frac{\partial (e+p) u r}{\partial r} + \frac{1}{r}\frac{\partial (u p_{rz} + v p_{\theta r} + w p_{rr} - q_r)}{\partial r} \\ &+ \frac{1}{r}\frac{\partial (e+p) v}{\partial \theta} + \frac{\partial (\rho (e+p) w)}{\partial z} \\ &+ \frac{1}{r}\frac{\partial (u p_{r\theta} + v p_{\theta\theta} + w p_{z\theta} - q_\theta)}{\partial \theta} + \frac{\partial (u p_{rz} + v p_{\theta z} + w p_{zz} - q_z)}{\partial z} = 0. \end{aligned} \tag{5.156}$$

where $\rho e = \rho(\epsilon + u^2/2) = \rho(c_v T + u^2/2)$.

As before, the stress tensor and heat flux vector are expanded in a series as

$$p_{ij} = p_{ij}^{Eu} + p_{ij}^{NS} + p_{ij}^{Bu} + p_{ij}^{SB} + O(Kn^{n+1}) \tag{5.157}$$

$$q_i = q_i^{Eu} + q_i^{NS} + q_i^{Bu} + q_i^{SB} + O(Kn^{n+1}) \tag{5.158}$$

where n represents the accuracy in terms of $O(Kn)$.

Note that the Euler order stress and heat flux terms are zero as seen in Sect. 5.4.1.

5.5.1 Navier–Stokes Stress and Heat Flux

The Navier–Stokes order stress and heat flux terms are included here for completeness:

$$p_{rr}^{NS} = -\mu\left(\delta_1\frac{\partial u}{\partial r} + \delta_2\frac{1}{r}\frac{\partial v}{\partial\theta} + \delta_2\frac{\partial w}{\partial z} + \delta_2\frac{v}{r}\right) \quad (5.159)$$

$$p_{\theta\theta}^{NS} = -\mu\left(\delta_2\frac{\partial u}{\partial r} + \delta_1\frac{1}{r}\frac{\partial v}{\partial\theta} + \delta_2\frac{\partial w}{\partial z} + \delta_1\frac{u}{r}\right) \quad (5.160)$$

$$p_{zz}^{NS} = -\mu\left(\delta_2\frac{\partial u}{\partial r} + \delta_2\frac{1}{r}\frac{\partial v}{\partial\theta} + \delta_1\frac{\partial w}{\partial z} + \delta_2\frac{u}{r}\right) \quad (5.161)$$

$$p_{r\theta}^{NS} = p_{\theta r}^{NS} = -\mu\left(\frac{\partial v}{\partial r} - \frac{v}{r} + \frac{1}{r}\frac{\partial u}{\partial\theta}\right) \quad (5.162)$$

$$p_{\theta z}^{NS} = p_{z\theta}^{NS} = -\mu\left(\frac{1}{r}\frac{\partial w}{\partial\theta} + \frac{\partial v}{\partial z}\right) \quad (5.163)$$

$$p_{zr}^{NS} = p_{zr}^{NS} = -\mu\left(\frac{\partial u}{\partial z} + \frac{\partial w}{\partial r}\right) \quad (5.164)$$

$$q_r^{NS} = -k\frac{\partial T}{\partial r} \quad (5.165)$$

$$q_\theta^{NS} = -k\frac{1}{r}\frac{\partial T}{\partial\theta} \quad (5.166)$$

$$q_z^{NS} = -k\frac{\partial T}{\partial z} \quad (5.167)$$

where the coefficients are given as: $\delta_1 = 1/3$ and $\delta_2 = 2/3$.

5.5.2 Burnett Stress and Heat Flux

Singh and Agrawal [130] used the coordinate transformation technique on stress and heat flux terms in Cartesian coordinates (Eqs. (5.146) and (5.147)) to derive the following Burnett order stress and heat flux terms in cylindrical coordinates (some of the typographical errors appeared in [126] have been corrected in the following equations).

5.5.2.1 Burnett Stress Terms

In this section, we document all the six components of the stress tensor (out of which five are independent given the traceless nature of the stress tensor).

$$
\begin{aligned}
p_{rr}^{Bu} = \frac{\mu^2}{p}\Bigg[& \left(\frac{2\omega_1}{3} - \frac{14\omega_2}{9} + \frac{2\omega_6}{9}\right)\left(\frac{\partial u}{\partial r}\right)^2 + \left(\frac{\omega_1}{3} + \frac{2\omega_2}{9} - \frac{2\omega_6}{9}\right)\left(\frac{u}{r}\frac{\partial u}{\partial r}\right) \\
& + \left(\frac{\omega_2}{3} + \frac{\omega_6}{12}\right)\left(\frac{1}{r^2}\right)\left(\frac{\partial u}{\partial \theta}\right)^2 - \left(\frac{2\omega_2}{3} + \frac{\omega_6}{6}\right)\left(\frac{v}{r^2}\frac{\partial u}{\partial \theta}\right) \\
& + \left(\frac{\omega_2}{3} + \frac{\omega_6}{12}\right)\left(\frac{\partial u}{\partial z}\right)^2 + \left(-\frac{\omega_1}{3} + \frac{7\omega_2}{9} - \frac{\omega_6}{9}\right)\left(\frac{u}{r}\right)^2 \\
& - \left(\frac{2\omega_2}{3} - \frac{\omega_6}{12}\right)\left(\frac{\partial v}{\partial r}\right)^2 - \left(\frac{\omega_1}{3} - \frac{7\omega_2}{9} + \frac{\omega_6}{9}\right)\left(\frac{1}{r}\frac{\partial v}{\partial \theta}\right)^2 \\
& + \left(\frac{\omega_2}{3} - \frac{\omega_6}{6}\right)\left(\frac{\partial v}{\partial z}\right)^2 + \left(\frac{2\omega_2}{3} - \frac{\omega_6}{6}\right)\left(\frac{v}{r}\frac{\partial v}{\partial r}\right) \\
& - \left(\frac{2\omega_1}{3} - \frac{14\omega_2}{9} + \frac{2\omega_6}{9}\right)\left(\frac{u}{r^2}\frac{\partial v}{\partial \theta}\right) + \left(\frac{\omega_2}{3} + \frac{\omega_6}{12}\right)\left(\frac{v}{r}\right)^2 \\
& + \left(\frac{\omega_2}{3} - \frac{\omega_6}{6}\right)\left(\frac{1}{r}\frac{\partial w}{\partial \theta}\right)^2 \\
& - \left(\frac{\omega_1}{3} - \frac{7\omega_2}{9} + \frac{\omega_6}{9}\right)\left(\frac{\partial w}{\partial z}\right)^2 - \left(\frac{2\omega_1}{3} + \frac{4\omega_2}{9} - \frac{4\omega_6}{9}\right)\left(\frac{\partial w}{\partial z}\frac{u}{r}\right) \\
& + \left(\frac{\omega_1}{3} + \frac{2\omega_2}{9} - \frac{2\omega_6}{9}\right)\left(\frac{\partial u}{\partial r}\frac{\partial w}{\partial z}\right) - \left(\frac{2\omega_2}{3} - \frac{\omega_6}{6}\right)\left(\frac{1}{r}\frac{\partial u}{\partial \theta}\frac{\partial v}{\partial r}\right) \\
& - \left(\frac{2\omega_1}{3} + \frac{4\omega_2}{9} - \frac{4\omega_6}{9}\right)\left(\frac{1}{r}\frac{\partial v}{\partial \theta}\frac{\partial w}{\partial z}\right) + \left(\frac{\omega_1}{3} + \frac{2\omega_2}{9} - \frac{2\omega_6}{9}\right)\left(\frac{1}{r}\frac{\partial u}{\partial r}\frac{\partial v}{\partial \theta}\right) \\
& - \left(\frac{2\omega_2}{3} - \frac{\omega_6}{6}\right)\left(\frac{\partial w}{\partial r}\frac{\partial u}{\partial z}\right) - \left(\frac{2\omega_2}{3} - \frac{\omega_6}{12}\right)\left(\frac{\partial w}{\partial r}\right)^2 \\
& + \left(\frac{4\omega_2}{3} - \frac{\omega_6}{3}\right)\left(\frac{1}{r}\frac{\partial w}{\partial \theta}\frac{\partial v}{\partial z}\right) - \frac{\omega_3}{3}\frac{R}{r}\frac{\partial T}{\partial r} + \frac{2\omega_3 R}{3}\frac{\partial^2 T}{\partial r^2} \\
& - \frac{\omega_3 R}{3}\frac{\partial^2 T}{\partial z^2} - \frac{\omega_3}{3}\frac{R}{r^2}\frac{\partial^2 T}{\partial \theta^2} \\
& + \frac{2\omega_4}{3}\frac{R}{p}\frac{\partial T}{\partial r}\frac{\partial p}{\partial r} - \frac{\omega_4}{3}\frac{R}{p}\frac{\partial T}{\partial z}\frac{\partial p}{\partial z} - \frac{\omega_4}{3}\frac{1}{r^2}\frac{R}{p}\frac{\partial T}{\partial \theta}\frac{\partial p}{\partial \theta} + \frac{2\omega_5}{3}\frac{R}{T}\left(\frac{\partial T}{\partial r}\right)^2
\end{aligned}
$$

$$
-\frac{\omega_5}{3}\frac{R}{T}\left(\frac{\partial T}{\partial z}\right)^2-\frac{\omega_5}{3}\frac{1}{r^2}\frac{R}{T}\left(\frac{\partial T}{\partial \theta}\right)^2-\frac{2\omega_2}{3}\frac{1}{\rho}\frac{\partial^2 p}{\partial r^2}+\frac{\omega_2}{3}\frac{1}{\rho}\frac{\partial^2 p}{\partial z^2}
+\frac{\omega_2}{3}\frac{1}{r^2}\frac{1}{\rho}\frac{\partial^2 p}{\partial \theta^2}+\frac{\omega_2}{3}\frac{1}{r}\frac{1}{\rho}\frac{\partial p}{\partial r}+\frac{2\omega_2}{3}\frac{1}{\rho^2}\frac{\partial p}{\partial r}\frac{\partial \rho}{\partial r}-\frac{\omega_2}{3}\frac{1}{\rho^2}\frac{\partial p}{\partial z}\frac{\partial \rho}{\partial z}
-\frac{\omega_2}{3}\frac{1}{r^2}\frac{1}{\rho^2}\frac{\partial p}{\partial \theta}\frac{\partial \rho}{\partial \theta}\Bigg] \tag{5.168}
$$

$$
\begin{aligned}
p^{Bu}_{\theta\theta} = \frac{\mu^2}{p}\Bigg[& -\left(\frac{\omega_1}{3}-\frac{7\omega_2}{9}+\frac{1\omega_6}{9}\right)\left(\frac{\partial u}{\partial r}\right)^2+\left(\frac{\omega_1}{3}+\frac{2\omega_2}{9}-\frac{2\omega_6}{9}\right)\left(\frac{u}{r}\frac{\partial u}{\partial r}\right)\\
&+\left(\frac{2\omega_1}{3}-\frac{14\omega_2}{9}+\frac{2\omega_6}{9}\right)\left(\frac{u}{r}\right)^2-\left(\frac{2\omega_2}{3}-\frac{\omega_6}{12}\right)\left(\frac{1}{r^2}\right)\left(\frac{\partial u}{\partial \theta}\right)^2\\
&+\left(\frac{4\omega_2}{3}-\frac{\omega_6}{6}\right)\left(\frac{v}{r^2}\frac{\partial u}{\partial \theta}\right)+\left(\frac{\omega_2}{3}-\frac{\omega_6}{6}\right)\left(\frac{\partial u}{\partial z}\right)^2+\left(\frac{\omega_2}{3}+\frac{\omega_6}{12}\right)\left(\frac{\partial v}{\partial r}\right)^2\\
&+\left(\frac{2\omega_2}{3}-\frac{\omega_6}{6}\right)\left(\frac{v}{r}\frac{\partial v}{\partial r}\right)-\left(\frac{2\omega_2}{3}-\frac{\omega_6}{12}\right)\left(\frac{v}{r}\right)^2\\
&+\left(\frac{2\omega_1}{3}-\frac{14\omega_2}{9}+\frac{2\omega_6}{9}\right)\left(\frac{1}{r}\frac{\partial v}{\partial \theta}\right)^2+\left(\frac{4\omega_1}{3}-\frac{28\omega_2}{9}+\frac{4\omega_6}{9}\right)\left(\frac{u}{r^2}\frac{\partial v}{\partial \theta}\right)\\
&+\left(\frac{\omega_2}{3}+\frac{\omega_6}{12}\right)\left(\frac{\partial v}{\partial z}\right)^2\\
&+\left(\frac{\omega_2}{3}-\frac{\omega_6}{6}\right)\left(\frac{\partial w}{\partial r}\right)^2-\left(\frac{2\omega_2}{3}-\frac{\omega_6}{12}\right)\left(\frac{1}{r}\frac{\partial w}{\partial \theta}\right)^2\\
&-\left(\frac{2\omega_1}{3}+\frac{4\omega_2}{9}-\frac{4\omega_6}{9}\right)\left(\frac{\partial u}{\partial r}\frac{\partial w}{\partial z}\right)+\left(\frac{4\omega_2}{3}-\frac{\omega_6}{3}\right)\frac{\partial u}{\partial z}\frac{\partial w}{\partial r}\\
&-\left(\frac{2\omega_2}{3}-\frac{\omega_6}{6}\right)\left(\frac{1}{r}\frac{\partial u}{\partial \theta}\frac{\partial v}{\partial r}\right)+\left(\frac{\omega_1}{3}+\frac{2\omega_2}{9}-\frac{2\omega_6}{9}\right)\left(\frac{1}{r}\frac{\partial v}{\partial \theta}\frac{\partial w}{\partial z}\right)\\
&+\left(\frac{\omega_1}{3}+\frac{2\omega_2}{9}-\frac{2\omega_6}{9}\right)\left(\frac{1}{r}\frac{\partial u}{\partial r}\frac{\partial v}{\partial \theta}\right)-\left(\frac{2\omega_2}{3}+\frac{\omega_6}{6}\right)\left(\frac{1}{r}\frac{\partial w}{\partial \theta}\frac{\partial v}{\partial z}\right)\\
&+\frac{2\omega_3}{3}\frac{R}{r}\frac{\partial T}{\partial r}-\frac{\omega_3 R}{3}\frac{\partial^2 T}{\partial r^2}-\frac{\omega_3 R}{3}\frac{\partial^2 T}{\partial z^2}+\frac{2\omega_3}{3}\frac{R}{r^2}\frac{\partial^2 T}{\partial \theta^2}-\frac{\omega_4}{3}\frac{R}{p}\frac{\partial T}{\partial r}\frac{\partial p}{\partial r}\\
&-\frac{\omega_4}{3}\frac{R}{p}\frac{\partial T}{\partial z}\frac{\partial p}{\partial z}+\frac{2\omega_4}{3}\frac{1}{p}\frac{R}{r^2}\frac{\partial T}{\partial \theta}\frac{\partial p}{\partial \theta}-\frac{\omega_5}{3}\frac{R}{T}\left(\frac{\partial T}{\partial r}\right)^2-\frac{\omega_5}{3}\frac{R}{T}\left(\frac{\partial T}{\partial z}\right)^2\\
&+\frac{2\omega_5}{3}\frac{R}{T}\frac{1}{r^2}\left(\frac{\partial T}{\partial \theta}\right)^2+\left(\frac{\omega_1}{3}+\frac{2\omega_2}{9}-\frac{2\omega_6}{9}\right)\left(\frac{\partial w}{\partial z}\frac{u}{r}\right)
\end{aligned}
$$

$$
\begin{aligned}
&-\left(\frac{\omega_1}{3}-\frac{7\omega_2}{9}+\frac{\omega_6}{9}\right)\left(\frac{\partial w}{\partial z}\right)^2+\frac{\omega_2}{3}\frac{1}{\rho}\frac{\partial^2 p}{\partial r^2}+\frac{\omega_2}{3}\frac{1}{\rho}\frac{\partial^2 p}{\partial z^2}\\
&-\frac{2\omega_2}{3}\frac{1}{\rho}\frac{1}{r^2}\frac{\partial^2 p}{\partial \theta^2}-\frac{2\omega_2}{3}\frac{1}{\rho}\frac{1}{r}\frac{\partial p}{\partial r}\\
&-\frac{\omega_2}{3}\frac{1}{\rho^2}\frac{\partial p}{\partial r}\frac{\partial \rho}{\partial r}-\frac{\omega_2}{3}\frac{1}{\rho^2}\frac{\partial p}{\partial z}\frac{\partial \rho}{\partial z}+\frac{2\omega_2}{3}\frac{1}{r^2}\frac{1}{\rho^2}\frac{\partial p}{\partial \theta}\frac{\partial \rho}{\partial \theta}\Bigg]
\end{aligned}
\tag{5.169}
$$

$$
\begin{aligned}
p_{zz}^{Bu}=\frac{\mu^2}{p}\Bigg[&-\left(\frac{\omega_1}{3}-\frac{7\omega_2}{9}+\frac{1\omega_6}{9}\right)\left(\frac{\partial u}{\partial r}\right)^2-\left(\frac{2\omega_1}{3}+\frac{4\omega_2}{9}-\frac{4\omega_6}{9}\right)\left(\frac{u}{r}\frac{\partial u}{\partial r}\right)\\
&-\left(\frac{\omega_1}{3}-\frac{7\omega_2}{9}+\frac{\omega_6}{9}\right)\left(\frac{u}{r}\right)^2+\left(\frac{\omega_2}{3}-\frac{\omega_6}{6}\right)\left(\frac{1}{r^2}\right)\left(\frac{\partial u}{\partial \theta}\right)^2\\
&-\left(\frac{2\omega_2}{3}-\frac{\omega_6}{3}\right)\left(\frac{v}{r^2}\frac{\partial u}{\partial \theta}\right)-\left(\frac{2\omega_2}{3}-\frac{\omega_6}{12}\right)\left(\frac{\partial u}{\partial z}\right)^2\\
&+\left(\frac{\omega_2}{3}-\frac{\omega_6}{6}\right)\left(\frac{\partial v}{\partial r}\right)^2+\left(\frac{\omega_2}{3}-\frac{\omega_6}{6}\right)\left(\frac{v}{r}\right)^2\\
&-\left(\frac{\omega_1}{3}-\frac{7\omega_2}{9}+\frac{\omega_6}{9}\right)\left(\frac{1}{r}\frac{\partial v}{\partial \theta}\right)^2-\left(\frac{2\omega_1}{3}-\frac{14\omega_2}{9}+\frac{2\omega_6}{9}\right)\left(\frac{u}{r^2}\frac{\partial v}{\partial \theta}\right)\\
&-\left(\frac{2\omega_2}{3}-\frac{\omega_6}{12}\right)\left(\frac{\partial v}{\partial z}\right)^2+\left(\frac{\omega_2}{3}+\frac{\omega_6}{12}\right)\left(\frac{\partial w}{\partial r}\right)^2\\
&+\left(\frac{\omega_2}{3}+\frac{\omega_6}{12}\right)\left(\frac{1}{r}\frac{\partial w}{\partial \theta}\right)^2+\left(\frac{2\omega_1}{3}-\frac{14\omega_2}{9}+\frac{2\omega_6}{9}\right)\left(\frac{\partial w}{\partial z}\right)^2\\
&+\left(\frac{4\omega_2}{3}-\frac{\omega_6}{3}\right)\left(\frac{1}{r}\frac{\partial u}{\partial \theta}\frac{\partial v}{\partial r}\right)+\left(\frac{\omega_1}{3}+\frac{2\omega_2}{9}-\frac{2\omega_6}{9}\right)\frac{\partial w}{\partial z}\frac{\partial u}{\partial r}\\
&+\left(\frac{\omega_1}{3}+\frac{2\omega_2}{9}-\frac{2\omega_6}{9}\right)\left(\frac{1}{r}\frac{\partial v}{\partial \theta}\frac{\partial w}{\partial z}\right)+\left(\frac{-2\omega_2}{3}+\frac{\omega_6}{6}\right)\frac{\partial u}{\partial z}\frac{\partial w}{\partial r}\\
&-\left(\frac{2\omega_1}{3}+\frac{4\omega_2}{9}-\frac{4\omega_6}{9}\right)\left(\frac{1}{r}\frac{\partial u}{\partial r}\frac{\partial v}{\partial \theta}\right)+\left(\frac{\omega_2}{3}+\frac{\omega_6}{3}\right)\frac{v}{r}\frac{\partial v}{\partial r}\\
&+\left(\frac{2\omega_2}{3}+\frac{\omega_6}{6}\right)\left(\frac{1}{r}\frac{\partial w}{\partial \theta}\frac{\partial v}{\partial z}\right)+\frac{\omega_4}{3}\frac{1}{p}\frac{R}{r^2}\frac{\partial T}{\partial \theta}\frac{\partial p}{\partial \theta}-\frac{\omega_5}{3}\frac{R}{T}\left(\frac{\partial T}{\partial r}\right)^2\\
&+\frac{2\omega_5}{3}\frac{R}{T}\left(\frac{\partial T}{\partial z}\right)^2-\frac{\omega_5}{3}\frac{R}{T}\frac{1}{r^2}\left(\frac{\partial T}{\partial \theta}\right)^2+\left(\frac{\omega_1}{3}+\frac{2\omega_2}{9}-\frac{2\omega_6}{9}\right)\left(\frac{\partial w}{\partial z}\frac{u}{r}\right)\\
&+\frac{\omega_2}{3}\frac{1}{\rho}\frac{\partial^2 p}{\partial r^2}-\frac{2\omega_2}{3}\frac{1}{\rho}\frac{\partial^2 p}{\partial z^2}+\frac{\omega_2}{3}\frac{1}{r^2}\frac{1}{\rho}\frac{\partial^2 p}{\partial \theta^2}+\frac{\omega_2}{3}\frac{1}{r}\frac{1}{\rho}\frac{\partial p}{\partial r}-\frac{\omega_2}{3}\frac{1}{\rho^2}\frac{\partial p}{\partial r}\frac{\partial \rho}{\partial r}
\end{aligned}
$$

$$+\frac{2\omega_2}{3}\frac{1}{\rho^2}\frac{\partial p}{\partial z}\frac{\partial \rho}{\partial z}-\frac{\omega_2}{3}\frac{1}{r^2}\frac{1}{\rho^2}\frac{\partial p}{\partial \theta}\frac{\partial \rho}{\partial \theta}-\frac{\omega_3}{3}\frac{R}{r}\frac{\partial T}{\partial r}-\frac{\omega_3}{3}R\frac{\partial^2 T}{\partial r^2}$$

$$+\frac{2\omega_3 R}{3}\frac{\partial^2 T}{\partial z^2}-\frac{\omega_3}{3}\frac{R}{r^2}\frac{\partial^2 T}{\partial \theta^2}-\frac{\omega_4}{3}\frac{R}{p}\frac{\partial T}{\partial r}\frac{\partial p}{\partial r}+\frac{2\omega_4}{3}\frac{R}{p}\frac{\partial T}{\partial z}\frac{\partial p}{\partial z}\Bigg] \quad (5.170)$$

$$p_{r\theta}^{Bu}=p_{\theta r}^{Bu}=\frac{\mu^2}{p}\Bigg[-\left(\frac{\omega_1}{2}-\frac{5\omega_2}{3}+\frac{\omega_6}{6}\right)\left(\frac{v}{r}\frac{\partial u}{\partial r}\right)$$

$$+\left(\frac{\omega_1}{2}-\frac{2\omega_2}{3}+\frac{\omega_6}{6}\right)\left(\frac{u}{r^2}\frac{\partial u}{\partial \theta}\right)-\left(\frac{\omega_1}{2}-\frac{2\omega_2}{3}-\frac{\omega_6}{6}\right)\left(\frac{uv}{r^2}\right)$$

$$+\left(\frac{\omega_1}{2}-\frac{5\omega_2}{3}+\frac{\omega_6}{6}\right)\left(\frac{u}{r}\frac{\partial v}{\partial r}\right)-\left(\frac{\omega_1}{2}-\frac{2\omega_2}{3}+\frac{\omega_6}{6}\right)\left(\frac{v}{r^2}\frac{\partial v}{\partial \theta}\right)$$

$$-\left(\frac{\omega_1}{2}+\frac{\omega_2}{3}-\frac{\omega_6}{3}\right)\left(\frac{v}{r}\frac{\partial w}{\partial z}\right)+\left(\frac{\omega_1}{2}+\frac{\omega_2}{3}-\frac{\omega_6}{3}\right)\left(\frac{1}{r}\frac{\partial u}{\partial \theta}\frac{\partial w}{\partial z}\right)$$

$$+\left(\frac{\omega_1}{2}-\frac{5\omega_2}{3}+\frac{\omega_6}{6}\right)\left(\frac{1}{r}\frac{\partial u}{\partial r}\frac{\partial u}{\partial \theta}\right)+\left(\frac{\omega_1}{2}+\frac{\omega_2}{3}-\frac{\omega_6}{3}\right)\left(\frac{\partial v}{\partial r}\frac{\partial w}{\partial z}\right)$$

$$+\left(\frac{\omega_1}{2}-\frac{2\omega_2}{3}+\frac{\omega_6}{6}\right)\left(\frac{\partial u}{\partial r}\frac{\partial v}{\partial r}\right)+\left(\frac{\omega_1}{2}-\frac{2\omega_2}{3}+\frac{\omega_6}{6}\right)$$

$$\times\left(\frac{1}{r^2}\frac{\partial u}{\partial \theta}\frac{\partial v}{\partial \theta}\right)+\left(\frac{\omega_1}{2}-\frac{5\omega_2}{3}+\frac{\omega_6}{6}\right)\left(\frac{1}{r}\frac{\partial v}{\partial r}\frac{\partial v}{\partial \theta}\right)$$

$$-\left(\omega_2-\frac{\omega_6}{4}\right)\left(\frac{\partial v}{\partial z}\frac{\partial w}{\partial r}\right)-\left(\omega_2-\frac{\omega_6}{4}\right)\left(\frac{1}{r}\frac{\partial u}{\partial z}\frac{\partial w}{\partial \theta}\right)-\left(\omega_2-\frac{\omega_6}{4}\right)$$

$$\times\left(\frac{1}{r}\frac{\partial w}{\partial r}\frac{\partial w}{\partial \theta}\right)+\omega_3\frac{R}{r}\frac{\partial^2 T}{\partial r\partial \theta}-\omega_3\frac{R}{r^2}\frac{\partial T}{\partial \theta}+\frac{\omega_4}{2}\frac{R}{r}\frac{1}{p}\frac{\partial p}{\partial \theta}\frac{\partial T}{\partial r}$$

$$+\frac{\omega_4}{2}\frac{1}{r}\frac{R}{p}\frac{\partial p}{\partial r}\frac{\partial T}{\partial \theta}+\omega_5\frac{1}{T}\frac{R}{r}\frac{\partial T}{\partial r}\frac{\partial T}{\partial \theta}+\frac{\omega_6}{4}\frac{\partial u}{\partial z}\frac{\partial v}{\partial z}+\omega_2\frac{1}{r^2}\frac{1}{\rho}\frac{\partial p}{\partial \theta}$$

$$-\omega_2\frac{1}{r}\frac{1}{\rho}\frac{\partial^2 p}{\partial r\partial \theta}+\frac{\omega_2}{2}\frac{1}{r}\frac{1}{\rho^2}\frac{\partial p}{\partial \theta}\frac{\partial \rho}{\partial r}+\frac{\omega_2}{2}\frac{1}{r}\frac{1}{\rho^2}\frac{\partial p}{\partial r}\frac{\partial \rho}{\partial \theta}\Bigg] \quad (5.171)$$

$$p_{\theta z}^{Bu}=p_{z\theta}^{Bu}=\frac{\mu^2}{p}\Bigg[\left(\omega_2-\frac{\omega_6}{4}\right)\left(\frac{v}{r}\frac{\partial u}{\partial z}\right)+\left(\frac{\omega_1}{2}-\frac{5\omega_2}{3}+\frac{\omega_6}{6}\right)\left(\frac{u}{r}\frac{\partial v}{\partial z}\right)$$

$$+\left(\omega_2-\frac{\omega_6}{4}\right)\left(\frac{v}{r}\frac{\partial w}{\partial r}\right)+\left(\frac{\omega_1}{2}-\frac{2\omega_2}{3}+\frac{\omega_6}{6}\right)\left(\frac{u}{r^2}\frac{\partial w}{\partial \theta}\right)$$

$$+\left(\frac{\omega_1}{2}-\frac{2\omega_2}{3}+\frac{\omega_6}{6}\right)\left(\frac{\partial v}{\partial z}\frac{\partial w}{\partial z}\right)+\left(\frac{\omega_1}{2}+\frac{\omega_2}{3}-\frac{\omega_6}{3}\right)\left(\frac{\partial u}{\partial r}\frac{\partial v}{\partial z}\right)$$

$$-\left(\omega_2-\frac{\omega_6}{4}\right)\left(\frac{1}{r}\frac{\partial u}{\partial\theta}\frac{\partial u}{\partial z}\right)-\left(\omega_2-\frac{\omega_6}{4}\right)\left(\frac{\partial u}{\partial z}\frac{\partial v}{\partial r}\right)$$
$$+\left(\frac{\omega_1}{2}-\frac{5\omega_2}{3}+\frac{\omega_6}{6}\right)\left(\frac{1}{r}\frac{\partial v}{\partial z}\frac{\partial v}{\partial\theta}\right)-\left(\omega_2-\frac{\omega_6}{4}\right)\left(\frac{1}{r}\frac{\partial u}{\partial\theta}\frac{\partial w}{\partial r}\right)$$
$$+\frac{\omega_6}{4}\frac{\partial v}{\partial r}\frac{\partial w}{\partial r}+\left(\frac{\omega_1}{2}-\frac{5\omega_2}{3}+\frac{\omega_6}{6}\right)\left(\frac{1}{r}\frac{\partial w}{\partial z}\frac{\partial w}{\partial\theta}\right)$$
$$+\left(\frac{\omega_1}{2}+\frac{\omega_2}{3}-\frac{\omega_6}{3}\right)\left(\frac{1}{r}\frac{\partial u}{\partial r}\frac{\partial w}{\partial\theta}\right)+\left(\frac{\omega_1}{2}-\frac{2\omega_2}{3}+\frac{\omega_6}{6}\right)$$
$$\times\left(\frac{1}{r^2}\frac{\partial v}{\partial\theta}\frac{\partial w}{\partial\theta}\right)+\omega_3\frac{R}{r}\frac{\partial^2 T}{\partial\theta\partial z}+\frac{\omega_4}{2}\frac{1}{r}\frac{R}{p}\frac{\partial T}{\partial z}\frac{\partial p}{\partial\theta}+\frac{\omega_4}{2}\frac{1}{r}\frac{R}{p}\frac{\partial T}{\partial\theta}\frac{\partial p}{\partial z}$$
$$+\omega_5\frac{1}{T}\frac{R}{r}\frac{\partial T}{\partial z}\frac{\partial T}{\partial\theta}-\omega_2\frac{1}{r}\frac{1}{\rho}\frac{\partial^2 p}{\partial\theta\partial z}+\frac{\omega_2}{2}\frac{1}{r}\frac{1}{\rho^2}\frac{\partial p}{\partial\theta}\frac{\partial\rho}{\partial z}$$
$$\left.+\frac{\omega_2}{2}\frac{1}{r}\frac{1}{\rho^2}\frac{\partial p}{\partial z}\frac{\partial\rho}{\partial\theta}\right] \tag{5.172}$$

$$p_{rz}^{Bu}=p_{zr}^{Bu}=\frac{\mu^2}{p}\left[\left(\frac{\omega_1}{2}+\frac{\omega_2}{3}-\frac{\omega_6}{3}\right)\left(\frac{u}{r}\frac{\partial u}{\partial z}\right)+\left(\omega_2-\frac{\omega_6}{4}\right)\left(\frac{v}{r}\frac{\partial v}{\partial z}\right)\right.$$
$$+\left(\frac{\omega_1}{2}+\frac{\omega_2}{3}-\frac{\omega_6}{3}\right)\left(\frac{u}{r}\frac{\partial w}{\partial r}\right)+\left(\frac{\omega_1}{2}-\frac{2\omega_2}{3}+\frac{\omega_6}{6}\right)\left(\frac{\partial u}{\partial z}\frac{\partial w}{\partial z}\right)$$
$$+\left(\frac{\omega_1}{2}-\frac{5\omega_2}{3}+\frac{\omega_6}{6}\right)\left(\frac{\partial u}{\partial r}\frac{\partial u}{\partial z}\right)-\left(\omega_2-\frac{\omega_6}{4}\right)\left(\frac{1}{r}\frac{\partial u}{\partial\theta}\frac{\partial v}{\partial z}\right)$$
$$-\left(\omega_2-\frac{\omega_6}{4}\right)\left(\frac{\partial v}{\partial z}\frac{\partial v}{\partial r}\right)+\left(\frac{\omega_1}{2}+\frac{\omega_2}{3}-\frac{\omega_6}{3}\right)\left(\frac{1}{r}\frac{\partial u}{\partial z}\frac{\partial v}{\partial\theta}\right)$$
$$+\left(\frac{\omega_1}{2}-\frac{5\omega_2}{3}+\frac{\omega_6}{6}\right)\left(\frac{\partial w}{\partial z}\frac{\partial w}{\partial r}\right)+\left(\frac{\omega_1}{2}-\frac{2\omega_2}{3}+\frac{\omega_6}{6}\right)$$
$$\times\left(\frac{\partial u}{\partial r}\frac{\partial w}{\partial r}\right)+\left(\frac{\omega_1}{2}+\frac{\omega_2}{3}-\frac{\omega_6}{3}\right)\left(\frac{1}{r}\frac{\partial v}{\partial\theta}\frac{\partial w}{\partial r}\right)-\frac{\omega_6}{4}\frac{v}{r^2}\frac{\partial w}{\partial\theta}$$
$$+\frac{\omega_6}{4}\frac{1}{r^2}\frac{\partial u}{\partial\theta}\frac{\partial w}{\partial\theta}-\left(\omega_2-\frac{\omega_6}{4}\right)\left(\frac{1}{r}\frac{\partial v}{\partial r}\frac{\partial w}{\partial\theta}\right)+\omega_3 R\frac{\partial^2 T}{\partial r\partial z}$$
$$+\frac{\omega_4}{2}\frac{R}{p}\frac{\partial T}{\partial r}\frac{\partial p}{\partial z}+\frac{\omega_4}{2}\frac{R}{p}\frac{\partial p}{\partial r}\frac{\partial T}{\partial z}+\omega_5\frac{R}{T}\frac{\partial T}{\partial r}\frac{\partial T}{\partial z}-\omega_2\frac{1}{\rho}\frac{\partial^2 p}{\partial r\partial z}$$
$$\left.+\frac{\omega_2}{2}\frac{1}{\rho^2}\frac{\partial p}{\partial z}\frac{\partial\rho}{\partial r}+\frac{\omega_2}{2}\frac{1}{\rho^2}\frac{\partial p}{\partial r}\frac{\partial\rho}{\partial z}\right] \tag{5.173}$$

5.5.2.2 Burnett Heat Flux Terms

In this section, we document the three components of the heat flux vector. As shown in Sect. 5.6, these terms are of order Knudsen square as also the stress tensor terms in the previous section.

$$
\begin{aligned}
q_r^{Bu} = \frac{\mu^2}{\rho}\Bigg[& -\left(\frac{2\theta_2}{3}-\frac{2\theta_4}{3}\right)\left(\frac{\partial^2 u}{\partial r^2}\right)+\frac{\theta_4}{2}\frac{\partial^2 u}{\partial z^2}+\frac{\theta_4}{2}\frac{1}{r^2}\frac{\partial^2 u}{\partial \theta^2} \\
& -\left(\frac{2\theta_2}{3}-\frac{2\theta_4}{3}\right)\left(\frac{1}{r}\frac{\partial u}{\partial r}\right)+\left(\frac{2\theta_2}{3}-\frac{2\theta_4}{3}\right)\left(\frac{u}{r^2}\right)-\left(\frac{2\theta_2}{3}-\frac{\theta_4}{6}\right) \\
& \times\left(\frac{1}{r}\frac{\partial^2 v}{\partial r\partial\theta}\right)+\left(\frac{2\theta_2}{3}-\frac{7\theta_4}{6}\right)\left(\frac{1}{r^2}\frac{\partial v}{\partial \theta}\right)-\left(\frac{2\theta_2}{3}-\frac{\theta_4}{6}\right)\left(\frac{\partial^2 w}{\partial r\partial z}\right) \\
& +\left(\theta_1-\frac{2\theta_2}{3}-\theta_5\right)\left(\frac{u}{r}\frac{1}{T}\frac{\partial T}{\partial r}\right)-\frac{3\theta_5}{2}\frac{v}{r^2}\frac{1}{T}\frac{\partial T}{\partial \theta}-\frac{\theta_3}{3}\frac{u}{r}\frac{1}{p}\frac{\partial p}{\partial r}-\frac{\theta_3}{2}\frac{v}{r^2}\frac{1}{p}\frac{\partial p}{\partial \theta} \\
& +\left(\theta_1-\frac{8\theta_2}{3}+2\theta_5\right)\left(\frac{1}{T}\frac{\partial u}{\partial r}\frac{\partial T}{\partial r}\right)+\frac{3\theta_5}{2}\frac{1}{r^2}\frac{1}{T}\frac{\partial u}{\partial \theta}\frac{\partial T}{\partial \theta}+\frac{3\theta_5}{2}\frac{1}{T}\frac{\partial u}{\partial z}\frac{\partial T}{\partial z} \\
& +\frac{\theta_3}{2}\frac{1}{p}\frac{\partial u}{\partial z}\frac{\partial p}{\partial z}+\frac{\theta_3}{2}\frac{1}{r^2}\frac{1}{p}\frac{\partial u}{\partial \theta}\frac{\partial p}{\partial \theta}-\frac{\theta_3}{3}\frac{1}{p}\frac{\partial w}{\partial z}\frac{\partial p}{\partial r} \\
& +\left(\theta_1-2\theta_2+\frac{3\theta_5}{2}\right)\left(\frac{1}{r}\frac{1}{T}\frac{\partial v}{\partial r}\frac{\partial T}{\partial \theta}\right)-\frac{\theta_3}{3}\frac{1}{r}\frac{1}{RT}\frac{\partial v}{\partial \theta}\frac{\partial p}{\partial r} \\
& +\left(\theta_1-\frac{2\theta_2}{3}-\theta_5\right)\frac{1}{r}\frac{1}{T}\frac{\partial T}{\partial r}\frac{\partial v}{\partial \theta}-\left(2\theta_2-\frac{3\theta_5}{2}\right)\left(\frac{1}{T}\frac{\partial w}{\partial r}\frac{\partial T}{\partial z}\right) \\
& +\frac{\theta_3}{2}\frac{1}{p}\frac{\partial w}{\partial r}\frac{\partial p}{\partial z}+\frac{1}{pr}\frac{\partial p}{\partial \theta}\frac{\partial v}{\partial r}\frac{\theta_3}{2}+\left(\theta_1-\frac{2\theta_2}{3}-\theta_5\right)\left(\frac{1}{T}\frac{\partial w}{\partial z}\frac{\partial T}{\partial r}\right) \\
& +\frac{1}{p}\frac{\partial p}{\partial r}\frac{\partial u}{\partial r}\frac{2\theta_3}{3}\Bigg]
\end{aligned}
\tag{5.174}
$$

$$
\begin{aligned}
q_\theta^{Bu} = \frac{\mu^2}{\rho}\Bigg[& \left(\frac{2\theta_2}{3}+\frac{7\theta_4}{6}\right)\left(\frac{1}{r^2}\frac{\partial u}{\partial \theta}\right)+\frac{\theta_4}{2}\frac{\partial^2 v}{\partial z^2}+\frac{\theta_4}{2}\frac{\partial^2 v}{\partial r^2} \\
& -\left(\frac{2\theta_2}{3}-\frac{2\theta_4}{3}\right)\left(\frac{1}{r^2}\frac{\partial^2 v}{\partial \theta^2}\right)+\frac{\theta_4}{2}\frac{1}{r}\frac{\partial v}{\partial r}+\frac{\theta_3}{2}\frac{1}{p}\frac{\partial v}{\partial r}\frac{\partial p}{\partial r}+\frac{\theta_4}{2}\frac{v}{r^2} \\
& +\left(\theta_1-\frac{2\theta_2}{3}-\theta_5\right)\left(\frac{u}{r^2}\frac{1}{T}\frac{\partial T}{\partial \theta}\right)-\frac{\theta_3}{2}\frac{v}{r}\frac{1}{p}\frac{\partial p}{\partial r} \\
& +\frac{2\theta_3}{3}\frac{u}{r^2}\frac{1}{p}\frac{\partial p}{\partial \theta}-\left(\frac{2\theta_2}{3}-\frac{\theta_4}{6}\right)\left(\frac{\partial^2 u}{\partial r\partial\theta}\right)+\left(\theta_1-\frac{2\theta_2}{3}-\theta_5\right)\left(\frac{1}{r}\frac{1}{T}\frac{\partial u}{\partial r}\frac{\partial T}{\partial \theta}\right)
\end{aligned}
$$

$$
\begin{aligned}
&-\frac{\theta_3}{3}\frac{1}{r}\frac{1}{p}\frac{\partial u}{\partial r}\frac{\partial p}{\partial \theta}+\frac{\theta_3}{2}\frac{1}{r}\frac{1}{p}\frac{\partial u}{\partial \theta}\frac{\partial p}{\partial r}+\frac{3\theta_5}{2}\frac{1}{T}\frac{\partial v}{\partial r}\frac{\partial T}{\partial r}+\frac{3\theta_5}{2}\frac{1}{T}\frac{\partial v}{\partial z}\frac{\partial T}{\partial z}\\
&+\frac{\theta_3}{2}\frac{1}{p}\frac{\partial v}{\partial z}\frac{\partial p}{\partial z}+\frac{2\theta_3}{3}\frac{1}{r^2}\frac{1}{RT}\frac{\partial v}{\partial \theta}\frac{\partial p}{\partial \theta}+\left(\theta_1-\frac{8\theta_2}{3}+2\theta_5\right)\left(\frac{1}{r^2}\frac{1}{T}\frac{\partial v}{\partial \theta}\frac{\partial T}{\partial \theta}\right)\\
&-\left(\frac{2\theta_2}{3}-\frac{\theta_4}{6}\right)\left(\frac{1}{r}\frac{\partial^2 w}{\partial z\partial \theta}\right)+\left(\theta_1-\frac{2\theta_2}{3}-\theta_5\right)\frac{1}{r}\frac{1}{T}\frac{\partial w}{\partial z}\frac{\partial T}{\partial \theta}\\
&-\left(2\theta_2-\frac{3\theta_5}{2}\right)\left(\frac{1}{r}\frac{1}{T}\frac{\partial w}{\partial \theta}\frac{\partial T}{\partial z}\right)-\frac{\theta_3}{3}\frac{1}{p}\frac{1}{r}\frac{\partial w}{\partial z}\frac{\partial p}{\partial \theta}+\frac{v}{r}\frac{1}{T}\frac{\partial T}{\partial r}\left(\frac{-3\theta_5}{2}+2\theta_2\right)\\
&+\left(-2\theta_2+\frac{3\theta_5}{2}\right)\frac{1}{r}\frac{1}{T}\frac{\partial T}{\partial r}\frac{\partial u}{\partial \theta}\Bigg]
\end{aligned}
\tag{5.175}
$$

$$
\begin{aligned}
q_z^{Bu}=\frac{\mu^2}{\rho}\Bigg[&-\left(\frac{2\theta_2}{3}-\frac{\theta_4}{6}\right)\left(\frac{1}{r}\frac{\partial u}{\partial z}\right)+\frac{\theta_4}{2}\frac{\partial^2 w}{\partial r^2}-\left(\frac{2\theta_2}{3}-\frac{2\theta_4}{3}\right)\left(\frac{\partial^2 w}{\partial z^2}\right)\\
&+\frac{\theta_4}{2}\frac{1}{r^2}\frac{\partial^2 w}{\partial \theta^2}+\left(\theta_1-\frac{2\theta_2}{3}-\theta_5\right)\left(\frac{u}{r}\frac{1}{T}\frac{\partial T}{\partial z}\right)-\frac{\theta_3}{3}\frac{u}{r}\frac{1}{p}\frac{\partial p}{\partial z}\\
&+\left(\theta_1-\frac{2\theta_2}{3}-\theta_5\right)\left(\frac{1}{T}\frac{\partial u}{\partial r}\frac{\partial T}{\partial z}\right)+\left(-2\theta_2+\frac{3\theta_5}{2}\right)\frac{1}{r}\frac{1}{T}\frac{\partial u}{\partial z}\frac{\partial T}{\partial r}\\
&-\frac{\theta_3}{3}\frac{1}{p}\frac{\partial u}{\partial r}\frac{\partial p}{\partial z}-\left(\frac{2\theta_2}{3}-\frac{\theta_4}{6}\right)\left(\frac{\partial^2 u}{\partial r\partial z}\right)+\frac{\theta_3}{2}\frac{1}{p}\frac{\partial u}{\partial z}\frac{\partial p}{\partial r}\\
&-\left(\frac{2\theta_2}{3}-\frac{\theta_4}{6}\right)\left(\frac{1}{r}\frac{\partial^2 v}{\partial z\partial \theta}\right)+\frac{\theta_3}{2}\frac{1}{r}\frac{1}{p}\frac{\partial v}{\partial z}\frac{\partial p}{\partial \theta}\\
&+\left(\theta_1-\frac{2\theta_2}{3}-\theta_5\right)\left(\frac{1}{r}\frac{1}{T}\frac{\partial v}{\partial \theta}\frac{\partial T}{\partial z}\right)-\frac{\theta_3}{3}\frac{1}{p}\frac{1}{r}\frac{\partial v}{\partial \theta}\frac{\partial p}{\partial z}+\frac{3\theta_5}{2}\frac{1}{T}\frac{\partial w}{\partial r}\frac{\partial T}{\partial r}\\
&+\frac{\theta_3}{2}\frac{1}{p}\frac{\partial w}{\partial r}\frac{\partial p}{\partial r}+\left(\theta_1-\frac{8\theta_2}{3}+2\theta_5\right)\left(\frac{1}{T}\frac{\partial w}{\partial z}\frac{\partial T}{\partial z}\right)+\frac{2\theta_3}{3}\frac{1}{p}\frac{\partial w}{\partial z}\frac{\partial p}{\partial z}\\
&+\frac{\theta_3}{2}\frac{1}{r^2}\frac{1}{p}\frac{\partial w}{\partial \theta}\frac{\partial p}{\partial \theta}+\frac{3\theta_5}{2}\frac{1}{r^2}\frac{1}{T}\frac{\partial w}{\partial \theta}\frac{\partial T}{\partial \theta}\\
&+\frac{\theta_4}{2}\frac{1}{r}\frac{\partial w}{\partial r}+\frac{1}{r}\frac{1}{T}\frac{\partial v}{\partial z}\frac{\partial v}{\partial \theta}\left(\frac{3\theta_5}{2}-2\theta_2\right)\Bigg]
\end{aligned}
\tag{5.176}
$$

The above equations involve 11 coefficients—six ω's and five θ's. The values of these coefficients for both Maxwell and hard sphere molecules are tabulated in Table 5.1.

The above Navier–Stokes and Burnett-order stress and heat flux terms when substituted into the conservation equations (Eqs. (5.152)–(5.156)) give the complete Burnett equation in cylindrical coordinates.

5.6 Order of Magnitude Analysis

An order of magnitude analysis can provide important insights about the various terms in an equation. Such an analysis can be utilized to remove the less significant terms from the equations, thereby simplifying the process of analysis. Specifically, we will examine the various terms in the Burnett-order stress tensor and heat flux vector, and compare them with the corresponding terms of the Navier–Stokes order.

We select U, ρ, and L as the velocity, density, and length scale, respectively. Pressure difference in a flow scales with ρU^2, while from ideal gas law, pressure scales as ρc^2 (where c is the speed of sound). In an analogous manner to pressure, the temperature difference scales as $\rho U^2/R$ while the absolute temperature scales as $\rho c^2/R$ where R is the gas constant.

Normalizing stress terms by ρU^2 and heat flux terms by ρU^3, we obtain

$$\frac{p_{ij}^{NS}}{\rho U^2} = -\frac{2\mu}{\rho U^2}\frac{\partial u_{\langle i}}{\partial x_{j\rangle}} \sim \frac{\mu}{\rho U^2}\frac{U}{L} = \frac{1}{Re} \sim \frac{Kn}{Ma}$$

$$\frac{q_i^{NS}}{\rho U^3} = -\frac{k}{\rho U^3}\frac{\partial T}{\partial x_i} \sim \frac{k}{\rho U^3}\frac{\rho U^2}{RL} \sim \frac{1}{Re Pr} \sim \frac{Kn}{Ma}$$

where Pr is Prandtl number, which is order unity for gases. It is therefore seen that both p_{ij}^{NS} and q_i^{NS} are of order Kn.

Repeating the above exercise for the various terms in Eq. (5.146) other than the external force term, we obtain

$$\frac{p_{ij,1}^{Bu}}{\rho U^2} = \frac{\omega_1}{\rho U^2}\frac{\mu^2}{p}\frac{\partial u_k}{\partial x_k} S_{ij} \sim \frac{\mu^2}{\rho U^2 \rho c^2}\frac{U^2}{L^2} = \frac{Ma^2}{Re^2} \sim Kn^2$$

$$\frac{p_{ij,2}^{Bu}}{\rho U^2} = -\frac{\omega_2}{\rho U^2}\frac{\mu^2}{p}\frac{\partial}{\partial x_{\langle i}}\left[\frac{1}{\rho}\frac{\partial p}{\partial x_{j\rangle}}\right] \sim \frac{\mu^2}{\rho U^2 \rho c^2}\frac{\rho U^2}{\rho L^2} = \frac{Ma^2}{Re^2} \sim Kn^2$$

$$\frac{p_{ij,3}^{Bu}}{\rho U^2} = -\frac{\omega_2}{\rho U^2}\frac{\mu^2}{p}\frac{\partial u_k}{\partial x_{\langle i}}\frac{\partial u_{j\rangle}}{\partial x_k} \sim \frac{\mu^2}{\rho U^2 \rho c^2}\frac{U^2}{L^2} = \frac{Ma^2}{Re^2} \sim Kn^2$$

$$\frac{p_{ij,4}^{Bu}}{\rho U^2} = -\frac{\omega_2}{\rho U^2}\frac{2\mu^2}{p}\frac{\partial u_k}{\partial x_{\langle i}} S_{j\rangle k} \sim \frac{\mu^2}{\rho U^2 \rho c^2}\frac{U^2}{L^2} = \frac{Ma^2}{Re^2} \sim Kn^2$$

$$\frac{p_{ij,5}^{Bu}}{\rho U^2} = \frac{\omega_3}{\rho U^2}\frac{\mu^2}{\rho T}\frac{\partial^2 T}{\partial x_{\langle i}\partial x_{j\rangle}} \sim \frac{\mu^2}{\rho U^2 \rho c^2}\frac{U^2}{L^2} = \frac{Ma^2}{Re^2} \sim Kn^2$$

$$\frac{p_{ij,6}^{Bu}}{\rho U^2} = \frac{\omega_4}{\rho U^2}\frac{\mu^2}{p\rho T}\frac{\partial T}{\partial x_{\langle i}}\frac{\partial p}{\partial x_{j\rangle}} \sim \frac{\mu^2}{\rho U^2 \rho c^2}\frac{\rho U^4}{\rho c^2 L^2} = \frac{Ma^4}{Re^2} \sim Ma^2 Kn^2$$

$$\frac{p_{ij,7}^{Bu}}{\rho U^2} = \frac{\omega_5}{\rho U^2}\frac{\mu^2}{\rho T^2}\frac{\partial T}{\partial x_{\langle i}}\frac{\partial T}{\partial x_{j\rangle}} \sim \frac{\mu^2}{\rho U^2 \rho c^2}\frac{U^4}{c^2 L^2} = \frac{Ma^4}{Re^2} \sim Ma^2 Kn^2$$

$$\frac{p_{ij,8}^{Bu}}{\rho U^2} = \frac{\omega_6}{\rho U^2}\frac{\mu^2}{p} S_{k\langle i}S_{j\rangle k} \sim \frac{\mu^2}{\rho U^2 \rho c^2}\frac{U^2}{L^2} = \frac{Ma^2}{Re^2} \sim Kn^2$$

Similarly, repeating the above exercise for the various terms in Eq. (5.147), we obtain

$$\frac{q_{i,1}^{Bu}}{\rho U^3} = \frac{\theta_1}{\rho U^3}\frac{\mu^2}{\rho T}\frac{\partial u_k}{\partial x_k}\frac{\partial T}{\partial x_i} \sim \frac{\mu^2}{\rho^2 U^3}\frac{U^3}{c^2 L^2} = \frac{Ma^2}{Re^2} \sim Kn^2$$

We decompose the first term in the curly braces of Eq. (5.147) into two terms

$$-\frac{2}{3}\frac{\partial}{\partial x_i}\left(T\frac{\partial u_k}{\partial x_k}\right) = -\frac{2}{3}T\frac{\partial^2 u_k}{\partial x_i x_k} - \frac{2}{3}\frac{\partial T}{\partial x_i}\frac{\partial u_k}{\partial x_k}$$

and proceed as

$$\frac{q_{i,2}^{Bu}}{\rho U^3} = -\frac{\theta_2}{\rho U^3}\frac{2}{3}\frac{\mu^2}{\rho}\frac{\partial^2 u_k}{\partial x_i \partial x_k} \sim \frac{\mu^2}{\rho^2 U^3}\frac{U}{L^2} = \frac{1}{Re^2} \sim \frac{Kn^2}{Ma^2}$$

$$\frac{q_{i,3}^{Bu}}{\rho U^3} = -\frac{\theta_2}{\rho U^3}\frac{2}{3}\frac{\mu^2}{\rho T}\frac{\partial u_k}{\partial x_k}\frac{\partial T}{\partial x_i} \sim \frac{\mu^2}{\rho^2 U^3}\frac{U^3}{c^2 L^2} = \frac{Ma^2}{Re^2} \sim Kn^2$$

$$\frac{q_{i,4}^{Bu}}{\rho U^3} = -2\frac{\theta_2}{\rho U^3}\frac{\mu^2}{\rho T}\frac{\partial u_k}{\partial x_i}\frac{\partial T}{\partial x_k} \sim \frac{\mu^2}{\rho^2 U^3}\frac{U^3}{c^2 L^2} = \frac{Ma^2}{Re^2} \sim Kn^2$$

$$\frac{q_{i,5}^{Bu}}{\rho U^3} = \frac{\theta_3}{\rho U^3}\frac{\mu^2}{\rho p} S_{ik}\frac{\partial p}{\partial x_k} \sim \frac{\mu^2}{\rho^2 U^3}\frac{\rho U^3}{\rho c^2 L^2} = \frac{Ma^2}{Re^2} \sim Kn^2$$

$$\frac{q_{i,6}^{Bu}}{\rho U^3} = \frac{\theta_4}{\rho U^3}\frac{\mu^2}{\rho}\frac{\partial S_{ik}}{\partial x_k} \sim \frac{\mu^2}{\rho^2 U^3}\frac{U}{L^2} = \frac{1}{Re^2} \sim \frac{Kn^2}{Ma^2}$$

$$\frac{q_{i,7}^{Bu}}{\rho U^3} = 3\frac{\theta_5}{\rho U^3}\frac{\mu^2}{\rho T} S_{ik}\frac{\partial T}{\partial x_k} \sim \frac{\mu^2}{\rho^2 U^3}\frac{U^3}{c^2 L^2} = \frac{Ma^2}{Re^2} \sim Kn^2$$

The above exercise clearly shows that all the Burnett terms are of order Kn^2. The stress terms in Eq. (5.146) contain eights terms, of which six terms are of $O(Kn^2)$ while two terms are of $O(Ma^2 Kn^2)$. Therefore, one can neglect the two terms of $O(Ma^2 Kn^2)$ at lower Mach numbers. The heat flux terms in Eq. (5.147) contain seven terms, of which five terms are of $O(Kn^2)$ while two terms are of $O(Kn^2/Ma^2)$. Therefore, one can neglect the two terms of $O(Kn^2/Ma^2)$ at higher values of Mach numbers.

5.7 Some Exact Solutions

Given the complex nature of the equations, it is not easy to find an analytical solution of the Burnett equations. Additionally, since the Burnett equations involve third-order derivatives, its solution requires additional boundary conditions. Prescribing the boundary conditions accurately is crucial, as the solution strongly depends on the boundary conditions. However, finding these additional boundary conditions is not easy and is therefore an additional impediment in obtaining analytical solution of the Burnett equations. Only recently, the Burnett equations have been solved analytically using the two approaches as given below.

5.7.1 *Iterative Approach of Agrawal and Singh*

It is apparent from Eqs. (5.157)–(5.158) that the Burnett equations is a superset of the Navier–Stokes equations. That is, all the terms of the Navier–Stokes equations are contained in the Burnett equations, along with additional terms being present in the latter equation. This observation was advantageously used by Agrawal and Singh to "build-up" solution of the Burnett equations starting from the solution of the Navier–Stokes equations for two different problems.

The primary steps of their approach are as follows (Singh et al. [132]—see also Fig. 5.6)

(i) Start with the solution of the Navier–Stokes equations and assume that this is a solution of the Burnett equations. (The approach is applicable only for cases for which the Navier–Stokes solution is available.)
(ii) Substitute this preliminary solution in the Burnett equations and evaluate the order of magnitude of various terms in the Burnett equations. (The fact that explicit expressions for density, velocity, pressure, etc. are available allows determination of the Burnett-order stress terms, which are essentially product of the gradients of these quantities.)
(iii) Compare the magnitude of each term with the highest-order term in the governing equations from which the preliminary solution has been obtained. Therefore, the significance of each of the additional term from the Burnett equations can be gauged, and allows identification of term(s) of significance. (For a physical problem with specified dimensions of the flow passage and gas properties, the magnitude of each term can be obtained and compared against the term in the equation with the highest magnitude.)
(iv) If the additional term(s) are significant, augment the governing equations with this additional term(s), and re-solve the augmented governing equations. (The approach works only if the intermediate equation obtained after augmenting the Navier–Stokes with the additional terms identified in this step can still be solved analytically. The order of the equation can also go up at this stage if

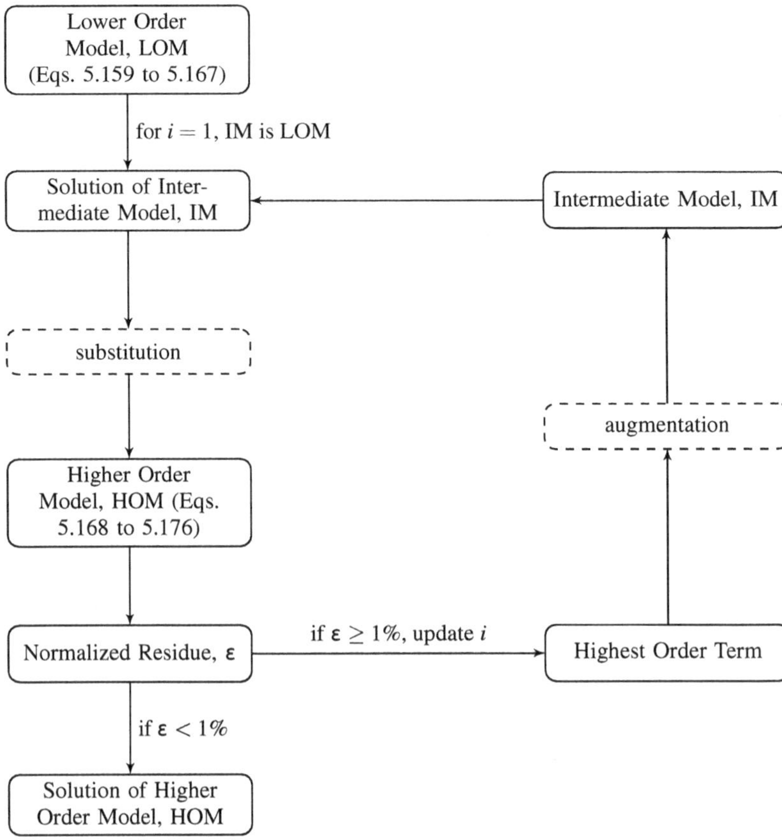

Fig. 5.6 Procedure for obtaining the solution of Burnett equations starting from the solution of Navier–Stokes equations, Singh and Agrawal [130]

a third-order derivative gets added to the governing equations, necessitating specification of additional boundary conditions.)

(v) Repeat steps (ii)–(iv) above till convergence is achieved. Convergence can be achieved either when no additional term remains in the Burnett equations or the remaining terms are insignificant. In the latter case, the remaining terms can be neglected and convergence is deemed to have been achieved.

Although the above procedure has been given in the context of Burnett and Navier–Stokes equations, it is general enough to be applicable to other non-linear partial differential equations, although the truth of this statement has to be established by applying it other equations.

Note that the procedure given above has to be applied at the point where the solution of the Burnett equations is expected to be substantially different than the solution for the Navier–Stokes equations, i.e., close to the wall. In practice, it is recommended to apply the above procedure at several points in the flow field.

Notice that per the above approach, additional boundary condition required for the solution of the Burnett equations is deferred till a third-order term gets added to the governing equation. That is, the procedure allows for a preliminary solution to be formulated, and defers prescription of the additional boundary conditions to the point that a third-order term gets added to the governing equations.

The above procedure was successfully applied by Singh et al. [132] to isothermal flow in a microchannel, and the following solution was obtained:

$$u(x,y) = \frac{Re\mu RT}{p(x)D_h}\left(\frac{y/H-(y/H)^2+2C_1Kn(z)+8C_2Kn^2(z)}{1/6+2C_1Kn(z)+8C_2Kn^2(z)}\right). \tag{5.177}$$

$$\begin{aligned}&\left(\frac{p}{p_0}\right)^2 - 1 + 24C_1Kn_0\left(\frac{p}{p_0}-1\right) + 96C_2Kn_0^2\log\frac{p}{p_0}\\ &+ 2Re^2\beta\chi\left[12C_1Kn_0\left(\frac{p_0}{p}-1\right)\right.\\ &\left.+ 24C_2Kn_0^2\left(\left(\frac{p_0}{p}\right)^2-1\right) - \log\frac{p}{p_0}\right] = -96Re\beta\frac{x-x_0}{D_h}\end{aligned} \tag{5.178}$$

where

$$\begin{aligned}\beta &= \frac{2Kn^2}{\pi}\chi = \frac{1}{A}\int_A\left(\frac{u}{\bar{u}}\right)^2 dA\\ &= \frac{1/30+\frac{2}{3}C_1Kn+4C_1^2Kn^2+\frac{8}{3}C_2Kn^2+64C_2^2Kn^4+32C_1C_2Kn^3}{(1/6+2C_1Kn+8C_2Kn^2)^2}.\end{aligned}$$

The above solution is same as that obtained in Sect. 2.4.2.2. This is because the procedure given above converged after the first step itself, and the magnitude of the all terms in the Burnett stress equation was negligible (<1% of the highest order term in the equation). Upon substituting the above Eqs. (5.177) and (5.178) in the conventional Burnett equations (Sect. 5.4.3), the normalized residue (difference in the left and right sides of the streamwise momentum equation normalized by the highest order term in the equation) was within a band of ±1% till Kn = 2.2 (Fig. 5.7a). (A finite residue is expected since convergence can happen where the magnitude of the remaining terms in the equation is small, but not exactly zero.) Since the proposed solution satisfies the Burnett equations and the boundary condition, it is deemed as a solution of the equations.

Similarly, Singh and Agrawal [130] obtained a solution for flow of gas in a microtube by solving Burnett equations in cylindrical coordinates (Sect. 5.5) using this iterative procedure. The solution is same as that given in Sreekanth [137] and the normalized residue was within a band of ±1% till Kn = 1.3 (Fig. 5.7b). Note that the solutions obtained here are applicable to the entire slip and a reasonable part of the transition regime.

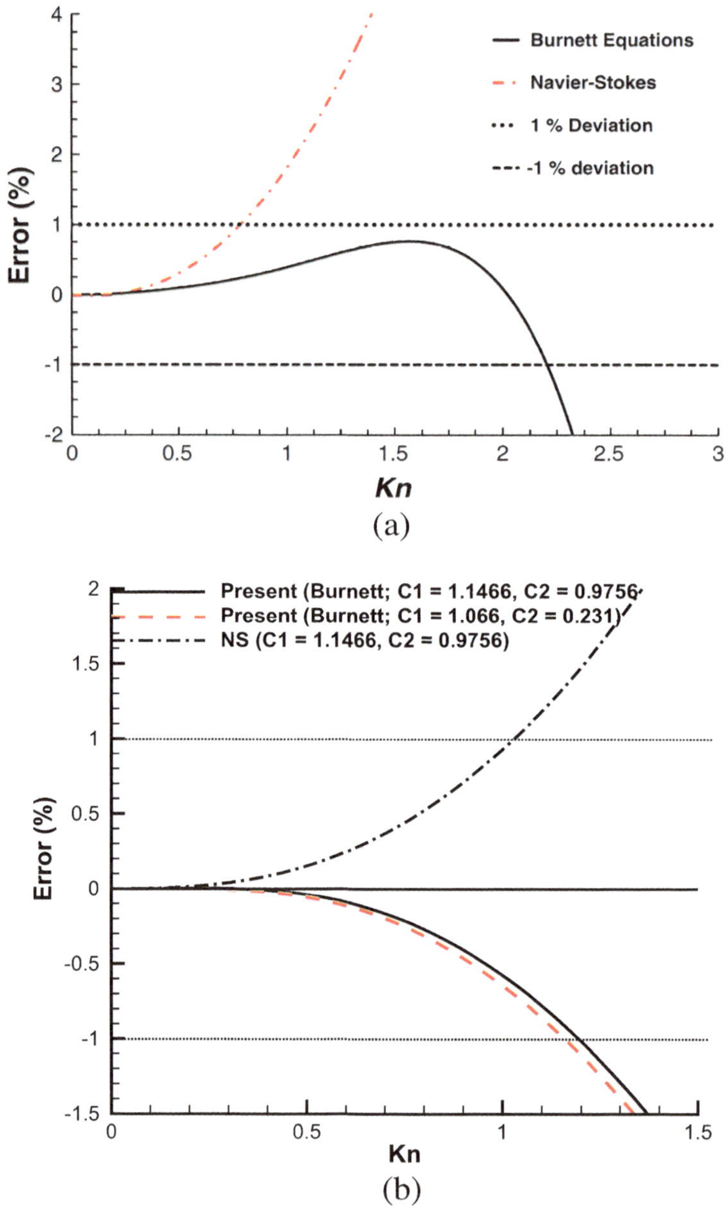

Fig. 5.7 Error (in %) versus Knudsen number for the full Burnett equations and the Navier–Stokes (NS) equations at $r \approx D/2$. Gas molecules assumed as Maxwellian with two different sets of slip coefficients. (**a**) Taken from [132] and (**b**) taken from [130]

5.7.2 Perturbation Based Solution

Rath et al. [120] recently performed a perturbation based analysis of steady, two-dimensional, isothermal gaseous slip flow in a long microchannel. The analysis is for the case when both the Reynolds and Mach numbers of the flow are small. Note that these are the same conditions under which Arkilic et al. [11] solved the Navier–Stokes equations using a perturbation based approach (Sect. 2.4.2.1).

Under the above-mentioned assumptions, the non-dimensional form of conventional Burnett equations can be written as

Continuity:

$$\varepsilon \frac{\partial (\rho u)}{\partial x} + \frac{\partial (\rho v)}{\partial y} = 0 \tag{5.179}$$

Streamwise (x) momentum equation:

$$\begin{aligned} &\frac{7}{3} \frac{k_2}{(p)^3} \frac{\varepsilon}{\mathrm{Re}} \frac{\partial p}{\partial y} \frac{\partial^2 p}{\partial x \partial y} + \frac{1}{3} \frac{k_2}{(p)^3} \frac{\varepsilon}{\mathrm{Re}} \frac{\partial^2 p}{\partial y \partial x} \frac{\partial^2 p}{\partial y^2} - \frac{2k_2}{(p)^4} \frac{\varepsilon}{\mathrm{Re}} \left(\frac{\partial p}{\partial y} \right)^2 \frac{\partial p}{\partial x} \\ &+ \frac{\mathrm{Re}\varepsilon}{Ma^2} \frac{1}{\gamma} \frac{\partial p}{\partial x} - \frac{2k_2}{3} \frac{\varepsilon}{\mathrm{Re}} \frac{1}{(p)^2} \frac{\partial^3 p}{\partial y^2 \partial x} = \frac{\partial^2 u}{\partial y^2} \end{aligned} \tag{5.180}$$

Cross-stream (y) momentum equation:

$$\frac{8}{3} \frac{k_2}{\mathrm{Re}} \frac{1}{(p)^3} \frac{\partial p}{\partial y} \frac{\partial^2 p}{\partial y^2} - \frac{2}{3} \frac{k_2}{\mathrm{Re}} \frac{1}{(p)^2} \frac{\partial^3 p}{\partial y^3} - \frac{k_2}{\mathrm{Re}} \frac{2}{(p)^4} \left(\frac{\partial p}{\partial y} \right)^3 + \frac{\mathrm{Re}}{Ma^2} \frac{1}{\gamma} \frac{\partial p}{\partial y} = 0 \tag{5.181}$$

The terms of the Navier–Stokes order and Burnett order can be easily recognized in the above equations: the velocity term on the right-hand side in Eq. (5.180) and the terms with γ (= ratio of specific heat) are the Navier–Stokes terms; rest of the terms are Burnett order terms.

Notice that Eq. (5.181) involves a single variable p. The solution of p once obtained can be substituted in Eq. (5.180) to obtain u. Finally, the solution of u and ρ (latter being proportional to p) can be substituted into Eq. (5.179) to obtain v, thereby completing the solution. That is, the assumption of small Mach and Reynolds numbers has helped by uncoupling the continuity and momentum equations. However, the pressure equation (Eq. (5.181)) is highly non-linear, and we resort to perturbation method for its solution.

Accordingly, we substitute

$$p = p_0(x) + p_1(x, y) \tag{5.182}$$

in Eq. (5.181), where p_0 is the solution of the Navier–Stokes equation ($\partial p/\partial y = 0$). Upon retaining only the first-order terms, we obtain the following simplified y-momentum equation:

$$\frac{(p_0)^4}{\gamma}\frac{\partial p_1}{\partial y} - \frac{2k_2'(p_0)^2}{3}\frac{\partial^3 p_1}{\partial y^3} = 0 \tag{5.183}$$

where $k_2' = \omega_2 Ma^2/\mathrm{Re}^2$ with ω_2 being a coefficient in the Burnett equations (Table 5.1).

The solution to (5.183) can now be easily obtained as

$$p_1(x, y) = c_1(x) + c_2(x)\cosh(p_0 A y) + c_3(x)\sinh(p_0 A y) \tag{5.184}$$

where the constants of integration c_1, c_2, and c_3 are in general functions of x, and $A = \sqrt{\frac{3}{2k_2'\gamma}}$. Thus, the complete solution can be written as

$$p(x, y) = p_0(x) + c_1(x) + c_2(x)\cosh(p_0 A y) + c_3(x)\sinh(p_0 A y)\,. \tag{5.185}$$

Proceeding further in order to obtain the constants of integration in (5.185) we require three boundary conditions for pressure along the y-direction.

Boundary Condition 1: Normal gradient of pressure at the channel centerline is zero due to symmetry:

$$\left.\frac{\partial p}{\partial y}\right|_{y=0} = 0 \tag{5.186}$$

Boundary Condition 2: Pressure at $y = 0$ is the centerline pressure (i.e. $p|_{y=0} = p_c$). The value of centerline pressure (p_c) therefore needs to be prescribed, for example, using DSMC data.

Boundary Condition 3: The third condition is also specified at the centerline as follows:

$$\left.\frac{d^2 p}{dy^2}\right|_{y=0} = a \tag{5.187}$$

where a is the curvature of the pressure curve at $y = 0$, which again needs to be prescribed. Note that all the boundary conditions have been imposed at the centerline. This treatment of the problem as an initial value problem is deemed to be superior than treating it as a boundary value problem as argued in Singh et al. [133].

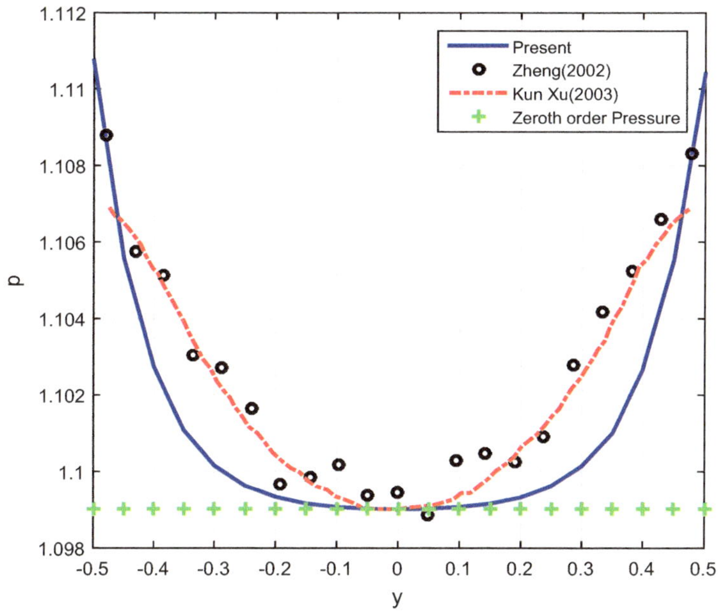

Fig. 5.8 Pressure distribution along the wall normal coordinate (with $y = 0$ at the centerline) at Knudsen number = 0.1, Rath et al. [120]

Applying the above-mentioned boundary conditions in Eq. (5.185) gives the value of all three coefficients. The solution finally becomes

$$p(x, y) = p_c + \frac{a}{(p_0 A)^2} \left(\cosh(p_0 A y) - 1\right). \tag{5.188}$$

Figure 5.8 represents the pressure distribution with respect to the wall-normal position at Knudsen number 0.1. The curvature of pressure field has been extracted from the pressure versus y graph obtained in Xu [170]. Comparison is further made with Zheng et al. [176] and Xu [170] who performed DSMC numerical simulations in long microchannel and compared the DSMC solution against solution of super-Burnett equation and the BGK Burnett equations. The comparison appears satisfactory, as the error is less than 0.3%.

The solution for streamwise velocity u is provided in Rath et al. [120] but not reproduced here for brevity. Figure 5.9 compares the lateral variation of stream wise velocity in the microchannel as obtained from the present approach against that of Arkilic [11] and Xu [170]. The velocity appears parabolic, i.e., the effect of correction on the velocity profile is minimal, at least for the present conditions.

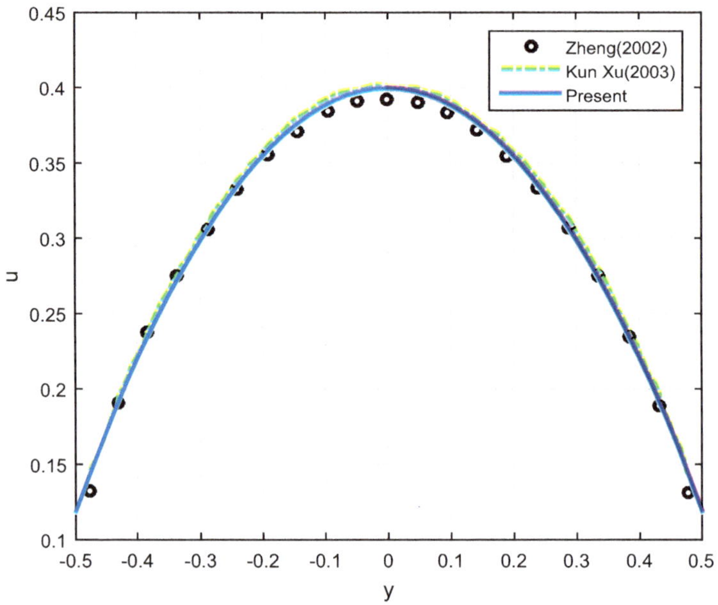

Fig. 5.9 Flow distribution in the cross-stream direction; u is the non-dimensionalized streamwise velocity and y is the non-dimensionalized wall normal coordinate (with $y = 0$ at the centerline). Knudsen number = 0.1, Rath et al. [120]

5.8 Stability Analysis of Burnett Equations

The linear stability analysis of Burnett equations was first undertaken by Bobylev [22] for Maxwellian molecules. The analysis was later extended by Uribe et al. [155] to other intermolecular potentials, and is followed here. In this section, we perform stability analysis of the conventional Burnett equations (Eqs. (5.146) and (5.147)), in a manner done in these earlier works. Towards this, we consider a one-dimensional wave and show that the Burnett equations are unstable to small wavelengths on applying small perturbations to the equilibrium solution.

Consider one-dimensional, time-varying flow; therefore, all flow variables are a function of (x, t) alone. The simplified equations for continuity, x-momentum and energy are

$$\frac{\partial \rho(x,t)}{\partial t} + \frac{\partial}{\partial x}\big(u(x,t)\, \rho(x,t)\big) = 0 \tag{5.189}$$

$$\frac{\partial u(x,t)}{\partial t} + u(x,t)\,\frac{\partial u(x,t)}{\partial x} = -\frac{1}{\rho(x,t)}\frac{\partial p_{xx}}{\partial x} \tag{5.190}$$

$$\frac{\partial T(x,t)}{\partial t} + u(x,t)\,\frac{\partial T(x,t)}{\partial x} = -\frac{2m}{3k_B\rho(x,t)}\left[p_{xx}(x,t)\frac{\partial u(x,t)}{\partial x} + \frac{\partial q_x(x,t)}{\partial x}\right] \tag{5.191}$$

The constitutive equations for p_{xx} and q_x are

$$\begin{aligned}
p_{xx} =& \frac{\rho k_B T}{m} - \frac{4}{3}\mu\frac{\partial u}{\partial x} + \frac{2}{3}\omega_1\mu^2\frac{m}{\rho k_B T}\left(\frac{\partial u}{\partial x}\right)^2 + \frac{2}{3}\omega_2\frac{\mu^2}{\rho T}\left[\frac{1}{\rho^2}\left(\frac{\partial \rho}{\partial x}\right)^2\right.\\
&\left.- \frac{1}{\rho}\frac{\partial \rho}{\partial x}\frac{\partial T}{\partial x} - \frac{1}{\rho^2}\frac{\partial^2 \rho}{\partial x^2} - \frac{\partial^2 T}{\partial x^2} - \frac{14}{9}\left(\frac{\partial u}{\partial x}\right)^2\right] + \frac{2}{3}\omega_3\frac{\mu^2}{\rho T}\frac{\partial^2 T}{\partial x^2}\\
&+ \frac{2}{3}\omega_4\frac{\mu^2 m}{\rho^2 k_B T^2}\left[\frac{k_B T}{m}\frac{\partial \rho}{\partial x}\frac{\partial T}{\partial x}\right.\\
&\left.+ \frac{\rho k_B}{m}\left(\frac{\partial T}{\partial x}\right)^2\right] + \frac{2}{3}\omega_5\frac{\mu^2}{\rho T^2}\left(\frac{\partial T}{\partial x}\right)^2 + \frac{2}{9}\omega_6\frac{m\mu^2}{\rho k_B T}\left(\frac{\partial u}{\partial x}\right)^2
\end{aligned} \tag{5.192}$$

$$\begin{aligned}
q_x =& -k\frac{\partial T}{\partial x} + \frac{\theta_1\mu^2}{\rho T}\frac{\partial u}{\partial x}\frac{\partial T}{\partial x} - \frac{8}{3}\frac{\theta_2\mu^2}{\rho T}\frac{\partial u}{\partial x}\frac{\partial T}{\partial x} - \frac{2}{3}\frac{\theta_2\mu^2}{\rho}\frac{\partial^2 u}{\partial x^2} + \frac{2}{3}\frac{\theta_3\mu^2 m}{\rho^2 k_B T}\\
&\left[\frac{k_B T}{m}\frac{\partial \rho}{\partial x} + \frac{\rho k_B}{m}\frac{\partial T}{\partial x}\right]\frac{\partial u}{\partial x} + \frac{2}{3}\frac{\theta_4\mu^2}{\rho}\frac{\partial^2 u}{\partial x^2} + 2\frac{\theta_5\mu^2}{\rho T}\frac{\partial T}{\partial x}\frac{\partial u}{\partial x}.
\end{aligned} \tag{5.193}$$

We now introduce a small perturbation around the base state: $u = 0$, $\rho = \rho_0$, $T = T_0$, where ρ_0 and T_0 are the uniform density and temperature in the flow in the undisturbed state. The total flow, which is sum of the base flow and the perturbation is therefore

$$\begin{aligned}
u(x,t) &= \sqrt{\frac{k_B T_0}{m}}\varepsilon u'(x,t)\\
\rho(x,t) &= \rho_0[1 + \varepsilon\rho'(x,t)]\\
T(x,t) &= T_0[1 + \varepsilon T'(x,t)]
\end{aligned} \tag{5.194}$$

Substituting u, ρ, and T from Eq. (5.194) into the governing equations (5.189)–(5.191) and retaining terms up to order ε, we obtain

$$\frac{\partial \rho'}{\partial t'} + \frac{\partial u'}{\partial s} = 0 \tag{5.195}$$

$$\frac{\partial u'}{\partial t} = -\frac{\partial}{\partial s}(\rho' + T') + \frac{4}{3}\frac{\partial^2 u'}{\partial s^2} + \frac{2}{3}\omega_2\frac{\partial^3 \rho'}{\partial s^3} - \frac{2}{3}(\omega_3 - \omega_2)\frac{\partial^3 T'}{\partial s^3} \tag{5.196}$$

$$\frac{\partial T'}{\partial t'} = -\frac{2}{3}\frac{\partial u'}{\partial s} + f\frac{\partial^2 T'}{\partial s^2} - \frac{4}{9}(\theta_4 - \theta_2)\frac{\partial^3 u'}{\partial s^3} \tag{5.197}$$

where $f = 2k_0/(3R\mu_0)$ is the Eucken factor, μ_0 is the viscosity, and k_0 is the thermal conductivity of the gas at temperature T_0. Further,

$$\lambda = \frac{\mu_0}{\rho_0}\sqrt{\frac{m}{k_B T_0}}, \quad s = \frac{x}{\lambda}, \quad t' = \frac{\rho_0 k_B T_0}{\mu_0 m}t \tag{5.198}$$

Since the equations admit exponential solutions in x and t, we assume normal modes of the form:

$$\begin{aligned} &\rho'(s, t') = \tilde{\rho}\ \exp(\Omega t' + i\kappa s),\ T'(s, t') = \tilde{T}\ \exp(\Omega t' + i\kappa s), \\ &u'(s, t') = \tilde{u}\ \exp(\Omega t' + i\kappa s),\ \mathrm{Im}(\kappa) = 0 \end{aligned} \tag{5.199}$$

Note that κ is a real variable, whereas Ω is complex. Substituting the above variables in Eqs. (5.195)–(5.197), we obtain the following dispersion relation:

$$18\Omega^3 + 69\Omega^2\kappa^2 + 97\Omega\kappa^4 - 14\Omega\kappa^6 + 30\Omega\kappa^2 + 45\kappa^4 + 60\kappa^6 = 0 \tag{5.200}$$

where we have assumed molecules to be Maxwellian. Accordingly, we employed

$$\theta_2 = \frac{45}{8}, \quad \theta_4 = 3, \quad \omega_2 = 2, \quad \omega_3 = 3$$

to obtain the above relation.

The real part of Ω determines the stability characteristics, as the disturbance grows exponentially in such case (refer Eq. (5.199)). It is apparent from Fig. 5.10 that there exists a critical wave number ($\kappa_c \approx 2.5$) beyond which one of the roots becomes positive, and the solution becomes unstable. This instability is usually referred to as "Bobylev's instability" and leads to difficulty when trying to solve these equations numerically on finer grids. A similar analysis on the Navier–Stokes equations leads to the conclusion that these equations are unconditionally stable, in contrast to our conclusion about the Burnett equations.

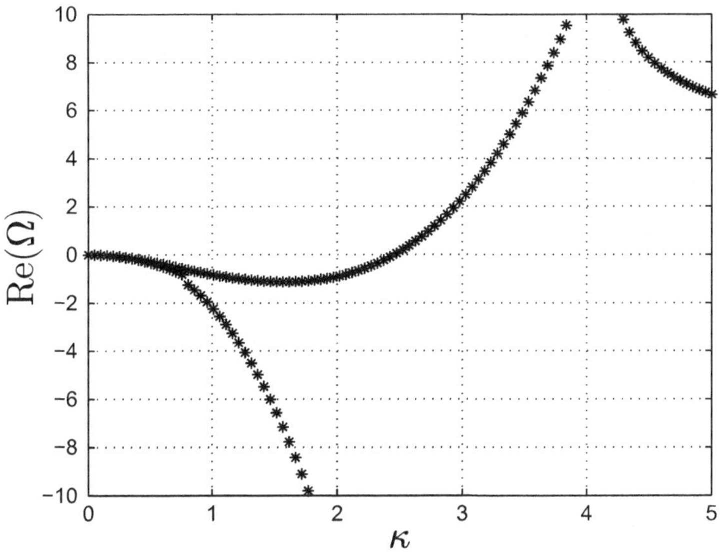

Fig. 5.10 Variation of real part of Ω with wave number κ, as obtained from Eq. (5.200)

5.9 Summary

In this chapter, a step-by-step derivation of the Navier–Stokes and Burnett equations starting from the Boltzmann equation has been presented. The approach involves expanding the distribution function involved in the Boltzmann equation in an infinite series with Knudsen number as the smallness parameter. A higher-order distribution function is expressed in terms of the lower-order distribution functions using the method of Enskog, which allows successive evaluation of higher-order distribution functions, starting from the Maxwell–Boltzmann distribution applicable to gas at rest. The approach leads to the correct form of the constitutive relations and the Navier–Stokes equations at first order, thereby lending credence to the approach. The constitutive equations at the next order become non-linear, involve second-order derivatives, and are complex in nature. The resulting equations from this approach—the Burnett equations have been presented here in both tensorial and cylindrical forms.

In order to better appreciate the Burnett equations, an order of magnitude analysis of the various terms in the equations has been undertaken. This analysis confirms that the Burnett stress and heat flux terms are of order Knudsen number-square. Analytical solutions of the Burnett equations, some obtained using a novel iterative approach, are also presented.

A stability analysis of the Burnett equations shows that these equations are unstable beyond a critical wave number. The issues of stability and violation of second law of thermodynamics have encouraged researchers to look for alternate forms of the Burnett equations. The expectation from these alternate equations is that they will enlarge the range of Knudsen number over which the equations will apply, yet not be plagued by the limitations of the Burnett equations. The variants of the Burnett equations are discussed in detail in Chap. 7.

Chapter 6
Grad Equations: Derivation and Analysis

In certain physical problems, the Navier–Stokes equations are insufficient to describe the flow physics accurately, making it necessary to take recourse to higher order continuum transport equations. Grad pioneered the moment method by which successive approximations of the microscopic Boltzmann equation can be obtained. Grad proposed to expand the particle distribution function $f(\mathbf{c}, \mathbf{x}, t)$ in terms of orthogonal Hermite polynomials, with an infinite set of Hermite coefficients being equivalent to the particle distribution function f itself. The number of Hermite coefficients to be considered in the expansion depends upon the degree of accuracy desired. The first few Hermite coefficients are the state variables, stress tensor and heat flux vector. These variables are not expressed in terms of other thermodynamic variables but are on par with them, and satisfy their own differential equation. This distinct approach of Grad along with the resulting equations is expounded in this chapter. The linearized form of the Grad equations is applied to two practical problems and the obtained solutions are compared against that obtained through the Navier–Stokes equations. The Grad equations can be used to obtain the Navier–Stokes equations and Cattaneo's equation as discussed here. The differences as well as the similarities between the Grad approach and the Chapman–Enskog approach (introduced in Chap. 5) are also brought forward.

6.1 Introduction

In the previous chapter, we presented the Chapman–Enskog solution of the Boltzmann equation and derived the hydrodynamic equations considering mass density ρ, bulk velocity u_i, and absolute temperature T as the primary variables. The closure was achieved by obtaining constitutive relationships for the pressure tensor and heat flux vector. In this chapter, we explain Grad moment-based method [62] to solve the Boltzmann equation which proved to be conceptually simpler yet powerful. The

A. Agrawal et al., *Microscale Flow and Heat Transfer*, Mechanical Engineering Series, https://doi.org/10.1007/978-3-030-10662-1_6

idea is to extend the primary variables (or moments) set, that is, in addition to density ρ, bulk velocity u_i, and absolute temperature T, pressure tensor P_{ij}, and heat flux vector q_i are included in the primary variables set and separate evolution equations are obtained for them. A bigger set of primary variables is expected to lead to higher accuracy of the solution, with a set of infinite moments being equivalent to the Boltzmann equation itself. We derive both 20-moment and 13-moment equations in this chapter before solving these equations to two specific problems.

6.2 Definition of Moments

We begin by defining moments of the distribution function in general terms. We will be confining our attention to first 20 moments and first 13 moments in Sects. 6.3 and 6.4, respectively. In the later sections, the evolution equations for these moments are discussed. In the present chapter, we try to follow the nomenclature that has been well established in the literature. Accordingly, a general nth order moment $I^a_{i_1 i_2 \ldots i_n}$ is defined as

$$I^a_{i_1 i_2 \ldots i_n} = m \int c^{2a} c_{i_1} c_{i_2} \cdots c_{i_n} f d\mathbf{c} \tag{6.1}$$

where c is the molecular velocity, and f is the particle distribution function of Eq. (5.1). Note that the superscript a associated with the term $I^a_{i_1 i_2 \ldots i_n}$ is appearing in the exponent of molecular speed inside the integral. These moments are defined with respect to the microscopic velocity and are generally called as convective moments [141].

As we will see in the following sections, it will be greatly beneficial to define the moments with respect to peculiar velocity $\mathbf{C}$ (where $\mathbf{C} = \mathbf{c} - \mathbf{u}$ with $\mathbf{u}$ being bulk velocity) as

$$U^a_{i_1 i_2 \ldots i_n} = m \int C^{2a} C_{i_1} C_{i_2} \cdots C_{i_n} f d\mathbf{c} \tag{6.2}$$

Accordingly, the second, third, and fourth order moments (P_{ij}, S_{ijk}, Q_{ijkl}) are defined as

$$P_{ij}(\mathbf{x}, t) = m \int C_i C_j f d\mathbf{C}, \tag{6.3}$$

$$S_{ijk}(\mathbf{x}, t) = m \int C_i C_j C_k f d\mathbf{C}, \tag{6.4}$$

$$Q_{ijkl}(\mathbf{x}, t) = m \int C_i C_j C_k C_l f d\mathbf{C} \tag{6.5}$$

Contracting the third order moment S_{ijk} as

$$S_{ikk} = m \int C_i C_k C_k f d\mathbf{C} = 2\frac{m}{2} \int C_i C^2 f d\mathbf{C} = 2q_i \tag{6.6}$$

where we have utilized definition of heat flux vector (Eq. (5.6)). Therefore, the contracted third order moment S_i turns out to be twice the heat flux vector.

The second order moment P_{ij} is identified as the pressure tensor and the contracted third order moment S_{ikk} as the heat flux vector as discussed in great detail in Sect. 5.2.2. However, other higher order moments does not seem to have particular physical significance. At best, according to the definition, complete third order moment S_{ijk} can be considered to be the flux of the pressure tensor. Likewise, Q_{ijkl} can be considered to be the flux of the third order moment S_{ijk}.

By definition, the higher order moments (P_{ij}, S_{ijk}, Q_{ijkl}) are symmetric in their subscripts. There are $\frac{1}{2}(n+1)(n+2)$ distinct moments of order n, so there are six distinct second order moments, ten distinct moments of third order, and fifteen distinct moments of fourth order. The divergence-less part of higher order moments (P_{ij}, S_{ijk}, Q_{ijkl}) are denoted by their respective small case alphabets as (p_{ij}, s_{ijk}, q_{ijkl}); the readers should make a note of this important convention which is followed in this chapter. So, when we make the second order moment P_{ij} divergence-less (trace $P_{ii} = 0$), we obtain stress tensor p_{ij} that has five distinct moments. In case of third order moment S_{ijk}, we can perform contraction as $i = j$, or $j = k$, or $i = k$; as such, there will be seven ($= 10 - 3$) distinct moments for the divergence-less part of the third order moment S_{ijk}, i.e., s_{ijk}. Likewise, for the fourth order moment Q_{ijkl}, its divergence-less part q_{ijkl} will have nine ($15 - 6$) distinct moments. In summary, a *symmetric traceless* second order moment has five distinct moments; similarly, third and fourth order moments have, respectively, seven and nine distinct moments.

The evolution equation for the nth order moment $I^a_{i_1 i_2 \ldots i_n}$ as defined in Eq. (6.1) can be generated by multiplying the Boltzmann equation (5.1) by $mc^{2a} c_{i_1} c_{i_2} \cdots c_{i_n}$ and integrating over the velocity space as

$$\begin{aligned} m \int c^{2a} c_{i_1} c_{i_2} \cdots c_{i_n} \frac{\partial f}{\partial t} d\mathbf{c} + m \int c^{2a} c_{i_1} c_{i_2} \cdots c_{i_n} c_k \frac{\partial f}{\partial x_k} d\mathbf{c} \\ = m \int c^{2a} c_{i_1} c_{i_2} \cdots c_{i_n} J(f, f_1) d\mathbf{c} \end{aligned} \tag{6.7}$$

$$\frac{\partial I^a_{i_1 i_2 \ldots i_n}}{\partial t} + \frac{\partial I^a_{i_1 i_2 \ldots i_n k}}{\partial x_k} = \bar{J}_{i_1 i_2 \cdots i_n} \tag{6.8}$$

where the production term is

$$\bar{J}_{i_1 i_2 \cdots i_n} = m \int c^{2a} c_{i_1} c_{i_2} \cdots c_{i_n} J(f, f_1) d\mathbf{c}.$$

Note that the forcing term in the Boltzmann equation has been neglected for simplicity while deriving the above equation. The Leibnitz rule and integration by parts are applied with the knowledge that **x**, **c**, and t are independent variables and the moment equation (6.8) is obtained.

The structure of the moment equation (6.8) is such that the evolution equation for nth order moment $I^a_{i_1 i_2 \ldots i_n}$ contains the next higher order $(n+1)$th moment $I^a_{i_1 i_2 \ldots i_n k}$ in the divergence term. Hence, this system of infinite coupled partial differential equations does not form a closed set. To achieve closure, Grad expanded the distribution function around Maxwell-Boltzmann distribution $f^{(0)}$ in terms of tensorial orthogonal Hermite polynomials as

$$
\begin{aligned}
f(\mathbf{x}, \mathbf{C}, t) &= f^{(0)} \sum_{n=0}^{\infty} \frac{1}{n!} a_i^{(n)}(\mathbf{x}, t) \mathscr{H}_i^{(n)}(\mathbf{C}) \\
&= f^{(0)} \left(a^{(0)} \mathscr{H}^{(0)} + a_i^{(1)} \mathscr{H}_i^{(1)} + \frac{1}{2!} a_{ij}^{(2)} \mathscr{H}_{ij}^{(2)} + \frac{1}{3!} a_{ijk}^{(3)} \mathscr{H}_{ijk}^{(3)} + \cdots \right)
\end{aligned}
\tag{6.9}
$$

This infinite series, however, needs to be truncated as it is possible to evaluate only a finite number of Hermite coefficients $a^{(n)}_{i_1 i_2 \cdots i_n}$. It turns out that the number of coefficients considered in the series must be equal to the primary variables considered so that the unknowns appearing in the equations can be evaluated in terms of the primary basic variables and the necessary closure is achieved. For example, for a full third order approximation, the primary variables set is represented as $\{\rho, u_i, \epsilon(= 3RT/2), P_{ij}, S_{ijk}\}$. Hence, the series (6.9) needs to be truncated after the fourth term by taking all the Hermite coefficients as zero for $n > 3$.

Notice that the functional form of the distribution function (Eq. (6.9)) is such that its dependence on space and time is only through the Hermite coefficients, $a_i^{(n)}$ and the molecular velocity dependence is through the Hermite polynomials, $\mathscr{H}^{(n)}_{i_1 i_2 \cdots i_n}$. Once the distribution function is determined, it can be used to compute the unknown higher order moments $I_{i_1 i_2 \ldots i_n k}$ appearing in the flux term and the production term $\bar{J}_{i_1 i_2 \cdots i_n}$, and the necessary closure is achieved.

6.3 20-Moment Approximation

In the present chapter, we first present a detailed derivation of the 20-moment equations, followed by the more popular 13-moment equations. A brief outline of the sections to follow is shown in Fig. 6.1. For the 20-moment equations, we adhere to the classic paper [62], so as to bring forth the essence of the moment method. However, it is more appropriate to consider a set of first 13 moments

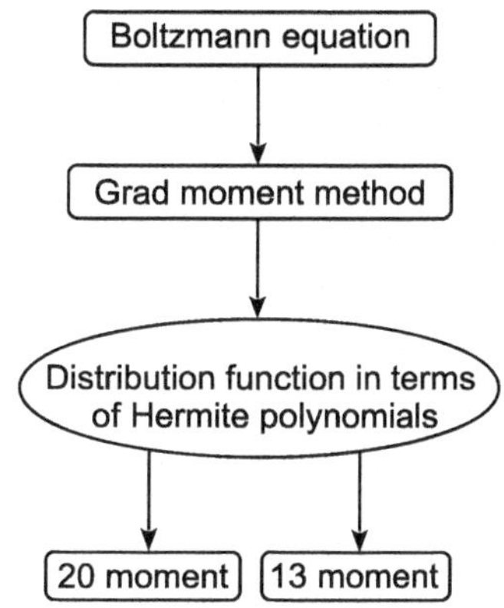

Fig. 6.1 20- and 13-moment equations from Boltzmann equation using Grad's distribution function

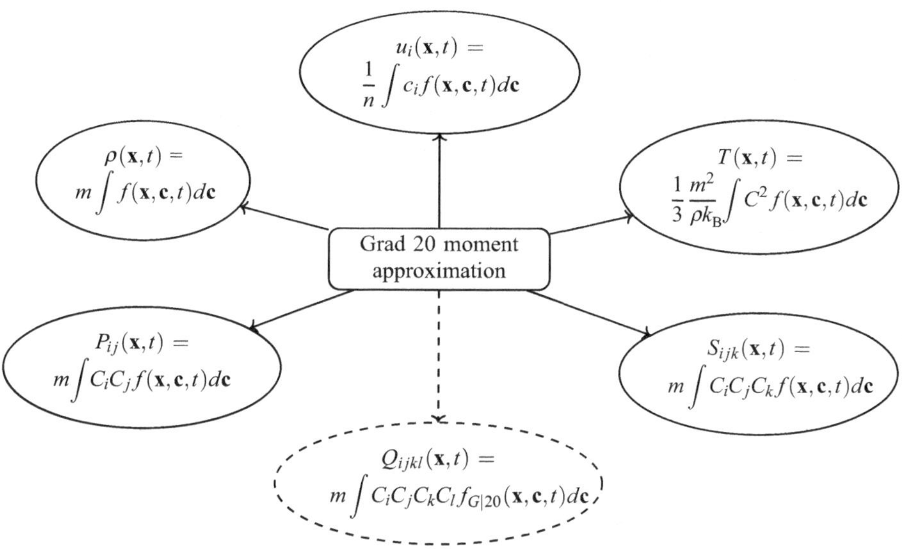

Fig. 6.2 Expressions for the variables considered in the Grad 20-moment approximation; variables in solid circles are the primary variables while those in dashed circles are the closure variables

(ρ, u_i, T, p_{ij}, and q_i) having physical intuitive meaning rather than the 20-moment approximation. The transition from 20-moment to 13-moment approximation can be done by contracting the third order moment S_{ijk} as S_{ikk} which actually is twice the heat flux vector (see Eq. (6.6)).

In 20-moment approximation, the primary variables considered are $\{\rho, u_i, \epsilon(= 3RT/2), P_{ij}, S_{ijk}\}$ as shown in Fig. 6.2. The higher order moments P_{ij} and S_{ijk} are treated on par with the thermodynamic variables and separate evolution equations are generated for these moments. However, in the evolution equation for S_{ijk}, a fourth order moment Q_{ijkl} is present in the flux term which needs to be evaluated in terms of primary variables to achieve the necessary closure.

6.3.1 Evolution Equations

The higher order moments P_{ij}, S_{ijk}, and Q_{ijkl} are defined with respect to peculiar velocity (Eqs. (6.3)–(6.5)). Hence, in order to generate the evolution equations for these moments, it is convenient to select peculiar velocity $\mathbf{C}$ as an independent variable. Accordingly, the Boltzmann equation (Eq. (5.1)) needs to be converted to some appropriate form in which peculiar velocity replaces the molecular velocity as an independent variable. The derivation of transformed Boltzmann equation $[f(\mathbf{x}, \mathbf{c}, t) \Rightarrow f(\mathbf{x}, \mathbf{C}, t)]$ is given in Appendix C.1 and is mathematically represented as

$$\frac{Df}{Dt} + C_k \frac{\partial f}{\partial x_k} + \left(F_k - \frac{Du_k}{Dt} \right) \frac{\partial f}{\partial C_k} - C_k \frac{\partial f}{\partial C_j} \frac{\partial u_j}{\partial x_k} = J(f, f_1) \tag{6.10}$$

To generate the evolution equation for higher order moment, say,

$$R^a_{i_1 i_2 \dots i_n} = m \int C^{2a} C_{i_1} C_{i_2} \dots C_{i_n} f(\mathbf{x}, \mathbf{C}, t) d\mathbf{C}$$

we multiply the transformed Boltzmann equation (Eq. (6.10)) by $mC^{2a} C_{i_1} C_{i_2} \dots C_{i_n}$ and integrate over the velocity space.

We have already generated the evolution equations for the first five variables $\{\rho, u_i, \epsilon(= 3RT/2)\}$ in Sect. 5.2.4 which read as

$$\frac{\partial \rho}{\partial t} + \frac{\partial (\rho u_k)}{\partial x_k} = 0, \tag{6.11}$$

$$\frac{\partial u_i}{\partial t} + u_k \frac{\partial u_i}{\partial x_k} + \frac{1}{\rho} \frac{\partial p}{\partial x_i} + \frac{1}{\rho} \frac{\partial p_{ik}}{\partial x_k} = F_i, \tag{6.12}$$

$$\frac{3}{2} R\rho \frac{\partial T}{\partial t} + \frac{3}{2} R\rho u_k \frac{\partial T}{\partial x_k} + p \frac{\partial u_i}{\partial x_i} + p_{ij} \frac{\partial u_i}{\partial x_j} + \frac{\partial q_k}{\partial x_k} = 0. \tag{6.13}$$

The evolution equations for the higher order moments are more involved and the detailed procedure is explained in the next section.

6.3.1.1 Evolution Equation for P_{ij}

In order to obtain the evolution equation for the pressure tensor P_{ij}, we multiply the transformed Boltzmann equation (Eq. (6.10)) by $mC_i C_j$ and then integrate over the velocity space as

$$m\int C_iC_j\frac{Df}{Dt}d\mathbf{C}+m\int C_iC_jC_k\frac{\partial f}{\partial x_k}d\mathbf{C}+m\int\left(F_k-\frac{Du_k}{Dt}\right)C_iC_j\frac{\partial f}{\partial C_k}d\mathbf{C}$$
$$-m\int C_iC_jC_k\frac{\partial f}{\partial C_l}\frac{\partial u_l}{\partial x_k}d\mathbf{C}=m\int C_iC_jJ(f,f_1)d\mathbf{C} \tag{6.14}$$

Since we have j as free index in Eq. (6.14), we changed the dummy index j in the fourth term of (Eq. (6.10)) to l. Applying rules of integration by parts and Leibnitz's rule to take out the differentiation outside the integral, we evaluate term by term as

$$\begin{aligned}(a)&=m\int C_iC_j\frac{Df}{Dt}d\mathbf{C}\\&=m\int\frac{DfC_iC_j}{Dt}d\mathbf{C}-m\int f\frac{DC_iC_j}{Dt}d\mathbf{C}\\&=\frac{D}{Dt}m\int fC_iC_jd\mathbf{C}-m\int f\left(C_i\frac{DC_j}{Dt}+C_j\frac{DC_i}{Dt}\right)d\mathbf{C}\\&=\frac{DP_{ij}}{Dt}-m\int f\,(0+0)\,d\mathbf{C}\qquad[\because \mathbf{C},t\text{ are independent variables}]\end{aligned}$$

$$\therefore\quad (a)=m\int C_iC_j\frac{Df}{Dt}d\mathbf{C}=\frac{DP_{ij}}{Dt} \tag{6.15}$$

$$\begin{aligned}(b)&=m\int C_iC_jC_k\frac{\partial f}{\partial x_k}d\mathbf{C}\\&=m\int\frac{\partial(fC_iC_jC_k)}{\partial x_k}d\mathbf{C}-m\int f\frac{\partial(C_iC_jC_k)}{\partial x_k}d\mathbf{C}\\&=\frac{\partial S_{ijk}}{\partial x_k}-0\qquad[\because \mathbf{C},\mathbf{x}\text{ are independent variables}]\end{aligned}$$

$$\therefore\quad (b)=m\int C_iC_jC_k\frac{\partial f}{\partial x_k}d\mathbf{C}=\frac{\partial S_{ijk}}{\partial x_k} \tag{6.16}$$

Assuming the external force to be independent of the peculiar velocity, the term in the rounded bracket can be taken outside the integral sign as

$$\begin{aligned}(c)&=m\int\left(F_k-\frac{Du_k}{Dt}\right)C_iC_j\frac{\partial f}{\partial C_k}d\mathbf{C}\\&=\left(F_k-\frac{Du_k}{Dt}\right)m\int C_iC_j\frac{\partial f}{\partial C_k}d\mathbf{C}\\&=\left(F_k-\frac{Du_k}{Dt}\right)m\int\frac{\partial C_iC_jf}{\partial C_k}d\mathbf{C}-\left(F_k-\frac{Du_k}{Dt}\right)m\int f\frac{\partial C_iC_j}{\partial C_k}d\mathbf{C}\end{aligned}$$

For the first term, we apply Gauss divergence theorem to convert the volume integral into a surface integral as

$$\int_{-\infty}^{+\infty} \frac{\partial}{\partial C_k}\left[\left(F_k - \frac{Du_k}{Dt}\right)mC_iC_jf\right]d\mathbf{C} = \oint_{C\to\infty}\left(F_k - \frac{Du_k}{Dt}\right)mC_iC_jfn_kdA \tag{6.17}$$

This surface integral needs to be performed over the control surface for $C \to \infty$. As there are no particles having infinite molecular velocities (or peculiar velocity), i.e., $f(C \to \infty) = 0$, this surface integral vanishes.

$$\begin{aligned}
\therefore\ (c) &= 0 - \left(F_k - \frac{Du_k}{Dt}\right)m\int f\frac{\partial C_iC_j}{\partial C_k}d\mathbf{C}\\
&= -\left(F_k - \frac{Du_k}{Dt}\right)m\int f\left(C_i\frac{\partial C_j}{\partial C_k} + C_j\frac{\partial C_i}{\partial C_k}\right)d\mathbf{C}\\
&= -\left(F_k - \frac{Du_k}{Dt}\right)m\int f\left(C_i\delta_{jk} + C_j\delta_{ik}\right)d\mathbf{C}\\
&= -\left(F_k - \frac{Du_k}{Dt}\right)\left[\delta_{jk}m\int fC_id\mathbf{C} + \delta_{ik}m\int fC_jd\mathbf{C}\right]\\
&= 0 \qquad\qquad [\because m\int fC_id\mathbf{C} = 0]
\end{aligned}$$

$$\therefore\quad (c) = m\int\left(F_k - \frac{Du_k}{Dt}\right)C_iC_j\frac{\partial f}{\partial C_k}d\mathbf{C} = 0 \tag{6.18}$$

$$\begin{aligned}
(d) &= -m\int C_iC_jC_k\frac{\partial f}{\partial C_l}\frac{\partial u_l}{\partial x_k}d\mathbf{C}\\
&= -\frac{\partial u_l}{\partial x_k}m\int C_iC_jC_k\frac{\partial f}{\partial C_l}d\mathbf{C}\\
&= -\frac{\partial u_l}{\partial x_k}m\int\frac{\partial C_iC_jC_kf}{\partial C_l}d\mathbf{C} + \frac{\partial u_l}{\partial x_k}m\int f\frac{\partial C_iC_jC_k}{\partial C_l}d\mathbf{C}
\end{aligned}$$

Converting the volume integral in the first term from to the surface integral and using the same arguments as used for first term in (c), the term vanishes.

$$\begin{aligned}
\therefore\ (d) &= 0 + \frac{\partial u_l}{\partial x_k}m\int f\left(C_iC_j\frac{\partial C_k}{\partial C_l} + C_iC_k\frac{\partial C_j}{\partial C_l} + C_jC_k\frac{\partial C_i}{\partial C_l}\right)d\mathbf{C}\\
&= \frac{\partial u_l}{\partial x_k}m\int f\left(C_iC_j\delta_{kl} + C_iC_k\delta_{jl} + C_jC_k\delta_{il}\right)d\mathbf{C}
\end{aligned}$$

$$= \frac{\partial u_l}{\partial x_k}\left[\delta_{kl} m \int f C_i C_j d\mathbf{C} + \delta_{jl} m \int C_i C_k d\mathbf{C} + \delta_{il} m \int C_j C_k d\mathbf{C}\right]$$

$$= P_{ij}\frac{\partial u_k}{\partial x_k} + P_{ik}\frac{\partial u_j}{\partial x_k} + P_{jk}\frac{\partial u_i}{\partial x_k}$$

$$\therefore \quad (d) = -m \int C_i C_j C_k \frac{\partial f}{\partial C_l}\frac{\partial u_l}{\partial x_k} d\mathbf{C} = P_{ij}\frac{\partial u_k}{\partial x_k} + P_{ik}\frac{\partial u_j}{\partial x_k} + P_{jk}\frac{\partial u_i}{\partial x_k} \tag{6.19}$$

For now, let's denote the production term as

$$m \int C_i C_j J(f, f_1) d\mathbf{C} = \bar{J}_{ij}^{(2)} \tag{6.20}$$

Substituting for the Lagrangian derivative in the first term (a) and combining results of all the terms, we obtain the evolution equation for the pressure tensor as

$$\frac{\partial P_{ij}}{\partial t} + \underline{u_k \frac{\partial P_{ij}}{\partial x_k}} + \frac{\partial S_{ijk}}{\partial x_k} + \underline{P_{ij}\frac{\partial u_k}{\partial x_k}} + P_{ik}\frac{\partial u_j}{\partial x_k} + P_{jk}\frac{\partial u_i}{\partial x_k} = \bar{J}_{ij}^{(2)}$$

$$\text{or,} \quad \frac{\partial P_{ij}}{\partial t} + \frac{\partial}{\partial x_k}(u_k P_{ij}) + \frac{\partial S_{ijk}}{\partial x_k} + P_{ik}\frac{\partial u_j}{\partial x_k} + P_{jk}\frac{\partial u_i}{\partial x_k} = \bar{J}_{ij}^{(2)} \tag{6.21}$$

Decomposing the pressure tensor P_{ij} into isotropic pressure p and stress tensor p_{ij} using the relation $p_{ij} = P_{ij} - p\delta_{ij}$, the above equation can be written as

$$\frac{\partial p_{ij}}{\partial t} + \underline{\delta_{ij}\frac{\partial p}{\partial t}} + \frac{\partial}{\partial x_k}(u_k p_{ij}) + \underline{\delta_{ij}\frac{\partial}{\partial x_k}(u_k p)} + \frac{\partial S_{ijk}}{\partial x_k}$$

$$+ p_{ik}\frac{\partial u_j}{\partial x_k} + \delta_{ik} p \frac{\partial u_j}{\partial x_k} + p_{jk}\frac{\partial u_i}{\partial x_k} + \delta_{jk} p \frac{\partial u_i}{\partial x_k} = \bar{J}_{ij}^{(2)} \tag{6.22}$$

For the underlined terms, we use Eq. (5.51)) (energy equation in terms of pressure) where dummy indices are changed accordingly to avoid conflict with the free indices,

$$\frac{\partial p}{\partial t} + \frac{\partial}{\partial x_k}(p u_k) = -\frac{2}{3} p \frac{\partial u_i}{\partial x_i} - \frac{2}{3} p_{ij}\frac{\partial u_i}{\partial x_j} - \frac{1}{3}\frac{\partial S_k}{\partial x_k} \tag{6.23}$$

Equation (6.22) for stress tensor then becomes

$$\frac{\partial p_{ij}}{\partial t} + \frac{\partial}{\partial x_k}(u_k p_{ij}) + \frac{\partial S_{ijk}}{\partial x_k} - \frac{2}{3}\delta_{ij} p_{kl}\frac{\partial u_k}{\partial x_l} - \frac{2}{3}\delta_{ij} p \frac{\partial u_k}{\partial x_k} - \frac{1}{3}\delta_{ij}\frac{\partial S_k}{\partial x_k}$$

$$+ p_{ik}\frac{\partial u_j}{\partial x_k} + p\frac{\partial u_j}{\partial x_i} + p_{jk}\frac{\partial u_i}{\partial x_k} + p\frac{\partial u_i}{\partial x_j} = \bar{J}_{ij}^{(2)} \tag{6.24}$$

Rearranging terms, we finally obtain the evolution equation for the stress tensor as

$$\frac{\partial p_{ij}}{\partial t} + \frac{\partial}{\partial x_k}(u_k p_{ij}) + \frac{\partial S_{ijk}}{\partial x_k} - \frac{1}{3}\delta_{ij}\frac{\partial S_k}{\partial x_k} + p_{ik}\frac{\partial u_j}{\partial x_k} + p_{jk}\frac{\partial u_i}{\partial x_k}$$
$$- \frac{2}{3}\delta_{ij} p_{kl}\frac{\partial u_k}{\partial x_l} + p\left(\frac{\partial u_i}{\partial x_j} + \frac{\partial u_j}{\partial x_i} - \frac{2}{3}\delta_{ij}\frac{\partial u_k}{\partial x_k}\right) = \bar{J}_{ij}^{(2)}. \tag{6.25}$$

6.3.1.2 Evolution Equation for S_{ijk}

Following a similar procedure as given above, we obtain the evolution equation for the third order moment S_{ijk}. For this, we multiply the transformed Boltzmann equation (Eq. (6.10)) by $mC_iC_jC_k$ and integrate over the velocity space to obtain

$$m\int C_iC_jC_k\frac{Df}{Dt}d\mathbf{C} + m\int C_iC_jC_kC_r\frac{\partial f}{\partial x_r}d\mathbf{C}$$
$$+ m\int\left(F_r - \frac{Du_r}{Dt}\right)C_iC_jC_k\frac{\partial f}{\partial C_r}d\mathbf{C}$$
$$- m\int C_iC_jC_kC_r\frac{\partial f}{\partial C_l}\frac{\partial u_l}{\partial x_r}d\mathbf{C} = m\int C_iC_jC_kJ(f, f_1)d\mathbf{C} \tag{6.26}$$

Using the same arguments, integration by parts and Leibnitz's rule, we evaluate term by term as

$$\begin{aligned}
(a) &= m\int C_iC_jC_k\frac{Df}{Dt}d\mathbf{C}\\
&= m\int\frac{DfC_iC_jC_k}{Dt}d\mathbf{C} - m\int f\frac{DC_iC_jC_k}{Dt}d\mathbf{C}\\
&= \frac{DS_{ijk}}{Dt} - 0\\
(b) &= m\int C_iC_jC_kC_r\frac{\partial f}{\partial x_r}d\mathbf{C}\\
&= m\int\frac{\partial(fC_iC_jC_kC_r)}{\partial x_r}d\mathbf{C} - m\int f\frac{\partial(C_iC_jC_kC_r)}{\partial x_r}d\mathbf{C}\\
&= \frac{\partial Q_{ijkr}}{\partial x_r} - 0\\
(c) &= m\int\left(F_r - \frac{Du_r}{Dt}\right)C_iC_jC_k\frac{\partial f}{\partial C_r}d\mathbf{C}\\
&= \left(F_r - \frac{Du_r}{Dt}\right)\left[m\int\frac{\partial fC_iC_jC_k}{\partial C_r}d\mathbf{C} - m\int f\frac{\partial C_iC_jC_k}{\partial C_r}d\mathbf{C}\right]
\end{aligned}$$

$$= 0 - \left(F_r - \frac{Du_r}{Dt}\right) m \int f \frac{\partial C_i C_j C_k}{\partial C_r} d\mathbf{C}$$

$$= -\left(F_r - \frac{Du_r}{Dt}\right) m \int f \left[C_i C_j \delta_{kr} + C_i C_k \delta_{jr} + C_j C_k \delta_{ir}\right] d\mathbf{C}$$

Using the momentum conservation equation (Eq. (5.40)),

$$\left(F_r - \frac{Du_r}{Dt}\right) = \frac{1}{\rho} \frac{\partial P_{rl}}{\partial x_l}$$

$$\therefore \quad (c) = -\frac{1}{\rho} \frac{\partial P_{rl}}{\partial x_l} \left[P_{ij} \delta_{kr} + P_{ik} \delta_{jr} + P_{jk} \delta_{ir}\right]$$

$$= -\frac{1}{\rho} \left[\frac{\partial P_{kl}}{\partial x_l} P_{ij} + \frac{\partial P_{jl}}{\partial x_l} P_{ik} + \frac{\partial P_{il}}{\partial x_l} P_{jk}\right]$$

$$(d) = -m \int C_i C_j C_k C_r \frac{\partial f}{\partial C_l} \frac{\partial u_l}{\partial x_r} d\mathbf{C}$$

$$= -m \int \frac{\partial}{\partial C_l} \left(f C_i C_j C_k C_r \frac{\partial u_l}{\partial x_r}\right) d\mathbf{C} + \frac{\partial u_l}{\partial x_r} m \int f \frac{\partial C_i C_j C_k C_r}{\partial C_l} d\mathbf{C}$$

$$= 0 + \frac{\partial u_l}{\partial x_r} m \int f \left[C_i C_j C_k \delta_{rl} + C_i C_r C_k \delta_{jl} + C_i C_j C_r \delta_{kl} + C_j C_k C_r \delta_{il}\right] d\mathbf{C}$$

$$= S_{ijk} \frac{\partial u_r}{\partial x_r} + S_{irk} \frac{\partial u_j}{\partial x_r} + S_{ijr} \frac{\partial u_k}{\partial x_r} + S_{jkr} \frac{\partial u_i}{\partial x_r} \tag{6.27}$$

Denoting the production term as

$$(e) = m \int C_i C_j C_k J(f, f_1) d\mathbf{C} = \bar{J}^{(3)}_{ijk}. \tag{6.28}$$

Combining and rearranging the terms, the evolution equation for third order moment S_{ijk} is obtained as

$$\frac{\partial S_{ijk}}{\partial t} + \frac{\partial}{\partial x_r}(u_r S_{ijk} + Q_{ijkr}) + \frac{\partial u_k}{\partial x_r} S_{ijr} + \frac{\partial u_j}{\partial x_r} S_{irk} + \frac{\partial u_i}{\partial x_r} S_{jkr}$$
$$- \frac{1}{\rho} \left[P_{ij} \frac{\partial P_{kr}}{\partial x_r} + P_{ik} \frac{\partial P_{jr}}{\partial x_r} + P_{jk} \frac{\partial P_{ir}}{\partial x_r}\right] = \bar{J}^{(3)}_{ijk}, \tag{6.29}$$

The evolution equations for pressure tensor P_{ij} (6.21), stress tensor p_{ij} (6.25), and third order moment S_{ijk} (6.29) are exact. However, the system of equations is not closed since fourth order moment Q_{ijkr} appears in the divergence term of the evolution equation for S_{ijk} (6.29).

6.3.2 Expansion in Hermite Polynomials

A brief introduction to the Hermite polynomials is given in Appendix C.2. Readers are advised to go through the Appendix before proceeding to this section.

Grad proposed to expand the function $\omega^{-1/2}g$ in orthogonal tensorial Hermite functions as (the superscript n represents an nth order tensor)

$$g(\mathbf{x}, \mathbf{v}, t) = \omega(\mathbf{v}) \sum_{n=0}^{\infty} \frac{1}{n!} a_i^{(n)}(\mathbf{x}, t) \mathcal{H}_i^{(n)}(\mathbf{v}), \tag{6.30}$$

where we have introduced the dimensionless microscopic velocity $\mathbf{v}$ and the dimensionless distribution function $g(\mathbf{x}, \mathbf{v}, t)$ as

$$\mathbf{v} = \frac{\mathbf{C}}{\sqrt{RT}}, \qquad g(\mathbf{x}, \mathbf{v}, t) = \frac{f(\mathbf{x}, \mathbf{C}, t)}{\left(\dfrac{n}{(RT)^{3/2}}\right)} \tag{6.31}$$

and the weight function in three-dimensional space ($N = 3$ in definition of weight function, Eq. (C.9)) is

$$\omega(\mathbf{v}) = \frac{1}{(2\pi)^{3/2}} \exp\left(\frac{-v^2}{2}\right) \tag{6.32}$$

The purpose of introducing the term $\omega^{-1/2}$ in the function $\omega^{-1/2}g$ is to get the Maxwellian distribution function around which the expansion is sought. Also, notice carefully the functional dependence of each of the variable, especially the Hermite coefficients $a_i^{(n)}$, which depend on the space coordinates and time and not on the microscopic velocity. It will become clear in the ensuing sections that the first few significant Hermite coefficients are actually the macroscopic variables.

In expanded notation, the dimensionless distribution function is represented as

$$g = \omega\left(a^{(0)}\mathcal{H}^{(0)} + a_i^{(1)}\mathcal{H}_i^{(1)} + \frac{1}{2!}a_{ij}^{(2)}\mathcal{H}_{ij}^{(2)} + \frac{1}{3!}a_{ijk}^{(3)}\mathcal{H}_{ijk}^{(3)} + \cdots\right) \tag{6.33}$$

The Maxwellian distribution $f^{(0)}$ (using $R = k_B/m$ in Eq. (5.90)) can be written in terms of the weight function ω as

$$\begin{aligned} f^{(0)} &= \frac{n}{(RT)^{3/2}} \frac{1}{(2\pi)^{3/2}} \exp\left(\frac{-C^2}{2RT}\right) \\ &= \frac{n}{(RT)^{3/2}} \frac{1}{(2\pi)^{3/2}} \exp\left(\frac{-v^2}{2}\right) \qquad \text{[Using Eq. (6.31)]} \end{aligned}$$

$$\therefore \; f^{(0)} = \frac{n}{(RT)^{3/2}}\omega \qquad \text{[Using Eq. (6.32)]} \tag{6.34}$$

Consequently, the expansion for distribution function f can be written in the form

$$\begin{aligned} f &= \frac{n}{(RT)^{3/2}} g \\ &= \frac{n}{(RT)^{3/2}} \omega(\mathbf{v}) \sum_{n=0}^{\infty} \frac{1}{n!} a_i^{(n)}(\mathbf{x}, t) \mathscr{H}_i^{(n)}(\mathbf{v}) \\ &= f^{(0)} \sum_{n=0}^{\infty} \frac{1}{n!} a_i^{(n)} \mathscr{H}_i^{(n)} \\ &= f^{(0)} \left(a^{(0)} \mathscr{H}^{(0)} + a_i^{(1)} \mathscr{H}_i^{(1)} + \frac{1}{2!} a_{ij}^{(2)} \mathscr{H}_{ij}^{(2)} + \frac{1}{3!} a_{ijk}^{(3)} \mathscr{H}_{ijk}^{(3)} + \cdots \right) \end{aligned} \tag{6.35}$$

where the Hermite coefficients $a_i^{(n)}$ are given by

$$\begin{aligned} a_i^{(n)} &= \int g \mathscr{H}_i^{(n)} d\mathbf{v} \\ &= \frac{1}{n} \int f \mathscr{H}_i^{(n)} d\mathbf{c} \qquad \text{[In dimensional form]} \\ &= \frac{1}{n} \int f \mathscr{H}_i^{(n)} d\mathbf{C} \qquad \text{[Using } d\mathbf{c} = d\mathbf{C}\text{]} \end{aligned} \tag{6.36}$$

The first few Hermite polynomials [61] are obtained in Appendix C.2 as

$$\mathscr{H}^{(0)} = 1, \tag{6.37}$$

$$\begin{aligned} \mathscr{H}_i^{(1)} &= v_i \\ &= \frac{C_i}{\sqrt{RT}}, \end{aligned} \tag{6.38}$$

$$\begin{aligned} \mathscr{H}_{ij}^{(2)} &= v_i v_j - \delta_{ij} \\ &= \frac{C_i C_j}{RT} - \delta_{ij}, \end{aligned} \tag{6.39}$$

$$\begin{aligned} \mathscr{H}_{ijk}^{(3)} &= v_i v_j v_k - (v_i \delta_{jk} + v_j \delta_{ik} + v_k \delta_{ij}) \\ &= \frac{C_i C_j C_k}{(RT)^{3/2}} - \frac{1}{\sqrt{RT}} (C_i \delta_{jk} + C_j \delta_{ik} + C_k \delta_{ij}), \end{aligned} \tag{6.40}$$

$$
\begin{aligned}
\mathscr{H}^{(4)}_{ijkl} &= v_i v_j v_k v_l - (v_i v_j \delta_{kl} + v_i v_k \delta_{jl} + v_i v_k \delta_{jl} + v_j v_k \delta_{il} + v_j v_l \delta_{ik} + v_k v_l \delta_{ij}) \\
&\quad + (\delta_{ij}\delta_{kl} + \delta_{ik}\delta_{jl} + \delta_{il}\delta_{jk}) \\
&= \frac{C_i C_j C_k C_l}{(RT)^2} \\
&\quad - \frac{1}{RT}(C_i C_j \delta_{kl} + C_i C_k \delta_{jl} + C_i C_k \delta_{jl} + C_j C_k \delta_{il} + C_j C_l \delta_{ik} + C_k C_l \delta_{ij}) \\
&\quad + (\delta_{ij}\delta_{kl} + \delta_{ik}\delta_{jl} + \delta_{il}\delta_{jk}) \qquad (6.41)
\end{aligned}
$$

Using the dimensional form of expression for Hermite coefficients (Eq. (6.36)), and substituting the expressions for Hermite polynomials (Eqs. (6.37)–(6.41)), the first few Hermite coefficients in terms of moments are obtained as

$$
\begin{aligned}
a^{(0)} &= \frac{1}{n}\int f \mathscr{H}^{(0)}_i d\mathbf{C} \\
&= \frac{1}{\rho}\int f d\mathbf{C} \\
&= 1, \qquad (6.42)
\end{aligned}
$$

$$
\begin{aligned}
a^{(1)}_i &= \frac{1}{n}\int f \mathscr{H}^{(1)}_i d\mathbf{C} \\
&= \frac{1}{\rho\sqrt{RT}}\int f C_i d\mathbf{C} \\
&= 0 \qquad (6.43)
\end{aligned}
$$

$$
\begin{aligned}
a^{(2)}_{ij} &= \frac{1}{n}\int f \mathscr{H}^{(2)}_i d\mathbf{C} \\
&= \frac{m}{\rho RT}\int f C_i C_j d\mathbf{C} - \delta_{ij}\frac{1}{n}\int f d\mathbf{C} \qquad [\because \rho = nm] \\
&= \frac{P_{ij}}{p} - \delta_{ij} \\
&= \frac{p_{ij}}{p}, \qquad (6.44)
\end{aligned}
$$

$$
\begin{aligned}
a^{(3)}_{ijk} &= \frac{1}{n(RT)^{3/2}}\int C_i C_j C_k f d\mathbf{C} - 0 \\
&= \frac{S_{ijk}}{p\sqrt{RT}}, \qquad (6.45)
\end{aligned}
$$

$$
\begin{aligned}
a^{(4)}_{ijkl} &= \frac{Q_{ijkl}}{pRT} - \frac{1}{p}(P_{ij}\delta_{kl} + P_{ik}\delta_{jl} + P_{il}\delta_{jk} + P_{jk}\delta_{il} + P_{jl}\delta_{ik} + P_{kl}\delta_{ij}) \\
&+ (\delta_{ij}\delta_{kl} + \delta_{ik}\delta_{jl} + \delta_{il}\delta_{jk}) \\
&= \frac{Q_{ijkl}}{pRT} - \frac{1}{p}(p_{ij}\delta_{kl} + p_{ik}\delta_{jl} + p_{il}\delta_{jk} + p_{jk}\delta_{il} + p_{jl}\delta_{ik} + p_{kl}\delta_{ij}) \\
&- (\delta_{ij}\delta_{kl} + \delta_{ik}\delta_{jl} + \delta_{il}\delta_{jk})
\end{aligned} \tag{6.46}
$$

where the usual definitions of mass density (Eq. (5.10)), pressure tensor (Eq. (5.5)), third order moment S_{ijk} (Eq. (6.4)), fourth order moment Q_{ijkl} (Eq. (6.5)), ideal gas equation ($p = \rho RT = nk_B T$) are used along with Eqs. (5.15) and (5.22).

The second order Hermite coefficient $a^{(2)}_{ij}$ (Eq. (6.44)) actually turns out to be a trace-free part of the pressure tensor, i.e., stress tensor p_{ij}. Similarly, when we contract third order Hermite coefficient $a^{(3)}_{ijk}$ (Eq. (6.45)), we obtain an expression in terms of heat flux vector as

$$
a^{(3)}_{ikk} = \frac{1}{n(RT)^{3/2}} \int C_i C^2 f d\mathbf{C} = \frac{m}{\rho(RT)^{3/2}} \int C_i C^2 f d\mathbf{C} = \frac{2q_i}{p\sqrt{RT}} \tag{6.47}
$$

6.3.3 Derivation of Distribution Function

In this section, we derive the full third order approximation of the distribution function $f_{G|20}$. We will utilize this form to evaluate the unknown higher order moment Q_{ijkl} appearing in the evolution equation of third order moment S_{ijk}. The basic idea of the third order approximation is to truncate the series (Eq. (6.35)) after the first four terms by assuming the higher order Hermite coefficients $a^{(n)}_i$ to be zero for $n > 3$. Substituting the expressions for Hermite coefficients $a^{(n)}_i$ (Eqs. (6.42)–(6.45)) and Hermite polynomials $\mathscr{H}^{(n)}_i$ (Eqs. (6.37)–(6.40)), we obtain

$$
\begin{aligned}
f_{G|20} &= f^{(0)} \left\{ a^{(0)}\mathscr{H}^{(0)} + a^{(1)}_i \mathscr{H}^{(1)}_i + \frac{1}{2!} a^{(2)}_{ij}\mathscr{H}^{(2)}_{ij} + \frac{1}{3!} a^{(3)}_{ijk}\mathscr{H}^{(3)}_{ijk} \right\} \\
&= f^{(0)} \left\{ 1 + \frac{1}{2} a^{(2)}_{ij}\mathscr{H}^{(2)}_{ij} + \frac{1}{6} a^{(3)}_{ijk}\mathscr{H}^{(3)}_{ijk} \right\} \qquad [\because a^{(0)} = \mathscr{H}^{(0)} = 1,\ a^{(1)}_i = 0] \\
&= f^{(0)} \left\{ 1 + \frac{1}{2}\frac{p_{ij}}{p}\left[\frac{C_i C_j}{RT} - \delta_{ij}\right] + \frac{1}{6}\frac{S_{ijk}}{p\sqrt{RT}}\left[\frac{C_i C_j C_k}{(RT)^{3/2}} \right.\right. \\
&\left.\left. - \frac{1}{\sqrt{RT}}(C_i\delta_{jk} + C_j\delta_{ik} + C_k\delta_{ij})\right]\right\}
\end{aligned}
$$

$$= f^{(0)}\left\{1 + \frac{1}{2}\frac{p_{ij}}{pRT}C_iC_j - \frac{1}{2}\frac{p_{ii}}{p} + \frac{S_{ijk}}{6p(RT)^2}C_iC_jC_k \right.$$
$$\left. - \frac{1}{6pRT}(C_iS_{ijj} + C_jS_{iji} + C_kS_{iik})\right\}$$
$$= f^{(0)}\left\{1 + \frac{p_{ij}}{2pRT}C_iC_j + \frac{S_{ijk}}{6p(RT)^2}C_iC_jC_k - \frac{S_i}{2pRT}C_i\right\} \quad [\because p_{ii} = 0] \tag{6.48}$$

Hence, the final form of the Grad 20-moment distribution function $f_{G|20}$ in terms of first 20 moments is obtained as

$$f_{G|20} = f^{(0)}\left\{1 + \frac{p_{ij}}{2pRT}C_iC_j + \frac{S_{ijk}}{6pR^2T^2}C_iC_jC_k - \frac{S_i}{2pRT}C_i\right\} \tag{6.49}$$

6.3.4 Closure

To obtain closure and reduce them to third order, Q_{ijkr} must be replaced by its third order equivalent, i.e., the unknown Q_{ijkr} must be expressed in terms of primary variables. For full third order approximation, the Hermite coefficients $a_i^{(n)}$ must be zero for $n > 3$. Recalling the equation for coefficient $a_{ijkr}^{(4)}$ (Eq. (6.46)) which we derived previously,

$$a_{ijkr}^{(4)} = \frac{Q_{ijkr}}{pRT} - \frac{1}{p}(p_{ij}\delta_{kr} + p_{ik}\delta_{jr} + p_{ir}\delta_{jk} + p_{jk}\delta_{ir} + p_{jr}\delta_{ik} + p_{kr}\delta_{ij})$$
$$- (\delta_{ij}\delta_{kr} + \delta_{ik}\delta_{jr} + \delta_{ir}\delta_{jk}) \tag{6.50}$$

and substituting $a_{ijkr}^{(4)} = 0$, we obtain the constitutive relationship for moment Q_{ijkr} and achieve the necessary closure as

$$Q_{ijkr} = RT(p_{ij}\delta_{kr} + p_{ik}\delta_{jr} + p_{ir}\delta_{jk} + p_{jk}\delta_{ir} + p_{jr}\delta_{ik} + p_{kr}\delta_{ij})$$
$$+ pRT(\delta_{ij}\delta_{kr} + \delta_{ik}\delta_{jr} + \delta_{ir}\delta_{jk}) \tag{6.51}$$

An important remark needs to be made here. We could have utilized the distribution function $f_{G|20}$ to evaluate the fourth order moment Q_{ijkr} as

$$Q_{ijkr} = m\int C_iC_jC_kC_r f_{G|20} d\mathbf{C} \tag{6.52}$$

which is however not done here. Rather, the knowledge of the higher order Hermite coefficients helped us to derive the necessary constitutive relationship for Q_{ijkr}.

6.3.5 20-Moment Equations

To obtain a closed set of equations, we substitute the constitutive relationship for Q_{ijkr} (Eq. (6.51)) in Eq. (6.29). Recalling the evolution equation for S_{ijk},

$$\frac{\partial S_{ijk}}{\partial t}+\frac{\partial u_r S_{ijk}}{\partial x_r}+\underbrace{\frac{\partial Q_{ijkr}}{\partial x_r}}_{(i)}+\frac{\partial u_k}{\partial x_r}S_{ijr}+\frac{\partial u_j}{\partial x_r}S_{irk}+\frac{\partial u_i}{\partial x_r}S_{rjk}$$
$$\underbrace{-\frac{1}{\rho}\left[P_{ij}\frac{\partial P_{kr}}{\partial x_r}+P_{ik}\frac{\partial P_{jr}}{\partial x_r}+P_{jk}\frac{\partial P_{ir}}{\partial x_r}\right]}_{(ii)}=\bar{J}^{(3)}_{ijk}, \tag{6.53}$$

Evaluating terms (i) and (ii) separately as

$$\begin{aligned}
(i)=\frac{\partial Q_{ijkr}}{\partial x_r}&=\frac{\partial(RTp_{ij})}{\partial x_k}+\frac{\partial(RTp_{ik})}{\partial x_j}+\delta_{jk}\frac{\partial(RTp_{ir})}{\partial x_r}+\frac{\partial(RTp_{jk})}{\partial x_i}\\
&+\delta_{ik}\frac{\partial(RTp_{jr})}{\partial x_r}+\delta_{ij}\frac{\partial(RTp_{kr})}{\partial x_r}+\frac{\partial}{\partial x_r}\left[pRT\left(\delta_{ij}\delta_{kr}+\delta_{ik}\delta_{jr}+\delta_{ir}\delta_{jk}\right)\right]\\
&=\frac{\partial(RTp_{ij})}{\partial x_k}+\frac{\partial(RTp_{ik})}{\partial x_j}+\frac{\partial(RTp_{jk})}{\partial x_i}+\delta_{ij}\frac{\partial(RTp_{kr})}{\partial x_r}+\delta_{jk}\frac{\partial(RTp_{ir})}{\partial x_r}\\
&+\delta_{ik}\frac{\partial(RTp_{jr})}{\partial x_r}+\delta_{ij}\frac{\partial(pRT)}{\partial x_k}+\delta_{ik}\frac{\partial(pRT)}{\partial x_j}+\delta_{jk}\frac{\partial(pRT)}{\partial x_i}\\
&=\frac{\partial(RTp_{ij})}{\partial x_k}+\frac{\partial(RTp_{ik})}{\partial x_j}+\frac{\partial(RTp_{jk})}{\partial x_i}+\frac{\partial RT}{\partial x_r}[p_{kr}\delta_{ij}+p_{ir}\delta_{jk}+p_{jr}\delta_{ik}]\\
&\underbrace{+RT\left[\delta_{ij}\frac{\partial p_{kr}}{\partial x_r}+\delta_{jk}\frac{\partial p_{ir}}{\partial x_r}+\delta_{ik}\frac{\partial p_{jr}}{\partial x_r}\right]}_{(i-a)}\\
&+p\left[\frac{\partial RT}{\partial x_k}\delta_{ij}+\frac{\partial RT}{\partial x_j}\delta_{ik}+\frac{\partial RT}{\partial x_i}\delta_{jk}\right]\underbrace{+RT\left[\delta_{ij}\frac{\partial p}{\partial x_k}+\delta_{ik}\frac{\partial p}{\partial x_j}+\delta_{jk}\frac{\partial p}{\partial x_i}\right]}_{(i-b)}
\end{aligned} \tag{6.54}$$

$$\begin{aligned}
(ii)&=-\frac{1}{\rho}\left[P_{ij}\frac{\partial P_{kr}}{\partial x_r}+P_{ik}\frac{\partial P_{jr}}{\partial x_r}+P_{jk}\frac{\partial P_{ir}}{\partial x_r}\right]\\
&=-\frac{1}{\rho}\left[(p_{ij}+p\delta_{ij})\frac{\partial P_{kr}}{\partial x_r}+(p_{ik}+p\delta_{ik})\frac{\partial P_{jr}}{\partial x_r}+(p_{jk}+p\delta_{jk})\frac{\partial P_{ir}}{\partial x_r}\right]
\end{aligned}$$

$$
\begin{aligned}
&= -\frac{1}{\rho}\left[p_{ij}\frac{\partial P_{kr}}{\partial x_r} + p_{ik}\frac{\partial P_{jr}}{\partial x_r} + p_{jk}\frac{\partial P_{ir}}{\partial x_r}\right] \\
&\quad - \frac{1}{\rho}\left[p\delta_{ij}\frac{\partial P_{kr}}{\partial x_r} + p\delta_{ik}\frac{\partial P_{jr}}{\partial x_r} + p\delta_{jk}\frac{\partial P_{ir}}{\partial x_r}\right] \\
&= -\frac{1}{\rho}\left[p_{ij}\frac{\partial(p_{kr} + p\delta_{kr})}{\partial x_r} + p_{ik}\frac{\partial(p_{jr} + p\delta_{jr})}{\partial x_r} + p_{jk}\frac{\partial(p_{ir} + p\delta_{ir})}{\partial x_r}\right] \\
&\quad - \frac{1}{\rho}\left[p\delta_{ij}\frac{\partial(p_{kr} + p\delta_{kr})}{\partial x_r} + p\delta_{ik}\frac{\partial(p_{jr} + p\delta_{jr})}{\partial x_r} + p\delta_{jk}\frac{\partial(p_{ir} + p\delta_{ir})}{\partial x_r}\right] \\
&= -\frac{1}{\rho}\left[p_{ij}\frac{\partial p_{kr}}{\partial x_r} + p_{ik}\frac{\partial p_{jr}}{\partial x_r} + p_{jk}\frac{\partial p_{ir}}{\partial x_r}\right] - \frac{1}{\rho}\left[p_{ij}\frac{\partial p}{\partial x_k} + p_{ik}\frac{\partial p}{\partial x_j} + p_{jk}\frac{\partial p}{\partial x_i}\right] \\
&\quad \underbrace{-RT\left[\delta_{ij}\frac{\partial p_{kr}}{\partial x_r} + \delta_{ik}\frac{\partial p_{jr}}{\partial x_r} + \delta_{jk}\frac{\partial p_{ir}}{\partial x_r}\right]}_{(ii-a)} \underbrace{-RT\left[\delta_{ij}\frac{\partial p}{\partial x_k} + \delta_{ik}\frac{\partial p}{\partial x_j} + \delta_{jk}\frac{\partial p}{\partial x_i}\right]}_{(ii-b)}
\end{aligned}
\tag{6.55}
$$

Substituting the expressions for (i) and (ii) in Eq. (6.53) and noting that the terms $(i-a)$ and $(i-b)$ of Eq. (6.54) cancel with the terms $(ii-a)$ and $(ii-b)$ of Eq. (6.55), respectively, and rearranging the terms, we obtain an equation for S_{ijk} as

$$
\begin{aligned}
&\frac{\partial S_{ijk}}{\partial t} + \frac{\partial}{\partial x_r}(u_r S_{ijk}) + \left(\frac{\partial u_k}{\partial x_r}S_{ijr} + \frac{\partial u_j}{\partial x_r}S_{irk} + \frac{\partial u_i}{\partial x_r}S_{rjk}\right) \\
&\quad + \left[\frac{\partial RTp_{ij}}{\partial x_k} + \frac{\partial RTp_{ik}}{\partial x_j} + \frac{\partial RTp_{jk}}{\partial x_i}\right] + \frac{\partial RT}{\partial x_r}[p_{kr}\delta_{ij} + p_{ir}\delta_{jk} + p_{jr}\delta_{ik}] \\
&\quad + p\left[\frac{\partial RT}{\partial x_k}\delta_{ij} + \frac{\partial RT}{\partial x_j}\delta_{ik} + \frac{\partial RT}{\partial x_i}\delta_{jk}\right] - \frac{1}{\rho}\left[p_{ij}\frac{\partial p_{kr}}{\partial x_r} + p_{ik}\frac{\partial p_{jr}}{\partial x_r} + p_{jk}\frac{\partial p_{ir}}{\partial x_r}\right] \\
&\quad - \frac{1}{\rho}\left[p_{ij}\frac{\partial p}{\partial x_k} + p_{ik}\frac{\partial p}{\partial x_j} + p_{jk}\frac{\partial p}{\partial x_i}\right] = \bar{J}^{(3)}_{ijk}
\end{aligned}
\tag{6.56}
$$

Notice that without explicit expressions for the production terms $\left[\bar{J}^{(2)}_{ij}, \bar{J}^{(3)}_{ijk}\right]$, the equation is still incomplete. The evaluation of the production terms is complicated. However, in the case of Maxwell molecules, even without knowing the full functional form of the distribution form, the exact values can be computed [149] and they read as

$$
\bar{J}^{(2)}_{ij} = -\frac{p}{\mu}p_{ij} \tag{6.57}
$$

$$
\bar{J}^{(3)}_{ijk} = -\frac{3}{2}\frac{p}{\mu}\left[S_{ijk} - \frac{1}{5}(S_i\delta_{jk} + S_j\delta_{ik} + S_k\delta_{ij})\right] \tag{6.58}
$$

We are now ready with the full 20-moment equations in closed form

$$\frac{\partial \rho}{\partial t} + \frac{\partial}{\partial x_i}(\rho u_i) = 0, \tag{6.59}$$

$$\frac{\partial u_i}{\partial t} + u_k \frac{\partial u_i}{\partial x_k} + \frac{1}{\rho}\frac{\partial p}{\partial x_i} + \frac{1}{\rho}\frac{\partial p_{ik}}{\partial x_k} = F_i, \tag{6.60}$$

$$\frac{3}{2}R\rho \frac{\partial T}{\partial t} + \frac{3}{2}R\rho u_k \frac{\partial T}{\partial x_k} + p\frac{\partial u_i}{\partial x_i} + p_{ij}\frac{\partial u_i}{\partial x_j} + \frac{\partial q_k}{\partial x_k} = 0, \tag{6.61}$$

$$\begin{aligned}&\frac{\partial p_{ij}}{\partial t} + \frac{\partial}{\partial x_r}(u_r p_{ij}) + \frac{\partial S_{ijr}}{\partial x_r} - \frac{2}{3}\delta_{ij}\frac{\partial q_k}{\partial x_k} + p_{ir}\frac{\partial u_j}{\partial x_r} + p_{jr}\frac{\partial u_i}{\partial x_r}\\ &- \frac{2}{3}\delta_{ij} p_{kl}\frac{\partial u_k}{\partial x_l} + p\left(\frac{\partial u_i}{\partial x_j} + \frac{\partial u_j}{\partial x_i} - \frac{2}{3}\delta_{ij}\frac{\partial u_k}{\partial x_k}\right) = -\frac{p}{\mu}p_{ij},\end{aligned} \tag{6.62}$$

$$\begin{aligned}&\frac{\partial S_{ijk}}{\partial t} + \frac{\partial}{\partial x_r}(u_r S_{ijk}) + \left(\frac{\partial u_k}{\partial x_r}S_{ijr} + \frac{\partial u_j}{\partial x_r}S_{irk} + \frac{\partial u_i}{\partial x_r}S_{rjk}\right)\\ &+ \left[\frac{\partial RTp_{ij}}{\partial x_k} + \frac{\partial RTp_{ik}}{\partial x_j} + \frac{\partial RTp_{jk}}{\partial x_i}\right] + \frac{\partial RT}{\partial x_r}[p_{kr}\delta_{ij} + p_{ir}\delta_{jk} + p_{jr}\delta_{ik}]\\ &+ p\left[\frac{\partial RT}{\partial x_k}\delta_{ij} + \frac{\partial RT}{\partial x_j}\delta_{ik} + \frac{\partial RT}{\partial x_i}\delta_{jk}\right] - \frac{1}{\rho}\left[p_{ij}\frac{\partial p_{kr}}{\partial x_r} + p_{ik}\frac{\partial p_{jr}}{\partial x_r} + p_{jk}\frac{\partial p_{ir}}{\partial x_r}\right]\\ &- \frac{1}{\rho}\left[p_{ij}\frac{\partial p}{\partial x_k} + p_{ik}\frac{\partial p}{\partial x_j} + p_{jk}\frac{\partial p}{\partial x_i}\right] = -\frac{3}{2}\frac{p}{\mu}\left[S_{ijk} - \frac{1}{5}(S_i\delta_{jk} + S_j\delta_{ik} + S_k\delta_{ij})\right]\end{aligned} \tag{6.63}$$

This set of 20 scalar equations determines the evolution of 20 third-order variables, viz., ρ, $\mathbf{u}$, T, p_{ij}, and S_{ijk} $(1 + 3 + 1 + 5 + 10 = 20)$ for Maxwellian molecules. The limitation of Maxwellian molecules arises because the production terms are known only for these molecules. This completes the derivation of Grad 20-moment equations.

6.4 The 13-Moment Approximation

In this section, we present the derivation of Grad 13-moment equations. The variables defining the state of the gas are ρ, $\mathbf{u}$, T, p_{ij}, and $\mathbf{q}$ which are thirteen $(1 + 3 + 1 + 5 + 3 = 13)$ in number. This set of Grad 13-moment equations is a more preferred choice over 20-moment equations since physical significance can be attributed to each of these variables. Remember, in 20-moment approximation, second order moment turned out to be the stress tensor, but, third order moment S_{ijk} does not seem to have any physical significance. However, the contracted third order moment S_i actually turns out to be twice the heat flux vector (see Eq. (6.6)) as

$$S_{ikk} = m \int C_i C^2 f d\mathbf{C} = 2q_i \tag{6.64}$$

Hence, it is more appropriate to consider moments having physical intuitive meaning rather than the full third order approximation. Accordingly, we include stress tensor p_{ij} and heat flux vector q_i in the primary variables set in addition to density ρ, bulk velocity u_i, and temperature T. The expressions for the primary variables and unknowns appearing in the 13-moment approximation are shown in Fig. 6.3.

The present section is outlined as follows: The evolution equations for stress tensor (p_{ij}) and heat flux vector (q_i) will be generated. The higher order moments appearing in these equations will be made divergence-free, a standard currently followed in the literature. The distribution function $f_{G|13}$ is then derived and utilized to evaluate the unknowns appearing in the evolution equations for p_{ij} and q_i to obtain a closed set of equations.

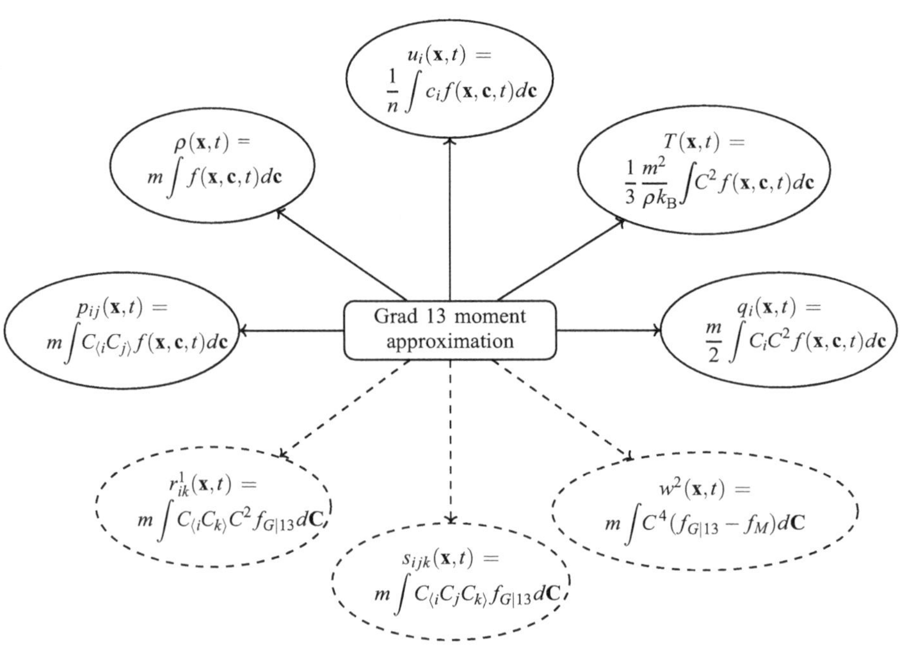

Fig. 6.3 Expression for the variables considered in the Grad 13-moment approximation; variables in solid circles are the primary variables while those in dashed circles are the closure variables evaluated using Grad 13-moment distribution function $f_{G|13}$

6.4.1 Evolution Equations for Stress Tensor and Heat Flux Vector

The evolution equation for stress tensor p_{ij} was derived in Sect. 6.3.1.1 as

$$\frac{\partial p_{ij}}{\partial t} + \frac{\partial}{\partial x_r}(u_r p_{ij}) + \underline{\frac{\partial S_{ijk}}{\partial x_k} - \frac{1}{3}\delta_{ij}\frac{\partial S_k}{\partial x_k}} + p_{ik}\frac{\partial u_j}{\partial x_k} + p_{jk}\frac{\partial u_i}{\partial x_k}$$
$$- \frac{2}{3}\delta_{ij}p_{kl}\frac{\partial u_k}{\partial x_l} + p\left(\frac{\partial u_i}{\partial x_j} + \frac{\partial u_j}{\partial x_i} - \frac{2}{3}\delta_{ij}\frac{\partial u_k}{\partial x_k}\right) = \bar{J}_{ij}^{(2)} \tag{6.65}$$

However, the third order moment S_{ijk} although symmetric is not divergence-free. To make S_{ijk} divergence-free, we use the following mathematical identity for the decomposition of any third order tensor W_{ijk},

$$w_{\langle ijk\rangle} = W_{(ijk)} - \frac{1}{5}\left[W_{(ill)}\delta_{jk} + W_{(jll)}\delta_{ik} + W_{(kll)}\delta_{ij}\right] \tag{6.66}$$

where the angular brackets denote the symmetric and trace-free part of the tensor and round brackets denote the symmetric part of the tensor which is obtained as

$$W_{(ijk)} = \frac{1}{6}\left(W_{ijk} + W_{kij} + W_{jki} + W_{ikj} + W_{jik} + W_{kji}\right) \tag{6.67}$$

Since S_{ijk} is already symmetric, we get rid of the round brackets and its trace-free part is then given as

$$s_{ijk} = S_{ijk} - \frac{1}{5}\left[S_{ill}\delta_{jk} + S_{jll}\delta_{ik} + S_{kll}\delta_{ij}\right] \tag{6.68}$$

With this decomposition, the underlined terms in Eq. (6.65) are evaluated as

$$\begin{aligned}
\frac{\partial S_{ijk}}{\partial x_k} - \frac{1}{3}\delta_{ij}\frac{\partial S_k}{\partial x_k} &= \frac{\partial s_{ijk}}{\partial x_k} + \frac{1}{5}\frac{\partial}{\partial x_k}\left[S_{ill}\delta_{jk} + S_{jll}\delta_{ik} + S_{kll}\delta_{ij}\right] - \frac{1}{3}\delta_{ij}\frac{\partial S_k}{\partial x_k} \\
&= \frac{\partial s_{ijk}}{\partial x_k} + \frac{1}{5}\frac{\partial S_i}{\partial x_j} + \frac{1}{5}\frac{\partial S_j}{\partial x_i} + \frac{1}{5}\delta_{ij}\frac{\partial S_k}{\partial x_k} - \frac{1}{3}\delta_{ij}\frac{\partial S_k}{\partial x_k} \\
&= \frac{\partial s_{ijk}}{\partial x_k} + \frac{2}{5}\frac{\partial q_i}{\partial x_j} + \frac{2}{5}\frac{\partial q_j}{\partial x_i} + \frac{2}{5}\delta_{ij}\frac{\partial q_k}{\partial x_k} - \frac{2}{3}\delta_{ij}\frac{\partial q_k}{\partial x_k} \quad [\because S_i = 2q_i] \\
&= \frac{\partial s_{ijk}}{\partial x_k} + \frac{4}{5}\left[\frac{1}{2}\left(\frac{\partial q_i}{\partial x_j} + \frac{\partial q_j}{\partial x_i}\right) - \frac{1}{3}\delta_{ij}\frac{\partial q_k}{\partial x_k}\right] \\
&= \frac{\partial s_{ijk}}{\partial x_k} + \frac{4}{5}\frac{\partial q_{\langle i}}{\partial x_{j\rangle}}
\end{aligned} \tag{6.69}$$

Introducing the Lagrangian derivative and using $\bar{J}_{ij}^{(2)} = -\frac{p}{\mu} p_{ij}$ for Maxwellian molecule (from Eq. (6.58)), the final evolution equation for the stress tensor is obtained as

$$\frac{Dp_{ij}}{Dt} + p_{ij}\frac{\partial u_k}{\partial x_k} + \frac{\partial s_{ijk}}{\partial x_k} + \frac{4}{5}\frac{\partial q_{\langle i}}{\partial x_{j\rangle}} + 2p_{r\langle i}\frac{\partial u_{j\rangle}}{\partial x_r} + 2p\frac{\partial u_{\langle i}}{\partial x_{j\rangle}} = -\frac{p}{\mu} p_{ij} \tag{6.70}$$

To obtain the evolution equation for the heat flux vector q_i, we multiply the transformed Boltzmann equation (Eq. (6.10)) by $\frac{m}{2} C_i C^2$ and integrate over the velocity space as (The argument for certain integrals being zero is same as given in Sect. 6.3.1.1 and therefore these arguments are not repeated here.)

$$\begin{aligned}&\frac{m}{2}\int C_i C^2 \frac{Df}{Dt} d\mathbf{C} + \frac{m}{2}\int C_i C^2 C_k \frac{\partial f}{\partial x_k} d\mathbf{C} + \frac{m}{2}\int \left(F_k - \frac{Du_k}{Dt}\right) C_i C^2 \frac{\partial f}{\partial C_k} d\mathbf{C}\\ &-\frac{m}{2}\int C_i C^2 C_k \frac{\partial f}{\partial C_j}\frac{\partial u_j}{\partial x_k} d\mathbf{C} = \frac{m}{2}\int C_i C^2 J(f, f_1) d\mathbf{C}\end{aligned} \tag{6.71}$$

Applying integration by parts and Leibnitz's rule to take out the differentiation outside the integral, we evaluate term by term as

$$\begin{aligned}(a) &= \frac{m}{2}\int C_i C^2 \frac{Df}{Dt} d\mathbf{C}\\ &= \frac{m}{2}\int \frac{Df C_i C^2}{Dt} d\mathbf{C} - \frac{m}{2}\int f \frac{DC_i C^2}{Dt} d\mathbf{C}\\ &= \frac{Dq_i}{Dt} - 0\end{aligned} \tag{6.72}$$

$$\begin{aligned}(b) &= \frac{m}{2}\int C_i C^2 C_k \frac{\partial f}{\partial x_k} d\mathbf{C}\\ &= \frac{m}{2}\int \frac{\partial f C_i C_k C^2}{\partial x_k} d\mathbf{C} - \frac{m}{2}\int f \frac{\partial C_i C^2 C_k}{\partial x_k} d\mathbf{C}\\ &= \frac{1}{2}\frac{\partial R_{ik}^1}{\partial x_k} - 0 \qquad \left[\text{Say } R_{ik}^1 = m\int f C_i C_k C^2 d\mathbf{C}\right]\end{aligned} \tag{6.73}$$

$$\begin{aligned}(c) &= \frac{m}{2}\int \left(F_k - \frac{Du_k}{Dt}\right) C_i C^2 \frac{\partial f}{\partial C_k} d\mathbf{C}\\ &= \left(F_k - \frac{Du_k}{Dt}\right)\left[\frac{m}{2}\int \frac{\partial f C_i C^2}{\partial C_k} d\mathbf{C} - \frac{m}{2}\int f \frac{\partial C_i C^2}{\partial C_k} d\mathbf{C}\right]\\ &= 0 - \left(F_k - \frac{Du_k}{Dt}\right)\frac{m}{2}\int f \frac{\partial C_i C_l C_l}{\partial C_k} d\mathbf{C} \qquad \left[\text{Since } C^2 = C_l C_l\right]\\ &= -\left(F_k - \frac{Du_k}{Dt}\right)\frac{m}{2}\int f\left[C_i C_l \delta_{lk} + C_i C_l \delta_{lk} + C_l C_l \delta_{ik}\right] d\mathbf{C}\end{aligned}$$

Using the following form of momentum conservation equation (Eq. (5.40)),

$$\left(F_k - \frac{Du_k}{Dt}\right) = \frac{1}{\rho}\frac{\partial P_{kl}}{\partial x_l} = \frac{1}{\rho}\frac{\partial p}{\partial x_k} + \frac{1}{\rho}\frac{\partial p_{kl}}{\partial x_l} \qquad [\because p_{kl} = P_{kl} - p\delta_{kl}]$$

$$\begin{aligned}
\therefore \quad (c) &= -\left(\frac{1}{\rho}\frac{\partial p}{\partial x_k} + \frac{1}{\rho}\frac{\partial p_{kl}}{\partial x_l}\right)\frac{m}{2}\int f\left[2C_iC_k + C^2\delta_{ik}\right]d\mathbf{C} \\
&= -\left(\frac{1}{\rho}\frac{\partial p}{\partial x_k} + \frac{1}{\rho}\frac{\partial p_{kl}}{\partial x_l}\right)\left[\frac{m}{2}\int f2C_iC_kd\mathbf{C} + \frac{m}{2}\int C^2\delta_{ik}d\mathbf{C}\right] \\
&= -\left(\frac{1}{\rho}\frac{\partial p}{\partial x_k} + \frac{1}{\rho}\frac{\partial p_{kl}}{\partial x_l}\right)\left[P_{ik} + \frac{3}{2}p\delta_{ik}\right] \qquad \left[\because p = \frac{m}{3}\int C^2 f d\mathbf{c}\right] \\
&= -\left(\frac{1}{\rho}\frac{\partial p}{\partial x_k} + \frac{1}{\rho}\frac{\partial p_{kl}}{\partial x_l}\right)\left[p_{ik} + p\delta_{ik} + \frac{3}{2}p\delta_{ik}\right] \\
&= -\left(\frac{1}{\rho}\frac{\partial p}{\partial x_k} + \frac{1}{\rho}\frac{\partial p_{kl}}{\partial x_l}\right)\left[p_{ik} + \frac{5}{2}p\delta_{ik}\right] \\
&= -\frac{p_{ik}}{\rho}\frac{\partial p}{\partial x_k} - \frac{5}{2}\frac{p}{\rho}\frac{\partial p}{\partial x_i} - \frac{p_{ik}}{\rho}\frac{\partial p_{kl}}{\partial x_l} - \frac{5}{2}\frac{p}{\rho}\frac{\partial p_{il}}{\partial x_l} \\
(d) &= -\frac{m}{2}\int C_iC^2C_k\frac{\partial f}{\partial C_j}\frac{\partial u_j}{\partial x_k}d\mathbf{C} = -\frac{m}{2}\int\frac{\partial}{\partial C_j}\left(fC_iC^2C_k\right)\frac{\partial u_j}{\partial x_k}d\mathbf{C} \\
&\quad + \frac{\partial u_j}{\partial x_k}\frac{m}{2}\int f\frac{\partial C_iC_kC_rC_r}{\partial C_j}d\mathbf{C} = 0 + \frac{\partial u_j}{\partial x_k}\frac{m}{2} \\
&\quad \times \int f\left[C_iC_kC_r\delta_{jr} + C_iC_kC_r\delta_{jr} + C_iC_rC_r\delta_{jk} + C_kC_rC_r\delta_{ij}\right]d\mathbf{C} \\
&= \frac{\partial u_j}{\partial x_k}\frac{m}{2}\int f\left[2C_iC_kC_j + C_iC^2\delta_{jk} + C_kC^2\delta_{ij}\right]d\mathbf{C} \\
&= \frac{\partial u_j}{\partial x_k}\frac{m}{2}\int 2fC_iC_kC_jd\mathbf{C} + \frac{\partial u_j}{\partial x_k}\delta_{jk}\frac{m}{2}\int fC_iC^2d\mathbf{C} \\
&\quad + \frac{\partial u_j}{\partial x_k}\delta_{ij}\frac{m}{2}\int fC_kC^2d\mathbf{C} = \frac{\partial u_j}{\partial x_k}S_{ijk} + \frac{\partial u_k}{\partial x_k}q_i + \frac{\partial u_i}{\partial x_k}q_k
\end{aligned} \tag{6.74}$$

To make S_{ijk} divergence-free, we use decomposition (Eq. (6.68)) for the first term $\frac{\partial u_j}{\partial x_k}S_{ijk}$ and proceed as

$$
\begin{aligned}
\frac{\partial u_j}{\partial x_k} S_{ijk} &= \frac{\partial u_j}{\partial x_k}\left[s_{ijk} + \frac{1}{5}(S_i\delta_{jk} + S_j\delta_{ik} + S_k\delta_{ij})\right] \\
&= \frac{\partial u_j}{\partial x_k}\left[s_{ijk} + \frac{2}{5}(q_i\delta_{jk} + q_j\delta_{ik} + q_k\delta_{ij})\right] \qquad [\because S_i = 2q_i] \\
&= \frac{\partial u_j}{\partial x_k} s_{ijk} + \frac{2}{5} q_i \frac{\partial u_k}{\partial x_k} + \frac{2}{5} q_j \frac{\partial u_j}{\partial x_i} + \frac{2}{5} q_k \frac{\partial u_i}{\partial x_k}
\end{aligned}
$$

Substituting the above expression in the (d) term, we have

$$
(d) = \frac{\partial u_j}{\partial x_k} s_{ijk} + \frac{7}{5} q_i \frac{\partial u_k}{\partial x_k} + \frac{7}{5} q_k \frac{\partial u_i}{\partial x_k} + \frac{2}{5} q_j \frac{\partial u_j}{\partial x_i} \tag{6.75}
$$

The production term for the Maxwellian molecules can be computed without knowing the actual form of the distribution function as

$$
(e) = \frac{m}{2} \int C_i C^2 J(f, f_1) d\mathbf{C} = -\frac{2}{3}\frac{p}{\mu} q_i \tag{6.76}
$$

Combining all the terms, the equation for heat flux vector q_i is obtained as

$$
\begin{aligned}
&\frac{Dq_i}{Dt} + \underbrace{\frac{1}{2}\frac{\partial R^1_{ik}}{\partial x_k}}_{(i)} \underbrace{- \frac{p_{ik}}{\rho}\frac{\partial p}{\partial x_k} - \frac{5}{2}\frac{p}{\rho}\frac{\partial p}{\partial x_i}}_{(ii)} - \frac{p_{ik}}{\rho}\frac{\partial p_{kl}}{\partial x_l} - \frac{5}{2}\frac{p}{\rho}\frac{\partial p_{il}}{\partial x_l} \\
&\quad + \frac{\partial u_j}{\partial x_k} s_{ijk} + \frac{7}{5} q_i \frac{\partial u_k}{\partial x_k} + \frac{7}{5} q_k \frac{\partial u_i}{\partial x_k} + \frac{2}{5} q_j \frac{\partial u_j}{\partial x_i} = -\frac{2}{3}\frac{p}{\mu} q_i,
\end{aligned} \tag{6.77}
$$

The higher order moment R^1_{ik} appearing in term (i) is, however, still not divergence-free. To make R^1_{ik} divergence-free, we decompose it as

$$
\begin{aligned}
(i) &= \frac{1}{2}\frac{\partial R^1_{ik}}{\partial x_k} \\
&= \frac{1}{2}\frac{\partial}{\partial x_k} m \int f C_i C_k C^2 d\mathbf{C} \\
&= \frac{1}{2}\frac{\partial}{\partial x_k} m \int f C_{\langle i} C_{k\rangle} C^2 d\mathbf{C} \\
&\quad + \frac{1}{2}\frac{\partial}{\partial x_k} m \int f \frac{1}{3} C^4 \delta_{ik} d\mathbf{C} \left[\because C_i C_k = C_{\langle i} C_{k\rangle} + \frac{C^2}{3}\delta_{ik}\right] \\
&= \frac{1}{2}\frac{\partial r^1_{ik}}{\partial x_k} + \frac{1}{6}\frac{\partial}{\partial x_i} m \int f C^4 d\mathbf{C} \qquad \left[\text{Say, } m \int f C_{\langle i} C_{k\rangle} C^2 d\mathbf{C} = r^1_{ik}\right]
\end{aligned} \tag{6.78}
$$

For term (ii) of Eq. (6.77), we use ideal gas equation $p = \rho RT$ to replace the pressure term as

$$
\begin{aligned}
(ii) &= -\frac{p_{ik}}{\rho}\frac{\partial p}{\partial x_k} - \frac{5}{2}\frac{p}{\rho}\frac{\partial p}{\partial x_i} \\
&= -\frac{p_{ik}}{\rho}\frac{\partial \rho RT}{\partial x_k} - \frac{5}{2}RT\frac{\partial \rho RT}{\partial x_i} \\
&= -p_{ik}R\frac{\partial T}{\partial x_k} - \frac{p_{ik}}{\rho}RT\frac{\partial \rho}{\partial x_k} - \frac{5}{2}\frac{\partial \rho (RT)^2}{\partial x_i} + \frac{5}{2}\rho RT\frac{\partial RT}{\partial x_i}
\end{aligned} \tag{6.79}
$$

The third term on the right-hand side of the above equation turns out to be

$$
\frac{5}{2}\frac{\partial \rho (RT)^2}{\partial x_i} = \frac{1}{6}\frac{\partial}{\partial x_i}m\int f_M C^4 d\mathbf{C}
$$

where f_M is a Maxwellian distribution function.

$$
\begin{aligned}
\therefore (ii) &= -\frac{p_{ik}}{\rho}\frac{\partial p}{\partial x_k} - \frac{5}{2}\frac{p}{\rho}\frac{\partial p}{\partial x_i} \\
&= -p_{ik}R\frac{\partial T}{\partial x_k} - \frac{p_{ik}}{\rho}RT\frac{\partial \rho}{\partial x_k} - \frac{1}{6}\frac{\partial}{\partial x_i}m\int f_M C^4 d\mathbf{C} + \frac{5}{2}\rho RT\frac{\partial RT}{\partial x_i}
\end{aligned} \tag{6.80}
$$

Combining the results for terms (i) (Eq. (6.78)) and (ii) (Eq. (6.80)), we proceed

$$
\begin{aligned}
(i) + (ii) &= \frac{1}{2}\frac{\partial R^1_{ik}}{\partial x_k} - \frac{p_{ik}}{\rho}\frac{\partial p}{\partial x_k} - \frac{5}{2}\frac{p}{\rho}\frac{\partial p}{\partial x_i} \\
&= \frac{1}{2}\frac{\partial r^1_{ik}}{\partial x_k} + \frac{1}{6}\frac{\partial w^2}{\partial x_i} - p_{ik}R\frac{\partial T}{\partial x_k} - \frac{p_{ik}}{\rho}RT\frac{\partial \rho}{\partial x_k} + \frac{5}{2}\rho RT\frac{\partial RT}{\partial x_i}
\end{aligned} \tag{6.81}
$$

where we have introduced $w^2 = m\int C^4(f - f_M)d\mathbf{C}$.

Substituting results for terms (i)+(ii) (Eq. (6.81)) into Eq. (6.77), we finally have the evolution equation for the heat flux vector along with stress tensor equation (for the sake of completeness) as

$$
\frac{Dp_{ij}}{Dt} + p_{ij}\frac{\partial u_k}{\partial x_k} + \frac{\partial s_{ijk}}{\partial x_k} + \frac{4}{5}\frac{\partial q_{\langle i}}{\partial x_{j\rangle}} + 2p_{r\langle i}\frac{\partial u_{j\rangle}}{\partial x_r} + 2p\frac{\partial u_{\langle i}}{\partial x_{j\rangle}} = -\frac{p}{\mu}p_{ij} \tag{6.82}
$$

$$\frac{Dq_i}{Dt} + \frac{1}{2}\frac{\partial r^1_{ik}}{\partial x_k} + \frac{1}{6}\frac{\partial w^2}{\partial x_i} - p_{ik}R\frac{\partial T}{\partial x_k} - \frac{p_{ik}}{\rho}RT\frac{\partial \rho}{\partial x_k} + \frac{5}{2}\rho RT\frac{\partial RT}{\partial x_i} - \frac{p_{ik}}{\rho}\frac{\partial p_{kl}}{\partial x_l}$$
$$- \frac{5}{2}RT\frac{\partial p_{ik}}{\partial x_k} + \frac{\partial u_j}{\partial x_k}s_{ijk} + \frac{7}{5}q_i\frac{\partial u_k}{\partial x_k} + \frac{7}{5}q_k\frac{\partial u_i}{\partial x_k} + \frac{2}{5}q_j\frac{\partial u_j}{\partial x_i} = -\frac{2}{3}\frac{p}{\mu}q_i, \tag{6.83}$$

This set of coupled equations does not form a closed set due to the unknowns

$$r^1_{ik} = m\int C_{\langle i}C_{k\rangle}C^2 f d\mathbf{C} \tag{6.84}$$

$$w^2 = m\int C^4(f - f_M)d\mathbf{C} \tag{6.85}$$

$$s_{ijk} = m\int C_{\langle i}C_jC_{k\rangle}C^2 f d\mathbf{C} \tag{6.86}$$

To evaluate these unknowns and achieve closure, we should know the distribution function which we derive in the next section.

6.4.2 Derivation of Distribution Function

To derive the distribution function $f_{G|13}$ for 13-moment approximation, we start with the distribution function for the Grad's 20-moment approximation $f_{G|20}$ (Eq. (6.49)) which reads as

$$f_{G|20} = f^{(0)}\left\{1 + \frac{p_{ij}}{2pRT}C_iC_j + \underline{\frac{S_{ijk}}{6pR^2T^2}C_iC_jC_k} - \frac{S_i}{2pRT}C_i\right\} \tag{6.87}$$

The third order symmetric tensor S_{ijk} is decomposed into divergence-less part as

$$s_{ijk} = S_{ijk} - \frac{1}{5}(S_{ill}\delta_{jk} + S_{jll}\delta_{ik} + S_{kll}\delta_{ij}) \tag{6.88}$$

Now, in order to obtain the closure, the number of coefficients considered in the series (Eq. (6.35)) must be equal to the primary basic variables considered. Hence, in order to reduce the 20-moment distribution function $f_{G|20}$ to 13-moment distribution function $f_{G|13}$, it becomes mandatory to take the divergence-less part of S_{ijk} (i.e., s_{ijk}) as zero. Hence, with $s_{ijk} = 0$, we proceed as

$$S_{ijk} = \frac{1}{5}(S_{ill}\delta_{jk} + S_{jll}\delta_{ik} + S_{kll}\delta_{ij}) \tag{6.89}$$

Substituting Eq. (6.89) into the underlined term of full third order distribution function (Eq. (6.87)),

$$\begin{aligned}\frac{S_{ijk}}{6pR^2T^2}C_iC_jC_k &= \frac{1}{30}\frac{(S_{ill}\delta_{jk}+S_{jll}\delta_{ik}+S_{kll}\delta_{ij})}{pR^2T^2}C_iC_jC_k \\ &= \frac{1}{30}\frac{S_{ill}C_iC^2+S_{jll}C_jC^2+S_{kll}C_kC^2}{pR^2T^2} \\ &= \frac{1}{10}\frac{S_iC_iC^2}{pR^2T^2} \\ &= \frac{1}{5}\frac{q_iC_iC^2}{pR^2T^2} \qquad [\because S_i = 2q_i]\end{aligned} \tag{6.90}$$

The final form of the distribution function $f_{G|13}$ in terms of first 13 moments is then obtained as

$$f_{G|13} = f^{(0)}\left\{1+\frac{p_{ij}}{2pRT}C_{\langle i}C_{j\rangle} - \frac{q_i}{pRT}C_i\left(1-\frac{C^2}{5RT}\right)\right\}$$
$$[\because p_{ij}C_iC_j = p_{ij}C_{\langle i}C_{j\rangle}] \tag{6.91}$$

According to Grad [62], the stresses do not alter the shape of distribution function $f_{G|13}$ radically; the spherical symmetry of $f^{(0)}$ is distorted into an ellipsoid and the heat flux vector introduces skewness in the distribution function $f_{G|13}$. Grad also believed that the expansion (Eq. (6.91)) describes the majority of flow problems in which the macroscopic variables do not vary by a large amount over a mean free path with sufficient accuracy. He believed that the expansion (6.91) is sufficient to study shock waves of medium strength, which unfortunately proved to be false in his later studies [63].

6.4.3 Closure

With the knowledge of the distribution function $f_{G|13}$ (Eq. (6.91)), we can now proceed to evaluate the unknowns as

$$r^1_{ik} = m\int f_{G|13}C_{\langle i}C_{k\rangle}C^2 d\mathbf{C} \tag{6.92}$$

$$w^2 = m\int C^4(f_{G|13}-f_M)d\mathbf{C} \tag{6.93}$$

$$s_{ijk} = m\int f_{G|13}C_{\langle i}C_jC_{k\rangle}d\mathbf{C} \tag{6.94}$$

Notice that replacement of f by $f_{G|13}$ makes the above equations approximate. As mentioned previously, it becomes necessary to take $s_{ijk} = 0$, hence, there is no further need to evaluate the third unknown. The procedure to evaluate the remaining two integrals is same as outlined in the previous chapter; hence, we do not show a step-by-step evaluation of these integrals. The final result can be obtained as

$$r_{ik}^{1} = 7RTp_{ik}, \qquad w^{2} = 0, \qquad s_{ijk} = 0. \tag{6.95}$$

6.4.4 13-Moment Equations

Substituting the expressions for the unknowns (Eq. 6.95) in the evolution equations for the stress tensor and heat flux vector (Eqs. (6.82) and (6.83)), closure is finally achieved. Below we present Grad's 13-moment equations along with the conservation laws:

$$\frac{\partial \rho}{\partial t} + \frac{\partial (\rho u_k)}{\partial x_k} = 0 \tag{6.96}$$

$$\frac{\partial u_i}{\partial t} + u_k \frac{\partial u_i}{\partial x_k} + \frac{1}{\rho}\frac{\partial p}{\partial x_i} + \frac{1}{\rho}\frac{\partial p_{ik}}{\partial x_k} = F_i, \tag{6.97}$$

$$\frac{3}{2}R\rho \frac{\partial T}{\partial t} + \frac{3}{2}R\rho u_k \frac{\partial T}{\partial x_k} + p\frac{\partial u_i}{\partial x_i} + p_{ij}\frac{\partial u_i}{\partial x_j} + \frac{\partial q_k}{\partial x_k} = 0, \tag{6.98}$$

$$\frac{Dp_{ij}}{Dt} + p_{ij}\frac{\partial u_k}{\partial x_k} + \frac{4}{5}\frac{\partial q_{\langle i}}{\partial x_{j\rangle}} + 2p_{r\langle i}\frac{\partial u_{j\rangle}}{\partial x_r} + 2p\frac{\partial u_{\langle i}}{\partial x_{j\rangle}} = -\frac{p}{\mu}p_{ij} \tag{6.99}$$

$$\begin{aligned}\frac{Dq_i}{Dt} &+ \frac{5}{2}p_{ik}R\frac{\partial T}{\partial x_k} - \frac{p_{ik}}{\rho}RT\frac{\partial \rho}{\partial x_k} + \frac{5}{2}\rho RT\frac{\partial RT}{\partial x_i} - \frac{p_{ik}}{\rho}\frac{\partial p_{kl}}{\partial x_l} + RT\frac{\partial p_{ik}}{\partial x_k}\\ &+ \frac{7}{5}q_i\frac{\partial u_k}{\partial x_k} + \frac{7}{5}q_k\frac{\partial u_i}{\partial x_k} + \frac{2}{5}q_j\frac{\partial u_j}{\partial x_i} = -\frac{2}{3}\frac{p}{\mu}q_i.\end{aligned} \tag{6.100}$$

The detailed flowchart of the Grad's moment method is shown in Fig. 6.4 for convenience. The distribution function is expanded in terms of orthogonal tensorial Hermite polynomials. Depending on the desired accuracy, the series is truncated at an appropriate order of approximation. The number of Hermite coefficients considered must be equal to the number of primary variables considered. The evolution equations for the primary variables are generated from the transformed Boltzmann equation. The unknowns appearing in these equations are then evaluated using the distribution function and the necessary closure is achieved.

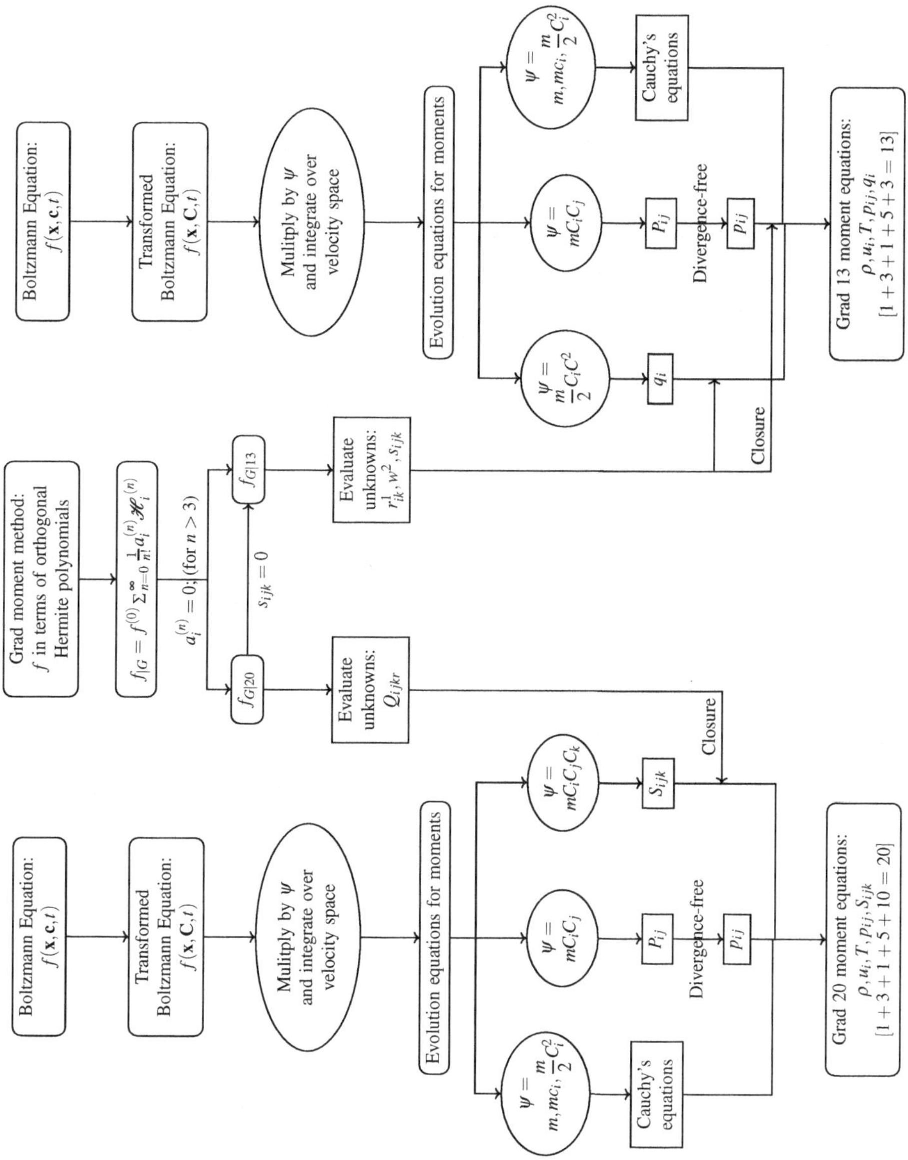

Fig. 6.4 Flowchart depicting the Grad moment method

6.5 Boundary Conditions for 13-Moment Equations

The equations derived above need appropriate number of boundary conditions for their solution. This is further required to bring the higher order continuum transport equations within the framework of computational fluid dynamics and for gaining mainstream adoption. In this section, we present the boundary conditions as given by Grad in his seminal work [62, 64]. Grad derived these boundary conditions for a two-dimensional flow problem purely on a mathematical basis. However, here we follow the method outlined by Yang [172] which has a more physical basis. The method is based on conservation laws and give the same results as that obtained by Grad.

When a gas molecule strikes a solid boundary, there are a number of ways that the interaction can occur: the gas molecule may be specularly or diffusively reflected; it may be adsorbed at the solid boundary; it may react with the wall molecules to form a chemical bond; it may simply dissociate or ionize; or it may even dislodge the solid molecule. This interaction may further get affected by the nature of the solid boundary and its surface finish, temperature, etc. Because of these numerous possibilities, an accurate prediction of the effective interaction of the gas molecules with the solid boundary is particularly difficult. This is the primary reason because of which an accurate prediction of the wall boundary conditions has remained largely elusive. As such, the development of this research topic is still in progress and the theory presented in this section should not be considered as the definite answer.

As higher order continuum transport equations are derived from the Boltzmann equation, it is natural that the boundary conditions should also originate from the Boltzmann equation. Consider a stream of gas molecules that is about to collide against a solid wall perpendicular to the x_1-direction as shown in Fig. 6.5. To simplify the problem, we make the following assumptions [28, 84]:

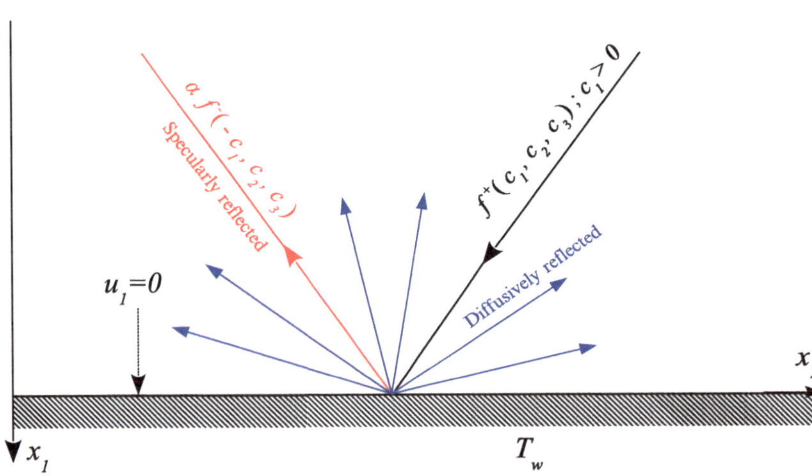

Fig. 6.5 Schematic showing the gas–wall interaction

1. The solid wall is stationary. Otherwise, the boundary conditions are formulated in a frame where the solid wall is at rest.
2. No chemical reaction takes place between the gas and wall molecules.
3. The gas molecule remains in contact with the wall for a much lesser time, i.e., the adsorption time is much smaller than the characteristic time scale.
4. After collision, the gas molecule is re-emitted into the flow from the same point.

With these assumptions, consider a molecule at time t striking the wall with a velocity between $\mathbf{c}^I$ and $\mathbf{c}^I + d\mathbf{c}^I$ at a certain point $\mathbf{x}$ and re-emitted at exactly the same point into the flow with velocity between $\mathbf{c}^R$ and $\mathbf{c}^R + d\mathbf{c}^R$. Let $f^+(\mathbf{x}, \mathbf{c}, t)$ and $f^-(\mathbf{x}, \mathbf{c}, t)$ be the incident and reflected distribution functions, respectively. Then, we can write the distribution function at the wall as a sum of these two contributions as

$$f(\mathbf{x}, \mathbf{c}, t) = f^+(\mathbf{x}, \mathbf{c}, t) + f^-(\mathbf{x}, \mathbf{c}, t). \tag{6.101}$$

Now, for the molecular stream, the molecules with x_1 component of molecular velocity less than zero will not collide with the solid boundary (see Fig. 6.5 for definition of the axis introduced). That is, these molecules do not belong to the incident stream. In a similar way, the molecules with x_1 component of molecular velocity greater than zero does not belong to the reflected stream. Mathematically, these conditions are represented as (functional dependence of distribution function on space and time is suppressed for convenience from now on)

$$f^+(\mathbf{c}) = \begin{cases} f(\mathbf{c}), & \text{for } c_1 > 0, \\ 0, & \text{for } c_1 < 0, \end{cases} \tag{6.102}$$

$$f^-(\mathbf{c}) = \begin{cases} 0, & \text{for } c_1 > 0, \\ f(\mathbf{c}), & \text{for } c_1 < 0, \end{cases} \tag{6.103}$$

The appropriate form of the boundary condition should be such that it specifies the unknown distribution f^- of the outgoing molecules from the known distribution of the incoming molecules f^+.

Now, upon collision, a certain fraction α of the incoming molecules is specularly reflected and the remaining molecules are perfectly absorbed by the wall and re-emitted with a Maxwellian distribution. Mathematically, this boundary condition can be represented as

$$\begin{aligned} f^+(c_1, c_2, c_3) &= \alpha f^-(-c_1, c_2, c_3) \\ &\quad + \frac{\rho_w}{m}\left(\frac{1}{2\pi R T_w}\right)^{3/2} \exp\left(-\frac{c^2}{2RT_w}\right), \qquad c_1 > 0 \\ &= f(c_1, c_2, c_3), \qquad c_1 < 0 \end{aligned} \tag{6.104}$$

Substituting the coefficient of the second term on the left-hand side as $k = \frac{\rho_w}{m}\left(\frac{1}{2\pi RT_w}\right)^{3/2}$, we write as

$$f^+(c_1, c_2, c_3) = \alpha f^-(-c_1, c_2, c_3) + k\exp\left(-\frac{c^2}{2RT_w}\right), \qquad c_1 > 0 \tag{6.105}$$

This coefficient k is determined in such a way that the wall does not collect molecules, i.e., $u_1 = 0$. Note that $\alpha = 1 - \sigma_v$, with σ_v being the tangential momentum accommodation coefficient (TMAC).

For a two-dimensional flow problem, the third component of velocity vector vanishes, i.e., $u_3 = 0$ and the stress tensor and heat flux vector reduce to

$$p_{ij} = \begin{bmatrix} p_{11} & p_{12} & 0 \\ p_{12} & p_{22} & 0 \\ 0 & 0 & 0 \end{bmatrix}, \qquad \text{with} \quad p_{11} + p_{22} = 0 \tag{6.106}$$

$$q_i = \left[q_1, q_2, 0\right]. \tag{6.107}$$

Specifying the distribution of the incoming molecules as the Grad's distribution function (Eq. (6.91)), i.e., $f^+ = f_{G|13}$ and expanding by considering $u_3 = 0$, we obtain

$$\begin{aligned} f^+(c_1, c_2, c_3) &= \frac{\rho}{m\,(2\pi RT)^{3/2}}\exp\left(-\frac{(c_1-u_1)^2 + (c_2-u_2)^2 + c_3^2}{2RT}\right) \\ &\times\left[1 + \frac{p_{11}(c_1-u_1)^2}{2pRT} - \frac{p_{11}(c_2-u_2)^2}{2pRT} + \frac{p_{12}(c_1-u_1)(c_2-u_2)}{pRT}\right. \\ &- \frac{q_1(c_1-u_1)}{pRT} - \frac{q_2(c_2-u_2)}{pRT} + \frac{q_1(c_1-u_1)^3}{5p(RT)^2} + \frac{q_1(c_1-u_1)(c_2-u_2)^2}{5p(RT)^2} \\ &\left.+ \frac{q_1(c_1-u_1)c_3^2}{5p(RT)^2} + \frac{q_2(c_1-u_1)^2(c_2-u_2)}{5p(RT)^2} + \frac{q_2(c_2-u_2)^3}{5p(RT)^2} + \frac{q_2(c_2-u_2)c_3^2}{5p(RT)^2}\right] \end{aligned} \tag{6.108}$$

For specular reflection, only the normal component of the molecular velocity changes sign while the tangential component remains unchanged. Hence, the molecular distribution function for the specularly reflected molecules can simply be obtained by replacing c_1 by $-c_1$ in Eq. (6.108). For molecules re-emitted with Maxwellian distribution with wall temperature T_w, we can write

$$f_w = k\exp\left(-\frac{c_1^2 + c_2^2 + c_3^2}{2RT_w}\right), \qquad \text{where} \quad k = \frac{\rho_w}{m}\left(\frac{1}{2\pi RT_w}\right)^{3/2} \tag{6.109}$$

Only those molecules having $c_1 > 0$ collide with the wall, hence, the mass of the molecules striking the unit area of the wall can be written as

$$m^+ = m\int_{-\infty}^{+\infty} dc_3 \int_{-\infty}^{+\infty} dc_2 \int_0^{+\infty} c_1 f^+(c_1, c_2, c_3)dc_1 \tag{6.110}$$

Substituting the distribution function for the incident molecules (Eq. (6.108)) in the above equation, and performing the integration with $u_1 = 0$ at the wall, we obtain the final expression for m^+ as

$$m^+ = m\frac{2p + p_{11}}{2\sqrt{2\pi RT}} \tag{6.111}$$

Similarly, using the distribution function for the reflected molecules, the mass of molecules leaving the unit area of wall in specular reflection is

$$\begin{aligned} \alpha m^- &= \alpha m \int_{-\infty}^{+\infty} dc_3 \int_{-\infty}^{+\infty} dc_2 \int_0^{\infty} c_1 f^-(-c_1, c_2, c_3)dc_1 \\ &= \alpha m\frac{2p + p_{11}}{2\sqrt{2\pi RT}} \end{aligned} \tag{6.112}$$

The mass of molecules leaving with Maxwellian distribution (Eq. (6.109)) can be written as

$$\begin{aligned} km_w &= km \int_{-\infty}^{+\infty} dc_3 \int_{-\infty}^{+\infty} dc_2 \int_0^{\infty} c_1 f_w(c_1, c_2, c_3)dc_1 \\ &= km \int_{-\infty}^{+\infty} \exp\left(-\frac{c_3^2}{2RT_w}\right) dc_3 \int_{-\infty}^{+\infty} \exp\left(-\frac{c_2^2}{2RT_w}\right) \\ &\quad \times dc_2 \int_0^{\infty} c_1 \exp\left(-\frac{c_1^2}{2RT_w}\right) dc_1 \\ &= km 2\pi (RT_w)^2 \end{aligned} \tag{6.113}$$

Since the solid wall does not accumulate any molecules ($u_1 = 0$), according to the law of conservation of mass, the mass of molecules leaving the wall (Eqs. (6.112) + (6.113)) should be exactly equal to the mass of molecules striking the wall (Eq. (6.110)). Accordingly, we get

$$\begin{aligned} \frac{2p + p_{11}}{2\sqrt{2\pi RT}} &= \alpha\frac{2p + p_{11}}{2\sqrt{2\pi RT}} + k2\pi(RT_w)^2 \\ \therefore \quad k &= \frac{1-\alpha}{2\pi(RT_w)^2}\frac{2p + p_{11}}{2\sqrt{2\pi RT}} \end{aligned} \tag{6.114}$$

This is an equation for k derived using the basic law of conservation of mass that ensures that the wall does not accumulate any gas molecule.

We now apply laws of conservation of momentum and energy at the wall to derive additional boundary conditions. As the incident molecules approach the wall, they bring tangential momentum to the unit area of the wall as

$$\begin{aligned} M^+ &= m\int_{-\infty}^{+\infty} dc_3 \int_{-\infty}^{+\infty} (c_2 - u_2)dc_2 \int_0^{+\infty} (c_1 - u_1) f^+(c_1, c_2, c_3)dc_1 \\ &= m\left(\frac{p_{12}}{2} + \frac{q_2}{5\sqrt{2\pi RT}}\right) \end{aligned} \tag{6.115}$$

For specularly reflected molecules, the tangential momentum carried away from the unit area of the wall is

$$\begin{aligned} \alpha M^- &= \alpha m\int_{-\infty}^{+\infty} dc_3 \int_{-\infty}^{+\infty} (c_2 - u_2)dc_2 \int_0^{\infty} (c_1 - u_1) f^-(-c_1, c_2, c_3)dc_1 \\ &= \alpha m\left(-\frac{p_{12}}{2} + \frac{q_2}{5\sqrt{2\pi RT}}\right) \end{aligned} \tag{6.116}$$

and for those molecules that are emitted with Maxwellian distribution (Eq. (6.109)) is

$$\begin{aligned} kM_w &= km\int_{-\infty}^{+\infty} dc_3 \int_{-\infty}^{+\infty} (c_2 - u_2)dc_2 \int_0^{\infty} (c_1 - u_1) f_w(c_1, c_2, c_3)dc_1 \\ &= km\int_{-\infty}^{+\infty} \exp\left(-\frac{c_3^2}{2RT_w}\right) dc_3 \int_{-\infty}^{+\infty} (c_2 - u_2) \exp\left(-\frac{c_2^2}{2RT_w}\right) dc_2 \\ &\quad \int_0^{\infty} (c_1 - u_1) \exp\left(-\frac{c_1^2}{2RT_w}\right) dc_1 \\ &= -km2\pi(RT_w)^2 u_2 \end{aligned} \tag{6.117}$$

According to law of conservation of momentum, the momentum brought to the unit area of the wall (Eq. (6.115)) is equal to sum of the momentum carried away by the specularly reflected molecules (Eq. (6.116)) and that by diffusively reflected molecules (Eq. (6.117)) as

$$M^+ = \alpha M^- + kM_w$$

$$\therefore \quad \frac{p_{12}}{2} + \frac{q_2}{5\sqrt{2\pi RT}} = \alpha\left(-\frac{p_{12}}{2} + \frac{q_2}{5\sqrt{2\pi RT}}\right) - k2\pi(RT_w)^2 u_2 \tag{6.118}$$

Substituting the expression for k (Eq. (6.114)), we obtain after simplification as

$$\frac{p_{12}}{p}+\frac{2(1-\alpha)u_2}{(1+\alpha)\sqrt{2\pi RT}}\left(1+\frac{p_{11}}{2p}\right)+\frac{2(1-\alpha)q_2}{5(1+\alpha)p\sqrt{2\pi RT}}=0 \tag{6.119}$$

Finally, we apply law of conservation of energy at the wall to obtain the final boundary condition. The stream of incident molecules brings kinetic energy to the unit area of the wall which is given as (with respect to the mean motion $\mathbf{u}$)

$$\begin{aligned} E^{+} &= \frac{1}{2}m\int_{-\infty}^{+\infty}dc_3\int_{-\infty}^{+\infty}dc_2\int_{0}^{+\infty}(c_1-u_1) \\ &\quad\times\left[(c_1-u_1)^2+(c_2-u_2)^2+c_3^2\right]f^{+}(c_1,c_2,c_3)dc_1 \\ &= m\left[\frac{q_1}{2}+\left(2p+\frac{3}{2}p_{11}\right)\left(\frac{RT}{2\pi}\right)^{1/2}\right] \end{aligned} \tag{6.120}$$

For specularly reflected molecules, the kinetic energy (with respect to the mean motion $\mathbf{u}$) carried away from the unit area of the wall is

$$\begin{aligned} \alpha E^{-} &= \frac{1}{2}\alpha m\int_{-\infty}^{+\infty}dc_3\int_{-\infty}^{+\infty}dc_2\int_{0}^{\infty}(c_1-u_1) \\ &\quad\times\left[(c_1-u_1)^2+(c_2-u_2)^2+c_3^2\right]f^{-}(-c_1,c_2,c_3)dc_1 \\ &= m\alpha\left[-\frac{q_1}{2}+\left(2p+\frac{3}{2}p_{11}\right)\left(\frac{RT}{2\pi}\right)^{1/2}\right] \end{aligned} \tag{6.121}$$

and for those molecules that are emitted with Maxwellian distribution (Eq. (6.109)) is

$$\begin{aligned} kE_w &= \frac{km}{2}\int_{-\infty}^{+\infty}dc_3\int_{-\infty}^{+\infty}dc_2\int_{0}^{\infty}(c_1-u_1) \\ &\quad\times\left[(c_1-u_1)^2+(c_2-u_2)^2+c_3^2\right]f_w(c_1,c_2,c_3)dc_1 \\ &= km\pi(RT_w)^2\left(4RT_w+u_2^2\right) \end{aligned} \tag{6.122}$$

According to law of conservation of energy, the kinetic energy brought to the unit area of the wall (Eq. (6.120)) is equal to sum of the kinetic energy carried away by the specularly reflected molecules (Eq. (6.121)) and that by diffusively reflected molecules (Eq. (6.122)) as

$$E^{+} = \alpha E^{-} + kE_w$$

$$\therefore \frac{q_1}{2} + \left(2p + \frac{3}{2}p_{11}\right)\left(\frac{RT}{2\pi}\right)^{1/2} = \alpha\left[-\frac{q_1}{2} + \left(2p + \frac{3}{2}p_{11}\right)\left(\frac{RT}{2\pi}\right)^{1/2}\right] + k\pi(RT_w)^2\left(4RT_w + u_2^2\right) \tag{6.123}$$

Substituting the expression for k (Eq. (6.114)), we obtain the final boundary condition after simplification as

$$\left(\frac{2\pi}{RT}\right)^{1/2}\frac{q_1}{p} + \frac{4(1-\alpha)}{1+\alpha}\left[1 - \frac{T_w}{T} + \frac{p_{11}}{2p}\left(\frac{3}{2} - \frac{T_w}{T}\right) - \left(1 + \frac{p_{11}}{2p}\right)\frac{(u_2/2)^2}{RT}\right] = 0 \tag{6.124}$$

The appropriate boundary conditions for the Grad's 13-moment equations can therefore be summarized as follows:

$$u_1 = 0 \tag{6.125}$$

$$\frac{p_{12}}{p} + \frac{2(1-\alpha)u_2}{(1+\alpha)\sqrt{2\pi RT}}\left(1 + \frac{p_{11}}{2p}\right) + \frac{2(1-\alpha)q_2}{5(1+\alpha)p\sqrt{2\pi RT}} = 0 \tag{6.126}$$

$$\left(\frac{2\pi}{RT}\right)^{1/2}\frac{q_1}{p} + \frac{4(1-\alpha)}{1+\alpha}\left[1 - \frac{T_w}{T} + \frac{p_{11}}{2p}\left(\frac{3}{2} - \frac{T_w}{T}\right) - \left(1 + \frac{p_{11}}{2p}\right)\frac{(u_2/2)^2}{RT}\right] = 0 \tag{6.127}$$

The first boundary condition (Eq. (6.125)) simply states that the solid wall does not accumulate or emit gas molecules. The stress boundary condition (Eq. (6.126)) states that the tangential stress (p_{12}) or tangential heat flux (q_2) at the wall boundary is related to tangential slip velocity of the stream. The third boundary condition (Eq. (6.127)) states that the normal component of the heat flux vector (q_1) is roughly associated with a temperature jump between the solid wall and the adjacent gas molecules.

For flows that are not too far from equilibrium, we can safely say that

$$p_{11} << p, \quad T_w \cong T, \quad \mathbf{u} \cong 0$$

and the last two boundary conditions get simplified.

6.6 Significance of 13-Moment Equations

In order to appreciate the significance of 13-moment equations, we apply these equations to simple problems as proposed by Grad [62]. We will focus on discerning the physics and obtaining precise results in these simpler problems.

As a first example, consider the behavior of the gas when the variables namely ρ, u_i, and T are constants, i.e., they do not vary with time and space, no external force acts on the gas, while the variables p_{ij} and q_i depend on time alone. With these assumptions, all the spatial gradient terms vanish and Grad 13-moment equations (6.99) and (6.100) reduce to

$$\begin{aligned} \frac{dp_{ij}}{dt} &= -\frac{p}{\mu} p_{ij} \\ \frac{dq_i}{dt} &= -\frac{2}{3}\frac{p}{\mu} q_i \end{aligned} \tag{6.128}$$

By integrating the above ordinary differential equations with respect to time, we obtain exact solutions as

$$\begin{aligned} p_{ij}(t) &= p_{ij}(0) \exp\{-t/(\mu/p)\} \\ q_i(t) &= q_i(0) \exp\{-t/(3\mu/2p)\} \end{aligned} \tag{6.129}$$

Hence, the viscous stresses and heat flow approach zero exponentially with periods μ/p and $3\mu/2p$, respectively.

Next, consider a steady one-dimensional heat flow in a gas at rest in the thirteen-moment approximation. With these assumptions, we have

$$\frac{\partial}{\partial t} = 0, \; q_2 = q_3 = 0, \; u_i = 0. \tag{6.130}$$

From the conservation equations ((6.97) and (6.98)), we obtain for $F_i = 0$

$$\frac{1}{\rho}\frac{\partial p}{\partial x_i} + \frac{1}{\rho}\frac{\partial p_{ij}}{\partial x_j} = 0 \Rightarrow \frac{\partial p}{\partial x_i} + \frac{\partial p_{ij}}{\partial x_j} = 0 \tag{6.131}$$

$$\text{and } \frac{\partial q_i}{\partial x_i} = 0 \Rightarrow \frac{dq_1}{dx_1} = 0 \tag{6.132}$$

From second equation, we can conclude that q_1 is a constant. With this result and using the assumptions stated above, all the terms on the left-hand side of the evolution equation for stress tensor (Eq. 6.99) become zero and we have

$$-\frac{p}{\mu} p_{ij} = 0 \Rightarrow p_{ij} = 0. \tag{6.133}$$

Therefore, in a steady state one-dimensional heat flow in a gas at rest which corresponds to heat conduction problem, we get the viscous stress tensor to be zero, a quite obvious result. Also, Eq. (6.100) gives

$$\frac{5}{2} pR \frac{dT}{\partial x_1} = -\frac{2}{3}\frac{p}{\mu} q_1 \quad \Rightarrow q_1 = -k \frac{dT}{\partial x_1} \qquad [\text{where } k = \frac{15}{4}\mu R]$$

The aim of introducing these simple examples was to verify whether the moment equations are able to explain the physics of these problems. If the moment equations had predicted results in stark contradiction with the fundamentals and understanding of the problems, the faith in these equations would have been shaken and then, there was no point in extending these equations to complicated problems. We now proceed to solve more complicated problems by these equations in the following sections.

6.7 Exact Solution of Stokes' First Problem

We will now consider the problem of an infinite plate suddenly set into motion at time $t = 0$ with a uniform velocity U (small with respect to sonic speed) in its own plane (see Fig. 6.6). The plate is assumed to be at rest before the arrival

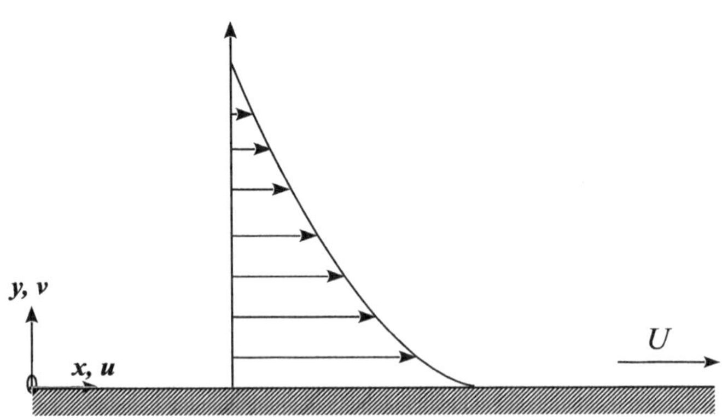

Fig. 6.6 Schematic of Stokes's first problem

of the impulse at $t = 0$. The x and y directions are along and normal to the plate, respectively; with u and v as the corresponding velocity components. The governing equations can be considerably simplified for this problem because the flow variables vary only with y and t, while $\partial()/\partial x = 0$. This problem is also referred to as Rayleigh's problem in the literature.

6.7.1 Navier–Stokes Based Solution

The problem can be easily solved within the framework of the Navier–Stokes equations [85, 102]. The two-dimensional continuity equation for incompressible flow reduces to $\partial v/\partial y = 0$ (because $\partial u/\partial x = 0$ from the assumption of long plate). Now, since $v = 0$ at the wall (wall being non-porous), we obtain that $v = 0$ everywhere.

The x- and y-momentum equations reduce to

$$\rho \frac{\partial u}{\partial t} = \mu \frac{\partial^2 u}{\partial y^2} \tag{6.134}$$

$$\frac{\partial p}{\partial y} = 0 \tag{6.135}$$

The y-momentum equation simply states that pressure is not a function of y, while the x-momentum equation needs to be solved under the following boundary conditions:

$$u(y, t = 0) = 0 \tag{6.136}$$

$$u(y = 0, t) = U \tag{6.137}$$

$$u(y \to \infty, t) = 0 \tag{6.138}$$

Note that the condition of no-slip is being applied at the wall ($y = 0$). Further, the fluid is assumed to be undisturbed at infinity.

Several methods (including Laplace transform and similarity variable approach) can be used to solve the x-momentum equation. The resulting solution is

$$u = U \operatorname{erfc}\left(\frac{y}{2\sqrt{\nu t}}\right) \tag{6.139}$$

from where the shear stress at the wall can be obtained as

$$p_{xy} = \frac{\rho U \nu^{1/2}}{(\pi t)^{1/2}} \exp\left(-\frac{y^2}{4\nu t}\right) \tag{6.140}$$

where $\nu = \mu/\rho$ and $\mathrm{erfc}(x)$ is complementary error function defined as

$$\mathrm{erfc}(x) = \frac{2}{\sqrt{\pi}} \int_x^\infty e^{-\eta^2} d\eta \tag{6.141}$$

The Navier–Stokes equations based solution however does not apply *just* after the arrival of the impulse which sets the plate into motion. This is because for small time ($t/\tau = o(1)$, where $\tau = c^2/\nu$ is the mean collision time, c is the speed of sound), the collision of the gas molecules with the plate is more important than inter-molecular collisions. (See below for further discussion on this point). The situation is therefore closer to that occurring in the free-molecular regime, which will eventually transit to the continuum regime (through the transition and slip regimes) with the passage of time. That is, the Navier–Stokes based solution applies for $t/\tau \gg 1$ while correction to the solution is required to understand the flow just after the arrival of the impulse.

6.7.2 Navier–Stokes Solution with Slip Boundary Condition

The above solution was improved by Schaaf with the use of slip boundary condition at the wall. The obtained solution as quoted in Gross and Jackson [67] is

$$u = U\mathrm{erfc}\left(\frac{y}{2\sqrt{\nu t}}\right) - U\exp\left(\frac{y}{\delta} + \frac{\nu t}{\delta^2}\right)\mathrm{erfc}\left(\frac{y}{2\sqrt{\nu t}} + \frac{\sqrt{\nu t}}{\delta}\right) \tag{6.142}$$

$$p_{xy} = \frac{\mu U}{\delta}\exp\left(\frac{y}{\delta} + \frac{\nu t}{\delta^2}\right)\mathrm{erfc}\left(\frac{y}{2\sqrt{\nu t}} + \frac{\sqrt{\nu t}}{\delta}\right) \tag{6.143}$$

where $\delta = C_1\lambda$ is a constant of the order of a mean free path, and C_1 is first order slip coefficient introduced earlier in Chap. 2.

6.7.3 13-Moment Equation Based Solution

We now follow Yang [172], and solve the linearized form of 13-moment equations for this problem. Yang and Lees solved the general case of compressible flow involving variation of temperature in the gas. The moment equations for the problem can be written as

Continuity:

$$\frac{\partial \rho}{\partial t} + v\frac{\partial \rho}{\partial y} + \rho\frac{\partial v}{\partial y} = 0 \tag{6.144}$$

Momentum:

$$\frac{\partial u}{\partial t} + v\frac{\partial u}{\partial y} + \frac{1}{\rho}\frac{\partial p_{xy}}{\partial y} = 0 \tag{6.145}$$

$$\frac{\partial v}{\partial t} + v\frac{\partial v}{\partial y} + \frac{1}{\rho}\frac{\partial p}{\partial y} + \frac{1}{\rho}\frac{\partial p_{yy}}{\partial y} = 0 \tag{6.146}$$

Energy (from Eq. 5.51):

$$\frac{\partial p}{\partial t} + v\frac{\partial p}{\partial y} + \frac{2}{3}p_{xy}\frac{\partial u}{\partial y} + \left(\frac{5}{3}p + \frac{2}{3}p_{yy}\right)\frac{\partial v}{\partial y} + \frac{2}{3}\frac{\partial q_y}{\partial y} = 0 \tag{6.147}$$

State:

$$p = \rho RT \tag{6.148}$$

Stress:

$$\frac{\partial p_{xx}}{\partial t} + v\frac{\partial p_{xx}}{\partial y} + \frac{4}{3}p_{xy}\frac{\partial u}{\partial y} + \left(p_{xx} - \frac{2}{3}p_{yy} - \frac{2}{3}p\right)\frac{\partial v}{\partial y} - \frac{4}{15}\frac{\partial q_y}{\partial y} + \frac{p}{\mu}p_{xx} = 0 \tag{6.149}$$

$$\frac{\partial p_{xy}}{\partial t} + v\frac{\partial p_{xy}}{\partial y} + (p + p_{yy})\frac{\partial u}{\partial y} + 2p_{xy}\frac{\partial v}{\partial y} + \frac{2}{5}\frac{\partial q_x}{\partial y} + \frac{p}{\mu}p_{xy} = 0 \tag{6.150}$$

$$\frac{\partial p_{yy}}{\partial t} + v\frac{\partial p_{yy}}{\partial y} + \left(\frac{4}{3}p + \frac{7}{3}p_{yy}\right)\frac{\partial v}{\partial y} - \frac{2}{3}p_{xy}\frac{\partial u}{\partial y} + \frac{8}{15}\frac{\partial q_y}{\partial y} + \frac{p}{\mu}p_{yy} = 0 \tag{6.151}$$

Heat:

$$\frac{\partial q_x}{\partial t} + v\frac{\partial q_x}{\partial y} + \frac{7}{5}q_y\frac{\partial u}{\partial y} + \frac{7}{5}q_x\frac{\partial v}{\partial y} + \frac{5}{2}\frac{p_{xy}}{\rho}\frac{\partial p}{\partial y} - \frac{7}{2}\frac{pp_{xx}}{\rho^2}\frac{\partial p}{\partial y} + \left(\frac{p}{\rho} - \frac{p_{xx}}{\rho}\right)\frac{\partial p_{xy}}{\partial y} - \frac{p_{xy}}{\rho}\frac{\partial p_{yy}}{\partial y} + \frac{2}{3}\frac{p}{\mu}q_x = 0 \tag{6.152}$$

$$\frac{\partial q_y}{\partial t} + v\frac{\partial q_y}{\partial y} + \frac{16}{5}q_y\frac{\partial v}{\partial y} + \frac{2}{5}q_x\frac{\partial u}{\partial y} + \frac{5}{2}\left(\frac{p}{\rho} + \frac{p_{yy}}{\rho}\right)\frac{\partial p}{\partial y} \tag{6.153}$$
$$- \left(\frac{5}{2}\frac{p^2}{\rho^2} + \frac{7}{2}\frac{pp_{yy}}{\rho^2}\right)\frac{\partial \rho}{\partial y} + \left(\frac{p}{\rho} - \frac{p_{yy}}{\rho}\right)\frac{\partial p_{yy}}{\partial y} - \frac{p_{xy}}{\rho}\frac{\partial p_{xy}}{\partial y} + \frac{2}{3}\frac{p}{\mu}q_y = 0$$

The above 10 equations involve the following 10-variables: ρ, p, u_x, u_y, T, p_{xx}, p_{xy}, p_{yy}, q_x, q_y.

The relevant initial conditions are: all components of velocity, stress, and heat flux are zero at the initial time, while the initial density, pressure, and temperature are constants:

$$\rho(0, y) = \rho_o; \quad p(0, y) = p_o; \quad T(0, y) = T_o. \tag{6.154}$$

The relevant boundary conditions are: all variables should be bounded as $y \to \infty$, while the following condition applies at the wall ($y = 0$):

$$v(t, 0) = 0 \tag{6.155}$$

$$\frac{p_{xy}(t)}{p(t)} + \frac{2(1-\alpha)}{(1+\alpha)(2\pi R)^{1/2}}\frac{u(t) - U}{(T(t))^{1/2}}\left(1 + \frac{p_{yy}(t)}{2p(t)}\right)$$
$$+ \frac{2(1-\alpha)}{5(1+\alpha)(2\pi R)^{1/2}}\frac{q_x(t)}{p(t)(T(t))^{1/2}} = 0 \tag{6.156}$$

$$\left(\frac{2\pi}{R}\right)^{1/2}\frac{1}{(T(t))^{1/2}}\frac{q_y(t)}{p(t)} + \frac{4(1-\alpha)}{1+\alpha}\left[1 - \frac{T_w}{T(t)} + \frac{p_{yy}(t)}{2p(t)}\left(\frac{3}{2} - \frac{T_w}{T(t)}\right) - \right.$$
$$\left.\left(1 + \frac{p_{yy}(t)}{2p(t)}\right)\frac{(u(t) - U)^2}{RT(t)}\right] = 0 \tag{6.157}$$

The above set of equations is difficult to solve analytically due to the non-linear and coupled nature of the equations. However, it is possible to obtain a solution upon linearization of the equations and boundary conditions. Linearization is possible for small impulse velocity U (with respect to speed of sound) of the plate, and small temperature difference between the plate and initial gas temperature, $T_w - T_o$. Writing:

$$u = 0 + u'; \quad v = 0 + v'; \quad \rho = \rho_o + \rho'; \quad p = p_o + p'; \quad T = T_o + T';$$
$$p_{xx} = 0 + p'_{xx}; \quad p_{xy} = 0 + p'_{xy}; \quad p_{yy} = 0 + p'_{yy};$$
$$q_x = 0 + q'_x; \quad q_y = 0 + q'_y \tag{6.158}$$

where the primed quantities denote small perturbations with respect to the undisturbed (initial) quantities. Since viscosity is a function of temperature, it is also a function of position, and can be described as

$$\frac{\mu}{\mu_o} = \left(\frac{T}{T_o}\right)^{\omega} \tag{6.159}$$

That is,

$$\mu = \mu_o \left(1 + \frac{T'}{T_o}\right)^{\omega} \approx \mu_o + \mu' \tag{6.160}$$

where $\mu' = \mu_o \omega (T'/T_o)$.

Upon substituting the above quantities in the governing equations and neglecting terms of higher order leads to the following set of equations:

$$\frac{\partial \rho'}{\partial t} + \rho_o \frac{\partial v'}{\partial y} = 0 \tag{6.161}$$

$$\frac{\partial u'}{\partial t} + \frac{1}{\rho_o} \frac{\partial p'_{xy}}{\partial y} = 0 \tag{6.162}$$

$$\frac{\partial v'}{\partial t} + \frac{1}{\rho_o} \frac{\partial p'}{\partial y} + \frac{1}{\rho_o} \frac{\partial p'_{yy}}{\partial y} = 0 \tag{6.163}$$

$$\frac{\partial p'}{\partial t} + \frac{5}{3} p_o \frac{\partial v'}{\partial y} + \frac{2}{3} \frac{\partial q'_y}{\partial y} = 0 \tag{6.164}$$

$$\frac{\partial p'_{xx}}{\partial t} - \frac{2}{3} p_o \frac{\partial v'}{\partial y} - \frac{4}{15} \frac{\partial q'_y}{\partial y} + \frac{p_o}{\mu_o} p'_{xx} = 0 \tag{6.165}$$

$$\frac{\partial p'_{xy}}{\partial t} + p_o \frac{\partial u'}{\partial y} + \frac{2}{5} \frac{\partial q'_x}{\partial y} + \frac{p_o}{\mu_o} p'_{xy} = 0 \tag{6.166}$$

$$\frac{\partial p'_{yy}}{\partial t} + \frac{4}{3} p_o \frac{\partial v'}{\partial y} + \frac{8}{15} \frac{\partial q'_y}{\partial y} + \frac{p_o}{\mu_o} p'_{yy} = 0 \tag{6.167}$$

$$\frac{\partial q'_x}{\partial t} + \frac{p_o}{\rho_o} \frac{\partial p'_{xy}}{\partial y} + \frac{2}{3} \frac{p_o}{\mu_o} q'_x = 0 \tag{6.168}$$

$$\frac{\partial q_y'}{\partial t}+\frac{5}{2}\frac{p_o}{\rho_o}\frac{\partial p'}{\partial y}-\frac{5}{2}\frac{p_o^2}{\rho_o^2}\frac{\partial p'}{\partial y}+\frac{p_o}{\rho_o}\frac{\partial p_{yy}'}{\partial y}+\frac{2}{3}\frac{p_o}{\mu_o}q_y'=0 \tag{6.169}$$

and the equation of state becomes

$$p' = R(\rho_o T' + T_o \rho') \tag{6.170}$$

or

$$T' = \frac{1}{R\rho_o}p' - \frac{T_o}{\rho_o}\rho' \tag{6.171}$$

Similarly, the boundary conditions at the wall simplify to:

$$v'(t,0)=0 \tag{6.172}$$

$$\frac{p_{xy}'(t,0)}{p_o}+\frac{2(1-\alpha)}{(1+\alpha)(2\pi R)^{1/2}}\frac{u'(t,0)-U}{(T_o)^{1/2}}+\frac{2(1-\alpha)}{5(1+\alpha)(2\pi R)^{1/2}}\frac{q_x'(t,0)}{p_o(T_o)^{1/2}}=0 \tag{6.173}$$

$$\left(\frac{2\pi}{RT_o}\right)^{1/2}\frac{q_y'(t,0)}{p_o}+\frac{4(1-\alpha)}{1+\alpha}\left[1-\frac{T_w}{T_o}+\frac{p_{yy}'(t,0)}{4p_o}\right]=0 \tag{6.174}$$

Note that the process of linearization uncouples the variables into two sets: with tangential velocity u', tangential shear stress p_{xy}', tangential heat flux q_x' being governed by Eqs. (6.162), (6.166), (6.168), and boundary condition (6.173) along with the condition that variables have to be bounded as y tends to infinity. The other variables are governed by the remaining seven equations and the boundary conditions (Eqs. (6.172) and (6.174)) along with the unboundedness condition. This uncoupling occurs due to neglecting terms such as $p_{xy}\partial u/\partial y$ (in Eqs. (6.147) and (6.151)), etc. upon linearization. Note that the velocity of the plate governs the solution for the former set, while the temperature difference $T_w - T_o$ governs the solution for the latter set. Here, for simplicity, we will consider, $T_w - T_o = 0$, such that $v' = p' = \rho' = T' = p_{xx}' = p_{yy}' = q_y' = 0$.

The remaining equations can be solved by using the standard technique of Laplace transform. The resulting solution for u, p_{xy} and q_x is provided for both short and long times. The solution for short time gives an idea of the flow behavior immediately after moving the plate. The solution for large time is useful for comparison with the corresponding solution from the Navier–Stokes equations.

For small values of time:

$$
\begin{aligned}
p_{xy}(t, y) &= p_o \frac{U}{a_o} \frac{1}{\left(\frac{1+\alpha}{1-\alpha}\sqrt{\frac{3\pi}{10}} + \frac{6\sqrt{21}}{35}\right)} \\
&\times \left\{ 1 - \left(\frac{\frac{61}{21\sqrt{35}}}{\frac{1+\alpha}{1-\alpha}\sqrt{\frac{\pi}{2}} + \frac{6}{\sqrt{35}}} \frac{p_o}{\mu_o} \times \left(t - \frac{5}{\sqrt{21}}\frac{y}{a_o}\right) + \frac{125}{42\sqrt{21}} \frac{p_o}{\mu_o}\frac{y}{a_o} \right) \right. \\
&+ \left[\frac{\frac{1+\alpha}{1-\alpha}\sqrt{\frac{\pi}{70}}\frac{421}{588} + 0.364}{\left(\frac{1+\alpha}{1-\alpha}\sqrt{\frac{\pi}{2}} + \frac{6}{\sqrt{35}}\right)^2} \left(\frac{p_o^2}{\mu_o^2}\right) \frac{\left(t - \frac{5}{\sqrt{21}}\frac{y}{a_o}\right)^2}{2} \right. \\
&\left.\left. + \left(\frac{0.318}{\frac{1+\alpha}{1-\alpha}\sqrt{\frac{\pi}{2}} + \frac{6}{\sqrt{35}}} + 0.139 \right) \frac{p_o^2}{\mu_o^2}\frac{y}{a_o} \left(t - \frac{5}{\sqrt{21}}\frac{y}{a_o}\right) + 0.211 \frac{p_o^2}{\mu_o^2}\frac{y^2}{a_o^2} \right] \right\} \\
&+ O\left(t^3\right), \quad t > \frac{5}{\sqrt{21}}\frac{y}{a_o}; \\
&= 0, \quad 0 < t < \frac{5}{\sqrt{21}}\frac{y}{a_o}
\end{aligned}
\tag{6.175}
$$

$$
\begin{aligned}
u(t, y) &= \frac{U}{\frac{1+\alpha}{1-\alpha}\sqrt{\frac{7\pi}{10}} + \frac{6}{5}} \left[1 + \left(\frac{\frac{1+\alpha}{1-\alpha}\sqrt{\frac{5\pi}{14}}\frac{5}{6} + \frac{2}{15}}{\frac{1+\alpha}{1-\alpha}\sqrt{\frac{7\pi}{10}} + \frac{6}{5}} \frac{p_o}{\mu_o} \left(t - \frac{5}{\sqrt{21}}\frac{y}{a_o}\right) \right. \right. \\
&\left. - \frac{125}{42\sqrt{21}} \frac{p_o}{\mu_o}\frac{y}{a_o} \right) + \left\{ \left(\frac{\left(\frac{1+\alpha}{1-\alpha}\sqrt{\frac{5\pi}{14}}\frac{5}{6} + \frac{2}{15}\right)^2}{\left(\frac{1+\alpha}{1-\alpha}\sqrt{\frac{7\pi}{10}} + \frac{6}{5}\right)^2} \right. \right. \\
&\left. - \frac{\left(\frac{1+\alpha}{1-\alpha}\sqrt{\frac{\pi}{70}}\frac{1699}{504} + \frac{4}{45}\right)}{\left(\frac{1+\alpha}{1-\alpha}\sqrt{\frac{7\pi}{10}} + \frac{6}{5}\right)} \right) \frac{p_o^2}{\mu_o^2} \frac{\left(t - \frac{5}{\sqrt{21}}\frac{y}{a_o}\right)^2}{2}
\end{aligned}
$$

$$+\left(0.139\frac{\frac{1+\alpha}{1-\alpha}\sqrt{\frac{5\pi}{14}}\frac{5}{6}+\frac{2}{15}}{\frac{1+\alpha}{1-\alpha}\sqrt{\frac{7\pi}{10}}+\frac{6}{5}}\frac{125}{42\sqrt{21}}\right)\frac{p_o^2}{\mu_o^2}\frac{y}{a_o}\left(t-\frac{5}{\sqrt{21}}\frac{y}{a_o}\right)$$

$$\left.\left.+0.211\frac{p_o^2}{\mu_o^2}\frac{y^2}{a_o^2}\right\}\right]+O\left(t^3\right),\quad t>\frac{5}{\sqrt{21}}\frac{y}{a_o}$$

$$=0,\quad 0<t<\frac{5}{\sqrt{21}}\frac{y}{a_o} \tag{6.176}$$

$$q_x(t,y)=\frac{p_oU}{\frac{1+\alpha}{1-\alpha}\sqrt{\frac{7\pi}{10}}+\frac{6}{5}}\left[1-\left(\frac{\frac{1+\alpha}{1-\alpha}\sqrt{\frac{\pi}{70}}\frac{1}{2}+\frac{2}{3}}{\frac{1+\alpha}{1-\alpha}\sqrt{\frac{7\pi}{10}}+\frac{6}{5}}\frac{p_o}{\mu_o}\left(t-\frac{5}{\sqrt{21}}\frac{y}{a_o}\right)\right.\right.$$

$$\left.+\frac{125}{42\sqrt{21}}\frac{p_o}{\mu_o}\frac{y}{a_o}\right)+\left\{\left(\frac{\left(\frac{1+\alpha}{1-\alpha}\sqrt{\frac{\pi}{70}}\frac{1}{2}+\frac{2}{3}\right)^2}{\left(\frac{1+\alpha}{1-\alpha}\sqrt{\frac{7\pi}{10}}+\frac{6}{5}\right)^2}-\frac{\left(\frac{1+\alpha}{1-\alpha}\sqrt{\frac{\pi}{70}}\frac{299}{504}\right)}{\left(\frac{1+\alpha}{1-\alpha}\sqrt{\frac{7\pi}{10}}+\frac{6}{5}\right)}\right)\right.$$

$$\times\frac{p_o^2}{\mu_o^2}\frac{\left(t-\frac{5}{\sqrt{21}}\frac{y}{a_o}\right)^2}{2}+\left(\frac{\frac{1+\alpha}{1-\alpha}\sqrt{\frac{\pi}{70}}\frac{1}{2}+\frac{2}{3}}{\frac{1+\alpha}{1-\alpha}\sqrt{\frac{7\pi}{10}}+\frac{6}{5}}\frac{125}{42\sqrt{21}}+0.139\right)\frac{p_o^2}{\mu_o^2}\frac{y}{a_o}$$

$$\left.\left.\times\left(t-\frac{5}{\sqrt{21}}\frac{y}{a_o}\right)+0.211\frac{p_o^2}{\mu_o^2}\frac{y^2}{a_o^2}\right\}\right]+O\left(t^3\right),\quad t>\frac{5}{\sqrt{21}}\frac{y}{a_o}$$

$$=0,\quad 0<t<\frac{5}{\sqrt{21}}\frac{y}{a_o} \tag{6.177}$$

For large values of time:

$$p_{xy}(t,y)=\frac{U}{a_o}\sqrt{\frac{5\mu_op_o}{3\pi t}}\left[1-\left\{\left(\left(\left(\frac{1+\alpha}{1-\alpha}\right)^2\frac{\pi}{4}-\frac{1}{4}\right)\frac{\mu_o}{p_o}+\frac{(1+\alpha)}{2(1-\alpha)}\sqrt{\frac{\pi}{2}}\sqrt{\frac{\mu_o}{p_o}}\frac{y}{\sqrt{\nu_o}}\right.\right.\right.$$

$$+\frac{1}{4}\frac{y^2}{\nu_o}\Bigg\}\frac{1}{t}+\Bigg\{\left(\left(\frac{1+\alpha}{1-\alpha}\right)^2\frac{9\pi}{16}+\frac{10713}{800}\right)\frac{\mu_o^2}{p_o^2}+\left(\frac{3}{4}\left(\frac{1+\alpha}{1-\alpha}\right)^3\left(\frac{\pi}{2}\right)^{3/2}\right.$$

$$\left.-\frac{3}{5}\frac{(1+\alpha)}{(1-\alpha)}\sqrt{\frac{\pi}{2}}\right)\left(\frac{\mu_o}{p_o}\right)^{3/2}\frac{y}{\sqrt{\nu_o}}+\left(\left(\frac{1+\alpha}{1-\alpha}\right)^2\frac{3\pi}{16}-\frac{3}{80}\right)\frac{\mu_o}{p_o}\frac{y^2}{\nu_o}$$

$$+\frac{(1+\alpha)}{8(1-\alpha)}\sqrt{\frac{\pi}{2}}\sqrt{\frac{\mu_o}{p_o}}\frac{y^3}{\nu_o^{3/2}}+\frac{y^4}{32\nu_o^2}\Bigg\}\frac{1}{t^2}+O\left(\frac{1}{t^3}\right)\Bigg] \tag{6.178}$$

$$u(t,y)=U\Bigg[1-\frac{1}{\sqrt{\pi t}}\Bigg[\frac{1+\alpha}{1-\alpha}\sqrt{\frac{\pi\mu_o}{2p_o}}+\frac{y}{\sqrt{\nu_o}}-\Bigg\{\left(\left(\frac{1+\alpha}{1-\alpha}\right)^3\left(\frac{\pi}{2}\right)^{3/2}\right.$$

$$\left.-\frac{(1+\alpha)}{5(1-\alpha)}\sqrt{\frac{\pi}{2}}\right)\left(\frac{\mu_o}{p_o}\right)^{3/2}+\left(\left(\frac{1+\alpha}{1-\alpha}\right)^2\frac{\pi}{4}-\frac{1}{20}\right)\frac{\mu_o}{p_o}\frac{y}{\sqrt{\nu_o}}$$

$$+\frac{(1+\alpha)}{4(1-\alpha)}\sqrt{\frac{\pi\mu_o}{2p_o}}\frac{y^2}{\nu_o}+\frac{1}{12}\frac{y^3}{\nu_o^{3/2}}\Bigg\}\frac{1}{t}$$

$$+\Bigg\{\left(\frac{(1+\alpha)}{(1-\alpha)}\sqrt{\frac{\pi}{2}}\frac{153}{200}-\frac{9}{20}\left(\frac{1+\alpha}{1-\alpha}\right)^3\left(\frac{\pi}{2}\right)^{3/2}\right)\left(\frac{\mu_o}{p_o}\right)^{5/2}$$

$$-\left(\left(\frac{1+\alpha}{1-\alpha}\right)^2\frac{33\pi}{80}+\frac{87}{400}\right)\frac{\mu_o^2}{p_o^2}\frac{y}{\sqrt{\nu_o}}+\frac{3}{8}\left(\frac{1+\alpha}{1-\alpha}\right)^3\left(\frac{\pi}{2}\right)^{3/2}\left(\frac{\mu_o}{p_o}\right)^{3/2}\frac{y^2}{\nu_o}$$

$$+\left(\left(\frac{1+\alpha}{1-\alpha}\right)^2\frac{\pi}{16}+\frac{3}{80}\right)\frac{\mu_o}{p_o}\frac{y^3}{\nu_o^{3/2}}+\frac{(1+\alpha)}{32(1-\alpha)}\sqrt{\frac{\pi\mu_o}{2p_o}}\frac{y^4}{\nu_o^2}+\frac{1}{160}\frac{y^5}{\nu_o^{5/2}}\Bigg\}$$

$$\times\frac{1}{t^2}+O\left(\frac{1}{t^3}\right)\Bigg]\Bigg] \tag{6.179}$$

$$q_x(t,y)=\frac{3}{4\sqrt{2}}p_oU\left(\frac{\mu_o}{p_ot}\right)^{3/2}\Bigg\{\frac{1+\alpha}{1-\alpha}+\Bigg[\left(\frac{1+\alpha}{1-\alpha}\frac{69}{20}-\frac{(1+\alpha)^3}{(1-\alpha)^3}\frac{3\pi}{4}\right)\frac{\mu_o}{p_o}+$$

$$\left(\frac{12}{5}\sqrt{\frac{2}{\pi}}-\frac{1+\alpha}{1-\alpha}\sqrt{\frac{\pi}{2}}\frac{3}{2}\right)\sqrt{\frac{\mu_o}{p_o}}\frac{y}{\sqrt{\nu_o}}-\frac{1}{4}\sqrt{\frac{2}{\pi}}\sqrt{\frac{p_o}{\mu_o}}\frac{y^3}{\nu_o^{3/2}}\Bigg]\frac{1}{t}+O\left(\frac{1}{t^2}\right)\Bigg\} \tag{6.180}$$

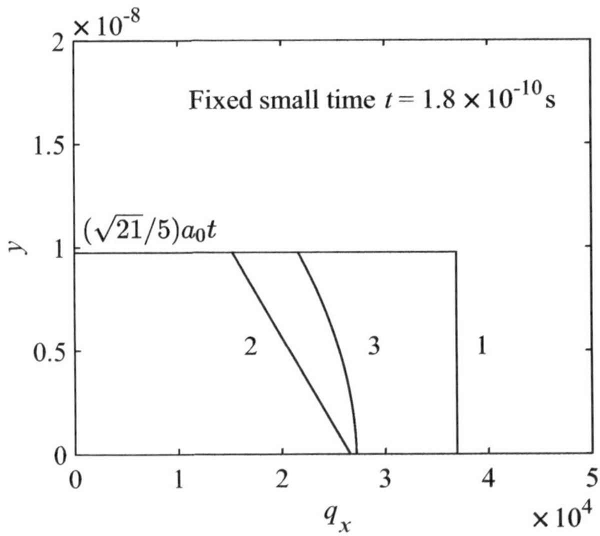

Fig. 6.7 Variation of q_x with y for small values of time; numbers 1, 2, and 3 denote the number of terms retained in evaluating the corresponding expression

where ν_o is coefficient of kinematic viscosity ($= \mu_o/p_o$), $a_o = \sqrt{\frac{5}{3}\frac{p_o}{\rho_o}}$, and $\alpha = 1 - \sigma_v$.

For small values of time, the above solution gives a step-function type of response for all the three quantities. For large time, the solutions are convergent only for sufficiently large values of time, $t/\tau > O(100)$. A diffusive behaviour for the three variables is seen, with the viscous layer growing as $\sqrt{t}$, in agreement with the solution suggested by the Navier–Stokes equations.

For fixed small values of time, the variation of tangential stress (p_{xy}), tangential flow energy (q_x), and tangential gas velocity (u) in the direction normal to the plate surface is shown in Figs. 6.7 and 6.8, where the numbers 1, 2, and 3 represent the number of terms considered in evaluating the functions. From Eqs. (6.175)–(6.177), a parameter in the small time solution is defined as

$$\frac{p_0 t}{\mu_0} = \frac{3}{5}\frac{Re}{M_0^2} \tag{6.181}$$

where Re is instantaneous Reynolds number and M_0 is Mach number given as

$$Re = \frac{\rho_0 U^2 t}{\mu_0}, \quad M_0 = \frac{U}{a_0} = \frac{U}{\sqrt{\frac{5 p_0}{3 \rho_0}}} \tag{6.182}$$

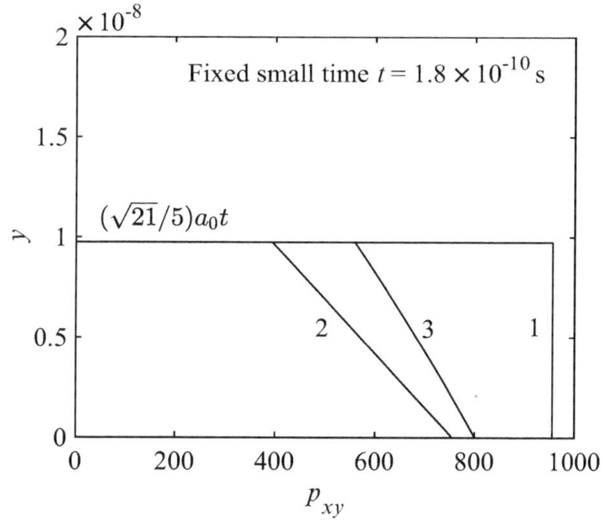

Fig. 6.8 Variation of p_{xy} with y for small values of time; numbers 1, 2, and 3 denote the number of terms retained in evaluating the corresponding expression

For small values of time,

$$\frac{p_0 t}{\mu_0} \sim \frac{a_0^2 t}{\nu} \sim \frac{t}{\tau} < 1$$

the solution in series expansions of the above defined parameter (Eq. (6.181)) is satisfactory.

From Figs. 6.7, 6.8, and 6.9, it is clear that, initially, the gas away from the plate does not feel any disturbance, because the gas molecules do not yet experience the motion of the impulsively accelerated plate. Understandably, the gas molecules near the plate are influenced by the plate motion. Inside the layer of thickness, $y < \sqrt{21}a_0 t/5$, (where $t = O(\tau)$), the gas moves with the plate at a fraction of the impulsive velocity U, as shown in Fig. 6.9. The corresponding tangential stress and tangential energy fluxes are shown in Figs. 6.7 and 6.8. The growth of thin layer is linear with time at a rate of $\sqrt{21}a_0/5$. The solution can also be plotted for larger values of time, $\frac{t}{\tau} > O(100)$. It is noted that the solutions given by Eqs. (6.178)–(6.180) are convergent.

The solution obtained here is not the most accurate solution known for the problem. Nonetheless, the above example illustrates the use of moment equations to a reasonably complex problem.

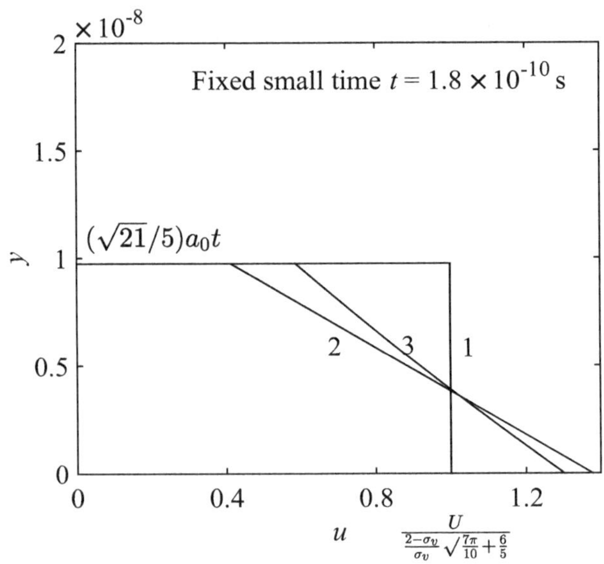

Fig. 6.9 Variation of u with y for small values of time; numbers 1, 2, and 3 denote the number of terms retained in evaluating the corresponding expression

6.7.4 *Kinetic Theory Based Solution*

The above solution has been improved by Gross and Jackson [67] using a kinetic model. They argued that the gas molecules which have interacted with the plate are affected by the movement of the plate. However, molecules which are yet to impinge on the plate remain unaffected by the movement of the plate (see Fig. 6.10). The distribution function for these two groups of molecules is different, and needs to be suitably accounted for, in order to obtain the correct dynamics immediately at the onset of motion of the plate. The following solution was obtained by Gross and Jackson [67] using kinetic theory:

For short time:

$$u = U\Big[0.139H(t - a_1 y)e^{-b_1 y} + 0.361H(t - a_2 y)e^{-b_1 y}\Big]$$
$$+U\Big[-6.10\Big(\frac{t - a_1 y}{\lambda}\Big)H(t - a_1 y)e^{-b_1 y} + 7.36\Big(\frac{t - a_2 y}{\lambda}\Big)H(t - a_2 y)e^{-b_2 y}\Big] \tag{6.183}$$

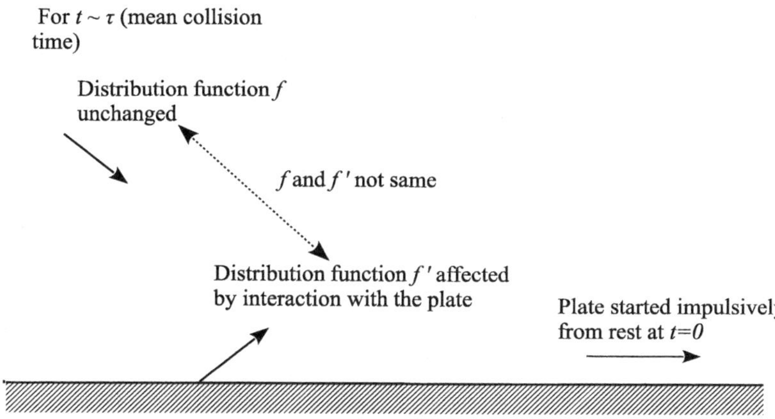

Fig. 6.10 Showing that the distribution function close to the plate is different from that away from the plate

$$p_{xy} = \frac{nU}{2(\pi\beta)^{1/2}}\left[0.616H(t-a_1y)e^{-b_1y} + 0.384H(t-a_2y)e^{-b_2y}\right] + \frac{nU}{2(\pi\beta)^{1/2}}$$
$$\left[-23.33\left(\frac{t-a_1y}{\lambda}\right)H(t-a_1y)e^{-b_1y} + 20.25\left(\frac{t-a_2y}{\lambda}\right)H(t-a_2y)e^{-b_2y}\right] \tag{6.184}$$

where $a_1 = 0.799$, $b_1 = 3.57/\lambda$, $a_2 = 3.33$, $b_2 = 7.42/\lambda$ (λ being mean free path of the gas).

For large time:

$$u(y,t) = U\mathrm{erfc}\left(\frac{yd}{2t^{1/2}}\right) - 1.24U\left(\frac{\lambda}{\pi t}\right)^{1/2} e^{\frac{-y^2d^2}{4t}} + 0.292U\left(\frac{\lambda}{\pi t}\right)^{1/2} e^{-cy} \tag{6.185}$$

$$p_{xy}(y,t) = \frac{nU}{2(\pi\beta)^{1/2}}\left[\frac{2}{d(t)^{1/2}}e^{\frac{-y^2d^2}{4t}} - 0.170\left(\frac{\lambda}{\pi t}\right)^{1/2} e^{-cy}\right] \tag{6.186}$$

where $c = 6.86/\lambda$ and $d = 1.54/(\lambda)^{1/2}$.

Note the presence of two step functions in the solution for small time (as opposed to a single step in the solution of Grad equation). The solution is further multiplied by $\exp(-by)$ (and therefore the strength of the step decreases on moving away from the plate); this multiplication is, however, missing in the Grad solution. This underscores the complexity of the problem in hand.

6.7.5 Comparison with Free-Molecular Regime Solution

The following solution applies to the problem in free-molecular regime [67]:

$$u(y,t) = \frac{U}{2}\text{erfc}\left(\frac{y\sqrt{\beta}}{t}\right) \tag{6.187}$$

$$p_{xy}(y,t) = \frac{\rho U}{2(\pi\beta)^{1/2}}\exp\left(\frac{-\beta y^2}{t^2}\right) \tag{6.188}$$

where $\beta = 1/(2RT)$. The free molecular solution suggests that as $t \to 0$, we get

$$u \to \frac{U}{2}; \quad p_{xy} \to \frac{\rho U}{2(\pi\beta)^{1/2}} \tag{6.189}$$

where we have utilized the following limiting properties of the complementary error function:

$$\begin{aligned} &\text{erfc}(x) \to 1 - \frac{2x}{\sqrt{\pi}}, && \text{as } x \to 0, \\ &\text{erfc}(x) \to \frac{\exp(-x^2)}{\sqrt{\pi}x}, && \text{as } x \to \infty. \end{aligned} \tag{6.190}$$

In contrast, the Navier–Stokes solution with no-slip boundary condition predicts $u \to U$ and $p_{xy} \to \infty$. The Grad equations predicts

$$u \to 0.373U; \quad p_{xy} \to 1.1\frac{\rho U}{2(\pi\beta)^{1/2}}. \tag{6.191}$$

That is, the predicted value of initial flow velocity from Grad equations is about 26% smaller than the true (free-molecular) velocity, while the shear stress is about 10% larger. The kinetic theory predicts the correct values of both the velocity and shear stress, although the variation of shear stress in space is not correctly captured even with the kinetic theory.

6.7.6 Significance

The solution of this problem is particularly interesting for understanding the more general case of arbitrary impulse on the surrounding fluid. The problem is also interesting as an analogy exists between the present problem and laminar flow over a flat plate. The boundary layer developing over the plate cannot be described by the Navier–Stokes equations at the leading-edge. This is because the gas molecules hitting the plate and the gas molecules slightly away from the plate, will equilibrate only after a certain distance away from the leading edge.

A similar situation arises for two plates placed close to each other (distance of the order of the mean free path). As discussed above (see Fig. 6.10), the distribution function of molecules impinging on a plate will primarily be governed by the velocity and temperature of the other plate, while the distribution function of molecules leaving the plate will primarily be governed by the velocity and temperature of the plate. The inability of the Navier–Stokes and Grad equations to treat the two distribution functions differently leads to error in the initial dynamics. However, as the density of the gas increases, the collision between the molecules increases, making the two distribution functions virtually identical; and the Navier–Stokes equations apply.

6.8 Exact Solution of Cylindrical Couette Flow

We will now consider the problem of steady cylindrical Couette flow, with outer cylinder (radius = b) stationary while the inner cylinder (radius = a) is rotating as shown in Fig. 6.11. We will assume that the cylinders are long such that there is no variation along the length of the cylinders. The steady state assumption allows setting $\frac{\partial}{\partial t} = 0$ while assumption of axisymmetry allows setting $\frac{\partial}{\partial \theta} = 0$. All the quantities are therefore a function of radial position r only, and the partial differential equations reduce to ordinary differential equations. The mean quantities involve r and θ components of velocity vector, u_r, u_θ; stress components, p_{rr}, $p_{r\theta}$, $p_{\theta\theta}$; and r and θ components of energy flux, q_r, q_θ; along with other flow variables, mass density ρ, pressure p, and temperature T. Under these assumptions, as shown by Ai [7], we can obtain a solution of the problem from moment equations with some additional simplifications introduced later. The Navier–Stokes based solution is already provided in Sect. 2.4.5.

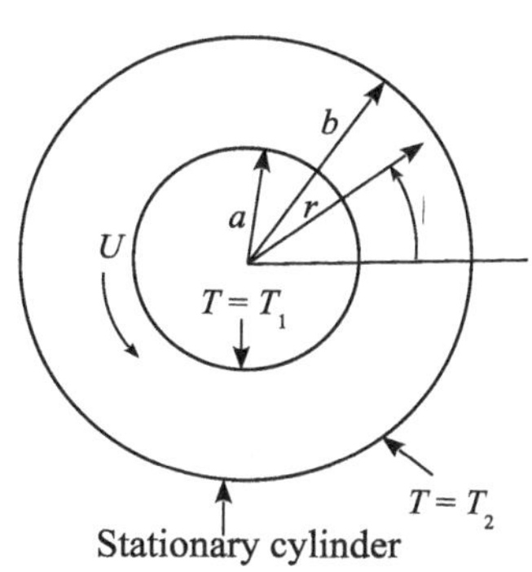

Fig. 6.11 Schematic of cylindrical Couette flow

6.8.1 Governing Equations and Boundary Conditions

13-moment equations in two-dimensional polar coordinate form are:

Continuity:

$$\frac{d}{dr}(\rho u_r r) = 0 \tag{6.192}$$

Momentum:

$$\frac{dp}{dr} + \frac{dp_{rr}}{dr} + \frac{p_{rr} - p_{\theta\theta}}{r} + \rho u_r \frac{du_r}{dr} - \rho \frac{u_\theta^2}{r} = 0 \tag{6.193}$$

$$\frac{dp_{r\theta}}{dr} + 2\frac{p_{r\theta}}{r} + \rho u_r \frac{du_\theta}{dr} + \rho \frac{u_\theta u_r}{r} = 0 \tag{6.194}$$

Energy:

$$\frac{5}{3}p\left(\frac{du_r}{dr} + \frac{u_r}{r}\right) + \frac{2}{3}\left(p_{rr}\frac{du_r}{dr} + p_{r\theta}\frac{du_\theta}{dr} + p_{\theta\theta}\frac{u_r}{r} - p_{r\theta}\frac{u_\theta}{r}\right)$$
$$+ \frac{2}{3}\left(\frac{dq_r}{dr} + \frac{q_r}{r}\right) + u_r \frac{dp}{dr} = 0 \tag{6.195}$$

Equation of State:

$$p = \rho RT \tag{6.196}$$

Stresses:

$$\frac{2}{3}p\left(2\frac{du_r}{dr} - \frac{u_r}{r}\right) + \frac{4}{15}\left(2\frac{dq_r}{dr} - \frac{q_r}{r}\right) + u_r \frac{dp_{rr}}{dr} - 2\frac{u_\theta}{r}p_{r\theta} + \frac{7}{3}p_{rr}\frac{du_r}{dr}$$
$$- \frac{2}{3}p_{r\theta}\frac{du_\theta}{dr} - \frac{4}{3}\frac{p_{r\theta}}{r}u_\theta - \frac{2}{3}\frac{p_{\theta\theta}u_r}{r} + \frac{p_{rr}u_r}{r} = -\frac{p}{\mu}p_{rr} \tag{6.197}$$

$$p\left(\frac{du_\theta}{dr} - \frac{u_\theta}{r}\right) + \frac{2}{5}\left(\frac{dq_\theta}{dr} - \frac{q_\theta}{r}\right) + u_r \frac{dp_{r\theta}}{dr} + u_\theta \frac{p_{rr} - p_{\theta\theta}}{r}$$
$$+ p_{rr}\frac{du_\theta}{dr} + 2p_{r\theta}\frac{u_r}{r} + 2p_{r\theta}\frac{du_r}{dr} - \frac{p_{\theta\theta}u_\theta}{r} = -\frac{p}{\mu}p_{r\theta} \tag{6.198}$$

$$-\frac{2}{3}p\left(\frac{du_r}{dr} - 2\frac{u_r}{r}\right) - \frac{4}{15}\left(\frac{dq_r}{dr} - 2\frac{q_r}{r}\right) + \frac{2}{r}u_\theta p_{r\theta} + u_r \frac{dp_{\theta\theta}}{dr} + \frac{4}{3}p_{r\theta}\frac{du_\theta}{dr}$$

$$-\frac{2}{3}\left(p_{rr}\frac{du_r}{dr}-p_{r\theta}\frac{u_\theta}{r}\right)+\frac{7}{3}p_{\theta\theta}\frac{u_r}{r}+p_{\theta\theta}\frac{du_r}{dr}=-\frac{p}{\mu}p_{\theta\theta} \tag{6.199}$$

Heat Flux:

$$\begin{aligned}
&\frac{5}{2}p\frac{dRT}{dr}+RT\left(\frac{dp_{rr}}{dr}+\frac{p_{rr}-p_{\theta\theta}}{r}\right)+q_r\left(\frac{du_r}{dr}+\frac{u_r}{r}\right)+u_r\frac{dq_r}{dr}-u_\theta\frac{q_\theta}{r}\\
&+\frac{11}{5}q_r\frac{du_r}{dr}+\frac{2}{5}\left(q_r\frac{u_r}{r}+q_\theta\frac{du_\theta}{dr}\right)-\frac{7}{5}q_\theta\frac{u_\theta}{r}-\frac{p_{rr}}{\rho}\frac{dp}{dr}+\frac{7}{2}p_{rr}\frac{dRT}{dr}\\
&-\frac{p_{rr}}{\rho}\left(\frac{dp_{rr}}{dr}+\frac{p_{rr}-p_{\theta\theta}}{r}\right)-\frac{p_{r\theta}}{\rho}\left(\frac{dp_{r\theta}}{dr}+2\frac{p_{r\theta}}{r}\right)=-\frac{2}{3}\frac{p}{\mu}q_r
\end{aligned} \tag{6.200}$$

$$\begin{aligned}
&RT\left(\frac{dp_{r\theta}}{dr}+2\frac{p_{r\theta}}{r}\right)+q_\theta\left(\frac{du_r}{dr}+\frac{u_r}{r}\right)+u_r\frac{dq_\theta}{dr}+q_r\frac{u_\theta}{r}+\frac{11}{5}q_\theta\frac{u_r}{r}\\
&+\frac{7}{5}q_r\frac{du_\theta}{dr}+\frac{2}{5}\left(q_\theta\frac{du_r}{dr}-q_r\frac{u_\theta}{r}\right)-\frac{p_{r\theta}}{\rho}\frac{dp}{dr}+\frac{7}{2}p_{r\theta}\frac{dRT}{dr}\\
&-\frac{p_{r\theta}}{\rho}\left(\frac{dp_{rr}}{dr}+\frac{p_{rr}-p_{\theta\theta}}{r}\right)-\frac{p_{\theta\theta}}{\rho}\left(\frac{dp_{r\theta}}{dr}+2\frac{p_{r\theta}}{r}\right)=-\frac{2}{3}\frac{p}{\mu}q_\theta
\end{aligned} \tag{6.201}$$

The above 10 equations involve the following 10-variables: ρ, p, T, u_r, u_θ, q_r, q_θ, p_{rr}, $p_{r\theta}$, $p_{\theta\theta}$.

The relevant boundary conditions are given below. At the outer stationary cylinder:

$$u_r(b)=0 \tag{6.202}$$

$$\begin{aligned}
&-\frac{p_{r\theta}(b)}{p(b)}+2\left(\frac{1-\alpha}{1+\alpha}\right)\frac{u_\theta(b)}{(2\pi RT(b))^{1/2}}\left(1+\frac{p_{rr}(b)}{2p(b)}\right)+\frac{2}{5}\left(\frac{1-\alpha}{1+\alpha}\right)\\
&\times\frac{q_\theta(b)}{(2\pi RT(b))^{1/2}p(b)}=0
\end{aligned} \tag{6.203}$$

$$\begin{aligned}
&-\left(\frac{2\pi}{RT(b)}\right)^{1/2}\frac{q_r(b)}{p(b)}+4\left(\frac{1-\alpha}{1+\alpha}\right)\left[1-\frac{T_2}{T(b)}-\frac{p_{rr}(b)}{2p(b)}\left(\frac{3}{2}-\frac{T_2}{T(b)}\right)\right.\\
&\left.+\left(1+\frac{p_{rr}(b)}{2p(b)}\right)\frac{(u_\theta(b))^2}{2RT(b)}\right]=0
\end{aligned} \tag{6.204}$$

In the above equations, $\alpha = 1 - \sigma_v$, with σ_v being the tangential momentum accommodation coefficient (TMAC). At the inner rotating cylinder ($r = a$)

$$u_r(a) = 0 \tag{6.205}$$

$$\frac{p_{r\theta}(a)}{p(a)} + 2\left(\frac{1-\alpha}{1+\alpha}\right)\frac{u_\theta(a) - U}{(2\pi RT(a))^{1/2}}\left(1 + \frac{p_{rr}(a)}{2p(a)}\right) + \frac{2}{5}\left(\frac{1-\alpha}{1+\alpha}\right) \times \frac{q_\theta(a)}{(2\pi RT(a))^{1/2}p(a)} = 0 \tag{6.206}$$

$$\left(\frac{2\pi}{RT(a)}\right)^{1/2}\frac{q_r(a)}{p(a)} + 4\left(\frac{1-\alpha}{1+\alpha}\right)\left[1 - \frac{T_1}{T(a)} + \frac{p_{rr}(a)}{2p(a)}\left(\frac{3}{2} - \frac{T_1}{T(a)}\right) + \left(1 + \frac{p_{rr}(a)}{2p(a)}\right)\frac{(u_\theta(a) - U)^2}{2RT(a)}\right] = 0 \tag{6.207}$$

Equation (6.192) can be solved with the boundary conditions Eq. (6.202) or (6.205) as $u_r = 0$, which helps get rid of several terms from the governing equations. Further, Eq. (6.194) can be readily integrated to obtain

$$p_{r\theta} = \frac{B}{r^2} \tag{6.208}$$

where B is a constant of integration. The shear stress can be converted into a torque per unit length by multiplying it by $2\pi r^2$. Therefore, Eq. (6.208) shows that the torque per unit length is constant across the annulus.

The above results can be used to simplify and integrate Eq. (6.195), and we obtain

$$r(q_r + p_{r\theta}u_\theta) = c \tag{6.209}$$

with c being another constant of integration. The equation therefore shows that the summation of heat flux and work done by the shear stress is a constant.

6.8.2 *Linearization of 13-Moment Equations*

The remaining equations are difficult to solve analytically due to the non-linear and coupled nature of the equations. However, as seen in the previous section, it is possible to obtain a solution upon linearization of the equations and the boundary conditions. Linearization is possible when the inner cylinder is rotating slowly and the temperature difference between the cylinder is kept small, i.e. $Ma = \frac{U}{\sqrt{\gamma R T_1}} << 1$ and $\frac{T_1 - T_2}{T_1} << 1$ (subscripts "1" and "2" denote quantities evaluated at the inner and outer cylinders, respectively).

Following the approach of the previous section, the variables ρ, p, and T are expressed as

$$\rho = \rho_1 + \rho^{'}, \quad p = p_1 + p^{'}, \quad T = T_1 + T^{'} \tag{6.210}$$

while for the remaining quantities (velocity, stress, and heat flux components), we have

$$Q = Q^{'} \tag{6.211}$$

Similarly, for the coefficient of viscosity we have

$$\mu = \mu_1 + \mu^{'} \tag{6.212}$$

where subscript "1" denotes quantities evaluated at the inner cylinder and the primed quantities denote small perturbations with respect to the undisturbed (initial) quantities.

Upon substituting the above quantities (Eqs. (6.210)–(6.211)) in the governing equations and neglecting terms of higher order leads to the following set of equations:
Momentum:

$$\frac{dp^{'}}{dr} + \frac{dp^{'}_{rr}}{dr} + \frac{p^{'}_{rr} - p^{'}_{\theta\theta}}{r} = 0 \tag{6.213}$$

Stresses:

$$\frac{4}{15}\left(2\frac{dq^{'}_r}{dr} - \frac{q^{'}_r}{r}\right) = -\frac{p_1}{\mu_1}p^{'}_{rr} \tag{6.214}$$

$$p_1\left(\frac{du^{'}_\theta}{dr} - \frac{u^{'}_\theta}{r}\right) + \frac{2}{5}\left(\frac{dq^{'}_\theta}{dr} - \frac{q^{'}_\theta}{r}\right) = -\frac{p_1}{\mu_1}p^{'}_{r\theta} \tag{6.215}$$

$$-\frac{4}{15}\left(\frac{dq^{'}_r}{dr} - 2\frac{q^{'}_r}{r}\right) = -\frac{p_1}{\mu_1}p^{'}_{\theta\theta} \tag{6.216}$$

Heat Flux:

$$\frac{5}{2}p_1\frac{dRT^{'}}{dr} - RT_1\frac{dp^{'}}{dr} = -\frac{2}{3}\frac{p_1}{\mu_1}q^{'}_r \tag{6.217}$$

$$q'_\theta = 0 \tag{6.218}$$

Equation of state:

$$p' = R(\rho_1 T' + T_1 \rho') \tag{6.219}$$

The linearized boundary conditions at $r = b$ are

$$-\frac{p'_{r\theta}(b)}{p_1} + 2\left(\frac{1-\alpha}{1+\alpha}\right)\frac{u'_\theta(b)}{(2\pi RT_1)^{1/2}} = 0 \tag{6.220}$$

$$-\left(\frac{2\pi}{RT_1}\right)^{1/2}\frac{q'_r(b)}{p_1} + 4\left(\frac{1-\alpha}{1+\alpha}\right)\left[\frac{T_1 - T_2}{T_1} + \frac{T_2}{T_1}\frac{T'(b)}{T_1} - \frac{p'_{rr}(b)}{4p_1}\right] = 0 \tag{6.221}$$

Similarly, the linearized boundary conditions at $r = a$ are

$$\frac{p'_{r\theta}(a)}{p_1} + 2\left(\frac{1-\alpha}{1+\alpha}\right)\frac{u'_\theta(a) - U}{(2\pi RT_1)^{1/2}} = 0 \tag{6.222}$$

$$\left(\frac{2\pi}{RT_1}\right)^{1/2}\frac{q'_r(a)}{p_1} + 4\left(\frac{1-\alpha}{1+\alpha}\right)\left[\frac{T'(a)}{T_1} + \frac{p'_{rr}(a)}{4p_1}\right] = 0 \tag{6.223}$$

6.8.3 *Solution*

Equation (6.215) (with the help of Eq. (6.218)) suggests that $(p'_{r\theta})$ is identical to the Navier–Stokes stress strain rate relation. Therefore, we obtain a similar expression for u'_θ as in the incompressible Navier–Stokes solution, i.e.,

$$u'_\theta = c_1 r + \frac{B}{2\mu_1 r} \tag{6.224}$$

Upon linearizing Eq. (6.209) we obtain

$$q'_r = \frac{c}{r} \tag{6.225}$$

Equations (6.214) and (6.216) can be solved in conjunction with the above equation, to obtain

$$p'_{rr} = \frac{4\mu_1}{5p_1}\frac{c}{r^2} \tag{6.226}$$

$$p'_{\theta\theta} = -\frac{4\mu_1}{5p_1}\frac{c}{r^2} \tag{6.227}$$

whereas these normal stresses are identically zero in the solution of the Navier–Stokes equation. Upon, substituting the solutions of p'_{rr} and $p'_{\theta\theta}$ into Eq. (6.213), we obtain $\frac{dp'}{dr} = 0$ or $p' =$ constant (which can be set as zero without loss of generality).

We can now turn to Eq. (6.217) and determine T', since the other terms in the equations have already been evaluated, as

$$T' = -\frac{4}{15}\left(\frac{c}{\mu_1 R}\right)\ln r + c_2 \tag{6.228}$$

Finally p' is obtained from the equation of state, as

$$p' = \frac{4}{15}\frac{\rho_1}{T_1}\left(\frac{c}{\mu_1 R}\right)\ln r - \frac{\rho_1}{T_1}c_2 \tag{6.229}$$

We now evaluate the constants B and c_1 appearing in $p_{r\theta}$ and u_θ by using the boundary conditions given by Eqs. (6.220) and (6.222). The obtained expressions for B and c_1 are

$$B = \frac{\mu_1 U \dfrac{2b^2}{b-a}\dfrac{Re}{Ma}}{\dfrac{1+\alpha}{1-\alpha}\sqrt{\dfrac{10\pi}{3}}\left(\dfrac{a}{b}+\dfrac{b^2}{a^2}\right)+\dfrac{a+b}{a}\dfrac{Re}{Ma}} \tag{6.230}$$

$$c_1 = \frac{\dfrac{1+\alpha}{1-\alpha}\sqrt{\dfrac{10\pi}{3}} - \dfrac{Re}{Ma}\dfrac{b}{b-a}}{\dfrac{1+\alpha}{1-\alpha}\sqrt{\dfrac{10\pi}{3}}\left(\dfrac{a}{b}+\dfrac{b^2}{a^2}\right)+\dfrac{a+b}{a}\dfrac{Re}{Ma}}\frac{U}{b} \tag{6.231}$$

For monatomic gas, Mach number and Reynolds number are defined based on conditions at the inner cylinder, and selecting annulus as length scale in the definition of Reynolds number,

$$Ma = \frac{U}{\sqrt{\frac{5}{3}RT_1}}, \qquad Re = \rho_1 U\frac{b-a}{\mu_1} \tag{6.232}$$

Hence, we obtain

$$p_{r\theta} = \frac{\mu_1 U \dfrac{2b^2}{b-a}\dfrac{Re}{Ma}}{\dfrac{1+\alpha}{1-\alpha}\sqrt{\dfrac{10\pi}{3}}\left(\dfrac{a}{b}+\dfrac{b^2}{a^2}\right)+\dfrac{a+b}{a}\dfrac{Re}{Ma}}\frac{1}{r^2} \tag{6.233}$$

$$\frac{u_\theta}{U} = \frac{\dfrac{1+\alpha}{1-\alpha}\sqrt{\dfrac{10\pi}{3}}\dfrac{r}{b}+\dfrac{Re}{Ma}\dfrac{b}{b-a}\left(\dfrac{b}{r}-\dfrac{r}{b}\right)}{\dfrac{1+\alpha}{1-\alpha}\sqrt{\dfrac{10\pi}{3}}\left(\dfrac{a}{b}+\dfrac{b^2}{a^2}\right)+\dfrac{a+b}{a}\dfrac{Re}{Ma}} \tag{6.234}$$

The drag coefficient, C_D is given by $C_D = \frac{p_{r\theta}}{\frac{1}{2}\rho_1 U^2}$. Therefore, at the stationary wall ($r = b$), we obtain

$$C_D Ma = \frac{4}{\dfrac{1+\alpha}{1-\alpha}\sqrt{\dfrac{10\pi}{3}}\left(\dfrac{a}{b}+\dfrac{b^2}{a^2}\right)+\dfrac{a+b}{a}\dfrac{Re}{Ma}} \tag{6.235}$$

For diffusively reflecting surfaces ($\alpha = 0$), the above expression simplifies to

$$\frac{1}{C_D Ma} = \frac{1}{4}\sqrt{\frac{10\pi}{3}}\left(\frac{a}{b}+\frac{b^2}{a^2}\right)+\frac{1}{2}\left(1+\frac{b-a}{2a}\right)\frac{Re}{Ma} \tag{6.236}$$

The above expression is in good agreement with several results in the literature, lending credibility to the above analysis, and particularly, the simplifying assumptions made in the process.

We can also determine the slip velocity at the surfaces of the cylinder. At the stationary outer cylinder, we obtain

$$\frac{u_\theta(b)}{U} = \frac{\dfrac{1+\alpha}{1-\alpha}\sqrt{\dfrac{10\pi}{3}}}{\dfrac{1+\alpha}{1-\alpha}\sqrt{\dfrac{10\pi}{3}}\left(\dfrac{a}{b}+\dfrac{b^2}{a^2}\right)+\dfrac{a+b}{a}\dfrac{Re}{Ma}} \tag{6.237}$$

whereas at the rotating inner cylinder we have

$$\frac{u_\theta(a)}{U} = \frac{u_\theta(b)}{U} + \frac{\dfrac{a+b}{a}\dfrac{Re}{Ma}}{\dfrac{1+\alpha}{1-\alpha}\sqrt{\dfrac{10\pi}{3}}\left(\dfrac{a}{b}+\dfrac{b^2}{a^2}\right)+\dfrac{a+b}{a}\dfrac{Re}{Ma}} \tag{6.238}$$

It is easy to verify from the above equations that the slip velocity at the surfaces goes to zero in the continuum limit ($Kn \to 0$ or $Re/Ma \to \infty$). In the other limit of free-molecular flow, $Re/Ma \to 0$, we obtain

$$\frac{u_\theta(b)}{U} = \frac{u_\theta(a)}{U} = \left(\frac{a}{b} + \frac{b^2}{a^2}\right)^{-1} \tag{6.239}$$

as the slip velocity at both the surfaces. If, further, the gap $(b-a)$ becomes negligibly small (i.e., $b/a \to 1$), the above expression reduces to

$$\frac{u_\theta(b)}{U} = \frac{u_\theta(a)}{U} = \frac{1}{2} \tag{6.240}$$

which corresponds to the result for plane flow case.

The other two constants c and c_2 are obtained from the remaining two boundary conditions (Eqs. (6.221), and (6.223)), as

$$c = \frac{\dfrac{T_1 - T_2}{T_1}}{\left(\dfrac{2\pi}{RT_1}\right)^{1/2} \dfrac{1}{4bp_1}\left(1 + \dfrac{b}{a}\dfrac{T_2}{T_1}\right)\left(\dfrac{1+\alpha}{1-\alpha}\right) + \dfrac{4}{15}\dfrac{T_2}{T_1}\dfrac{\ln(b/a)}{\mu_1 RT_1} + \dfrac{\mu_1}{5p_1^2 b^2}\left(1 + \dfrac{b^2}{a^2}\dfrac{T_2}{T_1}\right)} \tag{6.241}$$

and

$$c_2 = \frac{-(T_1 - T_2)\left[\left(\dfrac{2\pi}{RT_1}\right)^{1/2} \dfrac{1}{4ap_1}\left(\dfrac{1+\alpha}{1-\alpha}\right) - \dfrac{4}{15}\dfrac{1}{\mu_1 RT_1}\ln(a) + \dfrac{\mu_1}{5p_1^2 a^2}\right]}{\left(\dfrac{2\pi}{RT_1}\right)^{1/2} \dfrac{1}{4bp_1}\left(1 + \dfrac{b}{a}\dfrac{T_2}{T_1}\right)\left(\dfrac{1+\alpha}{1-\alpha}\right) + \dfrac{4}{15}\dfrac{T_2}{T_1}\dfrac{1}{\mu_1 RT_1}\ln(b/a) + \dfrac{\mu_1}{5p_1^2 b^2}\left(1 + \dfrac{b^2}{a^2}\dfrac{T_2}{T_1}\right)} \tag{6.242}$$

The temperature distribution across the annulus is given by

$$\frac{T_1 - T}{T_1 - T_2} = \frac{4}{15}\frac{\dfrac{Re}{Ma}\ln\left(1 + \dfrac{b-a}{a}\dfrac{r-a}{b-a}\right)}{\dfrac{1}{4}\sqrt{\dfrac{10\pi}{3}}\dfrac{b-a}{a}\left(1 + \dfrac{b}{a}\dfrac{T_2}{T_1}\right) + \dfrac{Re}{Ma}\dfrac{T_2}{T_1}\ln(b/a) + \dfrac{5}{4}\dfrac{Ma}{Re}\left(\dfrac{b-a}{b}\right)^2\left(1 + \dfrac{b^2}{a^2}\dfrac{T_2}{T_1}\right)}$$

$$+\frac{\frac{15}{16}\sqrt{\frac{10\pi}{3}}\frac{b-a}{a}+\frac{5}{4}\frac{Ma}{Re}\left(\frac{b-a}{a}\right)^2}{\frac{15}{16}\sqrt{\frac{10\pi}{3}}\frac{b-a}{b}\left(1+\frac{b}{a}\frac{T_2}{T_1}\right)+\frac{Re}{Ma}\frac{T_2}{T_1}\ln(b/a)+\frac{5}{4}\frac{Ma}{Re}\left(\frac{b-a}{b}\right)^2\left(1+\frac{b^2}{a^2}\frac{T_2}{T_1}\right)} \tag{6.243}$$

where we have assumed the surfaces to be diffuse ($\alpha = 0$) for simplicity. The gas temperature at the surface of the inner cylinder can now be determined as

$$T(a) = T_1 + \frac{\frac{b}{a}(T_2-T_1)\left(\sqrt{\frac{10\pi}{3}}+\frac{4}{3}\frac{1-\alpha}{1+\alpha}\frac{Ma}{Re}\frac{b-a}{a}\right)}{\sqrt{\frac{10\pi}{3}}\left(1+\frac{b}{a}\frac{T_2}{T_1}\right)+\frac{16}{15}\frac{1-\alpha}{1+\alpha}\frac{Re}{Ma}\frac{T_2}{T_1}\frac{b\ln(b/a)}{b-a}+\frac{4}{3}\frac{1-\alpha}{1+\alpha}\frac{Ma}{Re}\left(\frac{b-a}{b}\right)\left(1+\frac{b^2}{a^2}\frac{T_2}{T_1}\right)} \tag{6.244}$$

As before, we can obtain the temperature jump in the two limiting cases. When $\frac{Re}{Ma} \to 0$, we have

$$T_1 - T(a) = (T_1 - T_2)\left(\frac{a^2}{b^2} + \frac{T_2}{T_1}\right)^{-1} \tag{6.245}$$

Furthermore, when a approaches b and T_2 approaches T_1 we obtain $(T_1 - T(a)) = \frac{1}{2}(T_1 - T_2)$. On the other hand, for the case of $\frac{Re}{Ma} \to \infty$, we obtain $T_1 - T(a) = 0$, i.e., as expected there is no temperature jump in the limit of continuum flow.

At the surface of the outer cylinder we have in the limiting case of $\frac{Re}{Ma} \to 0$

$$T(b) - T_2 = (T_1 - T_2)\left(1 - \frac{b^2}{a^2}\frac{T_1 - T_2}{T_1}\right)\left(1 + \frac{b^2}{a^2}\frac{T_2}{T_1}\right)^{-1} \tag{6.246}$$

It is also instructive to write the ratio of the two temperature jumps, which is expressed as

$$\frac{T_1 - T(a)}{T(b) - T_2} = \frac{b^2}{a^2}\left(1 + \frac{b^2}{a^2}\frac{T_2}{T_1}\right)^{-1} \tag{6.247}$$

The above equation suggests that when $a/b \to 1$ and $T_2/T_1 \to 1$, we have

$$T_1 - T(a) = T(b) - T_2 = \frac{1}{2}(T_1 - T_2) \tag{6.248}$$

which is similar to the planar Couette flow result.

The Stanton number (St) is a useful non-dimensional number to define the analogy between fluid friction and heat transfer defined as

$$St = \frac{q_r}{\rho_1 U c_p (T_1 - T_2)}$$

We can obtain appropriate expressions for Stanton number for the present problem. At the wall of the inner cylinder, we have

$$StMa = \frac{1}{\frac{5}{2}\left[\frac{(1+\alpha)}{4(1-\alpha)}\frac{a}{b}\left(1+\frac{b}{a}\frac{T_2}{T_1}\right)\sqrt{\frac{10\pi}{3}}+\frac{T_2}{T_1}\frac{Re}{Ma}Pr\frac{a\ \ln(b/a)}{b-a}+\frac{5}{6}\frac{a}{b}\frac{b-a}{b}\frac{Ma}{Re}\left(1+\frac{b^2}{a^2}\frac{T_2}{T_1}\right)\right]} \tag{6.249}$$

Examining the limiting case of a approaches b at a fixed value of Re/Ma ($\gamma = 5/3$ for a monatomic gas), we obtain

$$\frac{1}{StMa} = \frac{5}{4}\frac{(1+\alpha)}{(1-\alpha)}\sqrt{\frac{10\pi}{3}} + \frac{Re}{Ma}Pr \tag{6.250}$$

which is again same as the result of planar Couette flow.

For arbitrary a and b, when $Re/Ma \to \infty$, we have

$$\frac{1}{StMa} \to \frac{T_2}{T_1}\frac{a}{(b-a)}\ln(b/a)\frac{Re}{Ma}Pr \tag{6.251}$$

which agrees with the classical solution of Fourier heat conduction for steady state. However, when $Re/Ma \to 0$, we obtain $StMa = 0$ which is different than the constant value given by a free molecule calculation.

6.8.4 Discussion on the Results

Using Eq. (6.234) the velocity profile u_θ/U across the annulus is plotted for different values of Re/Ma with a fixed value of $b/a = 10$, as shown in Fig. 6.12. Here, we considered the value of tangential momentum accommodation coefficient as unity. Figure 6.12 shows that the influence of rotation of inner cylinder on the gas reduces significantly with the radial distance.

Utilizing Eq. (6.238) the variation of Re/Ma with $u_\theta(a)/U$ for different values of b/a is shown in Fig. 6.13. Note that if the gap is very small, i.e. $b/a \to 0$, the results reduce to the case of planar Couette flow, as also noted from the corresponding expression. Figure 6.14 is drawn by using Eq. (6.236) to show the variation of $1/(C_D Ma)$ with Re/Ma for different values of b/a. It is observed that the quantity $1/(C_D Ma)$ increases linearly with increase in Re/Ma.

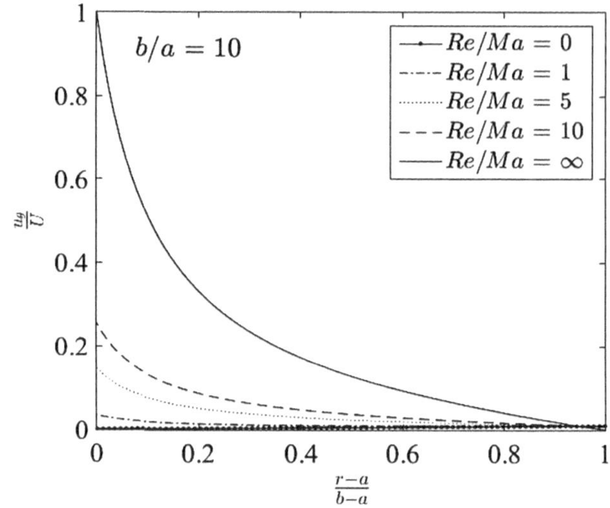

Fig. 6.12 Variation of u_θ/U with $(r-a)/(b-a)$

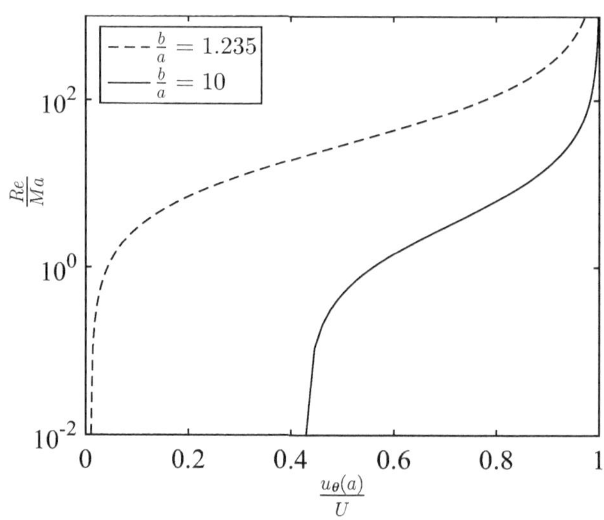

Fig. 6.13 Variation of Re/Ma with $u_\theta(a)/U$

Using Eq. (6.250), the variation of Re/Ma with the product of Stanton number and Mach number ($StMa$) for different set of the values of a and b is shown in Fig. 6.15. The value of Prandtl number (Pr) is considered to be 0.71, tangential momentum accommodation coefficient σ_v is considered to be unity and a constant

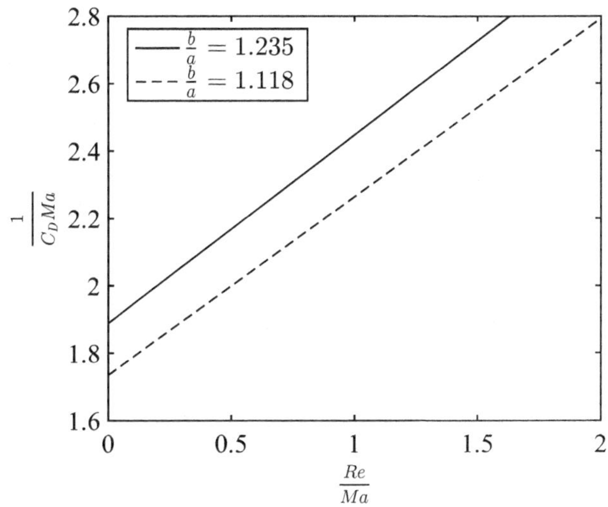

Fig. 6.14 Variation of $1/(C_D Ma)$ with Re/Ma

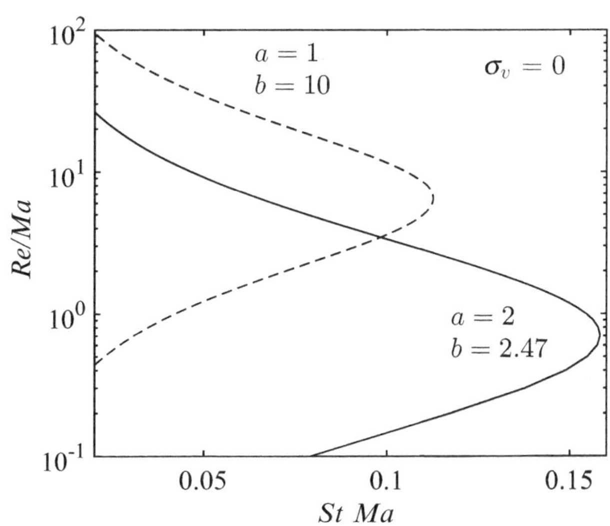

Fig. 6.15 Variation of Re/Ma with $StMa$

value of the ratio of outer cylinder temperature to the inner cylinder temperature (T_2/T_1) is considered. It is observed that with the increase in $StMa$, the increase in Re/Ma is more for the case $b/a = 10$ as compared to the case of $b/a = 1.235$.

6.9 Derivation of Navier–Stokes Equations from 13-Moment Equations

In this section, we show that by performing linearization, the Navier–Stokes equations can be extracted from the 13-moment equations. The linearized form of 13-moment equations read as

$$\frac{\partial \rho}{\partial t} + \frac{\partial(\rho u_k)}{\partial x_k} = 0, \tag{6.252}$$

$$\frac{\partial u_i}{\partial t} + u_k \frac{\partial u_i}{\partial x_k} + \frac{1}{\rho}\frac{\partial p}{\partial x_i} + \frac{1}{\rho}\frac{\partial p_{ik}}{\partial x_k} = F_i, \tag{6.253}$$

$$\frac{3}{2} R\rho \frac{\partial T}{\partial t} + \frac{3}{2} R\rho u_k \frac{\partial T}{\partial x_k} + p\frac{\partial u_i}{\partial x_i} + p_{ij}\frac{\partial u_i}{\partial x_j} + \frac{\partial q_k}{\partial x_k} = 0, \tag{6.254}$$

$$\frac{Dp_{ij}}{Dt} + \frac{4}{5}\frac{\partial q_{\langle i}}{\partial x_{j\rangle}} + \underline{2p\frac{\partial u_{\langle i}}{\partial x_{j\rangle}}} = \underline{-\frac{p}{\mu} p_{ij}} \tag{6.255}$$

$$\frac{Dq_i}{Dt} + \underline{\frac{5}{2}\rho RT \frac{\partial RT}{\partial x_i}} + RT\frac{\partial p_{ik}}{\partial x_k} = \underline{-\frac{2}{3}\frac{p}{\mu} q_i}. \tag{6.256}$$

For the five field variables: density, bulk velocity, and temperature, we can use the first three equations (which are actually five equations) given above. The remaining equations can be used to obtain the constitutive relations for these primary variables.

Since the stress tensor and heat flux vector vanish at equilibrium (i.e., $p_{ij} = 0$ and $q_i = 0$), Eqs. (6.255) and (6.256) can be substantially simplified with only the underlined terms in the above equations surviving. Hence, at equilibrium, we obtain

$$p_{ij} = -2\mu \frac{\partial u_{\langle i}}{\partial x_{j\rangle}}, \quad \text{and} \quad q_i = -k\frac{\partial T}{\partial x_i}, \tag{6.257}$$

where $k = (15/4)\mu R$ from kinetic theory.

Equation (6.257) represents the constitutive relations obtained for stress tensor and heat flux vector within the Navier–Stokes framework. As shown in Kogan [83], it is also possible to derive the Burnett equations from the Grad's equations. The Grad 13-moment equations can, therefore, be considered as a superset of both the Navier–Stokes and the Burnett equations.

6.10 Derivation of Cattaneo's Equation from 13-Moment Equations

The Cattaneo's equation is a non-Fourier heat conduction model introduced in Sect. 4.5. In this section, we derive the Cattaneo equation (4.14) starting from the Grad equations, following Agrawal [3]. The value of time delay which appeared as an additional unknown in the model can also be evaluated using this approach.

The derivation follows in a rather straightforward manner by setting the velocity and shear stress in the heat flux equation (6.100) as zero. This yields

$$\frac{\partial q_i}{\partial t} + \frac{5}{2}\rho RT \frac{\partial RT}{\partial x_i} = -\frac{2}{3}\frac{p}{\mu} q_i. \tag{6.258}$$

Simple algebric manipulation of the above equation (with the help of equation for state) yields

$$\begin{aligned} \frac{3}{2}\frac{\mu}{p}\frac{\partial q_i}{\partial t} + q_i &= -\frac{15}{4}\mu R \frac{\partial T}{\partial x_i} \\ &= -k\frac{\partial T}{\partial x_i} \end{aligned} \tag{6.259}$$

Note that $k = (15/4)\mu R$ has also been used in the above derivation. The above equation can be compared to the Cattaneo equation (Eq. (4.14)) introduced earlier. The relaxation term can be recognized by comparing Eqs. (4.14) and (6.259) and found to be $3\mu/(2p)$, where μ is viscosity and p is pressure. The value of relaxation time for air at standard condition is approximately 2.7×10^{-10} which is comparable to the value employed in the literature.

The derivation of Cattaneo's equation from Grad's equation should lend some credibility to the model.

6.11 Comparison of Chapman–Enskog and Grad Methods

Having examined both the Chapman–Enskog method and the Grad's moment method to solve the Boltzmann equation in detail in the previous and the present chapter, we now present some comments regarding the similarities and differences between the two approaches.

1. *Number of variables*: The Chapman–Enskog method involves the same set of variables as in the Navier–Stokes equations, whereas the Grad's approach is to expand the number of primary variables. Therefore, there are additional evolution equations involved in the Grad's approach; the ensuing 13-moment equations are substantially more involved and coupled.
2. *Initial and boundary conditions*: The requirement of additional initial and boundary conditions shows up for the complete solution, due to additional differential equations involved in the moment equations. The number of boundary conditions also increases in the Burnett equations—primarily because the constitutive relations obtained through the Chapman–Enskog method involve one order higher derivatives as compared to that present at the Navier–Stokes order. Specifically, the additional conditions have to be found for stress (p_{ij}) and

heat flux (q_i) in the Grad's approach, whereas additional boundary conditions for velocity and temperature are required in the Burnett equations.

3. *Requirement of distribution function*: Both the methods involve evaluation of the distribution function at some stage to evaluate the unknown terms appearing during the derivation process. In the Chapman–Enskog method, the number of terms retained in the expansion of the distribution function (in terms of Knudsen number) dictates the order of the ensuing equations. However, in the Grad moment method, the order of the various terms (in terms of Knudsen number) is not readily apparent.
4. *Derivation of distribution function:* The involvement of the Boltzmann equation in the derivation of distribution function in moment method is minimal. Recall that, in the Chapman–Enskog method, we expressed the distribution function as a perturbation series and substituted this form in the Boltzmann equation to obtain successive approximations for the distribution function. On the other hand, in moment method, the distribution function is expanded in terms of Hermite polynomials, and the series is truncated at some point. The Hermite coefficients are then evaluated and functional form of the distribution function is determined. Hence, it seems like the derivation of the distribution function is independent of the Boltzmann equation. However, note that the series is expanded about Maxwellian distribution function, which itself is a solution of Boltzmann equation for a gas at rest. Hence, the involvement of the Boltzmann equation in the derivation of the distribution function is minimal and involved only in generating the evolution equations for the moments.
5. *Comparison of distribution function:* In Chapman–Enskog method, the time dependence of the distribution function is not explicit, rather it is through that of temperature, pressure (or density), bulk velocity, and their gradients. However, in moment method, the time dependence is through the Hermite coefficients as well as temperature, pressure, and bulk velocity. The Chapman–Enskog form of the distribution function as obtained in first approximation, $f = f^{(0)} + f^{(1)}$ (see Eqs. (5.90) and (5.95)) can be retrieved from the Grad 13-moment form (Eq. (6.91)) by replacing the stress tensor and heat flux vector in Eq. (6.91) using the Navier–Stokes approximation as

$$p_{ij} = -2\mu \frac{\partial u_{\langle i}}{\partial x_{j\rangle}}, \qquad q_i = -k \frac{\partial T}{\partial x_i}$$

where absolute viscosity and thermal conductivity are given by Eqs. (5.128) and (5.134), respectively.

6. *Constitutive relation*: The accuracy of the constitutive relation affects the accuracy of the solution directly in the Chapman–Enskog method. However, the constitutive equation shows up for the higher moments (stress and heat flux) in the Grad method. Assuming that our primary interest is in lower moments

(velocity and temperature), any inaccuracy in the constitutive relation is expected to affect the lower moments of interest only indirectly (and hopefully more mildly!).

7. *Improvement in accuracy*: In principle, it is possible to increase the accuracy of the equations by employing still higher-order distribution function in the Enskog method. However, evaluation of the distribution function is intimately linked to the number of variables introduced in the Grad moment method. That is, an increase in accuracy requires increasing the moments further in the Grad's approach, which makes direct refinement of the moment equations difficult. Note that more moments means more governing equations and requirement of initial/boundary conditions for their solution.
8. *Complexity comparison:* The Grad moment method is conceptually more powerful and simpler than the Chapman–Enskog method even though we have to deal with a larger set of variables. This is evident from the fact that more advanced set of equations like 26-moment equations are available in the literature whereas super-Burnett equations are still not known in full three-dimensional non-linear form.

A relevant pertinent question to consider is the number of moments that should be retained in the Grad's approach. It is expected that more number of moments will make the equations more exact. However, ascribing a physical meaning to these higher moments becomes difficult. In the above section, 13 moments were considered; equations for 26 moments are also available in the literature (see Chap. 7). As can be readily appreciated, deriving and working with a larger number of equations is complicated and unattractive, especially when it is understood that the solution of the equations is intimately linked to the accuracy of the boundary conditions. However, as already argued, prescribing boundary conditions is not easy for the higher moments, especially when the physical meaning of these moments is itself unclear.

As a final remark, we note that closure relation/integral in Burnett and Grad equations depend on the inter-molecular potential/type of molecule (Maxwell, hard-sphere, etc.). So does it mean that different governing equations will apply to different gases? We have already seen in Sect. 2.3.4 that the wall boundary condition can differ between gases (monatomic versus non-monatomic, for example). The lower-order equations (Euler equations with zero stress and heat flux terms) do not differentiate between gases and liquids; similarly, the Navier–Stokes equations (with linear constitutive relation) do not differentiate between different types of gases. This universality however does not apply to higher-order equations! In other words, close to equilibrium, the nature of gas molecule does not matter, but the nature of the gas starts to matter if we are far away from equilibrium. Evidently, the large number of collisions in the near-equilibrium (continuum) regime between molecules masks the nature of the molecule.

6.12 Summary

In this chapter, Grad's idea of working with a larger number of primary variables is introduced. Based on this idea and representation of the distribution function in terms of Hermite polynomials, the Grad moment equations are derived. The boundary conditions are also derived for two-dimensional 13-moment case. The equations are used to solve the problem of an impulsively started plate, with the solution compared with that obtained from other equations.

The Grad equations are particularly important as the Navier–Stokes and Burnett equations can be obtained from them. Even certain non-Fourier models can be derived starting from Grad equations. Despite these features of the Grad equations, these equations suffer from several limitations as reviewed in the next chapter.

Chapter 7
Alternate Forms of Burnett and Grad Equations

In the previous two chapters, a formal derivation of the Burnett and Grad equations was presented. The derivation involved several novel ideas making the approach and the obtained equations invaluable. However, as reviewed in this chapter, both these equations have several limitations because of which working with alternate forms of these equations becomes necessary. Therefore, several variants of these equations have been proposed in the literature. The basic idea behind fixing the equations and the available variants are introduced in this chapter. Comparison of the results as obtained from the variants of the equations for a few specific cases is also presented.

7.1 Introduction

The Burnett and Grad equations derived in the previous two chapters have certain limitations because of which working with alternate forms of these equations becomes necessary. We first discuss these limitations, which helps appreciate the reason for existence of several variants of the equations in the literature. Most of the variants are however somewhat arbitrary, in that they cannot be derived from basic principles.

Recently, equations claiming to be thermodynamically more consistent have also been proposed. These equations have been derived from first principles and termed as Onsager-consistent equations. These OBurnett and O13 equations are also included, along with a test case on which these equations have been successfully applied.

A. Agrawal et al., *Microscale Flow and Heat Transfer*, Mechanical Engineering Series, https://doi.org/10.1007/978-3-030-10662-1_7

7.2 Limitations of Burnett Equations

The limitations of the Burnett equations are summarized herewith:

1. *Requirement of additional boundary conditions*: As seen in Chap. 5, the Burnett equations involve third-order derivatives (as opposed to second-order derivatives present in the Navier–Stokes equations), leading to the requirement of additional boundary conditions for the complete solution of the Burnett equations. Recall that the accuracy of solution of any differential equation strongly depends on the prescribed boundary conditions, making this issue particularly important. From experience, we know that while solving the Navier–Stokes equations there are no redundant boundary conditions that can be prescribed for any problem. Therefore, finding additional conditions for solving the Burnett equations is an issue. However, this issue does not, as such, invalidate the Burnett equations; only shows the difficulty in obtaining accurate flow solution using the equations. In order to accurately prescribe the additional boundary conditions, the interaction of the gas with the boundary needs careful attention. In Sect. 5.7.2 and several other works, the boundary value problem was converted into an initial value problem, and the problem of prescribing additional boundary conditions was circumvented by obtaining initial conditions from the DSMC simulations. Though the independent nature of the solution procedure was compromised to some extent, it was observed that the results of the Burnett equations were in reasonably good agreement with the DSMC results which enhances belief in these equations. But again, the challenge of prescribing additional boundary conditions is still an open problem, and without a definite answer, the Burnett equations will always be viewed with some suspicion.
2. *The convergence properties of the Chapman–Enskog series are not known*: The distribution function was expanded in an infinite series in Eq. (5.62) to capture far from equilibrium effects, with the assumption that the series would converge. Further, as evident from the detailed derivation given in Sect. 5.3.3, there are additional assumptions that are required in the process of derivation. It is therefore fair to claim that the derivation process is not based on physical arguments; assumptions are rather made to simplify the mathematical complexities.
   Note that the convergence (or otherwise) of the series has still not been proved and the assumptions involved are difficult to justify. Although some justification can be put forth in that the process leads to the correct equations (the Navier–Stokes) to first approximation, it suggests that the overall approach does not have as sound a mathematical footing as would be expected from fundamental equations in mechanics.
3. *The equations suffer from Bobylev's instability*: This point is already illustrated through stability analysis in Sect. 5.8 where it was shown that there exists a critical wave number beyond which the solution becomes unstable. An unconditional stability is expected from sound physical equations. These instabilities appear during transient processes, and manifest as unphysical oscillations in steady state processes.

4. *There is no proof that the Burnett equations follow positive definite entropy production*: This is unarguably the strongest criticism of these equations, as the equations have been shown to violate the second law of thermodynamics. This point is discussed at length by deGroot and Mazur [39] and Comeaux et al. [36], and the interested readers can refer to these sources for additional details and efforts made to mitigate this limitation.

7.3 Variants of the Burnett Equations

Two primary approaches have been adopted to overcome the above limitations: (1) Identify the source of difficulty and try fixing the issue or (2) Follow an alternative approach to derive the equations. The first approach involves either removing the terms that are the source of difficulty (i.e., obtain a subset of the Burnett equations) or compensating the effect of the difficult term with an additional term in the equation (i.e., form a superset of the Burnett equations). An order of magnitude analysis (of the type presented in Sect. 5.6) can help in identifying terms that are insignificant but can cause difficulties, and thus can be discarded. One can also resort to the super Burnett equations and selectively add terms from there to the Burnett equations in order to compensate the effect of the difficult term.

In all of the Burnett-variants, we have mass density (ρ), bulk velocity (u_i), and temperature (T) as the primary variables and closure is achieved by obtaining expressions for pressure tensor (P_{ij}) and heat flux vector (q_i) in terms of these primary variables and their gradients.

The basic conservation laws of mass, momentum, and energy as introduced earlier are being reproduced here for the sake of completeness and clarity.

$$\frac{\partial \rho}{\partial t}+\frac{\partial(\rho u_k)}{\partial x_k}=0 \tag{7.1}$$

$$\frac{\partial u_i}{\partial t}+u_k\frac{\partial u_i}{\partial x_k}+\frac{1}{\rho}\frac{\partial p}{\partial x_i}+\frac{1}{\rho}\frac{\partial p_{ik}}{\partial x_k}=F_i \tag{7.2}$$

$$\frac{3}{2}\rho R\frac{\partial T}{\partial t}+\frac{3}{2}\rho R u_k\frac{\partial T}{\partial x_k}+p\frac{\partial u_i}{\partial x_i}+p_{ij}\frac{\partial u_i}{\partial x_j}+\frac{\partial q_k}{\partial x_k}=0 \tag{7.3}$$

Note that at the zeroth and first approximation, all the variants give Euler and Navier–Stokes equations, respectively, i.e., the constitutive relationships for the pressure tensor and heat flux vector at the zeroth approximation read as

$$\begin{aligned} P_{ij}^{(Eu)} &= P_{ij}^{(0)} = \delta_{ij}p, \\ q_i^{(Eu)} &= q_i^{(0)} = 0 \end{aligned} \tag{7.4}$$

and at the first approximation,

$$P_{ij}^{(NS)} = p\delta_{ij} - 2\mu \frac{\partial u_{\langle i}}{\partial x_{j\rangle}}, \tag{7.5}$$

$$q_i^{(NS)} = -k \frac{\partial T}{\partial x_i} \tag{7.6}$$

The analysis differs for the second (and higher) approximations where, in all of the variants, we get different expressions for $P_{ij}^{(2)}$ and $q_i^{(2)}$ and accordingly, we get different sets of Burnett equations as

$$P_{ij}^{(Bu)} = P_{ij}^{(NS)} + P_{ij}^{(2)}$$

$$q_i^{(Bu)} = q_i^{(NS)} + q_i^{(2)}$$

Some important variants that were shown to overcome some of the limitations of Burnett equations are discussed in the next section.

7.3.1 Augmented Burnett Equations

As mentioned previously in Sect. 7.2, the conventional Burnett equations suffer from Bobylev's instability. In order to overcome this serious drawback, Zhong et al. [177] appended some higher order (super-Burnett) terms to the Burnett stress and heat terms, and termed new equations as augmented Burnett equations. The addition of these terms was not based on any fundamental principles but rather chosen in an arbitrary way with the sole purpose of making the equations stable without disturbing the second order accuracy of the equations. For one-dimensional case, the following linear terms of super-Burnett order were appended to $p_{xx}^{(2)}$ and $q_x^{(2)}$ respectively:

$$p_{xx}^{(3)} = \frac{\mu^3}{p^2} \left(\omega_7 RT \frac{\partial^3 u}{\partial x^3} \right) \tag{7.7}$$

$$q_1^{(3)} = \frac{\mu^3}{p\rho} \left(\theta_7 R \frac{\partial^3 T}{\partial x^3} + \theta_6 \frac{RT}{\rho} \frac{\partial^3 \rho}{\partial x^3} \right) \tag{7.8}$$

where the numerical values of the coefficients ω_7, θ_6, and θ_7 were properly *chosen* so that the new augmented Burnett equations are stable to disturbances of small wavelengths. For Maxwellian gas, following Wang Chang, the authors reported coefficients as: $\omega_7 = 2/9$, $\theta_6 = -5/8$, and $\theta_7 = 11/16$. However, from Eq. (5.151), and also noted by Struchtrup [141], the actual coefficient of term $\partial^3 T/\partial x^3$ is

$-157/16$, hence, θ_7 not only differs in magnitude but also in sign. This serious discrepancy in choosing the value of coefficient casts doubts about the validity of these equations.

These one-dimensional equations were then further generalized to three-dimensions by adding:

$$p_{ij}^{(3)} = \frac{\mu^3}{p^2}\left[\frac{3}{2}\omega_7 RT \frac{\partial^2}{\partial x_k \partial x_k}\left(\frac{\partial u_{\langle i}}{\partial x_{j\rangle}}\right)\right] \tag{7.9}$$

$$q_i^{(3)} = \frac{\mu^3}{p\rho}\left[\theta_7 R \frac{\partial}{\partial x_i}\left(\frac{\partial^2 T}{\partial x_k \partial x_k}\right) + \theta_6 \frac{RT}{\rho}\frac{\partial}{\partial x_i}\left(\frac{\partial^2 \rho}{\partial x_k \partial x_k}\right)\right] \tag{7.10}$$

to $p_{ij}^{(2)}$ and $q_i^{(2)}$, respectively. Regardless of these issues, the equations were found to be successful in computing the hypersonic shock structure and hypersonic flow past two-dimensional cylinder, but had only partial success in computing blunt body wakes and flat plate boundary layers.

7.3.2 BGK-Burnett Equations

In this section, we present the two sets of Burnett equations, augmented Burnett and BGK-Burnett equations in two-dimensional form as proposed by Agarwal et al. [2]. When only second order accurate terms are considered in the expressions of the stress tensor and heat flux vector, both these sets of equations have similar terms but with different values of coefficients. However, addition of the third order accurate terms has different basis and accordingly, we get two different sets of equations.

In two-dimensional form, the stress tensor and heat flux vector reduce to

$$p_{ij} = \begin{bmatrix} p_{xx}(x,y) & p_{xy}(x,y) & 0 \\ p_{xy}(x,y) & p_{yy}(x,y) & 0 \\ 0 & 0 & 0 \end{bmatrix}, \qquad \text{with} \quad p_{xx} + p_{yy} = 0 \tag{7.11}$$

$$q_i = \left[q_1(x,y)\ q_2(x,y)\ 0\right]. \tag{7.12}$$

The underlying principle in the derivation of these equations is similar to that of Chapman–Enskog method in that the stress tensor and heat flux vector are expanded as

$$p_{ij} = p_{ij}^{(0)} + p_{ij}^{(1)} + p_{ij}^{(2)} + p_{ij}^{(3)} \cdots = \sum_{r=0}^{\infty} p_{ij}^{(r)}$$

$$q_i = q_i^{(0)} + q_i^{(1)} + q_i^{(2)} + q_i^{(3)} + \cdots = \sum_{r=0}^{\infty} q_i^{(r)} \tag{7.13}$$

Similar to Chapman–Enskog method, we obtain Euler equations at the zeroth approximation and Navier–Stokes equations at the first approximation. At the second approximation, we consider the first three terms in the series (Eq. (7.13)), and the following expressions are obtained for the components of stress tensor and heat flux vector:

$$
\begin{aligned}
p_{xx}^{(2)} = \frac{\mu^2}{p}\Bigg\{ & \alpha_1\left(\frac{\partial u}{\partial x}\right)^2 + \alpha_2\frac{\partial u}{\partial x}\frac{\partial v}{\partial y} + \alpha_3\left(\frac{\partial v}{\partial y}\right)^2 + \alpha_4\frac{\partial u}{\partial x}\frac{\partial v}{\partial y} + \alpha_5\left(\frac{\partial u}{\partial y}\right)^2 \\
& + \alpha_6\left(\frac{\partial v}{\partial x}\right)^2 + \alpha_7 R\frac{\partial^2 T}{\partial x^2} + \alpha_8 R\frac{\partial^2 T}{\partial y^2} + \alpha_9\frac{RT}{\rho}\frac{\partial^2 \rho}{\partial x^2} + \alpha_{10}\frac{RT}{\rho}\frac{\partial^2 \rho}{\partial y^2} \\
& + \alpha_{11}\frac{RT}{\rho^2}\left(\frac{\partial \rho}{\partial x}\right)^2 + \alpha_{12}\frac{R}{\rho}\frac{\partial T}{\partial x}\frac{\partial \rho}{\partial x} + \alpha_{13}\frac{R}{T}\left(\frac{\partial T}{\partial x}\right)^2 \\
& + \alpha_{14}\frac{RT}{\rho^2}\left(\frac{\partial \rho}{\partial y}\right)^2 + \alpha_{15}\frac{R}{\rho}\frac{\partial T}{\partial y}\frac{\partial \rho}{\partial y} + \alpha_{16}\frac{R}{T}\left(\frac{\partial T}{\partial y}\right)^2\Bigg\}
\end{aligned} \tag{7.14}
$$

$$
\begin{aligned}
p_{yy}^{(2)} = \frac{\mu^2}{p}\Bigg\{ & \alpha_1\left(\frac{\partial v}{\partial y}\right)^2 + \alpha_2\frac{\partial u}{\partial x}\frac{\partial v}{\partial y} + \alpha_3\left(\frac{\partial u}{\partial x}\right)^2 + \alpha_4\frac{\partial u}{\partial y}\frac{\partial v}{\partial x} + \alpha_5\left(\frac{\partial v}{\partial x}\right)^2 \\
& + \alpha_6\left(\frac{\partial u}{\partial y}\right)^2 + \alpha_7 R\frac{\partial^2 T}{\partial y^2} + \alpha_8 R\frac{\partial^2 T}{\partial x^2} + \alpha_9\frac{RT}{\rho}\frac{\partial^2 \rho}{\partial y^2} + \alpha_{10}\frac{RT}{\rho}\frac{\partial^2 \rho}{\partial x^2} \\
& + \alpha_{11}\frac{RT}{\rho^2}\left(\frac{\partial \rho}{\partial y}\right)^2 + \alpha_{12}\frac{R}{\rho}\frac{\partial T}{\partial y}\frac{\partial \rho}{\partial y} + \alpha_{13}\frac{R}{T}\left(\frac{\partial T}{\partial y}\right)^2 \\
& + \alpha_{14}\frac{RT}{\rho^2}\left(\frac{\partial \rho}{\partial x}\right)^2 + \alpha_{15}\frac{R}{\rho}\frac{\partial T}{\partial x}\frac{\partial \rho}{\partial x} + \alpha_{16}\frac{R}{T}\left(\frac{\partial T}{\partial x}\right)^2\Bigg\}
\end{aligned} \tag{7.15}
$$

$$
\begin{aligned}
p_{xy}^{(2)} = p_{yx}^{(2)} = \frac{\mu^2}{p}\Bigg\{ & \beta_1\frac{\partial u}{\partial x}\frac{\partial u}{\partial y} + \beta_2\frac{\partial u}{\partial y}\frac{\partial v}{\partial y} + \beta_2\frac{\partial u}{\partial x}\frac{\partial v}{\partial x} \\
& + \beta_1\frac{\partial v}{\partial x}\frac{\partial v}{\partial y} + \beta_3 R\frac{\partial^2 T}{\partial x \partial y} + \beta_4\frac{RT}{\rho}\frac{\partial^2 \rho}{\partial x \partial y} + \beta_5\frac{R}{T}\frac{\partial T}{\partial x}\frac{\partial T}{\partial y} \\
& + \beta_6\frac{RT}{\rho^2}\frac{\partial \rho}{\partial x}\frac{\partial \rho}{\partial y} + \beta_7\frac{R}{\rho}\frac{\partial \rho}{\partial x}\frac{\partial T}{\partial y} + \beta_7\frac{R}{\rho}\frac{\partial T}{\partial x}\frac{\partial \rho}{\partial y}\Bigg\}
\end{aligned} \tag{7.16}
$$

$$q_x^{(2)} = \frac{\mu^2}{\rho}\left\{\gamma_1 \frac{1}{T}\frac{\partial T}{\partial x}\frac{\partial u}{\partial x} + \gamma_2 \frac{1}{T}\frac{\partial T}{\partial x}\frac{\partial v}{\partial y} + \gamma_3 \frac{\partial^2 u}{\partial x^2} + \gamma_4 \frac{\partial^2 u}{\partial y^2}\right.$$
$$+ \gamma_5 \frac{\partial^2 v}{\partial x \partial y} + \gamma_6 \frac{1}{T}\frac{\partial T}{\partial y}\frac{\partial v}{\partial x} + \gamma_7 \frac{1}{T}\frac{\partial T}{\partial y}\frac{\partial u}{\partial y} + \gamma_8 \frac{1}{\rho}\frac{\partial \rho}{\partial x}\frac{\partial u}{\partial x} + \gamma_9 \frac{1}{\rho}\frac{\partial \rho}{\partial x}\frac{\partial v}{\partial y}$$
$$\left. + \gamma_{10} \frac{1}{\rho}\frac{\partial \rho}{\partial y}\frac{\partial u}{\partial y} + \gamma_{11} \frac{1}{\rho}\frac{\partial \rho}{\partial y}\frac{\partial v}{\partial x}\right\} \tag{7.17}$$

$$q_y^{(2)} = \frac{\mu^2}{\rho}\left\{\gamma_1 \frac{1}{T}\frac{\partial T}{\partial y}\frac{\partial u}{\partial y} + \gamma_2 \frac{1}{T}\frac{\partial T}{\partial y}\frac{\partial v}{\partial x} + \gamma_3 \frac{\partial^2 v}{\partial y^2} + \gamma_4 \frac{\partial^2 v}{\partial x^2}\right.$$
$$+ \gamma_5 \frac{\partial^2 u}{\partial x \partial y} + \gamma_6 \frac{1}{T}\frac{\partial T}{\partial x}\frac{\partial u}{\partial y} + \gamma_7 \frac{1}{T}\frac{\partial T}{\partial x}\frac{\partial v}{\partial x} + \gamma_8 \frac{1}{\rho}\frac{\partial \rho}{\partial y}\frac{\partial v}{\partial y} + \gamma_9 \frac{1}{\rho}\frac{\partial \rho}{\partial y}\frac{\partial u}{\partial x}$$
$$\left. + \gamma_{10} \frac{1}{\rho}\frac{\partial \rho}{\partial x}\frac{\partial v}{\partial x} + \gamma_{11} \frac{1}{\rho}\frac{\partial \rho}{\partial x}\frac{\partial u}{\partial y}\right\} \tag{7.18}$$

The constitutive relationships for stress tensor and heat flux vector are similar for augmented Burnett and BGK-Burnett equations. However, the numerical values of the coefficients differ for these two sets and are given in Table 7.1.

Table 7.1 Numerical values of coefficients for stress and heat flux terms for augmented Burnett equations (for a hard sphere gas) and BGK-Burnett equations (for $\gamma = 1.4$)

Coefficient	Aug-Burnett	BGK-Burnett	Coefficient	Aug-Burnett	BGK-Burnett
α_1	1.199	−2.24	γ_2	0.407	−0.2
α_2	0.153	−40.48	γ_3	−2.269	−1.071
α_3	−0.600	0.56	γ_4	1.209	−2.0
α_4	−0.115	−1.20	γ_5	−3.478	−2.8
α_5	1.295	0.00	γ_6	−0.611	−7.5
α_6	−0.733	0.00	γ_7	11.033	−11.0
α_7	0.260	−19.6	γ_8	−2.060	−1.271
α_8	−0.130	−5.6	γ_9	1.030	1.0
α_9	−1.352	−1.6	γ_{10}	−1.545	−3.0
α_{10}	0.676	0.4	γ_{11}	−1.545	−3.0
α_{11}	1.352	1.6	β_1	−0.115	−1.4
α_{12}	−0.898	−19.6	β_2	1.913	−1.4
α_{13}	0.600	−18.0	β_3	0.390	0.00
α_{14}	−0.676	−0.4	β_4	−2.028	−2.0
α_{15}	0.449	−5.6	β_5	0.900	2.0
α_{16}	−0.300	−6.9	β_6	2.028	2.0
γ_1	10.830	−25.241	β_7	−0.676	0.00

7.3.2.1 Augmented Burnett Equations $O(Kn^3)$

Considering the first four terms in series (7.13), we obtain super-Burnett equations which are third order accurate in Knudsen number. However, in the case of augmented Burnett equations, the third-order terms are appended in a rather arbitrary way with the sole objective of obtaining stable numerical solutions without sacrificing the second-order accuracy of the solutions. The third-order super-Burnett terms in the augmented Burnett equations are given as

$$p_{xx}^{(3)^{(Aug)}} = \frac{\mu^3}{p^2}RT\left\{\alpha_{17}\frac{\partial^3 u}{\partial x^3} + \alpha_{17}\frac{\partial^3 u}{\partial x \partial y^2} + \alpha_{18}\frac{\partial^3 v}{\partial x^2 \partial y} + \alpha_{18}\frac{\partial^3 v}{\partial y^3}\right\} \tag{7.19}$$

$$p_{yy}^{(3)^{(Aug)}} = \frac{\mu^3}{p^2}RT\left\{\alpha_{17}\frac{\partial^3 v}{\partial y^3} + \alpha_{17}\frac{\partial^3 v}{\partial x^2 \partial y} + \alpha_{18}\frac{\partial^3 u}{\partial x \partial y^2} + \alpha_{18}\frac{\partial^3 u}{\partial x^3}\right\} \tag{7.20}$$

$$p_{xy}^{(3)^{(Aug)}} = p_{yx}^{(3)^{(Aug)}} = \frac{\mu^3}{p^2}RT\left\{\beta_8\frac{\partial^3 u}{\partial x^2 \partial y} + \beta_8\frac{\partial^3 u}{\partial y^3} + \beta_8\frac{\partial^3 v}{\partial x \partial y^2} + \beta_8\frac{\partial^3 v}{\partial x^3}\right\} \tag{7.21}$$

$$q_x^{(3)^{(Aug)}} = \frac{\mu^3}{p\rho}R\left\{\gamma_{12}\frac{\partial^3 T}{\partial x^3} + \gamma_{12}\frac{\partial^3 T}{\partial x \partial y^2} + \gamma_{13}\frac{T}{\rho}\frac{\partial^3 \rho}{\partial x^3} + \gamma_{13}\frac{T}{\rho}\frac{\partial^3 \rho}{\partial x \partial y^2}\right\} \tag{7.22}$$

$$q_y^{(3)^{(Aug)}} = \frac{\mu^3}{p\rho}R\left\{\gamma_{12}\frac{\partial^3 T}{\partial y^3} + \gamma_{12}\frac{\partial^3 T}{\partial x^2 \partial y} + \gamma_{13}\frac{T}{\rho}\frac{\partial^3 \rho}{\partial y^3} + \gamma_{13}\frac{T}{\rho}\frac{\partial^3 \rho}{\partial x^2 \partial y}\right\} \tag{7.23}$$

The superscript $(3)^{(Aug)}$ denotes the augmented Burnett terms which are third order accurate in Knudsen number to be added to the second order terms in Sect. 7.3.2. The coefficients in stress and heat flux terms are $\alpha_{17} = 0.2222, \alpha_{18} = -0.1111, \beta_8 = 0.1667, \gamma_{12} = 0.6875$, and $\gamma_{13} = -0.625$.

7.3.2.2 BGK-Burnett Equations $O(Kn^3)$

The super-Burnett order terms in the augmented Burnett equations are added in an ad hoc manner. However, in BGK-Burnett equations, the addition of the third order accurate terms has more fundamental basis. As seen in Sect. 5.3.3.2, the derivation of hydrodynamic equations becomes much simpler on replacing the non-linear collision integral in the Boltzmann equation by the much simpler BGK kinetic model. With this modification, it becomes possible to obtain a closed form of the distribution function to any order. Balakrishnan and Agarwal [17] considered first three terms in the Chapman–Enskog expansion (Eq. (5.62)) and derived the BGK-Burnett equations. The material derivatives in the first approximation to the distribution function ($f^{(1)}$), were replaced by the Euler equations to obtain the Navier–Stokes equations. In the next iteration, the material derivatives in the second approximation to the distribution function were replaced by the Navier–Stokes equations, and the resulting equations contain third order accurate Knudsen number (μ^3) terms. Had we replaced the material derivatives in the second approximation with the Euler equations, the resulting equations would have contained terms of Knudsen square (μ^2). Thus, the equations form a superset of Burnett equations which involved all the terms present in the stress tensor and heat flux expressions of conventional Burnett equations (Eqs. (5.144)–(5.145)) with third order derivatives terms similar to those present in super Burnett equations. The presence of these third order terms guaranteed the entropy consistency and unconditional stability of the equations for monatomic as well as polyatomic gases. The final form of equations for the components of stress tensor and heat flux vector read [2],

$$
\begin{aligned}
p_{xx}^{(3)(BGK)} &= \frac{\mu^3}{p^2} RT \left\{ \theta_1 \frac{\partial^3 u}{\partial x^3} + \theta_2 \frac{\partial^3 u}{\partial x \partial y^2} + \theta_3 \frac{\partial^3 v}{\partial x^2 \partial y} + \theta_4 \frac{\partial^3 v}{\partial y^3} \right\} \\
&- \frac{\mu^3}{p^2} \frac{RT}{\rho} \left\{ \theta_1 \frac{\partial \rho}{\partial x} \frac{\partial^2 u}{\partial x^2} \right. \\
&\left. + \theta_5 \frac{\partial \rho}{\partial x} \frac{\partial^2 v}{\partial x \partial y} + \theta_6 \frac{\partial \rho}{\partial x} \frac{\partial^2 u}{\partial y^2} + \theta_7 \frac{\partial \rho}{\partial y} \frac{\partial^2 v}{\partial x^2} + \theta_8 \frac{\partial \rho}{\partial y} \frac{\partial^2 u}{\partial x \partial y} + \theta_4 \frac{\partial \rho}{\partial y} \frac{\partial^2 v}{\partial y^2} \right\} \\
&+ \frac{\mu^3}{p^2} \left\{ \theta_9 \left(\frac{\partial u}{\partial x} \right)^3 + 3\theta_{10} \left(\frac{\partial u}{\partial x} \right)^2 \frac{\partial v}{\partial y} + \theta_{11} \frac{\partial u}{\partial x} \left(\frac{\partial v}{\partial y} \right)^2 - \theta_4 \frac{\partial u}{\partial x} \left(\frac{\partial u}{\partial y} \right)^2 \right. \\
&- 2\theta_4 \frac{\partial u}{\partial x} \frac{\partial u}{\partial y} \frac{\partial v}{\partial x} - \theta_4 \frac{\partial u}{\partial x} \left(\frac{\partial v}{\partial x} \right)^2 + \theta_{10} \left(\frac{\partial v}{\partial y} \right)^3 - \theta_{12} \frac{\partial v}{\partial y} \left(\frac{\partial u}{\partial y} \right)^2 \\
&\left. - 2\theta_{12} \frac{\partial u}{\partial y} \frac{\partial v}{\partial x} \frac{\partial v}{\partial y} - \theta_{12} \left(\frac{\partial v}{\partial x} \right)^2 \frac{\partial v}{\partial y} \right\} + \frac{\mu^3}{p^2} R \left\{ \theta_{13} \frac{\partial u}{\partial x} \frac{\partial^2 T}{\partial x^2} \right. \\
&\left. + \theta_{13} \frac{\partial u}{\partial x} \frac{\partial^2 T}{\partial y^2} + \theta_{14} \frac{\partial v}{\partial y} \frac{\partial^2 T}{\partial x^2} + \theta_{14} \frac{\partial v}{\partial y} \frac{\partial^2 T}{\partial y^2} \right\}
\end{aligned}
\tag{7.24}
$$

$$
\begin{aligned}
p_{yy}^{(3)(BGK)} &= \frac{\mu^3}{p^2} RT \left\{ \theta_1 \frac{\partial^3 v}{\partial y^3} + \theta_2 \frac{\partial^3 v}{\partial x^2 \partial y} + \theta_3 \frac{\partial^3 u}{\partial x \partial y^2} + \theta_4 \frac{\partial^3 u}{\partial x^3} \right\} \\
&- \frac{\mu^3}{p^2} \frac{RT}{\rho} \left\{ \theta_1 \frac{\partial \rho}{\partial y} \frac{\partial^2 v}{\partial y^2} \right. \\
&\left. + \theta_5 \frac{\partial \rho}{\partial y} \frac{\partial^2 u}{\partial x \partial y} + \theta_6 \frac{\partial \rho}{\partial y} \frac{\partial^2 v}{\partial x^2} + \theta_7 \frac{\partial \rho}{\partial x} \frac{\partial^2 u}{\partial y^2} + \theta_8 \frac{\partial \rho}{\partial x} \frac{\partial^2 v}{\partial x \partial y} + \theta_4 \frac{\partial \rho}{\partial x} \frac{\partial^2 u}{\partial x^2} \right\} \\
&+ \frac{\mu^3}{p^2} \left\{ \theta_9 \left(\frac{\partial v}{\partial y} \right)^3 + 3\theta_{10} \left(\frac{\partial v}{\partial y} \right)^2 \frac{\partial u}{\partial x} + \theta_{11} \frac{\partial v}{\partial y} \left(\frac{\partial u}{\partial x} \right)^2 - \theta_4 \frac{\partial v}{\partial y} \left(\frac{\partial v}{\partial x} \right)^2 \right.
\end{aligned}
$$

$$
\begin{aligned}
&- 2\theta_4 \frac{\partial v}{\partial y}\frac{\partial v}{\partial x}\frac{\partial u}{\partial y} - \theta_4 \frac{\partial v}{\partial y}\left(\frac{\partial u}{\partial y}\right)^2 + \theta_{10}\left(\frac{\partial u}{\partial x}\right)^3 - \theta_{12}\frac{\partial u}{\partial x}\left(\frac{\partial v}{\partial x}\right)^2 \\
&- 2\theta_{12}\frac{\partial v}{\partial x}\frac{\partial u}{\partial y}\frac{\partial u}{\partial x} - \theta_{12}\left(\frac{\partial u}{\partial y}\right)^2\frac{\partial u}{\partial x}\Bigg\} + \frac{\mu^3}{p^2}R\Bigg\{\theta_{13}\frac{\partial v}{\partial y}\frac{\partial^2 T}{\partial y^2} \\
&+ \theta_{13}\frac{\partial v}{\partial y}\frac{\partial^2 T}{\partial x^2} + \theta_{14}\frac{\partial u}{\partial x}\frac{\partial^2 T}{\partial y^2} + \theta_{14}\frac{\partial u}{\partial x}\frac{\partial^2 T}{\partial x^2}\Bigg\}
\end{aligned} \tag{7.25}
$$

$$
\begin{aligned}
p_{xy}^{(3)(BGK)} &= \frac{\mu^3}{p^2}RT\left\{\theta_{15}\frac{\partial^3 u}{\partial x^2 \partial y} + \frac{\partial^3 u}{\partial y^3} + \theta_{15}\frac{\partial^3 v}{\partial x \partial y^2} + \frac{\partial^3 v}{\partial x^3}\right\} - \frac{\mu^3}{p^2}\frac{RT}{\rho}\Bigg\{\theta_6\frac{\partial \rho}{\partial y}\frac{\partial^2 u}{\partial x^2} \\
&+ \theta_{16}\frac{\partial \rho}{\partial y}\frac{\partial^2 v}{\partial x \partial y} + \frac{\partial \rho}{\partial y}\frac{\partial^2 u}{\partial y^2} + \frac{\partial \rho}{\partial x}\frac{\partial^2 v}{\partial x^2} + \theta_{16}\frac{\partial \rho}{\partial x}\frac{\partial^2 u}{\partial x \partial y} + \theta_6\frac{\partial \rho}{\partial x}\frac{\partial^2 v}{\partial y^2}\Bigg\} \\
&- \frac{\mu^3}{p^2}\left(\frac{\partial u}{\partial y} + \frac{\partial v}{\partial x}\right)\Bigg\{\theta_4\left(\frac{\partial u}{\partial x}\right)^2 + 2\theta_{12}\frac{\partial u}{\partial x}\frac{\partial v}{\partial y} + 2\theta_7\frac{\partial u}{\partial y}\frac{\partial v}{\partial x} + \theta_7\left(\frac{\partial u}{\partial y}\right)^2 \\
&+ \theta_7\left(\frac{\partial v}{\partial x}\right)^2 + \theta_4\left(\frac{\partial v}{\partial y}\right)^2\Bigg\} + \frac{\mu^3}{p^2}R\Bigg\{\theta_{17}\frac{\partial u}{\partial y}\frac{\partial^2 T}{\partial x^2} + \theta_{17}\frac{\partial u}{\partial y}\frac{\partial^2 T}{\partial y^2} \\
&+ \theta_{17}\frac{\partial v}{\partial x}\frac{\partial^2 T}{\partial x^2} + \theta_{17}\frac{\partial v}{\partial x}\frac{\partial^2 T}{\partial y^2}\Bigg\}
\end{aligned} \tag{7.26}
$$

$$
\begin{aligned}
q_x^{(3)(BGK)} &= \frac{\mu^3}{p\rho}R\left\{\theta_{18}\frac{\partial^3 T}{\partial x^3} + \theta_{18}\frac{\partial^3 T}{\partial x \partial y^2} - \theta_{18}\frac{1}{\rho}\frac{\partial \rho}{\partial x}\frac{\partial^2 T}{\partial x^2} - \theta_{18}\frac{1}{\rho}\frac{\partial \rho}{\partial x}\frac{\partial^2 T}{\partial y^2}\right\} \\
&+ \frac{\mu^3}{p\rho}\Bigg\{\theta_{19}\frac{\partial u}{\partial x}\frac{\partial^2 u}{\partial x^2} + \theta_{20}\frac{\partial u}{\partial x}\frac{\partial^2 v}{\partial x \partial y} + \theta_6\frac{\partial u}{\partial x}\frac{\partial^2 u}{\partial y^2} + \theta_{21}\frac{\partial v}{\partial y}\frac{\partial^2 u}{\partial x^2} \\
&+ \theta_{22}\frac{\partial v}{\partial y}\frac{\partial^2 v}{\partial x \partial y} + \theta_7\frac{\partial v}{\partial y}\frac{\partial^2 u}{\partial y^2} + \theta_{23}\frac{\partial u}{\partial y}\frac{\partial^2 v}{\partial x^2} + \theta_{24}\frac{\partial u}{\partial y}\frac{\partial^2 u}{\partial x \partial y} + \theta_6\frac{\partial u}{\partial y}\frac{\partial^2 v}{\partial y^2} \\
&+ \theta_{23}\frac{\partial v}{\partial x}\frac{\partial^2 v}{\partial x^2} + \theta_{24}\frac{\partial v}{\partial x}\frac{\partial^2 u}{\partial x \partial y} + \theta_6\frac{\partial v}{\partial x}\frac{\partial^2 v}{\partial y^2}\Bigg\} \\
&- \frac{\mu^3}{p\rho}\left\{\frac{1}{\rho}\frac{\partial \rho}{\partial x} + \frac{1}{T}\frac{\partial T}{\partial x}\right\}\Bigg\{\theta_{13}\left(\frac{\partial u}{\partial x}\right)^2 \\
&+ 2\theta_{14}\frac{\partial u}{\partial x}\frac{\partial v}{\partial y} + 2\theta_{17}\frac{\partial u}{\partial y}\frac{\partial v}{\partial x} + \theta_{17}\left(\frac{\partial u}{\partial y}\right)^2 + \theta_{17}\frac{\partial v}{\partial x}^2 + \theta_{13}\left(\frac{\partial v}{\partial y}\right)^2\Bigg\} \\
&+ \frac{\mu^3 R}{p\rho T}\left\{\theta_{18}\frac{\partial T}{\partial x}\frac{\partial^2 T}{\partial x^2} + \theta_{18}\frac{\partial T}{\partial x}\frac{\partial^2 T}{\partial y^2}\right\}
\end{aligned} \tag{7.27}
$$

Table 7.2 Numerical values of coefficients for stress and heat flux terms for BGK-Burnett equations (for $\gamma = 1.4$)

Coefficient	Value	Coefficient	Value	Coefficient	Value
θ_1	2.56	θ_9	1.024	θ_{17}	1.4
θ_2	1.36	θ_{10}	−0.256	θ_{18}	4.9
θ_3	0.56	θ_{11}	1.152	θ_{19}	7.04
θ_4	−0.64	θ_{12}	0.16	θ_{20}	−0.16
θ_5	0.96	θ_{13}	2.24	θ_{21}	−1.76
θ_6	1.6	θ_{14}	−0.56	θ_{22}	4.24
θ_7	−0.4	θ_{15}	3.6	θ_{23}	3.8
θ_8	−0.24	θ_{16}	0.6	θ_{24}	3.4

$$
\begin{aligned}
q_y^{(3)(BGK)} &= \frac{\mu^3}{p\rho}R\left\{\theta_{18}\frac{\partial^3 T}{\partial y^3} + \theta_{18}\frac{\partial^3 T}{\partial x^2 \partial y} - \theta_{18}\frac{1}{\rho}\frac{\partial \rho}{\partial y}\frac{\partial^2 T}{\partial y^2} - \theta_{18}\frac{1}{\rho}\frac{\partial \rho}{\partial y}\frac{\partial^2 T}{\partial x^2}\right\} \\
&+ \frac{\mu^3}{p\rho}\left\{\theta_{19}\frac{\partial v}{\partial y}\frac{\partial^2 v}{\partial y^2} + \theta_{20}\frac{\partial v}{\partial y}\frac{\partial^2 u}{\partial x \partial y} + \theta_6\frac{\partial v}{\partial y}\frac{\partial^2 v}{\partial x^2} + \theta_{21}\frac{\partial u}{\partial x}\frac{\partial^2 v}{\partial y^2}\right. \\
&+ \theta_{22}\frac{\partial u}{\partial x}\frac{\partial^2 u}{\partial x \partial y} + \theta_7\frac{\partial u}{\partial x}\frac{\partial^2 v}{\partial x^2} + \theta_{23}\frac{\partial v}{\partial x}\frac{\partial^2 u}{\partial y^2} + \theta_{24}\frac{\partial v}{\partial x}\frac{\partial^2 v}{\partial x \partial y} + \theta_6\frac{\partial v}{\partial x}\frac{\partial^2 u}{\partial x^2} \\
&\left.+ \theta_{23}\frac{\partial u}{\partial y}\frac{\partial^2 u}{\partial y^2} + \theta_{24}\frac{\partial u}{\partial y}\frac{\partial^2 v}{\partial x \partial y} + \theta_6\frac{\partial u}{\partial y}\frac{\partial^2 u}{\partial x^2}\right\} \\
&- \frac{\mu^3}{p\rho}\left\{\frac{1}{\rho}\frac{\partial \rho}{\partial y} + \frac{1}{T}\frac{\partial T}{\partial y}\right\}\left\{\theta_{13}\left(\frac{\partial u}{\partial x}\right)^2\right. \\
&+ 2\theta_{14}\frac{\partial u}{\partial x}\frac{\partial v}{\partial y} + 2\theta_{17}\frac{\partial u}{\partial y}\frac{\partial v}{\partial x} + \theta_{17}\left(\frac{\partial u}{\partial y}\right)^2 + \theta_{17}\left(\frac{\partial v}{\partial x}\right)^2 \\
&\left.+ \theta_{13}\left(\frac{\partial v}{\partial y}\right)^2\right\} + \frac{\mu^3 R}{p\rho T}\left\{\theta_{18}\frac{\partial T}{\partial y}\frac{\partial^2 T}{\partial x^2} + \theta_{18}\frac{\partial T}{\partial y}\frac{\partial^2 T}{\partial y^2}\right\}
\end{aligned}
\tag{7.28}
$$

The superscript $(3)^{(BGK)}$ denotes third-order (in Knudsen number) super-Burnett stress and heat flux terms to be appended to their respective second order (in Knudsen number) terms in the BGK-Burnett equations in Sect. 7.3.2. The numerical values of the coefficients for $\gamma = 1.4$ are given in Table 7.2.

7.3.3 Other Forms of Burnett Equations

Instead of establishing another course to obtain Burnett equations, Woods [166] derived the generalized free path theory based on the mean free path approach to kinetic theory, initiated by Maxwell. This theory was neither based on Boltzmann's equation nor did it involve any power series expansion in terms of Knudsen number, making this theory interesting. The main crux of this theory was the phenomenological mean free path element, whose value was different depending on the mass, momentum, and energy transported by the molecules. Woods believed that this theory would be valid for any arbitrary state of the fluid, even for higher Knudsen numbers. Although this theory was successfully tested for propagation of sound waves in a rarefied monatomic gas [167], for reasons unknown, this theory failed to attain mainstream adoption in rarefied gas dynamics.

The other Burnett variants worth mentioning are the generalized Burnett equations [23] proposed by Bobylev and Burnett equations [37] proposed by Dadzie. Third order derivatives of velocity and temperature are absent in Bobylev's equations making them relatively simple to solve. These equations were shown to give better results in the shock waves flow problem than the Burnett equations at moderate Mach numbers [24]. The equations by Dadzie were shown to be thermo-mechanically consistent and unconditionally stable in both space and time for all values of transport coefficients.

7.4 Limitations of Grad Equations

The general critiques raised against the Grad equations are summarized below:

1. *Hyperbolic nature of equations:* Grad 13-moment equations form a hyperbolic set of equations, meaning that there is a finite speed of wave propagation. Although hyperbolic equations are preferable because of their mathematical simplicity, they give rise to discontinuous shocks. Grad, in his work of shock wave study [63] wrote ... *the solution breaks down completely, and no solution exists for stronger shocks (specifically, at Mach number $Ma = 1.65$).* However, in comparison, the Navier–Stokes equations yield results for infinite shock strength and only deviate from the 13-moment equations at $Ma = 1.2$. So, the range ($1.2 < Ma < 1.65$) over which Grad equations improve upon the Navier–Stokes equations is very narrow and does not justify the additional complexity that is involved in the Grad equations. This serious critique against 13-moment equations therefore put a question mark on its applicability. To overcome this critique, Weiss [164] increased the number of moments from 13 to 21, but this resulted in increase in the critical Ma from 1.65 to 1.887 only. Hence, a mere increase in the number of moments did not solve the problem at least for the shock wave flow problem.
2. *Unphysical negative values of distribution function:* The quantity $f(\mathbf{x}, \mathbf{c}, t)d\mathbf{x}d\mathbf{c}$ represents the number of particles in physical volume $\{\mathbf{x}, \mathbf{x} + d\mathbf{x}\}$ having molecular velocities in interval $\{\mathbf{c}, \mathbf{c} + d\mathbf{c}\}$ at time t. When normalized with total number of molecules (N), we obtain from Eq. (5.7),

$$1 = \iiint_V \iiint_{-\infty}^{\infty} \frac{f(\mathbf{x}, \mathbf{c}, t)}{N} d\mathbf{c} d\mathbf{x} \tag{7.29}$$

Hence, normalized distribution function actually gives the probability of finding a molecule in phase space volume $d\mathbf{x}d\mathbf{c}$, as such, its value must be strictly positive and should lie in the range [0, 1]. However, in one-dimensional steady plane shock wave flow problem [148], the 13-moment distribution function (Eq. (6.91)) in one-dimensional formulation assumes negative values at higher Mach numbers. This occurrence of negative values being unphysical leads to a question about the validity of the moment equations (and other set of equations that are based on Grad approach) at higher Mach numbers.

3. *Desired accuracy:* If we consider an infinite set of Hermite coefficients in the series [Eq. (6.35)], then there is no loss of kinetic information and we can say that the Grad description is equivalent to Boltzmann equation itself. However, it becomes mandatory to truncate the series at some point and obtain approximate solutions as only a finite number of coefficients can be evaluated. The accuracy of the solutions then largely depends upon the number of Hermite coefficients (and thereby the moments) considered. However, the number of moments considered cannot be increased arbitrarily since then the computational cost becomes comparable to that of DSMC simulations. In addition, the higher moments cannot be attributed any physical significance, hence, which moments to consider then becomes a serious question.
4. *Requirement of additional boundary conditions*: The Grad equations also suffer from the same issue that plaques the Burnett equations, i.e., the requirement of additional boundary conditions. This problem becomes more severe since, with the inclusion of the stress tensor and heat flux vector in the primary variables set, the boundary conditions need to be specified for these variables also.

7.5 Variants of Grad Equations

In order to overcome the limitations of the Grad equations, several variants of moment equations are proposed in the literature where the number of moments in the primary variables set is increased in a rather arbitrary way. Though increasing the number of moments helps to achieve somewhat more reasonable solutions, this remedy does not help to overcome all the limitations of the Grad equations as discussed in Sect. 7.4. Hence, a mere increase in the number of the moments does not suffice. In this section, we briefly discuss the two main variants proposed in the literature: regularized 13 (R13) and regularized 26 (R26) moment equations wherein the hyperbolic nature of the Grad equations is changed to parabolic by the process of regularization.

For the sake of completeness, the basic conservation laws of mass, momentum, and energy along with the evolution equations for the stress tensor and heat flux vector as derived in Sect. 6.4.1 are being reproduced here:

$$\frac{\partial \rho}{\partial t} + \frac{\partial (\rho u_k)}{\partial x_k} = 0. \tag{7.30}$$

$$\frac{\partial u_i}{\partial t} + u_k \frac{\partial u_i}{\partial x_k} + \frac{1}{\rho}\frac{\partial p}{\partial x_i} + \frac{1}{\rho}\frac{\partial p_{ik}}{\partial x_k} = F_i, \tag{7.31}$$

$$\frac{3}{2}\rho R \frac{\partial T}{\partial t} + \frac{3}{2}\rho R u_k \frac{\partial T}{\partial x_k} + p\frac{\partial u_i}{\partial x_i} + p_{ij}\frac{\partial u_i}{\partial x_j} + \frac{\partial q_k}{\partial x_k} = 0, \tag{7.32}$$

$$\frac{Dp_{ij}}{Dt} + p_{ij}\frac{\partial u_k}{\partial x_k} + \frac{\partial s_{ijk}}{\partial x_k} + \frac{4}{5}\frac{\partial q_{\langle i}}{\partial x_{j\rangle}} + 2p_{r\langle i}\frac{\partial u_{j\rangle}}{\partial x_r} + 2p\frac{\partial u_{\langle i}}{\partial x_{j\rangle}} = -\frac{p}{\mu}p_{ij} \tag{7.33}$$

$$\begin{aligned}\frac{Dq_i}{Dt} &+ \frac{1}{2}\frac{\partial r^1_{ik}}{\partial x_k} + \frac{1}{6}\frac{\partial w^2}{\partial x_i} - p_{ik}R\frac{\partial T}{\partial x_k} - \frac{p_{ik}}{\rho}RT\frac{\partial \rho}{\partial x_k} + \frac{5}{2}\rho RT\frac{\partial RT}{\partial x_i} - \frac{p_{ik}}{\rho}\frac{\partial p_{kl}}{\partial x_l}\\ &- \frac{5}{2}RT\frac{\partial p_{ik}}{\partial x_k} + \frac{\partial u_j}{\partial x_k}s_{ijk} + \frac{7}{5}q_i\frac{\partial u_k}{\partial x_k} + \frac{7}{5}q_k\frac{\partial u_i}{\partial x_k} + \frac{2}{5}q_j\frac{\partial u_j}{\partial x_i} = -\frac{2}{3}\frac{p}{\mu}q_i\end{aligned} \tag{7.34}$$

where the unknowns appearing in the equations are given as

$$\begin{aligned} r^1_{ik} &= m\int C_{\langle i}C_{k\rangle}C^2 f d\mathbf{C} \\ w^2 &= m\int C^4 (f - f_M) d\mathbf{C} \\ s_{ijk} &= m\int C_{\langle i}C_j C_{k\rangle}C^2 f d\mathbf{C} \end{aligned} \tag{7.35}$$

This set of equations forms the starting point in the derivation of R13 as well as R26 moment equations.

7.5.1 Regularized 13-Moment Equations

Torrilhon and Struchtrup [142, 148] combined the Chapman–Enskog and the Grad's moment method to bring together the advantages of both these methods and proposed regularized 13 (R13) moment equations. Basically, they proposed a new closure technique by obtaining approximation for higher order moments r^1_{ik}, w^2, and s_{ijk} (as given in Eq. (7.35)). Grad [64] himself had suggested to obtain such an approximation for these unknown moments. Recall that in the derivation of 13-moment equations, Grad obtained closure expressions for unknown moments as

$$r^1_{ik|G13} = 7RTp_{ik}, \qquad w^2_{|G13} = 0, \qquad s_{ijk|G13} = 0 \tag{7.36}$$

Just like we go one level higher from Euler equations to Navier–Stokes equations by obtaining constitutive relations for p_{ij} and q_i, Grad believed that performing this type of *iteration* for 13-moment equations is likely to increase the accuracy of the equations significantly. However, this idea suggested by Grad remained dormant for decades until Struchtrup and Torrilhon [142] pursued this idea and presented the regularized R13 moment equations.

7.5.1.1 Regularization Method

In the present section, we briefly touch upon the regularization method and present important steps in the derivation of the R13 equations. The moments involved in the R13 equations are depicted in Fig. 7.1 with their corresponding integral expressions in terms of distribution function. For detailed derivation, readers can follow [141, 142].

As we know that the Navier–Stokes equations are derived by means of Chapman–Enskog expansion around the equilibrium, a similar idea is applied by performing a first order Chapman–Enskog expansion around the Grad distribution. However, the authors assumed that the gas approaches the non-equilibrium manifold with a much faster time scale τ_ϵ which is assumed to be smaller than the mean free time τ. Hence, a gas initially in a non-equilibrium state first relaxes to the pseudo-equilibrium on a faster time scale and then proceeds towards the real equilibrium state on a slower time scale. There is no physical proof for the existence of the pseudo-equilibrium manifold; however, defining such a manifold greatly helps in improving the Grad moment system.

In R13 moment equations, we define deviations for unknowns, r^1_{ik}, w^2, and s_{ijk} as the difference between the true values and their approximated values calculated using $f_{G|13}$ as

$$Q_{ij} = r^1_{ij} - r^1_{ij|G13}, \qquad \Delta = w^2 - w^2_{|G13}, \qquad m_{ijk} = s_{ijk} - s_{ijk|G13}. \tag{7.37}$$

Note that the moment Q_{ij} is symmetric and trace-free even though it is denoted by a capital letter.

For the Grad closure, all the deviations of r^1_{ik}, w^2, and s_{ijk} (i.e., Δ, Q_{ij}, and m_{ijk}) are exactly zero. However, in R13 moment equations, the Grad's closure is defied and we seek non-zero approximations for the deviations of w^2, r^1_{ij}, and m_{ijk} as

$$\Delta = w^2 - w^2_{G|13} = w^2$$

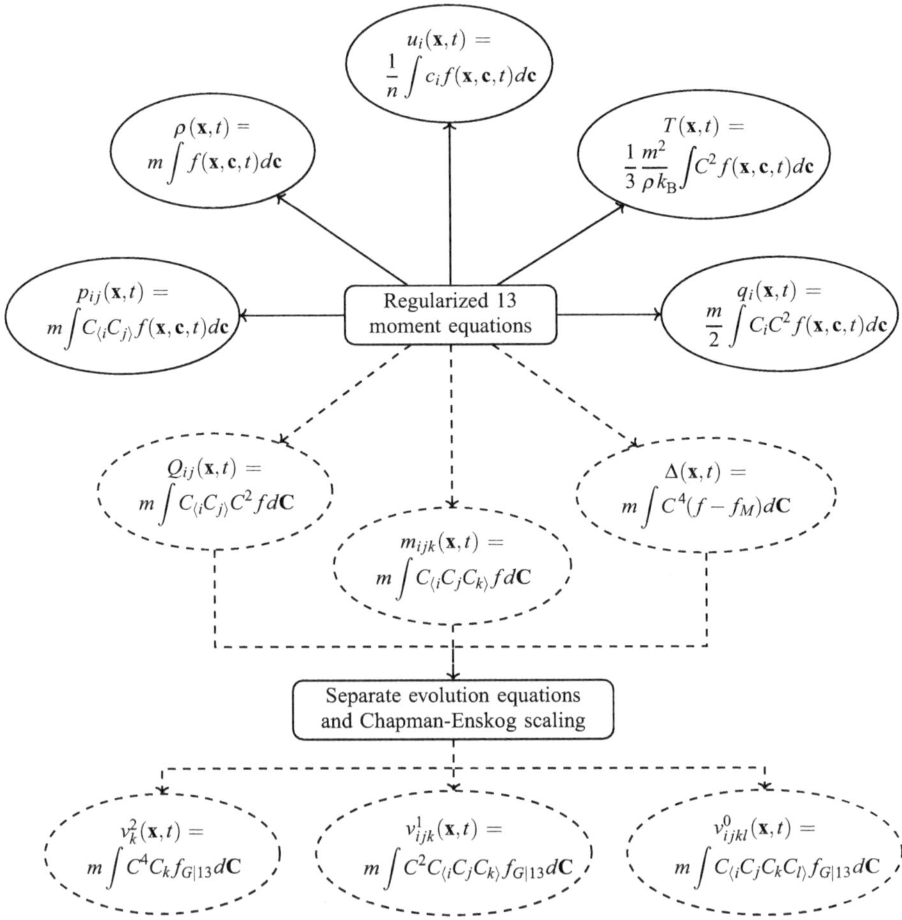

Fig. 7.1 Expressions for the variables considered in the regularized 13-moment equations; variables in solid circles are the primary variables while those in dashed circles are the closure variables; separate evolutions equations are generated for these closure variables and Chapman–Enskog scaling is then applied to obtain constitutive relationships for Δ, Q_{ij} and m_{ijk}

$$Q_{ij} = r^1_{ij} - r^1_{ijG|13} = r^1_{ij} - 7RTp_{ij} \tag{7.38}$$

$$m_{ijk} = s_{ijk} - s_{ijkG|13} = s_{ijk}$$

where we have utilized results in Eq. (7.36).

The first step in the process of regularization is to generate separate evolution equations for Δ, Q_{ij}, and m_{ijk} by multiplying the transformed Boltzmann equation (Eq. (6.10)) with $\psi = \{C^4, C^2C_{\langle i}C_{j\rangle}, C_{\langle i}C_jC_{k\rangle}\}$, respectively, and then integrating over the velocity space. After working out this step, the final result containing unknown moments v^2_k, v^1_{ijk}, v^0_{ijkl} (defined below) reads as

$$\frac{D\Delta}{Dt} - 20RT\frac{\partial q_k}{\partial x_k} + 8RTp_{kl}\frac{\partial u_k}{\partial x_l} + 4Q_{kl}\frac{\partial u_k}{\partial x_l} - 8q_k\frac{\partial RT}{\partial x_k}$$
$$- 8q_k RT\frac{\partial \ln\rho}{\partial x_k} - 8\frac{q_k}{\rho}\frac{\partial p_{kl}}{\partial x_l} + \frac{\partial v_k^2}{\partial x_k} + \frac{7}{3}\Delta\frac{\partial u_k}{\partial x_k} = -\frac{2}{3}\frac{p}{\mu}\Delta \tag{7.39}$$

$$\frac{DQ_{ij}}{Dt} + \frac{2}{5}\frac{\partial v^2_{\langle i}}{\partial x_{j\rangle}} - \frac{28}{5}RT\frac{\partial q_{\langle i}}{\partial x_{j\rangle}} - \frac{28}{5}q_{\langle i}\frac{\partial RT}{\partial x_{j\rangle}} - \frac{28}{5}RTq_{\langle i}\frac{\partial \ln\rho}{\partial x_{j\rangle}} - \frac{28}{5}\frac{q_{\langle i}}{\rho}\frac{\partial p_{j\rangle k}}{\partial x_k}$$
$$+ 4RT\left[p_{k\langle i}\frac{\partial u_k}{\partial x_{j\rangle}} + p_{k\langle i}\frac{\partial u_{j\rangle}}{\partial x_k} - \frac{2}{3}p_{ij}\frac{\partial u_k}{\partial x_k}\right] - \frac{14}{3}\frac{p_{ij}}{\rho}\frac{\partial q_k}{\partial x_k} - \frac{14}{3}\frac{p_{ij}p_{kl}}{\rho}p_{ij}\frac{\partial u_k}{\partial x_l}$$
$$- 7RT\frac{\partial m_{ijk}}{\partial x_k} - 2\frac{m_{ijk}}{\rho}\left(\frac{\partial p_{kl}}{\partial x_l} + \frac{\partial RT\rho}{\partial x_k}\right) + \frac{\partial v^1_{ijk}}{\partial x_k} + 2v^0_{ijkl}\frac{\partial u_k}{\partial x_l} + \frac{6}{7}Q_{\langle ij}\frac{\partial u_{k\rangle}}{\partial x_k}$$
$$+ \frac{4}{5}Q_{k\langle i}\frac{\partial u_k}{\partial x_{j\rangle}} + 2Q_{k\langle i}\frac{\partial u_{j\rangle}}{\partial x_k} + Q_{ij}\frac{\partial u_k}{\partial x_k} + \frac{14}{15}\Delta\frac{\partial u_{\langle i}}{\partial x_{j\rangle}} = -\frac{7}{6}\frac{p}{\mu}Q_{ij} \tag{7.40}$$

$$\frac{Dm_{ijk}}{Dt} - 3\frac{p_{\langle ij}}{\rho}\frac{\partial p_{k\rangle l}}{\partial x_l} - 3p_{\langle ij}RT\frac{\partial \ln\rho}{\partial x_{k\rangle}} + \frac{\partial v^0_{ijkl}}{\partial x_l} + \frac{3}{7}\frac{\partial Q_{\langle ij}}{\partial x_{k\rangle}}$$
$$+ 3RT\frac{\partial p_{\langle ij}}{\partial x_{k\rangle}} + 3m_{l\langle ij}\frac{\partial u_{k\rangle}}{\partial x_l} + m_{ijk}\frac{\partial u_l}{\partial x_l} + \frac{12}{5}q_{\langle i}\frac{\partial u_j}{\partial x_{k\rangle}} = -\frac{3}{2}\frac{p}{\mu}m_{ijk} \tag{7.41}$$

The unknown moments v_k^2, v^1_{ijk}, v^0_{ijkl} are defined as

$$v_k^2 = m\int C^4 C_k f d\mathbf{C}, \quad v^1_{ijk} = m\int C^2 C_{\langle i}C_j C_{k\rangle} f d\mathbf{C}, \quad v^0_{ijkl} = m\int C_{\langle i}C_j C_k C_{l\rangle} f d\mathbf{C}$$

After generating the transport equations for the higher moments $(\Delta, Q_{ij}, m_{ijk})$ by the moment method, we introduce the Chapman–Enskog scaling. Therefore, only the production terms of higher moments $(\Delta, Q_{ij}, m_{ijk})$ are scaled by ε while the production terms of stress tensor and heat flux vector remain unaffected. Accordingly, the terms on the RHS for the respective moment read as

$$-\frac{1}{\varepsilon}\frac{2}{3}\frac{p}{\mu}\Delta, \qquad -\frac{1}{\varepsilon}\frac{7}{6}\frac{p}{\mu}Q_{ij}, \qquad -\frac{1}{\varepsilon}\frac{3}{2}\frac{p}{\mu}m_{ijk} \tag{7.42}$$

while all the terms on the LHS remain same. In physical sense, by introducing a formal smallness parameter ε only in the higher moments $(\Delta, Q_{ij}, m_{ijk})$, we imply that these non-equilibrium moments change on a faster time scale than the non-equilibrium moments, p_{ij} and q_i.

The next step is to apply the Chapman–Enskog expansion for these higher moments with ε as the formal smallness parameter as

$$\Delta = \Delta^{(0)} + \varepsilon\Delta^{(1)} + \varepsilon^2\Delta^{(2)} + \cdots \tag{7.43}$$

$$Q_{ij} = Q_{ij}^{(0)} + \varepsilon Q_{ij}^{(1)} + \varepsilon^2 Q_{ij}^{(2)} + \cdots \tag{7.44}$$

$$m_{ijk} = m_{ijk}^{(0)} + \varepsilon m_{ijk}^{(1)} + \varepsilon^2 m_{ijk}^{(2)} + \cdots \tag{7.45}$$

Struchtrup and Torrilhon [142] truncated the series after the first two terms, thereby including only those terms which are of order $\mathcal{O}(\varepsilon^1)$. The authors believed that this approximation will yield stable equations, just like the first order Chapman–Enskog expansion resulted in the stable Navier–Stokes equations. Further notice that the zeroth order approximation produces the Grad 13-moment equations.

The next step is to obtain the constitutive relations for Δ, Q_{ij}, m_{ijk} in a way similar to that used for obtaining p_{ij} and q_i in the first order Chapman–Enskog expansion. To achieve this, we substitute series (7.44) in the evolution equations for Δ, Q_{ij}, m_{ijk} (Eqs. (7.39)–(7.41)) with the production terms replaced by values given in Eq. (7.42). The final result in short hand notation reads

$$\left[\frac{D\Delta}{Dt} + \{\cdots \text{space derivatives of moments} \cdots\}\right]_{f_{G|13}} = -\frac{2}{3}\frac{p}{\mu}\Delta^{(1)}$$

$$\left[\frac{DQ_{ij}}{Dt} + \{\cdots \text{space derivatives of moments} \cdots\}\right]_{f_{G|13}} = -\frac{7}{6}\frac{p}{\mu}Q_{ij}^{(1)}$$

$$\left[\frac{Dm_{ijk}}{Dt} + \{\cdots \text{space derivatives of moments} \cdots\}\right]_{f_{G|13}} = -\frac{3}{2}\frac{p}{\mu}m_{ijk}^{(1)}$$

The system of equations is still not closed yet, since we have higher order unknown moments sitting inside the curly brackets $\{\cdots\}$. The subscript $f_{G|13}$ attached to the left-hand side implies that all these unknown moments must be evaluated using the Grad's 13 moment distribution function $f_{G|13}$ (Eq. (6.91)). The final result has only one non-vanishing term with all the remaining unknowns turning out to be exactly zero, i.e.,

$$v^0_{ijkl|G13} = v^1_{ijk|G13} = 0, \quad v^2_{k|G13} = 28RTq_k \quad \text{and}$$
$$\Delta_{|G13} = Q_{ij|G13} = m_{ijk|G13} = 0 \tag{7.46}$$

With this closure, we can obtain the constitutive relations for the moments Δ, Q_{ij}, m_{ijk} which then can be substituted in the transport equations for stress tensor and heat flux vector.

Summarizing the result, we now present the transport equations for stress tensor and heat flux vector as

$$\frac{Dp_{ij}}{Dt} + p_{ij}\frac{\partial u_k}{\partial x_k} + \frac{4}{5}\frac{\partial q_{\langle i}}{\partial x_{j\rangle}} + 2p\frac{\partial u_{\langle i}}{\partial x_{j\rangle}} + 2p_{k\langle i}\frac{\partial u_{j\rangle}}{\partial x_k} + \frac{\partial m_{ijk}}{\partial x_k} = -\frac{p}{\mu}p_{ij} \tag{7.47}$$

$$\frac{Dq_i}{Dt}+\frac{5}{2}p\frac{\partial RT}{\partial x_i}+\frac{5}{2}p_{ik}\frac{\partial RT}{\partial x_k}+RT\frac{\partial p_{ik}}{\partial x_k}-RTp_{ik}\frac{\partial \ln\rho}{\partial x_k}+\frac{7}{5}q_k\frac{\partial u_i}{\partial x_k}+\frac{2}{5}q_k\frac{\partial u_k}{\partial x_i}$$
$$+\frac{7}{5}q_i\frac{\partial u_k}{\partial x_k}+\frac{1}{2}\frac{\partial Q_{ik}}{\partial x_k}+\frac{1}{6}\frac{\partial \Delta}{\partial x_i}+m_{ijk}\frac{\partial u_j}{\partial x_k}-\frac{p_{ij}}{\rho}\frac{\partial p_{jk}}{\partial x_k}=-\frac{2}{3}\frac{p}{\mu}q_i \tag{7.48}$$

and the constitutive relationships for the higher moments, Δ, Q_{ij} and m_{ijk} are given as

$$\Delta=-12\frac{\mu}{p}\left[RT\frac{\partial q_k}{\partial x_k}+\frac{5}{2}q_k\frac{\partial RT}{\partial x_k}-RTq_k\frac{\partial \ln\rho}{\partial x_k}-\frac{q_k}{\rho}\frac{\partial p_{kl}}{\partial x_l}+RTp_{kl}\frac{\partial u_k}{\partial x_l}\right] \tag{7.49}$$

$$Q_{ij}=-\frac{24}{5}\frac{\mu}{p}\left\{RT\frac{\partial q_{\langle i}}{\partial x_{j\rangle}}+q_{\langle i}\frac{\partial RT}{\partial x_{j\rangle}}-RTq_{\langle i}\frac{\partial \ln\rho}{\partial x_{j\rangle}}-\frac{q_{\langle i}}{\rho}\frac{\partial p_{j\rangle k}}{\partial x_k}\right.$$
$$\left.+\frac{5}{7}RT\left[p_{k\langle i}\frac{\partial u_{j\rangle}}{\partial x_k}+p_{k\langle i}\frac{\partial u_k}{\partial x_{j\rangle}}-\frac{2}{3}p_{ij}\frac{\partial u_k}{\partial x_k}\right]-\frac{5}{6}\frac{p_{ij}}{\rho}\frac{\partial q_k}{\partial x_k}-\frac{5}{6}\frac{p_{ij}p_{kl}}{\rho}\frac{\partial u_k}{\partial x_l}\right\} \tag{7.50}$$

$$m_{ijk}=-2\frac{\mu}{p}\left\{RT\frac{\partial p_{\langle ij}}{\partial x_{k\rangle}}-RTp_{\langle ij}\frac{\partial \ln\rho}{\partial x_{k\rangle}}+\frac{4}{5}q_{\langle i}\frac{\partial u_j}{\partial x_{k\rangle}}-\frac{p_{\langle ij}}{\rho}\frac{\partial p_{k\rangle l}}{\partial x_l}\right\} \tag{7.51}$$

The basic conservation laws (Eqs. (7.1)–(7.3)), the transport equations for stress tensor (Eq. (7.47)), and heat flux vector (Eq. (7.48)) along with the constitutive relations for the higher moments, Δ, Q_{ij} and m_{ijk} (Eqs. (7.49)–(7.51)) form a complete closed set of R13 moment equations.

7.5.1.2 Remarks on the R13 Equations

Some comments on the regularization procedure and the obtained R13 equations are in order.

1. Regularization seems to improve the governing equations, i.e., R13 equations are supposedly better than the Grad's equations. This is because we are essentially going one level up and then closing the equations. For example, instead of using the Grad's distribution function to evaluate the closure terms in Eq. (6.95), we instead use the Grad distribution function at the level of the evolution equation for the closure terms in Eq. (7.46). This helps improve the Knudsen number range over which the equations apply, but does not really solve all the problems.
2. Regularization also leads to the requirement of additional boundary conditions and makes the solution procedure even more complex as additional equations get involved.
3. Moreover, this approach takes the Grad's method for granted, and hence, any irregularities present in the Grad's method naturally propagate into the R13 equa-

tions. As such, this method has also faced critiques, and the authors themselves conceded in a later paper [140] *"While the properties of the R13 equations are highly desirable, their derivation by a method which stands between the methods of Grad and Chapman–Enskog did not convince all readers."*

To overcome the above critiques, Struchtrup [140] presented another set of R13 equations based on the order of magnitude of moments. This method does not take into account the Grad's approach but primarily considers the order of magnitude of the moments of the phase density. The resulting R13 equations were claimed to be a superior set of equations than the Burnett, Grad, and original R13 equations. However, in the work of Timokhin et al. [146], different R13 variants were applied to the shock structure problem. The authors reported that the newly derived R13 equations performed worse than the original R13 equations at supersonic velocities. Therefore, the ambiguity about accurate set of R13 equations still persists.

7.5.2 Regularized 26-Moment Equations

The regularization method introduced previously can be taken to the next level by considering more moments in the primary variables set. Gu and Emerson [68] proposed R26 moment equations on the similar lines of R13 moment equations in order to obtain more accurate solutions in the transition regime ($Kn > 0.1$). Accordingly, the variables set was expanded from 13 variables $\{\rho, u_i, T, p_{ij}, q_i\}$ to 26 variables $\{\rho, u_i, T, p_{ij}, q_i, m_{ijk}, Q_{ij}, \Delta\ (1+3+1+5+3+7+5+1 = 26 \text{ variables})\}$. The integral expressions for the primary variables along with the unknowns in terms of distribution function are shown in Fig. 7.2. Note that in R13 moment equations, we obtained constitutive relationships for m_{ijk}, Q_{ij}, and Δ. However, in R26 equations, these variables are treated on par with stress tensor and heat flux vector variables. The derivation procedure is similar to that R13 moment equations, here, we present only the important steps of the derivation of R26 moment equations.

Using a procedure similar to that outlined in Sect. 7.5.1, evolution equations for moments m_{ijk}, Q_{ij}, and Δ can be obtained from the transformed Boltzmann equation (6.10) as:

$$\frac{\partial m_{ijk}}{\partial t} + \frac{\partial u_l m_{ijk}}{\partial x_l} + \frac{\partial \phi_{ijkl}}{\partial x_l} = -\frac{3}{2}\frac{p}{\mu} m_{ijk} - 3p\frac{\partial (p_{\langle ij}/\rho)}{\partial x_{k\rangle}} + \mathscr{M}_{ijk} \tag{7.52}$$

$$\frac{\partial Q_{ij}}{\partial t} + \frac{\partial u_k Q_{ij}}{\partial x_k} + \frac{\partial \psi_{ijk}}{\partial x_k} = -\frac{7}{6}\frac{p}{\mu} Q_{ij} - \frac{28}{5}p\frac{\partial (q_{\langle i}/\rho)}{\partial x_{j\rangle}} + \mathscr{Q}_{ij} \tag{7.53}$$

$$\frac{\partial \Delta}{\partial t} + \frac{\partial \Delta u_i}{\partial x_i} + \frac{\partial \Omega_i}{\partial x_i} = -\frac{2}{3}\frac{p}{\mu}\Delta - 8p\frac{\partial (q_i/\rho)}{\partial x_i} + \aleph \tag{7.54}$$

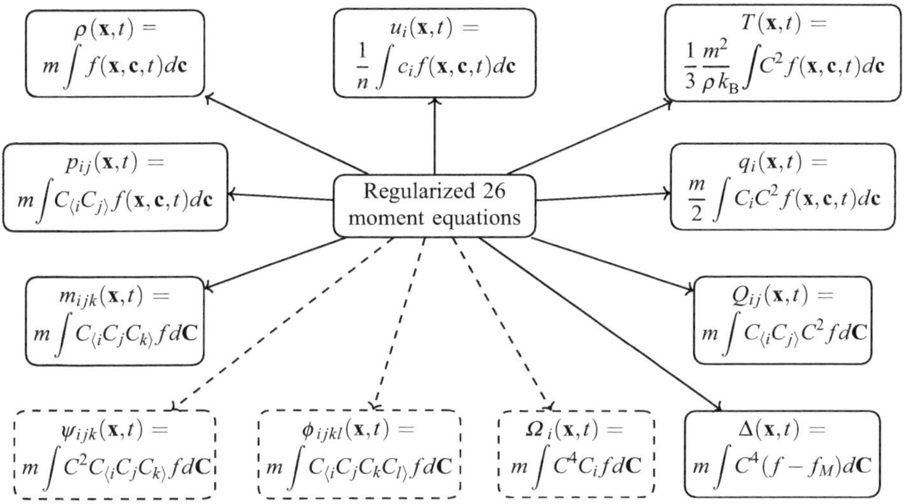

Fig. 7.2 Expressions for the variables considered in the regularized 26-moment equations; variables in solid rectangles are the primary variables while those in dashed rectangles are the closure variables; separate evolutions equations are generated for the closure variables ψ_{ijk}, ϕ_{ijkl}, and Ω_i and Chapman–Enskog scaling is then applied to obtain constitutive relationships for them

where

$$\mathscr{M}_{ijk} = 3\frac{p_{\langle ij}}{\rho}\frac{\partial p_{k\rangle l}}{\partial x_l} - \frac{12}{5}q_{\langle i}\frac{\partial u_j}{\partial x_{k\rangle}} - 3m_{l\langle ij}\frac{\partial u_{k\rangle}}{\partial x_l} - \frac{3}{7}\frac{\partial Q_{\langle ij}}{\partial x_{k\rangle}} \tag{7.55}$$

$$\begin{aligned}\mathscr{Q}_{ij} = &-\frac{2}{3}\frac{p}{\mu}\frac{p_{k\langle i}p_{j\rangle k}}{\rho} - \frac{28}{5}q_{\langle i}\frac{\partial RT}{\partial x_{j\rangle}} + \frac{28}{5}\frac{q_{\langle i}}{\rho}\frac{p_{j\rangle k}}{\partial x_k} + \frac{14}{3}\frac{p_{ij}}{\rho}\left(\frac{\partial q_k}{\partial x_k} + p_{kl}\frac{\partial u_k}{\partial x_l}\right)\\ &- 4RT\left(p_{k\langle i}\frac{\partial u_k}{\partial x_{j\rangle}} + p_{k\langle i}\frac{\partial u_{j\rangle}}{\partial x_k} - \frac{2}{3}p_{ij}\frac{\partial u_k}{\partial x_k}\right) - 2RT\frac{\partial m_{ijk}}{\partial x_k} - 9m_{ijk}\frac{\partial RT}{\partial x_k}\\ &- 2\phi_{ijkl}\frac{\partial u_k}{\partial x_l} + 2\frac{m_{ijk}}{\rho}\left(\frac{\partial p}{\partial x_k} + \frac{\partial p_{kl}}{\partial x_l}\right) - \left(\frac{6}{7}Q_{\langle ij}\frac{\partial u_{k\rangle}}{\partial x_k} + \frac{4}{5}Q_{k\langle i}\frac{\partial u_k}{\partial x_j}\right.\\ &\left. + 2Q_{k\langle i}\frac{\partial u_{j\rangle}}{\partial x_k}\right) - \frac{14}{15}\Delta\frac{\partial u_{\langle i}}{\partial x_{j\rangle}} - \frac{2}{5}\frac{\partial \Omega_{\langle i}}{\partial x_{j\rangle}}\end{aligned} \tag{7.56}$$

$$\aleph = -\frac{2}{3}\frac{p}{\mu}\frac{p_{ij}p_{ij}}{\rho} - 4(2RTp_{ij} + Q_{ij})\frac{\partial u_i}{\partial x_j} + 8\frac{q_i}{\rho}\frac{\partial p_{ij}}{\partial x_j} - 20q_i\frac{\partial RT}{\partial x_i} - \frac{4}{3}\Delta\frac{\partial u_i}{\partial x_i} \tag{7.57}$$

The unknowns appearing in the equations ϕ_{ijkl}, ψ_{ijk}, and Ω_i are the differences between the true value of the higher moments (v_{ijkl}, v^1_{ijk}, v^2_i) and their corresponding approximate value ($v_{ijklf_{G|26}}$, $v^1_{ijkf_{G|26}}$, $v^2_{if_{G|26}}$) evaluated using the

Grad's distribution function for the 26-moment approximation ($f_{G|26}$). Note that the unknowns ψ_{ijk} and Ω_i are single and double contracted moments of fifth order moment v_{ijklm}. These deviations are given as

$$\begin{aligned}\psi_{ijk} &= v^1_{ijk} - v^1_{ijkf_{G|26}}\\ &= m\int C^2 C_{\langle i}C_jC_{k\rangle} f d\mathbf{C} - m\int C^2 C_{\langle i}C_jC_{k\rangle} f_{G|26} d\mathbf{C}\end{aligned} \tag{7.58}$$

$$\begin{aligned}\Omega_i &= v^2_i - v^2_{if_{G|26}}\\ &= m\int C^4 C_i f d\mathbf{C} - m\int C^4 C_i f_{G|26} d\mathbf{C}\end{aligned} \tag{7.59}$$

$$\begin{aligned}\phi_{ijkl} &= v_{ijkl} - v_{ijklf_{G|26}}\\ &= m\int C_{\langle i}C_jC_kC_{l\rangle} f d\mathbf{C} - m\int C_{\langle i}C_jC_kC_{l\rangle} f_{G|26} d\mathbf{C}\end{aligned} \tag{7.60}$$

The functional form of the distribution function for Grad 26-moment approximation $f_{G|26}$ can be obtained in a similar way as that of $f_{G|20}$ (see Sect. 6.3.3). The final form reads as [141],

$$\begin{aligned}f_{G|26} = f^{(0)}\Bigg\{ & 1 + \frac{\Delta}{8pRT}\left(1 - \frac{2}{3}\frac{C^2}{RT} + \frac{1}{15}\frac{C^4}{R^2T^2}\right) + \frac{2}{5}\frac{q_k}{pRT}C_k\left(\frac{C^2}{2RT} - \frac{5}{2}\right)\\ &+ \frac{2RTp_{ij}}{4pR^2T^2}C_{\langle i}C_{j\rangle} + \frac{7RTp_{ij} - Q_{ij}}{4pR^2T^2}C_{\langle i}C_{j\rangle}\left(1 - \frac{1}{7}\frac{C^2}{RT}\right)\\ &+ \frac{m_{ijk}}{6\rho R^3T^3}C_{\langle i}C_jC_{k\rangle}\Bigg\}\end{aligned} \tag{7.61}$$

Utilizing this particular form of distribution function, the constitutive relationships for the unknowns can be obtained as

$$v_{ijklf_{G|26}} = 0; \quad v^1_{ijkf_{G|26}} = 9RTm_{ijk}; \quad v^2_{if_{G|26}} = 28RTq_i \tag{7.62}$$

Substituting this Grad closure in Eqs. (7.52)–(7.54), we obtain Grad 26-moment equations. However, we seek non-zero approximations for the deviations similar to R13 equations as

$$\begin{aligned}\psi_{ijk} &= v^1_{ijk} - v^1_{ijkf_{G|26}} = v^1_{ijk} - 9RTm_{ijk}\\ \Omega_i &= v^2_i - v^2_{if_{G|26}} = v^2_i - 28RTq_i\\ \phi_{ijkl} &= v_{ijkl} - v_{ijklf_{G|26}} = v_{ijkl}\end{aligned}$$

where the results of Eq. (7.62) are used.

The next step is to generate the evolution equations for these moments as (though this step is somewhat cumbersome and involves higher order tensor algebra, the equations can be generated in a straightforward manner)

$$\frac{\partial \phi_{ijkl}}{\partial t}+\frac{\partial u_m \phi_{ijkl}}{\partial x_m}+\frac{4}{9}\frac{\partial \psi_{\langle ijk}}{\partial x_{l\rangle}}-4\frac{m_{\langle ijk}}{\rho}\left(\frac{\partial p_{l\rangle m}}{\partial x_m}+\frac{\partial p}{\partial x_{l\rangle}}\right)$$
$$+4RT\frac{\partial m_{\langle ijk}}{\partial x_{l\rangle}}+4m_{\langle ijk}\frac{\partial RT}{\partial x_{l\rangle}}+4\phi_{m\langle ijk}\frac{\partial u_{l\rangle}}{\partial x_m}+\frac{12}{7}(Q_{\langle ij}+7RTp_{\langle ij})\frac{\partial u_k}{\partial x_{l\rangle}}$$
$$=-C_1\frac{p}{\mu}\phi_{ijkl}-C_2\frac{p}{\mu}\frac{p_{\langle ij}p_{kl\rangle}}{\rho} \tag{7.63}$$

$$\frac{\partial \psi_{ijk}}{\partial t}+\frac{\partial \psi_{ijk}u_m}{\partial x_m}+\frac{27}{7}RT\frac{\partial Q_{\langle ij}}{\partial x_{k\rangle}}+\frac{108}{5}RTq_{\langle i}\frac{\partial u_j}{\partial x_{k\rangle}}+8RTm_{\langle ijk}\frac{\partial u_{m\rangle}}{\partial x_m}$$
$$+\frac{27}{7}(2Q_{\langle ij}+7RTp_{\langle ij})\frac{\partial RT}{\partial x_{k\rangle}}+\frac{54}{7}RTm_{m\langle ij}\frac{\partial u_m}{\partial x_{k\rangle}}-6m_{ijk}RT\frac{\partial u_m}{\partial x_m}$$
$$-6\frac{m_{ijk}}{\rho}\left(\frac{\partial q_m}{\partial x_m}+p_{ml}\frac{\partial u_m}{\partial x_l}\right)-\frac{27}{7}\frac{Q_{\langle ij}}{\rho}\left(\frac{\partial p_{k\rangle m}}{\partial x_m}+\frac{\partial p}{\partial x_k}\right)+\frac{8}{9}\psi_{\langle ijk}\frac{\partial u_{m\rangle}}{\partial x_m}$$
$$+3\psi_{m\langle ij}\frac{\partial u_{k\rangle}}{\partial x_m}+\frac{6}{7}\psi_{m\langle ij}\frac{\partial u_m}{\partial x_{k\rangle}}+\frac{54}{35}\Omega_{\langle i}\frac{\partial u_j}{\partial x_{k\rangle}}-2\frac{\phi_{ijkm}}{\rho}\left(\frac{\partial p_{ml}}{\partial x_l}\frac{\partial p}{\partial x_m}\right)$$
$$-9RT\frac{\partial \phi_{ijkl}}{\partial x_l}=-Y_1\frac{p}{\mu}\psi_{ijk}-\frac{p}{\mu}\left(Y_2\frac{p_{\langle li}m_{jkl\rangle}}{\rho}+Y_3\frac{q_{\langle i}p_{jk\rangle}}{\rho}\right) \tag{7.64}$$

$$\frac{\partial \Omega_i}{\partial t}+\frac{\partial \Omega_i u_j}{\partial x_j}+\frac{7}{3}RT\frac{\partial \Delta}{\partial x_i}+4RT\frac{\partial Q_{ij}}{\partial x_j}-\frac{56}{3}\frac{q_i}{\rho}\left(\frac{\partial q_j}{\partial x_j}+p_{jk}\frac{\partial u_j}{\partial x_k}\right)$$
$$-4\frac{Q_{ij}}{\rho}\left(\frac{\partial p}{\partial x_j}+\frac{\partial p_{jk}}{\partial x_k}\right)+7\Delta\frac{\partial RT}{\partial x_i}-\frac{7}{3}\frac{\Delta}{\rho}\left(\frac{\partial p}{\partial x_i}+\frac{\partial p_{ij}}{\partial x_j}\right)$$
$$+2(9Q_{ij}+14RTp_{ij})\frac{\partial RT}{\partial x_j}+\frac{56}{5}RT\left(q_j\frac{\partial u_i}{\partial x_j}+q_j\frac{\partial u_j}{\partial x_i}\right)+4\psi_{ijk}\frac{\partial u_j}{\partial x_k}$$
$$+8RTm_{ijk}\frac{\partial u_j}{\partial x_k}+\left(\frac{5}{4}\Omega_i\frac{\partial u_j}{\partial x_j}+\frac{9}{5}\Omega_j\frac{\partial u_i}{\partial x_j}+\frac{4}{5}\Omega_j\frac{\partial u_j}{\partial x_i}\right)$$
$$=-\frac{p}{\mu}\Omega_i-\frac{2}{15}\frac{p}{\mu}\left(\frac{5m_{ijk}p_{jk}+14q_jp_{ij}}{\rho}\right) \tag{7.65}$$

where the production terms on the right-hand side of the equations are evaluated assuming Maxwellian molecules. Accordingly, the collision constants are

$$C_1=2.097,\ C_2=0.291,\ Y_1=1.698,\ Y_2=1.203,\ Y_3=0.854$$

Applying the Chapman–Enskog scaling as performed in R13 equations, all the terms on the LHS of evolution equations for ψ_{ijk}, Ω_i, and ϕ_{ijkl} are kept untouched, and only the production terms are scaled by ε as

$$\frac{\partial \phi_{ijkl}}{\partial t} + \{\ldots \text{space derivatives of moments} \ldots\} = -\frac{1}{\varepsilon} C_1 \frac{p}{\mu} \phi_{ijkl} - C_2 \frac{p}{\mu} \frac{p_{\langle ij} p_{kl\rangle}}{\rho} \tag{7.66}$$

$$\begin{aligned} &\frac{\partial \psi_{ijk}}{\partial t} + \{\ldots \text{space derivatives of moments} \ldots\} \\ &= -\frac{1}{\varepsilon} Y_1 \frac{p}{\mu} \psi_{ijk} - \frac{p}{\mu} \left(Y_2 \frac{p_{\langle li} m_{jkl\rangle}}{\rho} + Y_3 \frac{q_{\langle i} p_{jk\rangle}}{\rho} \right) \end{aligned} \tag{7.67}$$

$$\begin{aligned} &\frac{\partial \Omega_i}{\partial t} + \{\ldots \text{space derivatives of moments} \ldots\} \\ &= -\frac{1}{\varepsilon} \frac{p}{\mu} \Omega_i - \frac{2}{15} \frac{p}{\mu} \left(\frac{5 m_{ijk} p_{jk} + 14 q_j p_{ij}}{\rho} \right) \end{aligned} \tag{7.68}$$

Expanding the moments ψ_{ijk}, Ω_i, and ϕ_{ijkl} in terms of small parameter ε,

$$\begin{aligned} \phi_{ijkl} &= \phi_{ijkl}^{(0)} + \varepsilon \phi_{ijkl}^{(1)} + \varepsilon^2 \phi_{ijkl}^{(2)} + \cdots \\ \psi_{ijk} &= \psi_{ijk}^{(0)} + \varepsilon \psi_{ijk}^{(1)} + \varepsilon^2 \psi_{ijk}^{(2)} + \cdots \\ \Omega_i &= \Omega_i^{(0)} + \varepsilon \Omega_i^{(1)} + \varepsilon^2 \Omega_i^{(2)} + \cdots \end{aligned} \tag{7.69}$$

and substituting in Eqs. (7.66)–(7.68), the zeroth order approximations of ϕ_{ijkl}, ψ_{ijk} and Ω_i are obtained as

$$\phi_{ijkl}^{(0)} = 0, \quad \psi_{ijk}^{(0)} = 0, \quad \Omega_i^{(0)} = 0 \tag{7.70}$$

Considering the first two terms in series (7.69), the first order approximations of ϕ_{ijkl}, ψ_{ijk}, and Ω_i are obtained as

$$\begin{aligned} \phi_{ijkl} = &-\frac{4\mu}{C_1} \frac{\partial \left(m_{\langle ijk} / \rho \right)}{\partial x_{l\rangle}} - \frac{12}{C_1} \frac{\mu}{\rho} p_{\langle ij} \frac{\partial u_k}{\partial x_{l\rangle}} + \frac{4}{C_1} \frac{\mu}{p\rho} m_{\langle ijk} \frac{\partial p_{l\rangle m}}{\partial x_m} \\ &- \frac{12}{7} \frac{\mu}{C_1} \frac{Q_{\langle ij}}{p} \frac{\partial u_k}{\partial x_{l\rangle}} - \frac{C_2}{C_1} \frac{p_{\langle ij} p_{kl\rangle}}{\rho}, \end{aligned} \tag{7.71}$$

$$
\begin{aligned}
\psi_{ijk} = & -\frac{27\mu}{7Y_1}\frac{\partial\left(Q_{\langle ij}/\rho\right)}{\partial x_{k\rangle}} - \frac{27}{7}\frac{\mu}{Y_1 p}\left(Q_{\langle ij} + 7RTp_{\langle ij}\right)\frac{\partial RT}{\partial x_{k\rangle}} - \frac{108}{5Y_1}\frac{\mu}{\rho}q_{\langle i}\frac{\partial u_j}{\partial x_{k\rangle}} \\
& + \frac{27}{7}\frac{\mu}{Y_1}\frac{Q_{\langle ij}}{p\rho}\frac{\partial p_{k\rangle m}}{\partial x_m} + \frac{6\mu}{Y_1 p}\frac{m_{ijk}}{\rho}\left(\frac{\partial q_m}{\partial x_m} + p_{ml}\frac{\partial u_m}{\partial x_l}\right) \\
& - \frac{\mu}{Y_1\rho}\left(\frac{54}{7}m_{m\langle ij}\frac{\partial u_m}{\partial x_{k\rangle}} + 8m_{\langle ijk}\frac{\partial u_{m\rangle}}{\partial x_m} - 6m_{ijk}\frac{\partial u_m}{\partial x_m}\right) \\
& - \left(\frac{Y_2}{Y_1}\frac{p_{\langle li}m_{jkl\rangle}}{\rho} + \frac{Y_3}{Y_1}\frac{q_{\langle i}p_{jk\rangle}}{\rho}\right)
\end{aligned}
\tag{7.72}
$$

$$
\begin{aligned}
\Omega_i = & -\frac{7\mu}{3}\frac{\partial(\Delta\,\rho)}{\partial x_i} - 4\mu\frac{\partial(Q_{ij}/\rho)}{\partial x_j} - \frac{56}{5}\frac{\mu}{\rho}\left(q_j\frac{\partial u_i}{\partial x_j} + q_j\frac{\partial u_j}{\partial x_i}\right) - 8\frac{\mu}{\rho}m_{ijk}\frac{\partial u_j}{\partial x_k} \\
& - 14\frac{\mu}{p}(2RTp_{ij} + Q_{ij})\frac{\partial RT}{\partial x_j} + \frac{56}{3}\frac{\mu}{p}\frac{q_i}{\rho}\left(\frac{\partial q_j}{\partial x_j} + p_{jk}\frac{\partial u_j}{\partial x_k}\right) \\
& + 4\frac{\mu}{p}\frac{Q_{ij}}{\rho}\frac{\partial p_{jk}}{\partial x_k} + \frac{7}{3}\frac{\mu}{p}\Delta\left(\frac{\partial p_{ij}}{\rho\partial x_j} - 2\frac{\partial RT}{\partial x_i}\right) - \frac{2}{15}\left(\frac{5m_{ijk}p_{jk} + 14q_j p_{ij}}{\rho}\right)
\end{aligned}
\tag{7.73}
$$

This completes the closure for the regularized 26-moment equations. Summarizing the basic conservation laws equations (7.30)–(7.32), the transport equations for stress tensor equation (7.47) and heat flux vector equation (7.48), the transport equations for higher moments Δ, Q_{ij} and m_{ijk} Eqs. (7.52)–(7.54), along with the constitutive relations for the unknown higher moments, ϕ_{ijkl}, ψ_{ijk} and Ω_i Eqs. (7.71)–(7.73) forms a complete closed set of R26 moment equations.

Expanding the primary variables set by including higher order moments, R26 equations should help to achieve more accurate solutions into the transition regime than the R13 equations. For example, R13 equations fail to capture the Knudsen layer velocity profile at $Kn = 0.25$ in planar Couette flow. They also predict a larger slip velocity for $Kn > 0.4$, thereby overestimating the mass flow rate in force-driven Poiseuille flow. However, by virtue of expanded variables set, R26 equations can overcome these limitations and increase the Knudsen number envelope over which accurate solutions can be obtained. However, note that the increased accuracy comes at the cost of substantial complexity involved both in the theoretical formulation and in the numerical computation. Moreover, obtaining numerical solution of R26 equations for two-dimensional and three-dimensional flow problems seems extremely difficult at the moment.

7.6 Onsager Consistent Approach

The Chapman–Enskog approach and Grad approach to arrive at Burnett equations and moment equations, respectively, were presented in the previous chapters. These equations have a mathematical basis, but suffer from limitations as discussed in Sects. 7.2 and 7.4. In view of these limitations, Singh and Agrawal [131] presented a novel approach to obtain Burnett-*like* and Grad-*like* equations utilizing an alternate form of the distribution function wherein Onsager's reciprocity principle is additionally involved in the derivation. Instead of representing the distribution function in an infinite series in terms of Knudsen number as formal smallness parameter (ϵ) as done in Chapman–Enskog approach or in terms of orthogonal Hermite polynomials as done in Grad approach, the idea is to cast the distribution function in terms of thermodynamic forces and fluxes.

7.6.1 *Derivation of Distribution Function*

For any fluid flow transitioning from one conserved non-equilibrium state to another, the process is accompanied by entropy generation $\sigma(J_i, X_i)$ due to its thermodynamic force X_i and associated conjugate thermodynamic flux J_i. According to Onsager, the thermodynamic force is defined as derivative of entropy density σ with respect to thermodynamic variable α_i as

$$X_i = \frac{\partial \sigma}{\partial \alpha_i}, \quad i = 1, 2, 3, \ldots \tag{7.74}$$

The subscript i represents the type of thermodynamic force and is *not* a tensorial index. For example, thermodynamic force associated with stress tensor and heat flux vector are represented by X_τ and X_q, respectively.

The entropy generated in the process can be expressed in terms of thermodynamic forces and fluxes as

$$\sigma = \sum_{i=1}^{n} J_i X_i \tag{7.75}$$

Each flux is a linear homogeneous function of all the forces of the same tensorial order,

$$J_i = \sum_{k=1}^{n} L_{ik} X_k \tag{7.76}$$

where L_{ik} are phenomenological coefficients and obey Onsager's reciprocal relationship [107, 108], $L_{ik} = L_{ki}$ with $i, k = 1, 2, \ldots, n$.

7.6.1.1 First Order Correction to Distribution Function

The first order distribution function can be represented in terms of thermodynamic forces and microscopic fluxes around local Maxwellian distribution function f_0 as [39, 131]

$$f^{(1)} = f_0 - (\Upsilon_\tau : X_\tau + \Upsilon_q \cdot X_q) \tag{7.77}$$

where

$$\Upsilon_j \odot X_j = t_{r(j)} \left(\frac{\partial f_0}{\partial t} + \nabla_x . (\mathbf{c}\, f_0) \right)_{X_j = 0\ \forall j \neq i} \tag{7.78}$$

The macroscopic thermodynamic flux J_i can be obtained from its microscopic counterpart Υ_i as

$$J_i = \int \bar{\Upsilon}_j f d\mathbf{c} \tag{7.79}$$

where

$$\Upsilon_j = -f_0 t_{r(j)} \bar{\Upsilon}_j \tag{7.80}$$

$$\bar{\Upsilon}_\tau = -\left\{ \mathbf{C} \otimes \mathbf{C} - \frac{1}{2} \left[|\mathbf{C}|^2 (\gamma - 1) \right] \mathbf{I} \right\} \tag{7.81}$$

$$\bar{\Upsilon}_q = -\left(\frac{5}{2\beta} - |\mathbf{C}|^2 \right) \mathbf{C} \tag{7.82}$$

$$X_\tau = \beta \left[\nabla \otimes \mathbf{u} + (\nabla \otimes \mathbf{u})^T \right] \tag{7.83}$$

$$X_q = \nabla \beta \tag{7.84}$$

In the above equations, $\odot$ represents full contraction of the tensors of same order, $\otimes$ denotes the outer product, $t_{r(\tau)}$ and $t_{r(q)}$ are relaxation times for momentum [$= \mu/p$] and energy [$= k(\gamma - 1)/(R\gamma p)$] transport, respectively γ is ratio of specific heat, and $\beta = 1/(2RT)$. The standard BGK collision model is employed in the above formulation. However, note that the introduction of different time scales for momentum and thermal diffusion ensures the correct value of Prandtl number.

This first-order correction to the distribution function yields Navier–Stokes constitutive relationships but at the same time satisfies the Onsager's symmetry principle. However, the second order correction to the distribution function as obtained in Chapman–Enskog approach which yields Burnett equations does not satisfy Onsager's symmetry principle [95, 122]. This inconsistency with respect to symmetry principle may be the source of problems that plaque Burnett equations. With this form of distribution function in terms of forces and fluxes, we derive the second order correction to the distribution function in the next section.

7.6.1.2 Second Order Correction to Distribution Function

Keeping the same functional form of the distribution function in terms of thermodynamic forces and fluxes, the second-order correction to the distribution function as obtained by Singh and Agarwal [131] is

$$f_{|O} = f_0 - (\Upsilon_\tau : X_\tau + \Upsilon_q \cdot X_q + (\Upsilon'_{\tau\tau} \odot X_\tau) : X_\tau + (\Upsilon'_{qq} \odot X_q) \cdot X_q) \tag{7.85}$$

where expressions for $\Upsilon'_{\tau\tau} \odot X_\tau$ and $\Upsilon'_{qq} \odot X_q$ are

$$\begin{aligned}\bar{\Upsilon}'_{\tau\tau} \odot X_\tau &= \bar{\Upsilon}_{\tau\tau} \odot X_\tau + \frac{\Delta}{t_{r(\tau)}}(\Upsilon_q \,.\, X_q)\bar{\Upsilon}_\tau + \left(\frac{5}{2\beta} - |\mathbf{C}|^2\right) \\ &\times \Lambda\left\{\bar{\Upsilon}_\tau\left[[\varphi(\gamma-1)-\gamma]\nabla \,.\, \mathbf{u} + \frac{\varphi}{\beta}\mathbf{C} \,.\, \nabla\beta + \mathbf{C} \,.\, \nabla g\right]\right. \\ &\left.+ C_i\left[\mathbf{C} \otimes \frac{\partial \mathbf{u}}{\partial x_i} + \left(\mathbf{C} \otimes \frac{\partial \mathbf{u}}{\partial x_i}\right)^T\right]\right\}\end{aligned} \tag{7.86}$$

$$\bar{\Upsilon}'_{qq} \odot X_q = \bar{\Upsilon}_{qq} \odot X_q + \frac{\Delta}{t_{r(q)}}(\Upsilon_\tau : X_\tau)\bar{\Upsilon}_q + \left(\frac{5}{2\beta} - |\mathbf{C}|^2\right)(\Omega\mathbf{C} \,.\, \nabla\beta\mathbf{C}) \tag{7.87}$$

where Ω, Δ, Λ are defined in Eq. (7.91) and ϕ is viscosity exponent as in $\mu = \mu_0(T/T_0)^\phi$. Note that there are two parts in each of the above equations. The terms $\bar{\Upsilon}_{\tau\tau} \odot X_\tau$ and $\bar{\Upsilon}_{qq} \odot X_q$ as proposed by Mahendra and Singh [91] are second-order correction terms to the distribution function. However, explicit expressions for these terms and other terms on the right-hand side which are correction terms were obtained by Singh and Agarwal [131].

The terms $\bar{\Upsilon}_{\tau\tau} \odot X_\tau$ and $\bar{\Upsilon}_{qq} \odot X_q$ are

$$\begin{aligned}\bar{\Upsilon}_{\tau\tau} \odot X_\tau = &-C_i\left[\mathbf{C} \otimes \frac{\partial \mathbf{u}}{\partial x_i} + \left(\mathbf{C} \otimes \frac{\partial \mathbf{u}}{\partial x_i}\right)^T\right] + \frac{1}{2\beta}\left[\mathbf{C} \otimes \nabla g + (\mathbf{C} \otimes \nabla g)^T\right] \\ &- \left[\frac{1}{2\beta}(\gamma-1)\mathbf{C} \,.\, \nabla g - \frac{1}{2\beta}(\gamma-1)(\mathbf{C} \otimes \mathbf{C}) : X_\tau\right]\mathbf{I} \\ &- \bar{\Upsilon}_\tau\left(\frac{1}{t_{(r)\tau}}\sum_j \Upsilon_j \odot X_j\right) \\ &+ \bar{\Upsilon}_\tau\left\{[\varphi(\gamma-1)-\gamma]\nabla \,.\, \mathbf{u} + \frac{\varphi}{\beta}\mathbf{C} \,.\, \nabla\beta + \mathbf{C} \,.\, \nabla g\right\}\end{aligned} \tag{7.88}$$

$$\bar{\Upsilon}_{qq} \odot X_q = -\bar{\Upsilon}_q\left(\frac{1}{t_{(r)q}}\sum_j \Upsilon_j \odot X_j\right) - \left\{\frac{1}{\beta}[\mathbf{C}\,.\,\nabla g] - \frac{1}{\beta}(\mathbf{C}\otimes\mathbf{C})\colon X_\tau\right.$$
$$\left. + \frac{5}{2\beta}(\gamma-1)\nabla\,.\,\mathbf{u} + \frac{5}{2\beta^2}(\mathbf{C}\,.\,\nabla\beta)\right\}\mathbf{C} - \left[\frac{5}{2\beta} - |\mathbf{C}|^2\right]\left[\frac{1}{2\beta}\nabla g\right]$$
$$- \left[\frac{5}{2\beta} - |\mathbf{C}|^2\right]\left[\mathbf{C}\otimes\nabla\mathbf{u}\right]$$
$$+ \bar{\Upsilon}_q\left\{[\varphi(\gamma-1)-\gamma]\nabla\,.\,\mathbf{u} + \frac{\varphi}{\beta}\mathbf{C}\,.\,\nabla\beta + \mathbf{C}\,.\,\nabla g\right\} \tag{7.89}$$

where β and g are given as

$$\beta = \frac{1}{2RT}, \qquad g = \log\left(\frac{\rho}{\beta}\right) \tag{7.90}$$

However, since the distribution function in terms of $\bar{\Upsilon}_{\tau\tau}\odot X_\tau$ and $\bar{\Upsilon}_{qq}\odot X_q$ does not satisfy the compatibility conditions (Eq. (5.67)), it becomes necessary to modify the distribution function. This task is not trivial and Singh and Agrawal [131] adopted a correction procedure that appended additional terms to the expressions $\bar{\Upsilon}_{\tau\tau}\odot X_\tau$ and $\bar{\Upsilon}_{qq}\odot X_q$. Note that the added terms are not appended on an ad hoc basis, these terms are necessary in satisfying the compatibility conditions without breaking the symmetry principle. The coefficients Ω, Λ, and Δ appearing in Eqs. (7.86) and (7.87) are given as

$$\Omega = \varphi + 2; \quad \Lambda = -\frac{2}{5}; \quad \Delta = \frac{1}{2}\left[-\varphi\frac{t_{r(\tau)}}{t_{r(q)}} - \frac{t_{r(q)}}{t_{r(\tau)}} + 2\right] \tag{7.91}$$

The final form of the distribution function (Eq. (7.85)) satisfies the linearized Boltzmann equation and compatibility conditions without breaking the Onsager's symmetry principle. This particular form of distribution function $f_{|O}$ (where the subscript O stands for Onsager) is termed as "Onsager consistent distribution function." This form is then utilized to derive Onsager–Burnett (OBurnett) equations and Onsager-13 (O13) equations as shown in Fig. 7.3.

7.6.2 *Onsager–Burnett Equations*

In this section, we utilize the Onsager consistent distribution function (Eq. (7.85)) to derive expressions for pressure tensor and heat flux vector as

$$P_{ij} = p\delta_{ij} + p_{ij} = m\int C_iC_j f_{|O}d\mathbf{C}, \quad q_i = \frac{m}{2}\int C^2C_i f_{|O}d\mathbf{C} \tag{7.92}$$

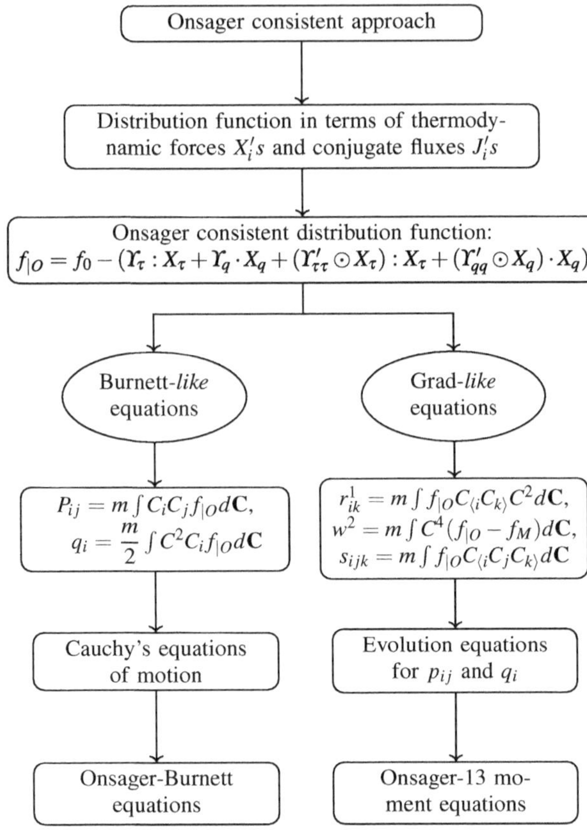

Fig. 7.3 Flowchart showing the detailed procedure of Onsager consistent approach to obtain OBurnett and O13 equations

The expression for the Onsager consistent distribution function (Eq. (7.85)) contains a large number of terms. After evaluating the integrals, the constitutive relationships for normal stress p_{xx}, shear stress p_{xy}, and heat flux component q_x are obtained as [136]

$$
\begin{aligned}
p_{xx} &= p_{xx}^{NS} + p_{xx}^{B} \\
&= \mu \left(\delta_1 \frac{\partial u}{\partial x} + \delta_2 \frac{\partial v}{\partial y} + \delta_2 \frac{\partial w}{\partial z} \right) + 4 \frac{\mu^2 \beta}{\rho} \left\{ \alpha_1 \left(\frac{\partial u}{\partial x} \right)^2 + \alpha_2 \left(\frac{\partial u}{\partial y} \right)^2 \right. \\
&\quad + \alpha_3 \left(\frac{\partial u}{\partial z} \right)^2 + \alpha_4 \frac{\partial u}{\partial y} \frac{\partial v}{\partial x} + \alpha_5 \frac{\partial u}{\partial z} \frac{\partial w}{\partial x} + \alpha_6 \left(\frac{\partial w}{\partial x} \right)^2 + \alpha_7 \left(\frac{\partial v}{\partial x} \right)^2 \\
&\quad + \alpha_8 \frac{\partial u}{\partial x} \frac{\partial v}{\partial y} + \alpha_9 \left(\frac{\partial v}{\partial y} \right)^2 + \alpha_{10} \left(\frac{\partial w}{\partial z} \right)^2 + \alpha_{11} \frac{\partial v}{\partial y} \frac{\partial w}{\partial z} + \alpha_{12} \frac{\partial u}{\partial x} \frac{\partial w}{\partial z} \\
&\quad \left. + \alpha_{13} \frac{\partial v}{\partial z} \frac{\partial w}{\partial y} + \alpha_{14} \left(\frac{\partial w}{\partial y} \right)^2 + \alpha_{15} \left(\frac{\partial v}{\partial z} \right)^2 \right\}
\end{aligned} \tag{7.93}
$$

$$
\begin{aligned}
p_{xy} &= p_{xy}^{NS} + p_{xy}^{B} \\
&= \mu\delta_3 \frac{\partial u}{\partial y} + \mu\delta_3 \frac{\partial v}{\partial x} + 4\frac{\mu^2\beta}{\rho}\left\{\beta_1 \frac{\partial u}{\partial x}\frac{\partial u}{\partial y} + \beta_2 \frac{\partial v}{\partial x}\frac{\partial v}{\partial y} + \beta_3 \frac{\partial u}{\partial z}\frac{\partial v}{\partial z}\right. \\
&+ \beta_4 \frac{\partial u}{\partial x}\frac{\partial v}{\partial x} + \beta_5 \frac{\partial u}{\partial y}\frac{\partial v}{\partial y} + \beta_6 \frac{\partial w}{\partial x}\frac{\partial w}{\partial y} + \beta_7 \frac{\partial v}{\partial z}\frac{\partial w}{\partial x} + \beta_8 \frac{\partial u}{\partial z}\frac{\partial w}{\partial y} \\
&\left.+ \beta_9 \frac{\partial u}{\partial y}\frac{\partial w}{\partial z} + \beta_{10} \frac{\partial v}{\partial x}\frac{\partial w}{\partial z}\right\}
\end{aligned} \tag{7.94}
$$

$$
\begin{aligned}
q_x &= q_x^{NS} + q_x^{B} \\
&= \delta_4 k \frac{1}{2R\beta^2}\frac{\partial \beta}{\partial x} + 4\frac{\mu^2\beta}{\rho}\left\{\gamma_1 \frac{1}{\beta}\frac{\partial g}{\partial x}\frac{\partial u}{\partial x} + \gamma_2 \frac{1}{\beta^2}\frac{\partial \beta}{\partial x}\frac{\partial v}{\partial y} + \gamma_3 \frac{1}{\beta^2}\frac{\partial \beta}{\partial x}\frac{\partial w}{\partial z}\right. \\
&+ \gamma_4 \frac{1}{\beta}\frac{\partial g}{\partial y}\frac{\partial u}{\partial y} + \gamma_5 \frac{1}{\beta}\frac{\partial g}{\partial y}\frac{\partial v}{\partial x} + \gamma_6 \frac{1}{\beta}\frac{\partial g}{\partial z}\frac{\partial w}{\partial x} + \gamma_7 \frac{1}{\beta^2}\frac{\partial \beta}{\partial x}\frac{\partial u}{\partial x} + \gamma_8 \frac{1}{\beta^2}\frac{\partial \beta}{\partial y}\frac{\partial u}{\partial y} \\
&+ \gamma_9 \frac{1}{\beta^2}\frac{\partial \beta}{\partial z}\frac{\partial u}{\partial z} + \gamma_{10} \frac{1}{\beta^2}\frac{\partial \beta}{\partial y}\frac{\partial v}{\partial x} + \gamma_{11} \frac{1}{\beta^2}\frac{\partial \beta}{\partial z}\frac{\partial w}{\partial x} + \gamma_{12} \frac{1}{\beta}\frac{\partial g}{\partial x}\frac{\partial v}{\partial y} \\
&\left.+ \gamma_{13} \frac{1}{\beta}\frac{\partial g}{\partial x}\frac{\partial w}{\partial z}\right] + \left(\frac{2k(\gamma-1)}{R\gamma}\right)^2 \frac{1}{\rho\beta}\left[\gamma_{14} \frac{\partial \beta}{\partial y}\frac{\partial v}{\partial x} + \gamma_{15} \frac{\partial \beta}{\partial z}\frac{\partial w}{\partial x}\right. \\
&\left.+ \gamma_{16} \frac{\partial \beta}{\partial x}\frac{\partial u}{\partial x} + \gamma_{17} \frac{\partial \beta}{\partial y}\frac{\partial u}{\partial y} + \gamma_{18} \frac{\partial \beta}{\partial z}\frac{\partial u}{\partial z} + \gamma_{19} \frac{\partial \beta}{\partial x}\frac{\partial v}{\partial y} + \gamma_{20} \frac{\partial \beta}{\partial x}\frac{\partial w}{\partial z}\right\}
\end{aligned} \tag{7.95}
$$

where g and β are defined through Eq. (7.90). The coefficients, $\alpha's$, $\beta's$, $\gamma's$, and $\delta's$, with numerical subscripts are functions of the type of gas and the interaction potential between the molecules. The constants appearing in front of the derivatives of flow field variables are

$$
\delta_1 = -\frac{4}{3},\ \delta_2 = \frac{2}{3},\ \alpha_1 = \left(\frac{125\gamma^2 - 576\gamma + \varphi(110 - 160\gamma + 50\gamma^2) + 643}{40}\right),
$$

$$
\alpha_2 = \frac{4}{5},\ \alpha_3 = \frac{4}{5},\ \alpha_4 = \frac{1}{5},\ \alpha_5 = \frac{1}{5},\ \alpha_6 = -\frac{3}{5},\ \alpha_7 = -\frac{3}{5},
$$

$$
\alpha_8 = \left(\frac{100\gamma^2 - 484\gamma + \varphi(90 - 140\gamma + 50\gamma^2) + 459}{20}\right),
$$

$$
\alpha_9 = \left(\frac{125\gamma^2 - 392\gamma + \varphi(70 - 120\gamma + 50\gamma^2) + 291}{40}\right),
$$

$$
\alpha_{10} = \alpha_9,\ \alpha_{11} = \alpha_9 + \frac{18}{40},\ \alpha_{12} = \alpha_{11},\ \alpha_{13} = -\frac{2}{5},
$$

$$\alpha_{14} = -\frac{1}{5}, \ \alpha_{15} = \alpha_{14},$$

$$\delta_3 = -1, \ \beta_1 = \frac{23\gamma + 5\gamma\varphi - 5\varphi - 37}{10}, \ \beta_2 = \beta_1, \beta_3 = -1, \ \beta_4 = \beta_1 - \frac{14}{10},$$
$$\beta_5 = \beta_4, \ \beta_6 = 4\beta_4, \ \beta_7 = -\frac{3}{10}, \ \beta_8 = \beta_7, \ \beta_9 = \beta_1$$

$$\delta_4 = 1, \ \gamma_1 = \left(\frac{-47 + 25\gamma}{8}\right), \gamma_2 = \varphi\left(\frac{49 - 35\gamma}{8}\right), \gamma_3 = \gamma_2,$$
$$\gamma_4 = -\frac{1}{2}, \gamma_5 = \gamma_4, \gamma_6 = \gamma_4, \ \gamma_7 = +\varphi\left(\frac{77 - 35\gamma}{8}\right), \ \gamma_8 = \frac{7\varphi}{4},$$
$$\gamma_9 = \gamma_8, \ \gamma_{10} = \gamma_9, \gamma_{11} = \gamma_8, \ \gamma_{12} = \left(\frac{-39 + 25\gamma}{8}\right),$$
$$\gamma_{13} = \gamma_{12}, \ \gamma_{14} = -1, \ \gamma_{15} = \gamma_{14}, \ \gamma_{16} = \left(\frac{-77 + 35\gamma + 10\varphi(-1 + \gamma)}{8}\right),$$
$$\gamma_{17} = -\frac{7}{4}, \ \gamma_{18} = \gamma_{17}, \ \gamma_{19} = \frac{-59 + 35\gamma + 10\varphi(-1 + \gamma)}{8}, \ \gamma_{20} = \gamma_{19}.$$

The Burnett contribution for other components of stress tensor and heat flux vector can be obtained by applying suitable change of variables in an appropriate base equation as given in Table 7.3. Substitution of these stress and heat flux terms in Cauchy's equations of motion and energy conservation equation completes the derivation of the OBurnett equations.

Table 7.3 Base equation and change of variables to be followed while evaluating Burnett order contribution for other components of stress tensor and heat flux vector

Variable	Base equation	Change of variables
p^B_{yy}	p^B_{xx}	$u \to v, x \to y, v \to u$, and $y \to x$
p^B_{zz}	p^B_{xx}	$u \to w, x \to z, w \to u$, and $z \to x$
p^B_{yz}	p^B_{xy}	$u \to v, x \to y, v \to w, y \to z, w \to u$, and $z \to x$
p^B_{zx}	p^B_{xy}	$u \to w, x \to z, v \to u, y \to x, w \to v$, and $z \to y$
q^B_y	q^B_x	$u \to v, x \to y, v \to u$, and $y \to x$
q^B_z	q^B_x	$u \to w, x \to z, w \to u$, and $z \to x$

7.6.2.1 Linear Stability Analysis

A linear stability analysis of the OBurnett equations is performed for a one-dimensional wave about the following equilibrium state: $u = 0$; $\rho = \rho_0$ and $T = T_0$. Assuming small perturbations around equilibrium, we can write

$$\begin{aligned} u &= u'(x, t), \\ \rho &= \rho_0 + \rho'(x, t), \\ T &= T_0 + T'(x, t) \end{aligned} \tag{7.96}$$

where $'$ denotes small quantities away from equilibrium. Substituting Eq. (7.96) in the Cauchy's equations of motion and energy conservation equation, the following set of equations is obtained upon linearization:

$$\frac{\partial \rho'}{\partial t} + \rho_0 \frac{\partial u'}{\partial x} = 0, \tag{7.97}$$

$$\rho_0 \frac{\partial u'}{\partial t} + \rho_0 R \frac{\partial T'}{\partial x} + R T_0 \frac{\partial \rho'}{\partial x} + \frac{4}{3} \mu \frac{\partial^2 u'}{\partial x^2} = 0, \tag{7.98}$$

$$\frac{3}{2} \rho_0 R \frac{\partial T'}{\partial t} - k \frac{\partial^2 T'}{\partial x^2} + p_0 \frac{\partial u'}{\partial x} = 0. \tag{7.99}$$

Notice that there are no Burnett-order terms in the above equations and the form of the equations is similar to that of perturbed Navier–Stokes equations. Since perturbed Navier–Stokes equations are unconditionally stable, OBurnett equations are also unconditionally stable which in itself is a remarkable feature considering the fact that the Burnett equations are plagued with stability issues. This stability feature is due to the fact that Eqs. (7.93)–(7.95) involve only the cross-product of derivative of velocity, temperature, and pressure. Therefore, Burnett order contribution to the stress and heat flux components (p_{xx}^B, p_{xy}^B, and q_x^B) becomes identically zero for small perturbation. In contrast, the presence of second-derivatives of velocity and temperature in stress tensor and heat flux vector in Burnett equations lead to the problem of stability.

7.6.2.2 Features of OBurnett Equations

The OBurnett equations have several attractive features:

1. *Fewer boundary conditions:* The constitutive relationships for OBurnett equations do not contain second and higher order derivatives of velocity as compared to conventional Burnett equations (which contain third order derivatives of velocity). Therefore, the OBurnett equations need the same number of boundary

conditions as the Navier–Stokes equations. Hence, the need of prescribing additional boundary conditions is mitigated.

2. *Unconditionally stable:* As per the linear stability analysis, the equations are unconditionally stable. This point is already demonstrated in Sect. 7.6.2.1.
3. *Reduced coupling:* The absence of temperature gradient terms in the constitutive relationships for stress tensor reduces the amount of coupling between the momentum and energy equations. Hence, in certain cases like incompressible forced convection, rarefied gas flow problems, one can obtain velocity field before solving the energy equation.
4. *Correct value of Prandtl number:* The formulation of distribution function involves two different relaxation times for momentum and energy transport which ensures that the correct value of Prandtl number can be recovered for any gas.
5. *Larger Knudsen number envelope:* The derivation of the OBurnett equations does not involve Chapman–Enskog like expansion, hence the OBurnett equations are expected to give accurate solutions in a larger Knudsen number envelope.

7.6.3 Onsager-13 Moment Equations

In this section, we present the derivation of Onsager-13 (O13) moment equations utilizing the Onsager consistent distribution function (Eq. (7.85)). The derivation of O13 moment equations can be outlined as: The evolution equations for the stress tensor and heat flux vector are generated as derived in Sect. 6.4.1. However, instead of using Grad distribution function ($f_{G|13}$), we employ the Onsager-consistent distribution function ($f_{|O}$) to evaluate the unknowns, r^1_{ik}, w^2 and s_{ijk}, appearing in the evolution equations to achieve the closure as

$$\begin{aligned} r^1_{ik} &= m\int f_{|O} C_{\langle i} C_{k\rangle} C^2 d\mathbf{C} \\ &= 7\mu \frac{p}{\rho}\frac{\partial u_{\langle i}}{\partial x_{k\rangle}} + \frac{1}{20}\frac{\rho}{\beta^2}\left(\frac{\mu}{p}\right)^2 \Bigg[[342 + 35\varphi \\ &\quad - 7\gamma(31 + 5\varphi)] \frac{\partial u_{\langle i}}{\partial x_{k\rangle}}\frac{\partial u_l}{\partial x_l} + 27\frac{\partial u_{\langle i}}{\partial x_l}\frac{\partial u_l}{\partial x_{k\rangle}} + 45\frac{\partial u_{\langle i}}{\partial x_l}\frac{\partial u_{k\rangle}}{\partial x_l} - 18\frac{\partial u_l}{\partial x_{\langle i}}\frac{\partial u_l}{\partial x_{k\rangle}} \Bigg] \\ &\quad + \frac{7}{8}\frac{\rho}{\beta^5}\left(\frac{k(\gamma-1)}{R\gamma p}\right)^2 \left[2\frac{\partial \beta}{\partial x_{\langle i}}\frac{\partial \beta}{\partial x_{k\rangle}} + \beta \frac{\partial \beta}{\partial x_{\langle i}}\frac{\partial g}{\partial x_{k\rangle}} \right] \end{aligned} \tag{7.100}$$

$$w^2 = m\int C^4 (f_{|O} - f_M) d\mathbf{C}$$

$$= (50-30\gamma)\mu\frac{p}{\rho}\frac{\partial u_j}{\partial x_j} + \frac{5}{2}\frac{\rho}{\beta^5}\left(\frac{k(\gamma-1)}{R\gamma p}\right)^2\left[\frac{7}{2}\frac{\partial \beta}{\partial x_j}\frac{\partial \beta}{\partial x_j} + \beta^2\frac{\partial g}{\partial x_j}\frac{\partial g}{\partial x_j}\right]$$
$$+ \frac{\rho}{\beta^2}\left(\frac{\mu}{p}\right)^2\frac{7}{16}(-5+3\gamma)(-63-10\varphi+5\gamma(7+2\varphi))\left[\left(\frac{\partial u_j}{\partial x_j}\right)^2\right] \tag{7.101}$$

$$s_{ijk} = m\int f_{|O}C_{\langle i}C_jC_{k\rangle}d\mathbf{C}$$
$$= \frac{3}{2}\frac{\rho}{\beta^3}\left\{\beta\left(\frac{\mu}{p}\right)^2\left[\frac{\partial g}{\partial x_{\langle i}}\frac{\partial u_j}{\partial x_{k\rangle}}\right] + \left[\varphi\left(\frac{k(\gamma-1)}{R\gamma p}\right)^2 - \left(\frac{\mu}{p}\right)^2\right]\left[\frac{\partial \beta}{\partial x_{\langle i}}\frac{\partial u_j}{\partial x_{k\rangle}}\right]\right\} \tag{7.102}$$

Substituting the expressions for higher order moments r^1_{ik}, w^2, and s_{ijk} (Eq. (7.102)) in the evolution equations for stress tensor and heat flux vector, we obtain O13 moment equations as

$$\frac{\partial p_{ij}}{\partial t} + \frac{3}{2}\frac{\partial}{\partial x_k}\left\{\beta\left(\frac{\mu}{p}\right)^2\left[\frac{\partial g}{\partial x_{\langle i}}\frac{\partial u_j}{\partial x_{k\rangle}}\right]\right.$$
$$\left. + \frac{\rho}{\beta^3}\left[\varphi\left(\frac{k(\gamma-1)}{R\gamma p}\right)^2 - \left(\frac{\mu}{p}\right)^2\right]\left[\frac{\partial \beta}{\partial x_{\langle i}}\frac{\partial u_j}{\partial x_{k\rangle}}\right]\right\}$$
$$+ u_k\frac{\partial p_{ij}}{\partial x_k} + \frac{4}{5}\frac{\partial q_{\langle i}}{\partial x_{j\rangle}} + 2p_{k\langle i}\frac{\partial u_{j\rangle}}{\partial x_k} + 2p\frac{\partial u_{\langle i}}{\partial x_{j\rangle}} + p_{ij}\frac{\partial u_k}{\partial x_k} = -\frac{p}{\mu}p_{ij}, \tag{7.103}$$

$$\frac{\partial q_i}{\partial t} + u_k\frac{\partial q_i}{\partial x_k} + \frac{5}{2}\left(\frac{p}{\rho}\frac{\partial p}{\partial x_i} - \frac{p^2}{\rho^2}\frac{\partial \rho}{\partial x_i}\right) - \frac{5}{2}\frac{p}{\rho}\frac{\partial p_{ik}}{\partial x_k} - \frac{p_{ik}}{\rho}\frac{\partial p}{\partial x_k} + 7\frac{\partial}{\partial x_k}\left(\mu\frac{p}{\rho}\frac{\partial u_{\langle i}}{\partial x_{k\rangle}}\right)$$
$$+ \frac{7}{5}q_k\frac{\partial u_i}{\partial x_k} + \frac{\partial}{\partial x_k}\left\{\frac{1}{20}\frac{\rho}{\beta^2}\left(\frac{\mu}{p}\right)^2\left[[342+35\varphi-7\gamma(31+5\varphi)]\frac{\partial u_{\langle i}}{\partial x_{k\rangle}}\frac{\partial u_l}{\partial x_l} + 27\frac{\partial u_{\langle i}}{\partial x_l}\frac{\partial u_l}{\partial x_{k\rangle}}\right.\right.$$
$$\left.\left. + 45\frac{\partial u_{\langle i}}{\partial x_l}\frac{\partial u_{k\rangle}}{\partial x_l} - 18\frac{\partial u_l}{\partial x_{\langle i}}\frac{\partial u_l}{\partial x_{k\rangle}}\right] + \frac{7}{8}\frac{\rho}{\beta^5}\left(\frac{k(\gamma-1)}{R\gamma p}\right)^2\left[2\frac{\partial \beta}{\partial x_{\langle i}}\frac{\partial \beta}{\partial x_{k\rangle}} + \beta\frac{\partial \beta}{\partial x_{\langle i}}\frac{\partial g}{\partial x_{k\rangle}}\right]\right\}$$
$$+ \frac{7}{5}q_i\frac{\partial u_k}{\partial x_k} + \frac{2}{5}q_k\frac{\partial u_k}{\partial x_i} + \frac{1}{6}\frac{\partial}{\partial x_i}\left\{(50-30\gamma)\mu\frac{p}{\rho}\frac{\partial u_j}{\partial x_j} + \frac{\rho}{\beta^2}\left(\frac{\mu}{p}\right)^2\frac{7}{16}(-5+3\gamma)\right.$$
$$\times(-63-10\varphi+5\gamma(7+2\varphi))\left[\left(\frac{\partial u_j}{\partial x_j}\right)^2\right] + \frac{5}{2}\frac{\rho}{\beta^5}\left(\frac{k(\gamma-1)}{R\gamma p}\right)^2\left[\frac{7}{2}\frac{\partial \beta}{\partial x_j}\frac{\partial \beta}{\partial x_j}\right.$$

$$+\beta^2\frac{\partial g}{\partial x_j}\frac{\partial g}{\partial x_j}\Bigg]\Bigg\} - \frac{p_{ij}}{\rho}\frac{\partial p_{jk}}{\partial x_k} + \frac{3}{2}\frac{\rho}{\beta^3}\frac{\partial u_j}{\partial x_k}\Bigg\{\beta\left(\frac{\mu}{p}\right)^2\left[\frac{\partial g}{\partial x_{\langle i}}\frac{\partial u_j}{\partial x_{k\rangle}}\right]$$

$$+\left[\varphi\left(\frac{k(\gamma-1)}{R\gamma p}\right)^2 - \left(\frac{\mu}{p}\right)^2\right]\left[\frac{\partial \beta}{\partial x_{\langle i}}\frac{\partial u_j}{\partial x_{k\rangle}}\right]\Bigg\} = -\frac{2}{3}\frac{p}{\mu}q_i \tag{7.104}$$

with β, g, γ, ϕ as defined in Sect. 7.6.1.2. The basic conservation laws (Eqs. (7.1)–(7.3)), the transport equations for stress tensor (Eq. (7.103)) and heat flux vector (Eq. (7.104)) form a complete closed set of O13 moment equations.

7.6.3.1 Comparison with Grad Closure

In this section, Onsager consistent closure is compared with the Grad closure. Most of the terms in the expressions for the unknowns r^1_{ik}, w^2, and s_{ijk} (Eqs. (7.100)–(7.102)) have $(\mu/p)^2$ as their coefficient, indicating that these are higher order terms. The coefficient $[k(\gamma-1)/R\gamma p]^2$ can be written as $1/Pr^2(\mu/p)^2$, signifying that the terms with this coefficient are also higher order terms. With this result, Onsager consistent closure to the first order can be written as

$$r^1_{ik} = 7\mu\frac{p}{\rho}\frac{\partial u_{\langle i}}{\partial x_{k\rangle}} \tag{7.105}$$

$$w^2 = (50-30\gamma)\mu\frac{p}{\rho}\frac{\partial u_j}{\partial x_j} \tag{7.106}$$

$$s_{ijk} = 0 \tag{7.107}$$

The comparison of Onsager consistent closure to the first order with the Grad closure is shown in Table 7.4. Note that the term $\mu\partial u_{\langle i}/\partial x_{k\rangle}$ simplifies to p_{ik} to the first order and unknown w^2 reduces to zero using $\gamma = 5/3$ for a monatomic gas. Therefore, Onsager consistent closure to the first order agrees with the Grad closure which enhances confidence in the O13 moment equations.

Including second order terms [$(\mu/p)^2$ terms] in the constitutive relationships for r^1_{ik}, w^2, and s_{ijk}, we get additional terms in the O13 moment equations when compared with Grad 13-moment equations. These terms include products of

Table 7.4 Comparison of Onsager consistent closure to the first order with the Grad closure. Note that the Grad equations are for monatomic gases

Unknowns	Grad closure	Onsager consistent closure
r^1_{ik}	$7RTp_{ik}$	$7\mu\frac{p}{\rho}\frac{\partial u_{\langle i}}{\partial x_{k\rangle}} = 7RTp_{ik}$
w^2	0	$(50-30\gamma)\mu\frac{p}{\rho}\frac{\partial u_j}{\partial x_j} = 0$ for $\gamma = \frac{5}{3}$
s_{ijk}	0	0

derivatives of velocity, temperature, and density and with inclusion of these higher order terms, the applicability of O13 equations is expected to be higher than Grad 13-moment equations.

Since we have first order derivatives of stress tensor and heat flux vector (as opposed to second order derivatives in R13 equations), we need fewer boundary conditions as compared to R13 equations. This point becomes clear when we compare the constitutive relationships obtained for both the cases. With respect to R13 moment equations, we note the presence of stress and heat flux terms in the expressions for Δ, Q_{ij}, and m_{ijk} (Eqs. (7.49)–(7.51)). However, in case of O13 moment equations, we have only first order derivatives of velocity, pressure, and temperature in the expressions for r^1_{ik}, w^2, and s_{ijk}, thereby, reducing the need of prescribing additional boundary conditions.

7.6.4 *Summary*

We presented a derivation of Onsager consistent distribution function which satisfies the compatibility conditions, H-theorem, and linearized Boltzmann equation. The form of the distribution function is such that the Onsager's symmetry principle is preserved, an important and rigid constraint that was somehow ignored in the previous derivations. This form of distribution function is then utilized to derive OBurnett and O13 equations. With all the significant features the Onsager consistent distribution function offer, it is safe to say that the derivation of OBurnett and O13 moment equations has strong roots in the fundamental principles of the non-equilibrium thermodynamics and is believed to capture the strong non-equilibrium effects at higher Knudsen numbers.

7.7 Comparison of Various Forms of Equations

In this section, we compare some of the available variants of Burnett and Grad equations for two relatively simple problems for which analytical (or semi-analytical) solution is possible.

7.7.1 *Couette Flow*

The various forms of the Burnett equations have been compared for some specific problems. Here we will follow the analysis of Singh et al. [133] and compare the results for the Couette flow problem. The solution is further compared with the Grad's equations and R13 equations.

Table 7.5 Value of A in Eq. (7.108)

Higher order continuum model	A
Burnett equations	0
Augmented Burnett equations	$\frac{1}{6}Kn^2$
Super Burnett equations (Agarwal et al. [2])	Kn^2
Super Burnett equations (Shavaliyev [129])	$-\frac{2}{3}Kn^2$
Grad's equations	0
R13 equations	$\frac{9}{5}Kn^2$

The non-dimensional governing equations involving the shear stress (p_{xy}) can be written in the following general form:

$$p_{xy} = -\frac{du}{dy} + A\frac{d^3u}{dy^3} \tag{7.108}$$

where the value of A varies with the type of the equation; its value is given in Table 7.5. The above equation was solved by setting the value of velocity u and its curvature to be zero at the centerline (due to symmetry), and with a slope of α at the centerline. The value of α was determined from DSMC calculation.

The solution of the above equation can be readily obtained as:

$$\begin{aligned} u &= p_{xy}y + \sqrt{A}(p_{xy} + \alpha)\sinh\left(\frac{y}{\sqrt{A}}\right) \quad \text{for } A \geq 0 \\ u &= p_{xy}y + \sqrt{A}(p_{xy} + \alpha)\sin\left(\frac{y}{\sqrt{A}}\right) \quad \text{for } A < 0. \end{aligned} \tag{7.109}$$

In the above equation, p_{xy} is non-dimensional shear stress whose value is evaluated as the harmonic mean of values in the continuum and free-molecular regimes, as

$$p_{xy} = \left(1 + 2\left(\frac{2-\sigma_v}{\sigma_v}\sqrt{\frac{2}{\pi}}Kn\right)\right)^{-1} \quad \text{(where } \sigma_v = 1\text{)}. \tag{7.110}$$

The value of p_{xy} and α depends only on Kn. Therefore, the solution from different equations can be readily compared for different values of Knudsen number. The DSMC data serves as the benchmark against which the obtained analytical solutions are compared.

Figures 7.4 and 7.5 show such comparison for different sets of equations. Note that the equations are not able to capture accurately the non-linearity in the velocity profile (Knudsen layer) observed in the near wall region.

7.7.2 *Force-Driven Poiseuille Flow*

We now consider flow between two parallel plates, with the flow driven by a constant external force, a [78]. Normally such flows are pressure driven; however,

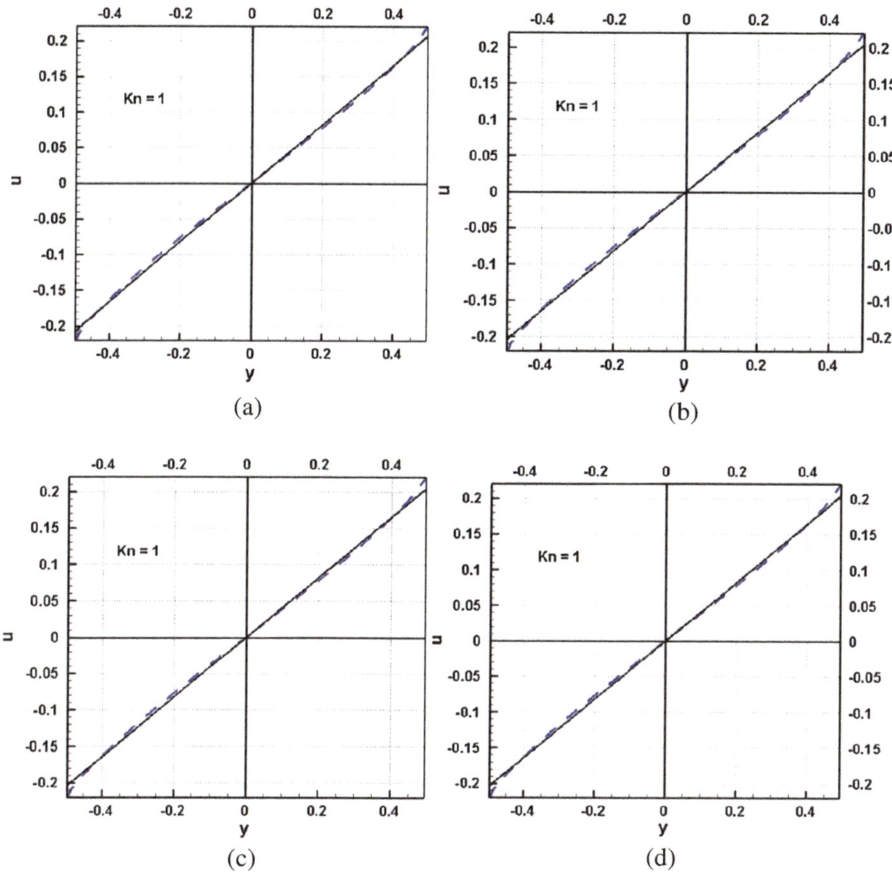

Fig. 7.4 Velocity distribution across the channel for Couette flow at $Kn = 1$ for (**a**) augmented Burnett equations [177], (**b**) super-Burnett equations [2], (**c**) super-Burnett equations [129] and (**d**) R13 equations; dashed line represents the DSMC data (all figures taken from Singh et al. [133])

the solution of pressure driven flow is difficult to solve analytically (pressure being the function of two space-coordinates, x and y), and therefore we resort to an external force driving the flow. As in Poiseuille flows, we consider the plates to be infinitely long and the flow to be steady. See Fig. 7.6 for additional details of the problem setup.

The analysis proceeds by making the assumptions suggested by Uribe and Garcia [154]: the velocity normal to the stationary walls, $v = 0$, and further that the flow variables are functions of y-direction only. The latter assumption can be justified due to the flow being fully developed. With these assumptions, the basic conservation laws reduce to

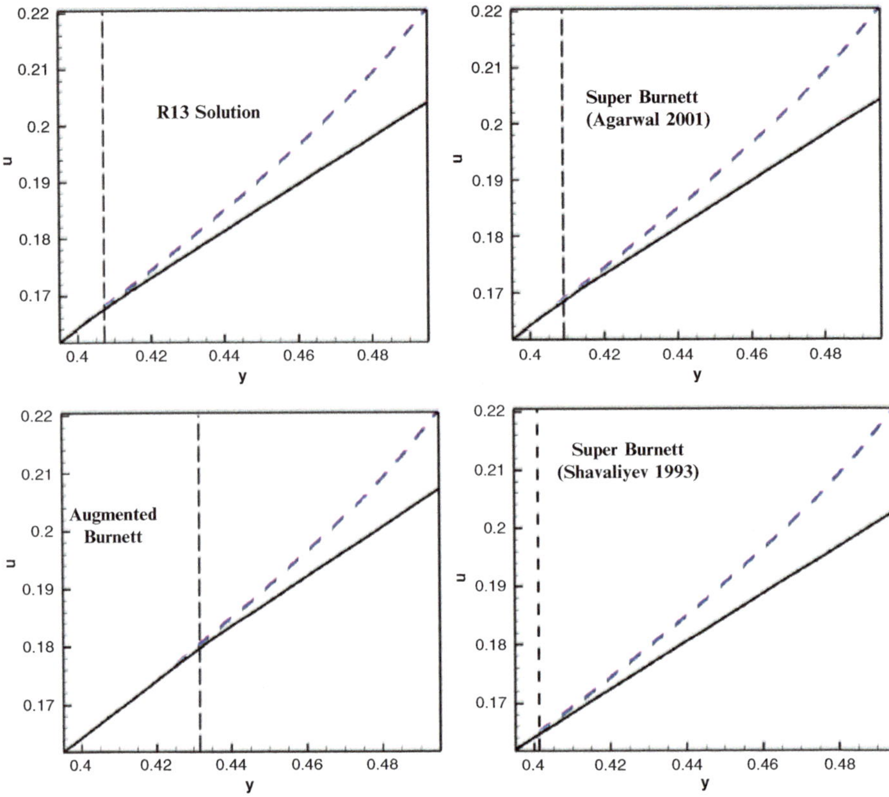

Fig. 7.5 Zoomed in view of the velocity profile in the near wall region according to R13 equations, super-Burnett equations [2], augmented Burnett equations [177] and super-Burnett equations [129]; dashed line represents the DSMC data (all figures taken from Singh et al. [133])

Fig. 7.6 Schematic of plane Poiseuille flow driven by an external force, "a"

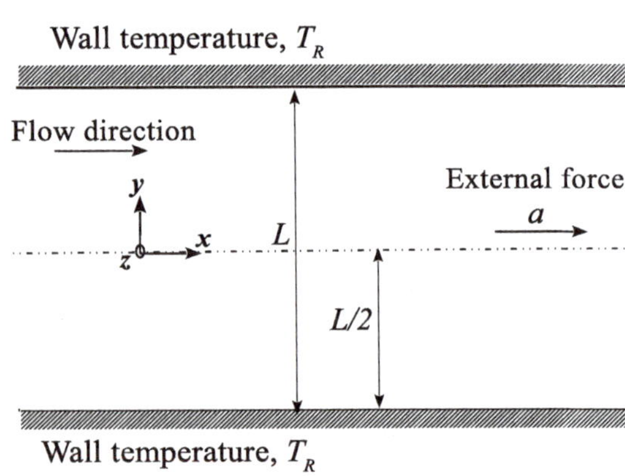

$$\frac{dp_{xy}}{dy} = \rho a, \qquad \text{[}x\text{-momentum equation]} \tag{7.111}$$

$$\frac{dp}{dy} + \frac{dp_{yy}}{dy} = 0, \qquad \text{[}y\text{-momentum equation]} \tag{7.112}$$

$$\frac{dq_2}{dy} + p_{xy}\frac{du}{dy} = 0. \qquad \text{[energy equation]} \tag{7.113}$$

Since the constitutive relationships are different for the Navier–Stokes, conventional Burnett, and OBurnett equations, the analysis differs from this point onwards.

1. Navier–Stokes framework: According to the linear constitutive relationships of the Navier–Stokes equations, we obtain

$$p_{xy} = -\mu\frac{du}{dy}, \quad p_{yy} = 0, \quad q_y = -k\frac{dT}{dy}$$

   Hence, the closed form of the equations is obtained as

$$-\frac{d}{dy}\left(\mu\frac{du}{dy}\right) = \frac{p}{RT}a,$$

$$\frac{dp}{dy} = 0,$$

$$\frac{d}{dy}\left(k\frac{dT}{dy}\right) + \mu\left(\frac{du}{dy}\right)^2 = 0,$$

   where we have utilized the ideal gas equation $p = \rho RT$ to replace density in the x-momentum equation.

2. Conventional Burnett equations: In conventional Burnett equations, the x-momentum and the energy equation remain unchanged. However, because of non-zero Burnett contribution to normal stress p_{yy} in Eq. (7.112), we get closed form of equations as

$$-\frac{d}{dy}\left(\mu\frac{du}{dy}\right) = \frac{p}{RT}a, \tag{7.114}$$

$$\frac{dp}{dy} + \frac{dp_{yy}^{CBu}}{dy} = 0, \tag{7.115}$$

$$\frac{d}{dy}\left(k\frac{dT}{dy}\right) + \mu\left(\frac{du}{dy}\right)^2 = 0. \tag{7.116}$$

where p_{yy}^{CBu} (with superscript depicting conventional Burnett equations) is given as

$$p_{yy}^{CBu} = \frac{2}{3}\frac{\omega_2\mu^2}{p\rho}\frac{dp}{dy}\frac{d\rho}{dy} - \frac{2}{3}\frac{\omega_2\mu^2}{p\rho}\frac{d^2p}{dy^2} - \frac{2}{3}\frac{\omega_2\mu^2}{p}\left(\frac{du}{dy}\right)^2 + \frac{2}{3}\frac{\omega_3\mu^2}{\rho T}\frac{d^2T}{dy^2}$$
$$+ \frac{2}{3}\frac{\omega_4\mu^2}{pT\rho}\frac{dp}{dy}\frac{dT}{dy} + \frac{2}{3}\frac{\omega_5\mu^2}{\rho T^2}\left(\frac{dT}{dy}\right)^2 + \frac{1}{12}\frac{\omega_6\mu^2}{p}\left(\frac{du}{dy}\right)^2 \tag{7.117}$$

Note that the Burnett contribution to shear stress p_{xy} and normal heat flux q_y is zero.

3. OBurnett equations: Again, in case of OBurnett equations, the x-momentum and the energy equation remain unchanged. However, the Burnett contribution to the normal stress p_{yy} is particularly different as compared to that of conventional Burnett equations and given as

$$p_{yy}^{OBu} = 4\frac{\mu^2\beta}{\rho}\left[\alpha_7\left(\frac{du}{dy}\right)^2\right]$$

(with superscript depicting OBurnett equations). The closed set of equations is then obtained as [78]

$$-\frac{d}{dy}\left(\mu\frac{du}{dy}\right) = \frac{p}{RT}a, \tag{7.118}$$

$$\frac{dp}{dy} + \frac{d}{dy}\left[2\alpha_7\frac{\mu^2}{p}\left(\frac{du}{dy}\right)^2\right] = 0, \tag{7.119}$$

$$\frac{d}{dy}\left(k\frac{dT}{dy}\right) + \mu\left(\frac{du}{dy}\right)^2 = 0. \tag{7.120}$$

The analysis is performed by considering the temperature dependence of the transport coefficients. Accordingly, the dynamic viscosity (μ) and thermal conductivity (k), for a dilute gas of rigid spheres are given as [31],

$$\mu = \frac{5c_\mu}{16\sigma^2}\left(\frac{mkT(y)}{\pi}\right)^{1/2}, \quad k = \frac{75c_\lambda}{64\sigma^2}\left(\frac{k^3T(y)}{\pi m}\right)^{1/2}, \tag{7.121}$$

where σ is the particle diameter and accurate values of $c_\mu = 1.016034$ and $c_\lambda = 1.02513$ are known.

The variables are non-dimensionalized as

$$p^*(s) = \frac{p(y)}{p(0)}, \quad T^*(s) = \frac{T(y)}{T_R}, \quad u^*(s) = \frac{u(y)}{(2kT_R/m)^{1/2}}, \text{ where } s = \frac{y}{L}. \tag{7.122}$$

Taking into account the temperature dependence of the transport coefficients and performing non-dimensionalization, we obtain coupled ordinary differential equations as

1. Navier–Stokes framework:

$$\frac{d^2u^*}{ds^2} = -\frac{1}{2T^*}\frac{dT^*}{ds}\frac{du^*}{ds} + \frac{b_0' p^*}{T^{*3/2}}, \tag{7.123}$$

$$\frac{dp^*}{ds} = 0 \tag{7.124}$$

$$\frac{d^2T^*}{ds^2} = -\frac{8c_\mu}{15c_\lambda}\left(\frac{du^*}{ds}\right)^2 - \frac{1}{2T^*}\left(\frac{dT^*}{ds}\right)^2. \tag{7.125}$$

where the coefficient b_0' is given as

$$b_0' = -\frac{8}{5}\frac{L^2 a\sqrt{2\pi}\,p(0)m\sigma^2}{c_\mu k^2 T_R^2}, \tag{7.126}$$

2. Conventional Burnett equations:

$$\frac{d^2u^*}{ds^2} = -\frac{1}{2T^*}\frac{dT^*}{ds}\frac{du^*}{ds} + \frac{b_0' p^*}{T^{*3/2}}, \tag{7.127}$$

$$\begin{aligned}\frac{d^2p^*}{ds^2} = {} & \frac{b_1 p^{*2}}{T^{*2}} + \frac{b_2 p^{*2}}{T^{*2}} - \frac{b_1 p^{*3}}{T^{*2}} + \frac{1}{p^*}\left(\frac{dp^*}{ds}\right)^2 - \frac{1}{T^*}\frac{dp^*}{ds}\frac{dT^*}{ds} \\ & - \frac{2p^*}{T^*}\left(\frac{du^*}{ds}\right)^2 - \frac{8\omega_3 p^* c_\mu}{15\omega_2 T^* c_\lambda}\left(\frac{du^*}{ds}\right)^2 - \frac{\omega_3 p^*}{2\omega_2 T^{*2}}\left(\frac{dT^*}{ds}\right)^2 \\ & + \frac{\omega_4}{\omega_2 T^*}\frac{dp^*}{ds}\frac{dT^*}{ds} + \frac{\omega_5 p^*}{\omega_2 T^{*2}}\left(\frac{dT^*}{ds}\right)^2 + \frac{\omega_6 p^*}{4\omega_2 T^*}\left(\frac{du^*}{ds}\right)^2 \end{aligned} \tag{7.128}$$

$$\frac{d^2T^*}{ds^2} = -\frac{8c_\mu}{15c_\lambda}\left(\frac{du^*}{ds}\right)^2 - \frac{1}{2T^*}\left(\frac{dT^*}{ds}\right)^2. \tag{7.129}$$

where the coefficients b_0', b_1, and b_2 are given as

$$b_0' = -\frac{8}{5}\frac{L^2 a\sqrt{2\pi}\,p(0)m\sigma^2}{c_\mu k^2 T_R^2}, \quad b_1 = -\frac{384}{25}\frac{L^2 p(0)^2\pi\sigma^4}{\omega_2 c_\mu^2 k^2 T_R^2},$$

$$b_2 = \frac{L^2 \left.\frac{d^2p}{dy^2}\right|_{s=0} T(0)^2}{p(0)T_R^2} \tag{7.130}$$

3. OBurnett equations:

$$\frac{d^2u^*}{ds^2} = -\frac{1}{2T^*}\frac{dT^*}{ds}\frac{du^*}{ds} + \frac{b_0' p^*}{T^{*3/2}}, \tag{7.131}$$

$$\frac{dp^*}{ds} = \frac{\left\{ -\frac{2T^*}{p^*}\frac{du^*}{ds}\left[-\frac{1}{2T^*}\frac{dT^*}{ds}\frac{du^*}{ds} + \frac{b_o' p^*}{T^{*3/2}} \right] + \left(\frac{du^*}{ds}\right)^2 \frac{1}{p^*}\frac{dT^*}{ds} \right\}}{\left\{ b_2' - \frac{T^*}{p^{*2}}\left(\frac{du^*}{ds}\right)^2 \right\}}, \tag{7.132}$$

$$\frac{d^2T^*}{ds^2} = -\frac{8c_\mu}{15c_\lambda}\left(\frac{du^*}{ds}\right)^2 - \frac{1}{2T^*}\left(\frac{dT^*}{ds}\right)^2. \tag{7.133}$$

The coefficients are

$$b_0' = -\frac{8}{5}\frac{L^2 a\sqrt{2\pi}\,p(0)m\sigma^2}{c_\mu k^2 T_R^2}, \quad b_2' = \frac{64}{25}\left(\frac{p(0)L}{kT_R}\frac{\sigma^2}{c_\mu}\right)^2 \frac{\pi}{\alpha_7}, \tag{7.134}$$

Solving this problem of coupled differential equations as an initial value problem, the results of Navier–Stokes equations, conventional Burnett equations, OBurnett equations, and NCCR theory [52, 103] are validated against the direct simulation Monte Carlo (DSMC) and molecular dynamics (MD) simulations results as shown in Fig. 7.7. An important point to note is that the Burnett equations (which are second order accurate in Knudsen number) are not able to capture the temperature dip at the center which is a super-Burnett effect. Another important observation is the bi-modal pressure profile predicted by the conventional Burnett equations and confirmed by the DSMC results. However, we get monotonic pressure profile according to OBurnett equations, NCCR theory, and confirmed by more accurate MD results. Hence, we note that, even for a simpler looking one-dimensional flow problem like force-driven Poiseuille flow, there are many subtleties in the flow physics and the results of different equations set sometimes give qualitatively different results.

Once we obtain the solution for velocity, temperature, and pressure, the results for stresses and heat fluxes can be readily obtained and they are shown in Fig. 7.8. For more mathematical details of these variables, interested readers can refer to Uribe and Garcia [154] in case of conventional Burnett equations, Jadhav et al. [78] in case of OBurnett equations, and Myong [104] in case of NCCR equations. For stresses and heat fluxes, the results of the conventional Burnett equations, OBurnett equations, and NCCR theory are qualitatively similar and in reasonable agreement with the DSMC and MD results but discrepancies are observed in the near-wall region. It is important to emphasize the failure of the Navier–Stokes equations which essentially predict the normal stresses (p^*_{xx}, p^*_{yy}, p^*_{zz}) and tangential heat flux (q^*_1) to be zero, as clearly evident from Figs. 7.8b, c, d and f.

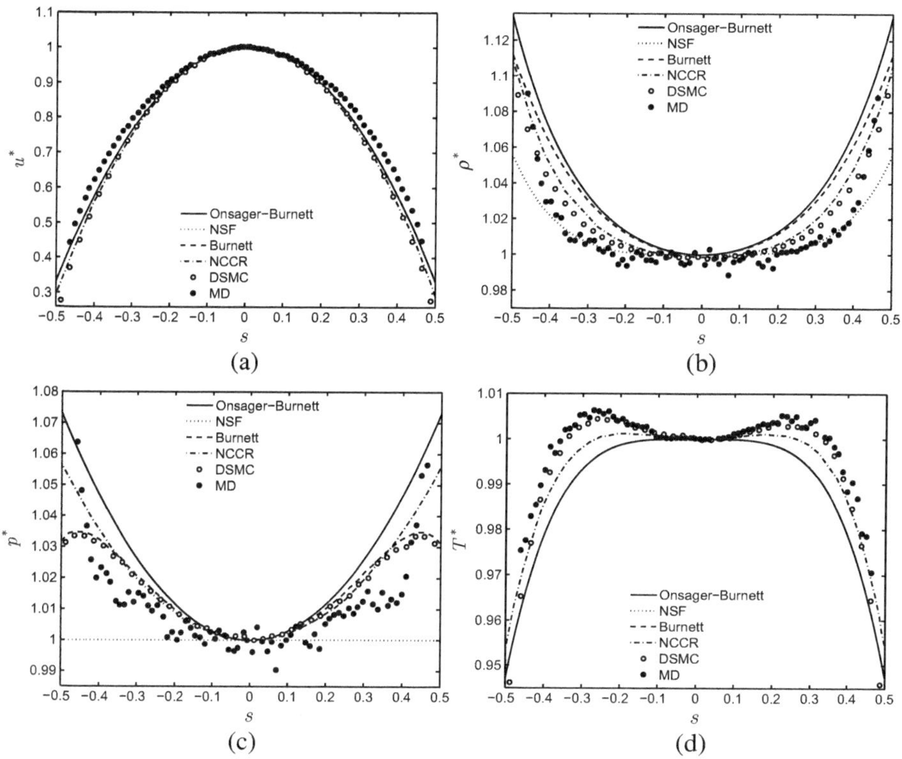

Fig. 7.7 Variation of (**a**) velocity, (**b**) density, (**c**) pressure, and (**d**) temperature in the cross stream direction in force-driven compressible plane Poiseuille flow at Kn = 0.1. Comparison is made for OBurnett equations, Navier–Stokes equations, conventional Burnett equations, and NCCR equations and validation is performed against the DSMC and MD results. Figures taken from Jadhav et al. [78]

7.8 Summary

In this chapter, the need for looking at variants of the Burnett and Grad equations is first explained and available variants of these equations are included. While several of these variants are obtained in a somewhat ad hoc manner, thermodynamically consistent Burnett-*like* and Grad-*like* equations have also been recently proposed.

These new equations being consistent with Onsager's principle are called OBurnett and O13 equations, respectively. The equations have been applied to two specific cases and satisfactory agreement of the available variants against benchmark data is noted. However, the results from these cases cannot be generalized in that if a particular variant works better in a given case than the others, it does not imply that variant is actually better. In fact, we urge the readers to apply the various available variants for their specific problem, and also try to arrive at better versions of these equations.

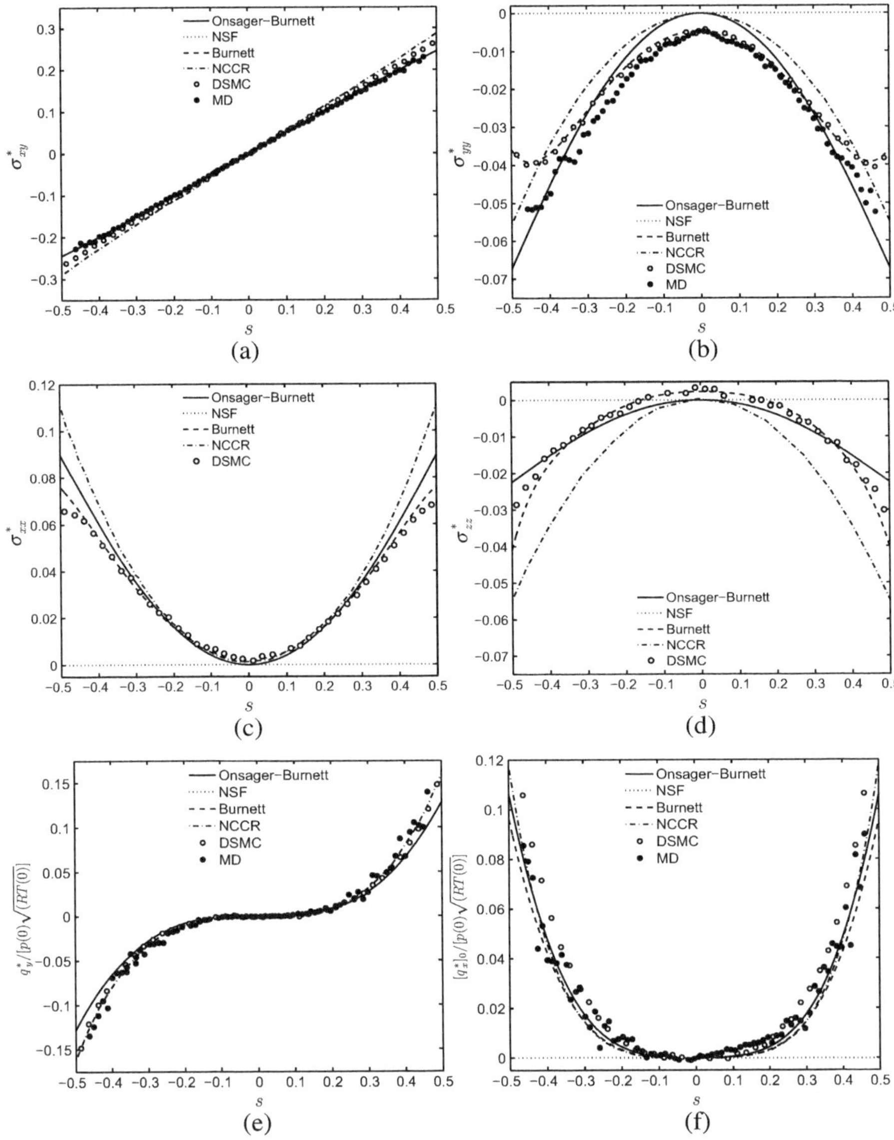

Fig. 7.8 Variation of (**a**) shear stress p^*_{xy}, (**b**) normal stress p^*_{yy}, (**c**) normal stress p^*_{xx} (**d**) normal stress p^*_{zz}, (**e**) normal heat flux q^*_2, and (**f**) tangential heat flux q^*_1 in the cross stream direction in force-driven compressible plane Poiseuille flow at Kn = 0.1. Comparison is made for OBurnett equations, Navier–Stokes equations, conventional Burnett equations, and NCCR equations and validation is performed against the DSMC and MD results. Note that stresses are represented as σ's. Figures taken from Jadhav et al. [78]

Chapter 8
Overview to Numerical and Experimental Techniques

In the previous chapters, the governing equation and analytical solution of the equation was presented for flows in relatively simple geometries. However, problems encountered in most practical situations are much more difficult, involving complex geometry, unclear boundary conditions, transition from one flow regime to another, and other complications. Obtaining an analytical solution in these cases is virtually not possible. One therefore resorts to either numerical methods for solution of the governing equation or experiments. In this chapter, we briefly comment on these techniques, while details can be found in specialized books and papers on the subject.

8.1 Numerical Approach

In this section, we provide a brief overview of the popular DSMC and molecular dynamics techniques. The discussion on these techniques is far from exhaustive.

8.1.1 Direct Simulation Monte Carlo Method

The Direct Simulation Monte Carlo (DSMC) method is the preferred choice for numerically simulating flow in microchannels, and as seen in previous chapters, analytical results in the high Knudsen number range are often benchmarked against DSMC data. In fact, it will perhaps not be an overstatement, that DSMC data virtually enjoys the same status as experimental data does in any other field! This statement also simultaneously underscores the absence of high-quality detailed experimental data in rarefied gas flows. DSMC is a particle based probabilistic numerical technique invented by Bird [21] and is particularly suited for simulation

A. Agrawal et al., *Microscale Flow and Heat Transfer*, Mechanical Engineering Series, https://doi.org/10.1007/978-3-030-10662-1_8

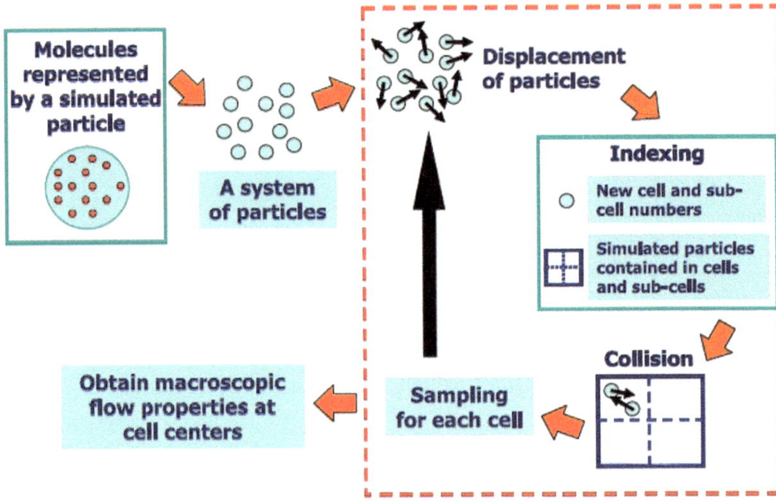

Fig. 8.1 The molecular direct simulation Monte Carlo method (DSMC) for low-Reynolds-number flows in the transition regime, Zhao et al. [175]

of dilute gases. Unlike conventional numerical techniques, DSMC does not solve a set of partial differential equations, rather simulates the movement and collision of particles on a grid. The model therefore has some similarity with the molecular nature of matter, with each simulated molecule representing a number of real molecules. The ratio of real to simulated molecules is a simulation variable, and has to be fine-tuned for each simulation case. Similar is the case with the number of cells in which the entire computational domain is divided into. Each cell is further sub-divided into sub cells such that the particles which are nearer to each other and within a distance of one mean free path are chosen for collision, so as to achieve physically consistent simulations while reducing the simulation time for choosing the particle pair for collision. A step by step implementation of DSMC method is shown in Fig. 8.1.

Note that the movement and collision of the particles in this method is uncoupled. That is, all particles are moved simultaneously to the next location as per the corresponding particle velocity, and then all collisions happen between the particle pairs selected on probabilistic basis. The time step involved in the computation is a fraction of the mean collision time between particles. Similarly, the cell size should be a fraction (typically one-third) of the mean free path of the gas. The gas–wall interaction can be modelled as either diffuse or specular (or some fraction thereof).

The physical variables (density, velocity, temperature) and even higher order moments (stress and heat flux) can be obtained by taking time averaging of the appropriate quantities (i.e., using the equations for these quantities in terms of distribution function introduced in Chap. 5) for steady flows. Unsteady flow simulations however tend to become expensive as ensemble averaging of various quantities is required, for which multiple simulations have to be performed with different (random) initial conditions.

It has been shown that DSMC solves the Boltzmann equation in the limit of small time step, small grid size, and large number of simulated molecules. It is further claimed [21] that the method does not have some of the restrictions of the Boltzmann equation, and can therefore yield physical effects that cannot be modelled even within the framework of the Boltzmann equation.

8.1.2 Alternative Techniques

Another popular alternative is Molecular Dynamics (MD) method. Unlike DSMC, MD is a deterministic method particularly suited for dense gases. The method involves solution of the Newton laws of motion as applied to gas particles, with the force acting on the particles being determined by the inter-molecular potential between the particles.

The governing equations provided in Chaps. 5–7 can be solved using standard numerical simulation techniques of computational fluid dynamics. Since these models involve higher-order terms of Knudsen number, the envelope of Knudsen number over which these equations apply is larger than the Navier-Stokes equations. These numerical results can eventually supplement DSMC data especially in the slip and early transition regimes, regimes where DSMC computations are relatively expensive. On the other hand, DSMC computations are particularly well suited to simulate flows in the Knudsen number range of 0.01–10, and most simulations in the literature are in this range of Knudsen number.

8.2 Experimental Approach

As seen earlier, microscale gaseous flows are characterized by relatively high Knudsen numbers. In practice, high Knudsen number can be realized either by reducing the length scale or by increasing the mean free path of the gas. Both these approaches have been followed in the literature while performing experiments on these flows.

8.2.1 Measurement at the Microscale

A typical microchannel would typically have a width of 60–200 μm, depth of 15–100 μm, and length of 300–2000 μm. The ability to first fabricate such microchannels and then instrument them are the two important prerequisites for performing measurements at the microscale. Note that in order to find the variation of any quantity along either the depth or width, the spatial resolution of the instrument measuring that quantity should be about an order of magnitude smaller than these dimensions. Further, needless to emphasize, the probe introduced should not disturb/alter the flow.

Therefore, measurement of mass flow rate, pressure, and temperature is more straightforward as these can be done outside the microchannel, while measurement of local flow velocity or wall shear stress is challenging as the probe has to be introduced into the microchannel in these cases. Again, measurement of pressure or temperature at the inlet and outlet reservoirs is much more straightforward, as compared to measurement of local pressure or temperature in the microchannel.

In order to measure the local pressure, taps can be added along the microchannel length as shown in Fig. 8.2a. Figure 2.12c shows the obtained pressure values from six pressure taps introduced by Pong et al. [88]. This data was subsequently employed by numerous researchers to benchmark their numerical and theoretical results. Wholefield pressure in the microchannel has also been measured using pressure sensitive paint by Huang et al. [76]. The arrangement of their setup is shown in Fig. 8.2b and a sample result in Fig. 8.2c.

Thermocouples remain the preferred choice for measurement of temperature. Thermocouples are typically mounted in the inlet/outlet reservoirs for obtaining the inlet and outlet temperatures of the fluid, and on the outer wall of the microchannel for measuring the local temperature. Of course, these measurements require extremely fine thermocouples beads; 25 μm bead size thermocouples are now commercially available. Use of infrared camera, which can provide whole-field temperature information, besides being non-intrusive, is another commonly employed measurement technique. The surface whose temperature is to be measured is coated with a paint of known emissivity for obtaining reliable data.

Micro-PIV (particle image velocimetry) is the preferred technique for obtaining velocity field in a microchannel. Here, fine oil droplets are added to the flow; these droplets scatter light (usually from twin Nd:YAG pulsed lasers), which is collected by a camera (or CCD). Two such images are collected and correlated to obtain the displacement of the droplets. Finally, knowing the time delay between the images gives the velocity of the droplets. From the assumption that the velocity of the droplet is equal to velocity of the fluid (an assumption that is satisfied when the Stokes number of the flow is small), one obtains the wholefield velocity information. However, although micro-PIV is a well-established experimental technique, there are only a few studies reporting measurement of gas flow in a microchannel, primarily because of the difficulty in carriage of oil-droplets by the flow at these small scales. However, measurement of gas flow rate is much more standard, and involves the use of mass flow controller of appropriate range and accuracy.

The microchannels can be fabricated using wet-etching and photolithography processes. An example process flow involved in the fabrication process is shown in Fig. 8.3. The material of the substrate on which microchannels are fabricated varies from thermally conducting materials such as silicon and copper to nonconducting materials such as polydimethylsiloxane (PDMS) and poly(methyl methyl acrylate) (PMMA). Good thermal conductivity of the substrate is obviously required in heat transfer studies. Closing of the microchannel is another aspect which has to be carefully handled. For this, plasma-bonding and PDMS-PDMS bonding are some of the options.

Fig. 8.2 (a) A close-up picture of transition section of a tested device with an included angle of 180°, Lee et al. [86]. (b) Schematic showing pressure sensitive paint (PSP) measurement, Huang et al. [76]. (c) Pressure map inside a microchannel with pressure ratio 4.74, Huang et al. [76]

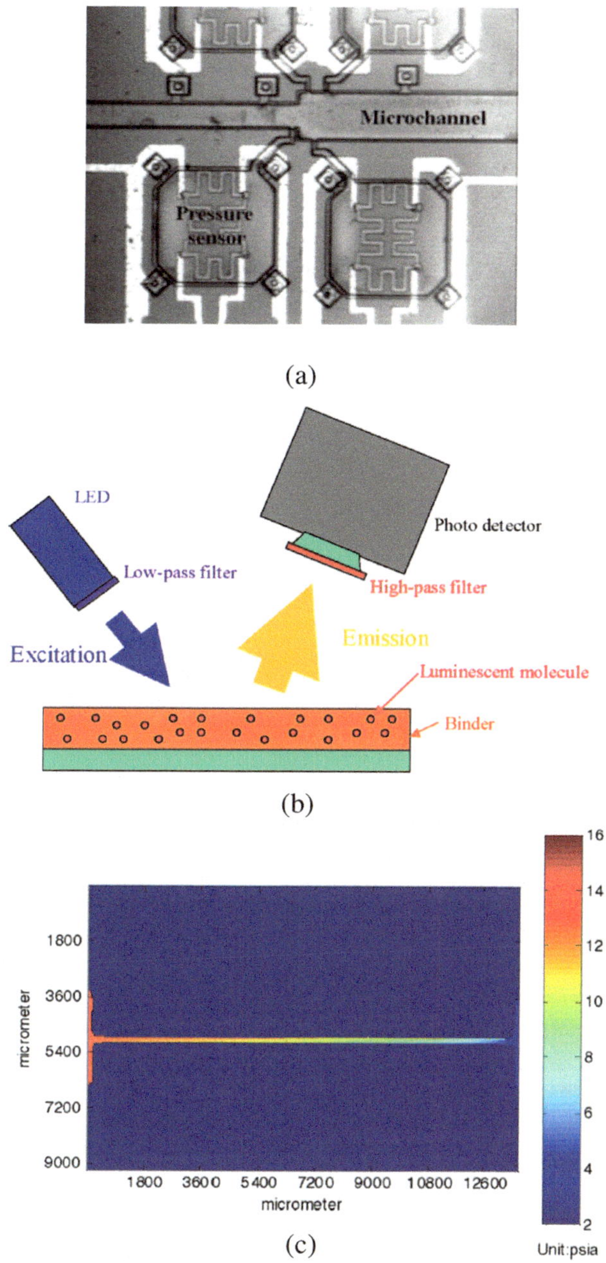

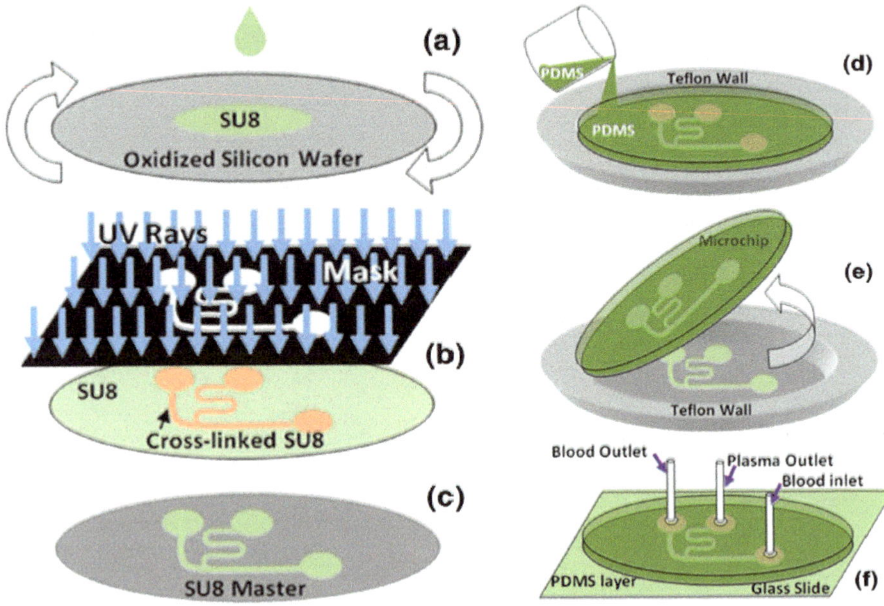

Fig. 8.3 Fabrication procedure for microdevices, (**a**) SU8 spin coating on oxidized silicon wafer, and pre-exposure bake, (**b**) exposure, and post-exposure bake and (**c**) development, (**d**) casting of PDMS on SU8 master, (**e**) peeling off PDMS chip from SU8 master, (**f**) spin coating of PDMS on glass slide, Prabhakar et al. [118]

8.2.2 Measurement in Rarefied Gas Flows

Working with reduced pressure (to increase the mean free path of the gas) while employing a scaled up facility (flow passage in conventional size range) makes measurements relatively easier [41, 137, 159]. Instruments for measurement of low pressure and low flow rate along with vacuum pumps required in such experiments are well developed and easily available in various ranges. The accuracy of vacuum pressure gauges and mass flow controllers can be very high, which helps in obtaining high-quality repeatable data.

Leakage of ambient air into the test section is a concern that has to be carefully addressed in order to obtain reliable data. There are various procedures that can be used to estimate the amount of leakage—including use of helium detectors, monitoring the rate of increase of pressure in the system after shutting off all the valves, etc.

The use of pitot tube is a cheaper and more convenient alternative than micro-PIV, although it is an intrusive and point-wise measurement technique. Pitot tube based measurements basically rely on the Bernoulli equation for determining the velocity from the measured pressures; however, the use of the Bernoulli equation in the case of rarefied gas can lead to substantial error. This is because the relatively

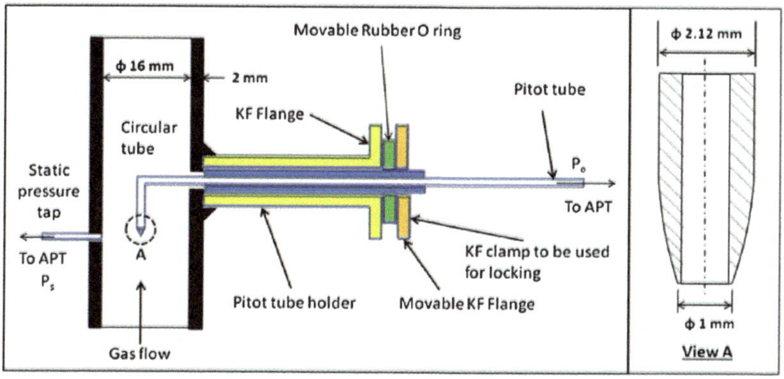

(a)

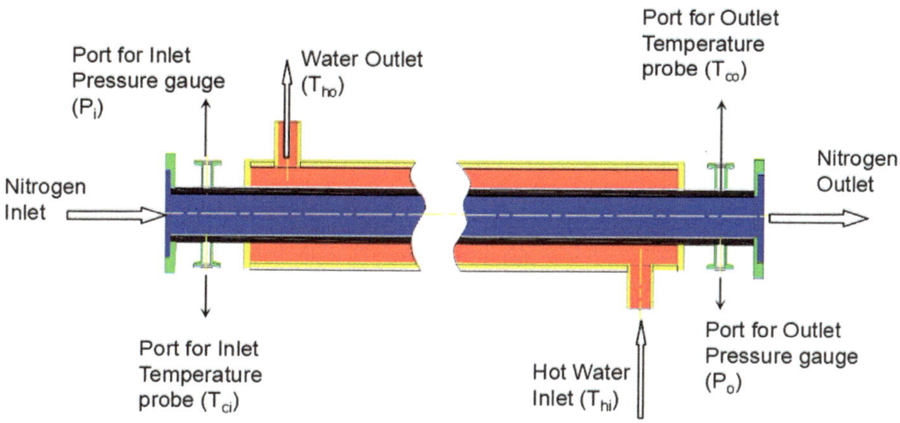

(b)

Fig. 8.4 (**a**) Schematic diagram of a pitot tube, Varade et al. [159]. (**b**) Schematic of a tube-in-tube heat exchanger, Demsis et al. [41]

low Reynolds number of the flow alters the flow pattern around the tube and slip at the walls, which change the pressure field at the tip of the tube. Recently, Varade et al. [159] showed that reliable measurements are possible with appropriate corrections to the Bernoulli equation for these viscous and rarefaction effects. The arrangement of pitot tube employed in this work is shown in Fig. 8.4a.

Measurement of local pressure [137] and local temperature [73, 156–158] is possible because of the relatively large passage size. A tube-in-tube counter-flow heat exchanger for measurement of heat transfer coefficient is shown in Fig. 8.4b. Attempts have also been made to obtain the local flow velocity by employing a hot wire. Providing an optical window and undertaking laser based measurements can provide detailed information about the flow, although such measurements have rarely been reported.

8.3 Summary

The purpose of this brief chapter is to provide a quick overview of the available options for performing numerical or physical experiments, with flows in the high Knudsen number regime. We noted that DSMC is a well-established numerical technique, while numerical solution of the higher-order models can further provide another compelling alternative way to study rarefied gas flows in complex situations. There is lot of possibility for improving the instrumentation for local flow measurements with substantially higher spatial resolution. There is also a rich possibility of developing optical based measurement techniques for measurement in low pressure gas flows.

Chapter 9
Summary and Future Research Directions

In the various chapters of this book, we have covered several different aspects of gas flow and heat transfer in the slip and transition regimes. In this last chapter, we present a broad summary of what is well understood and what remains to be better explored. We also identify some important research problems which can help in further development of this subject.

We have considered flow in simple (straight) and complex (involving change in cross sectional, flow direction, etc.) passages using both analytical and/or experimental/simulation tools. Therefore, there are four broad categories in which these various scenarios can be classified: *slip* or *transition* regime, *simple* or *complex* passages, *fluid flow* or *heat transfer* aspects, and *analytical* or *experimental/simulation* tools. We first summarize what is known (or reasonably well known) about these categories:

- *Analytical* solution for *flow* in *simple* passages in *slip* regime is available.
- *Experimental* and *DSMC* data for *flow* in *simple* passages in *slip* and *transition* regimes is available to a reasonable extent.
- *Analytical* solution for *heat transfer* in *simple* passages in *slip* regime is available. However, there is scope for considering several complicating effects together.
- Several *theoretical* models for describing *flow* and *heat transfer* in the *transition* regime are available. However, the reliability of these models has to be thoroughly tested and a final word on the superset of the Navier-Stokes equations has to be proclaimed.

We next summarize the major gaps (at least there is enough scope for further work in these areas):

- Reliable *experimental* data for *heat transfer* in *slip* regime for *simple* passages is not available.
- There is lot of scope (and need) for obtaining *experimental flow* and *heat transfer* data in the *slip* and *transition* regimes for *complex* passages.

A. Agrawal et al., *Microscale Flow and Heat Transfer*, Mechanical Engineering Series, https://doi.org/10.1007/978-3-030-10662-1_9

We now elaborate upon the above identified gaps by suggesting certain specific problems, which are particularly crucial for further development of this important subject.

1. *More data on slip coefficients*: As reviewed in Sect. 2.3.4, the value of tangential momentum accommodation coefficient or σ_v is available for a number of cases. However, such information has added to the need to resolve the dependency of slip coefficients (C_1 and C_2) on various parameters (esp. Knudsen number).
2. *Do slip coefficients depend on the gas (and surface material)?* This is a particularly crucial question, which has not received the attention that it should. A dependence of slip coefficient on the nature of the gas suggests that the friction factor and Nusselt number will additionally depend on the gas that we are working with, and will therefore substantially complicate the analysis of slip flows. Clearly the question is of fundamental interest as well as relevant from the viewpoint of developing sensors (and other applications) based on this difference.
3. *Data for alternate cross sections*: Although analysis and data are available for flow in straight passages, such information is limited to relatively simple cross sections (such as circular and rectangular). However, cross section of practical microchannels and microdevices is not limited to these simple shapes, rather may be trapezoidal, triangular, rhombic, and even irregular. Therefore, there is scope for obtaining reliable experimental and DSMC data for other cross sections of the flow passage and with surface modifications (esp. surface roughness).
4. *Single effect taken at a time in heat transfer problems*: As already stated, there is scope for considering several complicating effects (temperature jump, viscous dissipation, property variation, axial conduction, conjugate effect, end effects) together to make heat transfer analysis more realistic.
5. *Paucity of experimental heat transfer data*: The need for reliable experimental data is particularly felt in heat transfer because it not *a priori* obvious which of the multiple effects that co-exist can be neglected. This may be reason for the poor agreement between available theoretical and experimental findings. Further, the overall heat transfer coefficient has to be supplemented with local heat transfer coefficient measurements wherever possible.
6. *Development of miniature sensors for local flow and heat transfer measurements*: Having reliable local information is advantageous and therefore this requirement is obvious. The spatial resolution of most current flow and temperature sensors is at best several hundreds of microns; the resolution needs to be reduced to a few microns for local measurements to be meaningful.
7. *Development of sensors for measurement of additional variables*: As mentioned in Chap. 8, the current measurements are limited to overall behavior and a limited number of flow/thermal parameters. The measurements need to be expanded to include variables such as: velocity, wall shear stress, heat flux, etc.

8. *Generation of database of DSMC computations*: It would be useful to perform highly accurate DSMC computations for a range of problems and flow conditions. Any proposed new equations should then be tested against this database. This database is also useful because variability in DSMC computations from various groups exists. The theoretical research group can then focus exclusively on deriving new equations. The same requirement goes for generating high quality experimental data (for corresponding to DSMC database) and even other problems.
9. *Equations should be benchmarked against more involved problems*: Newly proposed equations need to be tested over a wider range of problems, and a wider range of parameter values to establish confidence in them. With the generation of comprehensive database mentioned above in point 8, this should become possible.
10. *Establishment of accurate boundary conditions for higher order models*: As seen in Chaps. 5–7, all higher order models (other than OBurnett equations) involve higher order derivatives and therefore need more number of boundary conditions than the Navier-Stokes equations. Procedure to obtain additional boundary conditions is therefore required.
11. *Dependency of solution on the type of gas needs testing*: As mentioned above, a final word on the correct form of the superset of the Navier-Stokes equations still needs to be proclaimed. Moreover a dependence of the form of the governing equation on the gas is felt (through the inter-molecular potential). Therefore the dependence of boundary condition (commented in point 2 above) and governing equation on the gas needs to be unambiguously resolved.

Appendix A
Appendix to Basic Tensor Algebra

Tensor notation greatly helps to represent equations in compact form. In this brief introduction to tensor algebra, we present the basic rules to be followed along with the general operations performed on tensors. Although this section aims to aid understanding of the tensor algebra employed in the book, the section is a standalone introduction to tensors and therefore other readers should also find it useful.

A.1 Introduction to Scalar, Vector, and Tensor

A scalar quantity can be considered as a zeroth order tensor that is completely defined by *magnitude* only and is independent of coordinate system. A vector is a first order tensor that has *magnitude* and a *direction* and requires three components along with three specified coordinate directions for its complete description. Generally, it is represented by a boldface symbol, for example, **x** for position vector, **u** for bulk velocity, **q** for heat flux vector, etc. In three-dimensional physical space, a vector **x** can be represented as

$$\mathbf{x} = x_1\mathbf{e_1} + x_2\mathbf{e_2} + x_3\mathbf{e_3}$$

where x_1, x_2, x_3 are the components of the vector **x** and $\mathbf{e_1}, \mathbf{e_2}, \mathbf{e_3}$ are the unit vectors in the direction of three mutually perpendicular axes. In tensorial notation, vector **x** is represented as x_i where i is known as the free index and goes from 1 to 3 in a three-dimensional Cartesian coordinate system. Similarly, a unit vector e_i refers to any one of the unit vectors $\mathbf{e_1}$, $\mathbf{e_2}$, or $\mathbf{e_3}$.

According to formal definition, a vector is a quantity whose components change like the components of a position vector under the rotation of the coordinate system [85]. To illustrate this, consider a coordinate system in two dimensions, $O\ 12$ which is rotated to obtain a new coordinate system $O\ 1'2'$ as shown in Fig. A.1. Let the

A. Agrawal et al., *Microscale Flow and Heat Transfer*, Mechanical Engineering Series, https://doi.org/10.1007/978-3-030-10662-1

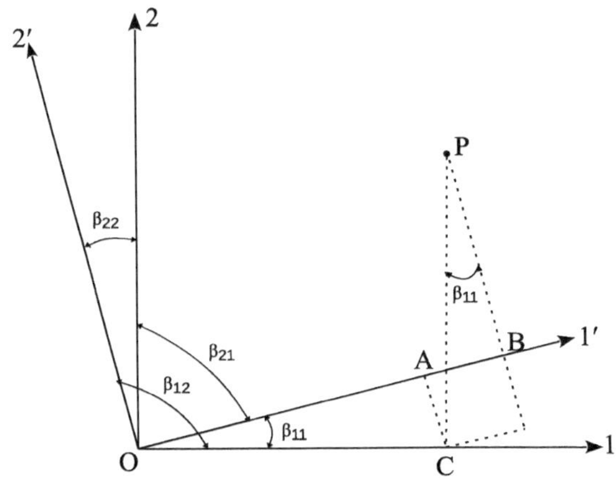

Fig. A.1 Coordinate system O 12 is rotated to O 1′2′

components of a position vector $\mathbf{x}$ in the old and new coordinate systems be x_i and x_i' where i goes from 1 to 2. The cosine of the angle β's between the old i and new j axes is represented by C_{ij}. In matrix form, C_{ij} can be represented as

$$C_{ij} = \begin{bmatrix} C_{11} & C_{12} \\ C_{21} & C_{22} \end{bmatrix} = \begin{bmatrix} \cos\beta_{11} & \cos\beta_{12} \\ \cos\beta_{21} & \cos\beta_{22} \end{bmatrix} = \cos\beta_{ij} \tag{A.1}$$

The first index of a matrix C_{ij} refers to the *old* axes while the second index refers to the *new* axes. Note that C_{ij} is not equal to C_{ji}. The components in the new coordinate system are related to the components in the old system as

$$\begin{aligned} x_1' &= \text{OB} = \text{OA} + \text{AB} \\ &= \text{OC}\cos\beta_{11} + \text{PC}\sin\beta_{11} \qquad [\because \text{AB} = \text{CD}] \\ &= x_1\cos\beta_{11} + x_2\sin\beta_{11} \\ &= x_1\cos\beta_{11} + x_2\cos\beta_{21} \qquad [\beta_{11} = \pi/2 - \beta_{21}, \ \therefore \sin\beta_{11} = \cos\beta_{21}] \\ &= x_1 C_{11} + x_2 C_{21} \end{aligned} \tag{A.2}$$

Similarly, for the second component,

$$x_2' = x_2\cos\beta_{22} + x_1\cos\beta_{12} = x_2 C_{22} + x_1 C_{12} \tag{A.3}$$

where we have used $\beta_{11} = \beta_{22} = \beta_{12} - \pi/2$ and $\sin\beta_{11} = \sin(\beta_{12} - \pi/2) = -\cos\beta_{12}$.

Generalizing to three dimensions, we can write the three components in the new coordinate system as

$$x_1' = x_1 C_{11} + x_2 C_{21} + x_3 C_{31} = \sum_{i=1}^{3} C_{i1} x_i$$

$$x'_2 = x_1 C_{12} + x_2 C_{22} + x_3 C_{32} = \sum_{i=1}^{3} C_{i2} x_i$$

$$x'_3 = x_1 C_{13} + x_2 C_{23} + x_3 C_{33} = \sum_{i=1}^{3} C_{i3} x_i$$

In compact tensorial notation, the above three equations can be represented in a single equation as

$$x'_j = \sum_{i=1}^{3} C_{ij} x_i \tag{A.4}$$

where the index j goes from 1 to 3 and called as free index. The index i is repeated in the same term of the right-hand side and summation operation is carried out for all values of this repeated index. Such an index is called as dummy index and choice of letter for a dummy index is immaterial as long as it is different from a free index. A generally accepted rule is to suppress the summation symbol in a term where an index is appearing twice. Such a rule is called as Einstein summation convention wherein for a repeated index, summation over all three values of that index is implied. With this convention, we can simply write as

$$x'_j = C_{ij} x_i \tag{A.5}$$

Equation (A.5) gives transformation for a position vector in a new coordinate system under the rotation operation. In a similar way, any quantity can be called as a Cartesian vector when its components transform like a position vector under the rotation of the coordinate system. Hence, **u** is a vector if its components transform as

$$u'_j = C_{ij} u_i \tag{A.6}$$

A second order tensor has nine components and two directions associated with it. The stress tensor, strain rate tensor, and velocity gradient tensor are frequently encountered second order tensors in fluid mechanics. In matrix form, a second order tensor can be represented as

$$\tau_{ij} = \begin{bmatrix} \tau_{11} & \tau_{12} & \tau_{13} \\ \tau_{21} & \tau_{22} & \tau_{23} \\ \tau_{31} & \tau_{32} & \tau_{33} \end{bmatrix} \tag{A.7}$$

Hence, for complete specification of a second order tensor, all these nine components must be determined. Similar to the definition of a vector, a transformation rule can be defined for a second order tensor as

$$\tau'_{ij} = C_{kj} C_{lj} \tau_{kl} \tag{A.8}$$

Further generalizing for an nth-order tensor, it has 3^n components which needs to be completely determined for complete specification of the tensor and n directions associated with it. The transformation rule for an nth order tensor can be written as

$$A'_{i_1 i_2 \ldots i_n} = C_{j_1 i_1} C_{j_2 i_2} \ldots C_{j_n i_n} A_{j_1 j_2 \ldots j_n} \tag{A.9}$$

Note that in transformation of a second order tensor, $\cos^2$, product of cos and sin, $\sin^2$ terms are involved. This is because there are two directions for a second order tensor, and transformation of each direction involves a cos or sin term. Similarly, transformation of an nth-order tensor involves product of n cos and sin terms.

A.2 Kronecker Delta

The Kronecker delta δ_{ij} is defined as

$$\delta_{ij} = \begin{cases} 1, & \text{if } i = j, \\ 0, & \text{if } i \neq j. \end{cases} \tag{A.10}$$

In matrix notation, we can write as

$$\delta_{ij} = \begin{bmatrix} \delta_{11} & \delta_{12} & \delta_{13} \\ \delta_{21} & \delta_{22} & \delta_{23} \\ \delta_{31} & \delta_{32} & \delta_{33} \end{bmatrix} = \begin{bmatrix} 1 & 0 & 0 \\ 0 & 1 & 0 \\ 0 & 0 & 1 \end{bmatrix} \tag{A.11}$$

According to the above definition, it can be shown that the components of δ_{ij} remain unchanged by a rotation of the coordinate system. Hence, δ_{ij} is a special isotropic tensor in the sense that it is the only isotropic tensor of second order. The Kronecker delta δ_{ij} establishes the orthonormality property of the unit vectors such that

$$\mathbf{e_i}\, \mathbf{e_j} = \delta_{ij} \tag{A.12}$$

The Kronecker delta is applied extensively in the following operation: For a term containing δ_{ij} with one of its indices being repeated, then we can simply replace that repeated index by the other index of δ_{ij}. For example,

$$q_i \delta_{ij} = q_1 \delta_{1j} + q_2 \delta_{2j} + q_3 \delta_{3j} \tag{A.13}$$

The right-hand side is q_1 when $j = 1$, q_2 when $j = 2$, and q_3 when $j = 3$, thus

$$q_i \delta_{ij} = q_j \tag{A.14}$$

Alternatively, the Kronecker delta can be used to change the index of an expression as

$$r_i = r_j \delta_{ij} \tag{A.15}$$

$$c_{ik} = c_{ij} \delta_{jk} \tag{A.16}$$

A.3 Some Important Results

Some important results are summarized below for easy reference:

1. $\dfrac{\partial x_i}{\partial x_j} = \delta_{ij}$ (A.17)

To prove this identity, we simply expand in the regular matrix notation as

$$\frac{\partial x_i}{\partial x_j} = \begin{bmatrix} \frac{\partial x_1}{\partial x_1} & \frac{\partial x_1}{\partial x_2} & \frac{\partial x_1}{\partial x_3} \\ \frac{\partial x_2}{\partial x_1} & \frac{\partial x_2}{\partial x_2} & \frac{\partial x_2}{\partial x_3} \\ \frac{\partial x_3}{\partial x_1} & \frac{\partial x_3}{\partial x_2} & \frac{\partial x_3}{\partial x_3} \end{bmatrix} = \begin{bmatrix} 1 & 0 & 0 \\ 0 & 1 & 0 \\ 0 & 0 & 1 \end{bmatrix} = \delta_{ij}$$

where we have utilized the fact that x_1, x_2, and x_3 are independent variables so that we can set terms like $\partial x_1/\partial x_3$ to 0.

2. $\delta_{ii} = 3$ (A.18)

Since the index i is repeated, a summation is implied over i as

$$\delta_{ii} = \sum_{i=1}^{3} \delta_{ii} = \delta_{11} + \delta_{22} + \delta_{33} = 1 + 1 + 1 = 3$$

where we have used definition of Kronecker delta (Eq. (A.10)).

3. $\delta_{ij}\delta_{jk} = \delta_{ik}$ (A.19)

The index j is repeated, so simply replacing the index j by other index k, we get the required expression (i.e. using the result in Eq. (A.14)).

4. $\dfrac{\partial x^2}{\partial x_i} = 2x_i$ (A.20)

Introducing k as dummy variable,

$$\frac{\partial x^2}{\partial x_i} = \frac{\partial x_k x_k}{\partial x_i} = x_k \frac{\partial x_k}{\partial x_i} + x_k \frac{\partial x_k}{\partial x_i} = 2x_k \delta_{ik} = 2x_i$$

where we have used results obtained in Eqs. (A.14) and (A.17).

The most useful takeaways from this section are the ideas of free index and repeated index. Note that in any term of a tensorial equation, index cannot occur more than twice. If it occurs, that equation is incorrect in tensorial structure. Further, each term of an equation should have exactly the same free indices for that equation to be tensorially consistent.

A.4 Decomposition of Tensors

Most of the tensors that we deal with are symmetric and trace-free; we therefore examine the procedure for making tensors symmetric and trace-free in this section. Consider a second order tensor A_{ij} and form a new tensor, say A'_{ji} by interchanging indices i and j. If the components of A'_{ji} are equal to A_{ij}, then A_{ij} is called symmetric in the indices i and j, i.e., $A_{ij} = A_{ji}$.

Given any nth order tensor, its symmetric part can be obtained as

$$A_{(i_1 i_2 \cdots i_n)} = \frac{1}{n!} \left(A_{i_1 i_2 \cdots i_n} + A_{i_2 i_1 \cdots i_n} + \cdots \text{all possible permutations of indices} \right) \tag{A.21}$$

where rounded brackets denote the symmetric part of the tensor.

Accordingly, the symmetric part of a second order tensor can be obtained as

$$A_{(ij)} = \frac{1}{2} \left(A_{ij} + A_{ji} \right) \tag{A.22}$$

For the third order tensor, its symmetric part is

$$A_{(ijk)} = \frac{1}{6} \left(A_{ijk} + A_{kij} + A_{jki} + A_{ikj} + A_{jik} + A_{kji} \right) \tag{A.23}$$

There are nine and twenty seven independent components for a general non-symmetric second order and third order tensor, respectively. After making symmetric an nth order tensor, there are $\frac{1}{2}(n+1)(n+2)$ independent components of the tensor. As such, there are six independent components for a second order tensor, ten independent components for a third order tensor, and fifteen independent components for a fourth order tensor.

A second order tensor A_{ij} is called trace-free (or divergence-free) if the sum of the diagonal elements is zero, i.e., $A_{ii} = 0$. The trace-free part of a symmetric first, second, and third order tensor can be obtained as

$$A_{\langle i\rangle} = A_i \tag{A.24}$$

$$A_{\langle ij\rangle} = A_{(ij)} - \frac{1}{3}A_{kk}\delta_{ij} \tag{A.25}$$

$$A_{\langle ijk\rangle} = A_{(ijk)} - \frac{1}{5}(A_{(ill)}\delta_{jk} + A_{(jll)}\delta_{ik} + A_{(kll)}\delta_{ij}) \tag{A.26}$$

where symmetric parts $A_{(ij)}$ and $A_{(ijk)}$ are given by Eqs. (A.22) and (A.23), respectively.

To illustrate in matrix form, a general second order tensor A_{ij} can be represented as

$$A_{ij} = \begin{bmatrix} A_{11} & A_{12} & A_{13} \\ A_{21} & A_{22} & A_{23} \\ A_{31} & A_{32} & A_{33} \end{bmatrix} \tag{A.27}$$

Using Eq. (A.22), its symmetric part $A_{(ij)}$ can be represented in matrix form as

$$A_{(ij)} = \begin{bmatrix} A_{11} & (A_{12}+A_{21})/2 & (A_{13}+A_{31})/2 \\ (A_{21}+A_{12})/2 & A_{22} & (A_{23}+A_{32})/2 \\ (A_{31}+A_{13})/2 & (A_{32}+A_{23})/2 & A_{33} \end{bmatrix} \tag{A.28}$$

Using Eq. (A.25), symmetric and trace-free part $A_{\langle ij\rangle}$ can be obtained as

$$A_{\langle ij\rangle} = \begin{bmatrix} A_{11}-(A_{11}+A_{22}+A_{33})/3 & (A_{12}+A_{21})/2 & (A_{13}+A_{31})/2 \\ (A_{21}+A_{12})/2 & A_{22}-(A_{11}+A_{22}+A_{33})/3 & (A_{23}+A_{32})/2 \\ (A_{31}+A_{13})/2 & (A_{32}+A_{23})/2 & A_{33}-(A_{11}+A_{22}+A_{33})/3 \end{bmatrix} \tag{A.29}$$

A general second order tensor A_{ij} has nine (3^n where $n = 2$) independent components while its symmetric part $A_{(ij)}$ has six independent components. Further, making it trace-free, $A_{\langle ij\rangle}$ has five independent components, three off-diagonal components, A_{12}, A_{13}, and A_{23} plus two of the three diagonal components. Notice that the third diagonal component can be determined by the relation $A_{\langle 11\rangle} + A_{\langle 22\rangle} + A_{\langle 33\rangle} = 0$.

A third order tensor A_{ijk} has twenty-seven (3^n where $n = 3$) independent components and can be represented in matrix form as

$$A_{ijk} = \begin{bmatrix} \underline{A_{111}} & \underline{A_{112}} & \underline{A_{113}} \\ A_{121} & \underline{A_{122}} & \underline{A_{123}} \\ A_{131} & A_{132} & \underline{A_{133}} \end{bmatrix} \begin{bmatrix} A_{211} & A_{212} & A_{213} \\ A_{221} & \underline{A_{222}} & \underline{A_{223}} \\ A_{231} & A_{232} & \underline{A_{233}} \end{bmatrix} \begin{bmatrix} A_{311} & A_{312} & A_{313} \\ A_{321} & A_{322} & A_{323} \\ A_{331} & A_{332} & \underline{A_{333}} \end{bmatrix} \tag{A.30}$$

The symmetric part of third order tensor $A_{(ijk)}$ has ten independent components (underlined terms) and can be obtained using Eq. (A.23) as

$$
\begin{aligned}
A_{(111)} &= \frac{1}{6}\left(A_{111} + A_{111} + A_{111} + A_{111} + A_{111} + A_{111}\right) = A_{111} \\
A_{(112)} &= \frac{1}{6}\left(A_{112} + A_{211} + A_{121} + A_{121} + A_{112} + A_{211}\right) \\
&= \frac{1}{3}\left(A_{112} + A_{211} + A_{121}\right) \\
A_{(113)} &= \frac{1}{6}\left(A_{113} + A_{311} + A_{131} + A_{131} + A_{113} + A_{311}\right) \\
&= \frac{1}{3}\left(A_{113} + A_{311} + A_{131}\right) \\
A_{(122)} &= \frac{1}{6}\left(A_{122} + A_{212} + A_{221} + A_{212} + A_{221} + A_{122}\right) \\
&= \frac{1}{3}\left(A_{122} + A_{212} + A_{221}\right) \\
A_{(123)} &= \frac{1}{6}\left(A_{123} + A_{312} + A_{231} + A_{132} + A_{213} + A_{321}\right) \\
A_{(133)} &= \frac{1}{6}\left(A_{133} + A_{313} + A_{331} + A_{313} + A_{331} + A_{133}\right) \\
&= \frac{1}{3}\left(A_{133} + A_{313} + A_{331}\right) \\
A_{(222)} &= \frac{1}{6}\left(A_{222} + A_{222} + A_{222} + A_{222} + A_{222} + A_{222}\right) = A_{222} \\
A_{(223)} &= \frac{1}{6}\left(A_{223} + A_{322} + A_{232} + A_{232} + A_{223} + A_{322}\right) \\
&= \frac{1}{3}\left(A_{223} + A_{322} + A_{232}\right) \\
A_{(233)} &= \frac{1}{6}\left(A_{233} + A_{323} + A_{332} + A_{323} + A_{332} + A_{233}\right) \\
&= \frac{1}{3}\left(A_{233} + A_{323} + A_{332}\right) \\
A_{(333)} &= \frac{1}{6}\left(A_{333} + A_{333} + A_{333} + A_{333} + A_{333} + A_{333}\right) = A_{333}
\end{aligned}
$$

The non-underlined terms (in Eq. (A.30)) can be obtained from the underlined terms as

$$
\begin{aligned}
A_{(112)} &= A_{(211)} = A_{(121)} \\
A_{(113)} &= A_{(311)} = A_{(131)} \\
A_{(122)} &= A_{(212)} = A_{(221)} \\
A_{(123)} &= A_{(312)} = A_{(231)} = A_{(132)} = A_{(213)} = A_{(321)} \\
A_{(133)} &= A_{(313)} = A_{(331)} \\
A_{(223)} &= A_{(322)} = A_{(232)} \\
A_{(233)} &= A_{(323)} = A_{(332)}
\end{aligned} \tag{A.31}
$$

Further to make trace-free, we can perform contraction as $i = j$, or $j = k$, or $k = i$; as such, there will be seven (= 10 − 3) independent components for the trace-free part of third order tensor $A_{\langle ijk \rangle}$. The elements of the three diagonals are related as

$$\underline{A_{\langle 111 \rangle}} + \underline{A_{\langle 122 \rangle}} + \underline{A_{\langle 133 \rangle}} = 0 \quad \Rightarrow \quad A_{\langle 133 \rangle} = -\underline{A_{\langle 111 \rangle}} - \underline{A_{\langle 122 \rangle}} \tag{A.32}$$

$$A_{\langle 211 \rangle} + \underline{A_{\langle 222 \rangle}} + \underline{A_{\langle 233 \rangle}} = 0 \quad \Rightarrow \quad A_{\langle 222 \rangle} = -A_{\langle 211 \rangle} - \underline{A_{\langle 233 \rangle}} \tag{A.33}$$

$$A_{\langle 311 \rangle} + A_{\langle 322 \rangle} + \underline{A_{\langle 333 \rangle}} = 0 \quad \Rightarrow \quad A_{\langle 333 \rangle} = -A_{\langle 311 \rangle} - A_{\langle 322 \rangle} \tag{A.34}$$

Hence, the components $A_{\langle 133 \rangle}$, $A_{\langle 222 \rangle}$, and $A_{\langle 333 \rangle}$ are not exactly independent since they can be determined from the remaining components. As a result, we have $A_{\langle 111 \rangle}$, $A_{\langle 112 \rangle}$, $A_{\langle 113 \rangle}$, $A_{\langle 122 \rangle}$, $A_{\langle 123 \rangle}$, $A_{\langle 223 \rangle}$, and $A_{\langle 233 \rangle}$ as seven independent components of trace free third order tensor. Using Eq. (A.26), components of $A_{\langle ijk \rangle}$ can be obtained as

$$
\begin{aligned}
A_{\langle 111 \rangle} &= A_{(111)} - \frac{1}{5}\left[A_{(1ll)}\delta_{11} + A_{(1ll)}\delta_{11} + A_{(1ll)}\delta_{11}\right] \\
&= A_{(111)} - \frac{3}{5}\left[A_{(111)} + A_{(122)} + A_{(133)}\right] \\
&= \frac{2}{5}A_{111} - \frac{1}{5}\left[A_{122} + A_{212} + A_{221} + A_{133} + A_{313} + A_{331}\right] \\
A_{\langle 112 \rangle} &= A_{(112)} - \frac{1}{5}\left[A_{(1ll)}\delta_{12} + A_{(1ll)}\delta_{12} + A_{(2ll)}\delta_{11}\right] \\
&= A_{(112)} - \frac{1}{5}\left[A_{(211)} + A_{(222)} + A_{(233)}\right] \\
&= \frac{4}{15}\left(A_{112} + A_{211} + A_{121}\right) - \frac{1}{5}\left[A_{222} + \frac{1}{3}\left(A_{233} + A_{323} + A_{332}\right)\right]
\end{aligned}
$$

$$
\begin{aligned}
A_{\langle 113\rangle} &= A_{(113)} - \frac{1}{5}\left[A_{(1ll)}\delta_{13} + A_{(1ll)}\delta_{13} + A_{(3ll)}\delta_{11}\right] \\
&= A_{(113)} - \frac{1}{5}\left[A_{(311)} + A_{(322)} + A_{(333)}\right] \\
&= \frac{4}{15}\left(A_{113} + A_{311} + A_{131}\right) - \frac{1}{5}\left[\frac{1}{3}\left(A_{322} + A_{232} + A_{223}\right) + A_{333}\right] \\
A_{\langle 122\rangle} &= A_{(122)} - \frac{1}{5}\left[A_{(1ll)}\delta_{22} + A_{(2ll)}\delta_{12} + A_{(2ll)}\delta_{21}\right] \\
&= A_{(122)} - \frac{1}{5}\left[A_{(111)} + A_{(122)} + A_{(133)}\right] \\
&= \frac{4}{15}\left(A_{122} + A_{212} + A_{221}\right) - \frac{1}{5}\left[A_{111} + \frac{1}{3}\left(A_{133} + A_{313} + A_{331}\right)\right] \\
A_{\langle 123\rangle} &= A_{(123)} - \frac{1}{5}\left[A_{(1ll)}\delta_{23} + A_{(2ll)}\delta_{13} + A_{(3ll)}\delta_{21}\right] \\
&= \frac{1}{6}\left(A_{123} + A_{312} + A_{231} + A_{132} + A_{213} + A_{321}\right) \\
A_{\langle 222\rangle} &= A_{(222)} - \frac{1}{5}\left[A_{(2ll)}\delta_{22} + A_{(2ll)}\delta_{22} + A_{(2ll)}\delta_{22}\right] \\
&= A_{(222)} - \frac{3}{5}\left[A_{(211)} + A_{(222)} + A_{(233)}\right] \\
&= \frac{2}{5}A_{222} - \frac{1}{5}\left[A_{211} + A_{121} + A_{112} + A_{233} + A_{323} + A_{332}\right] \\
A_{\langle 223\rangle} &= A_{(223)} - \frac{1}{5}\left[A_{(2ll)}\delta_{23} + A_{(2ll)}\delta_{23} + A_{(3ll)}\delta_{22}\right] \\
&= A_{(223)} - \frac{1}{5}\left[A_{(311)} + A_{(322)} + A_{(333)}\right] \\
&= \frac{4}{15}\left(A_{223} + A_{322} + A_{232}\right) - \frac{1}{5}\left[\frac{1}{3}\left(A_{311} + A_{131} + A_{113}\right) + A_{333}\right]
\end{aligned}
$$

Consider a third order tensor $\dfrac{\partial Q_{ij}}{\partial x_k}$ which is formed by taking the gradient of a symmetric and trace-free second order tensor Q_{ij}. Its symmetric part is obtained as

$$
\begin{aligned}
\frac{\partial Q_{(ij}}{\partial x_{k)}} &= \frac{1}{6}\left(\frac{\partial Q_{ij}}{\partial x_k} + \frac{\partial Q_{ki}}{\partial x_j} + \frac{\partial Q_{jk}}{\partial x_i} + \frac{\partial Q_{ik}}{\partial x_j} + \frac{\partial Q_{ji}}{\partial x_k} + \frac{\partial Q_{kj}}{\partial x_i}\right) \\
&= \frac{1}{6}\left(2\frac{\partial Q_{ij}}{\partial x_k} + 2\frac{\partial Q_{ki}}{\partial x_j} + 2\frac{\partial Q_{jk}}{\partial x_i}\right) \qquad [\because Q_{ij} = Q_{ji}] \\
&= \frac{1}{3}\left(\frac{\partial Q_{ij}}{\partial x_k} + \frac{\partial Q_{ki}}{\partial x_j} + \frac{\partial Q_{jk}}{\partial x_i}\right) \qquad \text{(A.35)}
\end{aligned}
$$

Its trace-free and symmetric part is then obtained using Eq. (A.26) as

$$\begin{aligned}
\frac{\partial Q_{\langle ij}}{\partial x_{k\rangle}} &= \frac{1}{3}\left(\frac{\partial Q_{ij}}{\partial x_k}+\frac{\partial Q_{ki}}{\partial x_j}+\frac{\partial Q_{jk}}{\partial x_i}\right)-\frac{1}{15}\left\{\left(\frac{\partial Q_{il}}{\partial x_l}+\frac{\partial Q_{il}}{\partial x_l}+\frac{\partial Q_{ll}}{\partial x_i}\right)\delta_{jk}\right.\\
&\left.+\left(\frac{\partial Q_{jl}}{\partial x_l}+\frac{\partial Q_{jl}}{\partial x_l}+\frac{\partial Q_{ll}}{\partial x_j}\right)\delta_{ik}+\left(\frac{\partial Q_{kl}}{\partial x_l}+\frac{\partial Q_{kl}}{\partial x_l}+\frac{\partial Q_{ll}}{\partial x_k}\right)\delta_{ij}\right\}\\
&= \frac{1}{3}\left(\frac{\partial Q_{ij}}{\partial x_k}+\frac{\partial Q_{ki}}{\partial x_j}+\frac{\partial Q_{jk}}{\partial x_i}\right)-\frac{2}{15}\left\{\frac{\partial Q_{il}}{\partial x_l}\delta_{jk}+\frac{\partial Q_{jl}}{\partial x_l}\delta_{ik}+\frac{\partial Q_{kl}}{\partial x_l}\delta_{ij}\right\}
\end{aligned} \tag{A.36}$$

where we have used $Q_{ll}=0$ (since Q_{ij} is trace-free and symmetric).

Likewise, for the fourth order symmetric tensor $A_{(ijkl)}$, there are fifteen independent components $[1/2(n+1)(n+2);\ n=4]$. For the trace-free part, we can perform contraction as $i=j$, or $i=k$ or $i=l$ or $j=k$ or $j=l$, or $k=l$ (six traces); as such, its trace-free part, $A_{\langle ijkl\rangle}$ will have nine $(15-6)$ independent components.

A.5 Some Examples

To illustrate the tensor operations, we consider three examples as below:

1. We first consider expression for viscous dissipation Φ, (Eq. (2.7)) and using tensorial operations, show that Φ is *always* a positive quantity (a result of profound significance!).

$$\begin{aligned}
\Phi &= 2\mu e_{ij}e_{ij}-\frac{2}{3}\mu\left(\frac{\partial u_k}{\partial x_k}\right)^2\\
&= 2\mu\left[e_{ij}e_{ij}-\frac{1}{3}\frac{\partial u_k}{\partial x_k}\frac{\partial u_l}{\partial x_l}\right]\\
&= 2\mu\left[e_{ij}e_{ij}+\frac{1}{9}\frac{\partial u_k}{\partial x_k}\frac{\partial u_l}{\partial x_l}3-\frac{2}{3}\frac{\partial u_k}{\partial x_k}\frac{\partial u_l}{\partial x_l}\right]\\
&= 2\mu\left[e_{ij}e_{ij}+\frac{1}{9}\frac{\partial u_k}{\partial x_k}\frac{\partial u_l}{\partial x_l}\delta_{ij}\delta_{ij}-\frac{2}{3}\frac{1}{2}\left(\frac{\partial u_i}{\partial x_i}+\frac{\partial u_i}{\partial x_i}\right)\frac{\partial u_l}{\partial x_l}\right]\\
&\qquad [\because \delta_{ij}\delta_{ij}=\delta_{ii}=3]\\
&= 2\mu\left[e_{ij}e_{ij}+\frac{1}{9}\frac{\partial u_k}{\partial x_k}\frac{\partial u_l}{\partial x_l}\delta_{ij}\delta_{ij}-\frac{2}{3}\frac{1}{2}\left(\frac{\partial u_i}{\partial x_j}\delta_{ij}+\frac{\partial u_j}{\partial x_i}\delta_{ij}\right)\frac{\partial u_l}{\partial x_l}\right]\\
&= 2\mu\left[e_{ij}e_{ij}+\frac{1}{9}\frac{\partial u_k}{\partial x_k}\frac{\partial u_l}{\partial x_l}\delta_{ij}\delta_{ij}-\frac{2}{3}e_{ij}\delta_{ij}\frac{\partial u_l}{\partial x_l}\right]\\
&= 2\mu\left[e_{ij}-\frac{1}{3}\frac{\partial u_k}{\partial x_k}\delta_{ij}\right]^2
\end{aligned} \tag{A.37}$$

where $e_{ij} = \frac{1}{2}\left(\frac{\partial u_i}{\partial x_j} + \frac{\partial u_j}{\partial x_i}\right)$ is strain rate tensor. Note the use of Kronecker delta in changing the index of the term in step 5. For example, term $\partial u_i/\partial x_i$ is transformed to e_{ij} by introducing δ_{ij}.

2. As a next exercise, we expand Φ in terms of velocity gradients using $e_{ij} = \frac{1}{2}\left(\frac{\partial u_i}{\partial x_j} + \frac{\partial u_j}{\partial x_i}\right)$ as

$$\begin{aligned}
\Phi &= 2\mu e_{ij}e_{ij} - \frac{2}{3}\mu\left(\frac{\partial u_k}{\partial x_k}\right)^2 \\
&= 2\mu\frac{1}{2}\left(\frac{\partial u_i}{\partial x_j} + \frac{\partial u_j}{\partial x_i}\right)\frac{1}{2}\left(\frac{\partial u_i}{\partial x_j} + \frac{\partial u_j}{\partial x_i}\right) - \frac{2}{3}\mu\left(\frac{\partial u_k}{\partial x_k}\right)^2 \\
&= \frac{1}{2}\mu\left(\frac{\partial u_i}{\partial x_j} + \frac{\partial u_j}{\partial x_i}\right)\left(\frac{\partial u_i}{\partial x_j} + \frac{\partial u_j}{\partial x_i}\right) - \frac{2}{3}\mu\left(\frac{\partial u_k}{\partial x_k}\right)^2 \\
&= \frac{1}{2}\mu\left(\frac{\partial u_i}{\partial x_j}\frac{\partial u_i}{\partial x_j} + 2\frac{\partial u_i}{\partial x_j}\frac{\partial u_j}{\partial x_i} + \underbrace{\frac{\partial u_j}{\partial x_i}\frac{\partial u_j}{\partial x_i}}_{i\to j;\, j\to i}\right) - \frac{2}{3}\mu\left(\frac{\partial u_k}{\partial x_k}\right)^2 \\
&= \mu\left(\frac{\partial u_i}{\partial x_j}\frac{\partial u_i}{\partial x_j} + \frac{\partial u_i}{\partial x_j}\frac{\partial u_j}{\partial x_i}\right) - \frac{2}{3}\mu\left(\frac{\partial u_k}{\partial x_k}\right)^2
\end{aligned} \tag{A.38}$$

Note that there are two repeated indices in the first two terms which implies summation over i (with i going from 1 to 3) and summation over j (with j going from 1 to 3) giving a total of nine terms. Performing summation first for j and then for i (the sequence of summation does not matter),

$$\begin{aligned}
(a) &= \sum_{i=1}^{3}\sum_{j=1}^{3}\frac{\partial u_i}{\partial x_j}\frac{\partial u_i}{\partial x_j} \\
&= \sum_{i=1}^{3}\left(\frac{\partial u_i}{\partial x_1}\frac{\partial u_i}{\partial x_1} + \frac{\partial u_i}{\partial x_2}\frac{\partial u_i}{\partial x_2} + \frac{\partial u_i}{\partial x_3}\frac{\partial u_i}{\partial x_3}\right) \\
&= \left[\frac{\partial u_1}{\partial x_1}\frac{\partial u_1}{\partial x_1} + \frac{\partial u_2}{\partial x_1}\frac{\partial u_2}{\partial x_1} + \frac{\partial u_3}{\partial x_1}\frac{\partial u_3}{\partial x_1}\right] + \left[\frac{\partial u_1}{\partial x_2}\frac{\partial u_1}{\partial x_2} + \frac{\partial u_2}{\partial x_2}\frac{\partial u_2}{\partial x_2} + \frac{\partial u_3}{\partial x_2}\frac{\partial u_3}{\partial x_2}\right] \\
&\quad + \left[\frac{\partial u_1}{\partial x_3}\frac{\partial u_1}{\partial x_3} + \frac{\partial u_2}{\partial x_3}\frac{\partial u_2}{\partial x_3} + \frac{\partial u_3}{\partial x_3}\frac{\partial u_3}{\partial x_3}\right] \\
&= \left(\frac{\partial u_1}{\partial x_1}\right)^2 + \left(\frac{\partial u_2}{\partial x_1}\right)^2 + \left(\frac{\partial u_3}{\partial x_1}\right)^2 + \left(\frac{\partial u_1}{\partial x_2}\right)^2 + \left(\frac{\partial u_2}{\partial x_2}\right)^2 + \left(\frac{\partial u_3}{\partial x_2}\right)^2 \\
&\quad + \left(\frac{\partial u_1}{\partial x_3}\right)^2 + \left(\frac{\partial u_2}{\partial x_3}\right)^2 + \left(\frac{\partial u_3}{\partial x_3}\right)^2
\end{aligned} \tag{A.39}$$

$$(b) = \sum_{i=1}^{3}\sum_{j=1}^{3} \frac{\partial u_i}{\partial x_j}\frac{\partial u_j}{\partial x_i}$$

$$= \sum_{i=1}^{3}\left(\frac{\partial u_i}{\partial x_1}\frac{\partial u_1}{\partial x_i} + \frac{\partial u_i}{\partial x_2}\frac{\partial u_2}{\partial x_i} + \frac{\partial u_i}{\partial x_3}\frac{\partial u_3}{\partial x_i}\right)$$

$$= \left[\frac{\partial u_1}{\partial x_1}\frac{\partial u_1}{\partial x_1} + \frac{\partial u_2}{\partial x_1}\frac{\partial u_1}{\partial x_2} + \frac{\partial u_3}{\partial x_1}\frac{\partial u_1}{\partial x_3}\right] + \left[\frac{\partial u_1}{\partial x_2}\frac{\partial u_2}{\partial x_1} + \frac{\partial u_2}{\partial x_2}\frac{\partial u_2}{\partial x_2} + \frac{\partial u_3}{\partial x_2}\frac{\partial u_2}{\partial x_3}\right]$$

$$+ \left[\frac{\partial u_1}{\partial x_3}\frac{\partial u_3}{\partial x_1} + \frac{\partial u_2}{\partial x_3}\frac{\partial u_3}{\partial x_2} + \frac{\partial u_3}{\partial x_3}\frac{\partial u_3}{\partial x_3}\right]$$

$$= \left(\frac{\partial u_1}{\partial x_1}\right)^2 + \frac{\partial u_2}{\partial x_1}\frac{\partial u_1}{\partial x_2} + \frac{\partial u_3}{\partial x_1}\frac{\partial u_1}{\partial x_3} + \frac{\partial u_1}{\partial x_2}\frac{\partial u_2}{\partial x_1} + \left(\frac{\partial u_2}{\partial x_2}\right)^2 + \frac{\partial u_3}{\partial x_2}\frac{\partial u_2}{\partial x_3}$$

$$+ \frac{\partial u_1}{\partial x_3}\frac{\partial u_3}{\partial x_1} + \frac{\partial u_2}{\partial x_3}\frac{\partial u_3}{\partial x_2} + \left(\frac{\partial u_3}{\partial x_3}\right)^2 \tag{A.40}$$

$$\text{and} \quad (c) = \left(\sum_{k=1}^{3}\frac{\partial u_k}{\partial x_k}\right)^2 = \left(\frac{\partial u_1}{\partial x_1} + \frac{\partial u_2}{\partial x_2} + \frac{\partial u_3}{\partial x_3}\right)^2 \tag{A.41}$$

Combining all the terms (a), (b), and (c), and completing squares for some of the terms, one can easily obtain

$$\Phi = 2\mu\left[\left(\frac{\partial u_1}{\partial x_1}\right)^2 + \left(\frac{\partial u_2}{\partial x_2}\right)^2 + \left(\frac{\partial u_3}{\partial x_3}\right)^2\right] + \mu\left(\frac{\partial u_1}{\partial x_2} + \frac{\partial u_2}{\partial x_1}\right)^2$$
$$+ \mu\left(\frac{\partial u_2}{\partial x_3} + \frac{\partial u_3}{\partial x_2}\right)^2 + \mu\left(\frac{\partial u_3}{\partial x_1} + \frac{\partial u_1}{\partial x_3}\right)^2 - \frac{2}{3}\mu\left(\frac{\partial u_1}{\partial x_1} + \frac{\partial u_2}{\partial x_2} + \frac{\partial u_3}{\partial x_3}\right)^2 \tag{A.42}$$

On replacing (u_1, u_2, u_3) by (u, v, w), and (x_1, x_2, x_3) by (x, y, z), Eq. (A.42) is same as that of Eq. (2.24).

3. Consider a complex equation (an evolution equation for symmetric and trace-free third order tensor $m_{\langle ijk \rangle}$) in index notation as

$$\frac{Dm_{\langle ijk\rangle}}{Dt} - 3\frac{p_{\langle ij}}{\rho}\frac{\partial p_{k\rangle l}}{\partial x_l} - 3p_{\langle ij}RT\frac{\partial \ln\rho}{\partial x_{k\rangle}} + \frac{\partial v_{ijkl}}{\partial x_l} + \frac{3}{7}\frac{\partial Q_{\langle ij}}{\partial x_{k\rangle}}$$
$$+ 3RT\frac{\partial p_{\langle ij}}{\partial x_{k\rangle}} + 3m_{l\langle ij}\frac{\partial u_{k\rangle}}{\partial x_l} + m_{\langle ijk\rangle}\frac{\partial u_l}{\partial x_l} + \frac{12}{5}q_{\langle i}\frac{\partial u_j}{\partial x_{k\rangle}} = -\frac{3}{2}\frac{p}{\mu}m_{\langle ijk\rangle} \tag{A.43}$$

We list some of the common rules as:

a. Each term in an equation must be of the same tensorial order. In Eq. (A.43), all the individual terms are third order tensors. In other words, the three free indices i, j, and k must appear in all the terms of the equation.
b. In the term, $\frac{\partial v_{ijkl}}{\partial x_l}$, l is a dummy index. Since the choice of a dummy index is immaterial, we can write $\frac{\partial v_{ijkl}}{\partial x_l} = \frac{\partial v_{ijkr}}{\partial x_r}$. However, note that the choice of dummy index should not clash with the free indices, i.e., dummy index should be selected other than i, j, and k for the above example considered.
c. An index cannot appear more than twice, for example, the term $\frac{\partial v_{ijll}}{\partial x_l}$ is incorrect, since index l is appearing thrice in a given term.

In order to illustrate the usefulness of the tensor algebra, it is worthwhile mentioning that Eq. (A.43) represents the evolution equations of seven independent components of $m_{\langle ijk \rangle}$. Recall that $m_{\langle ijk \rangle}$ is symmetric and trace-free and therefore has seven independent components: $m_{\langle 111 \rangle}, m_{\langle 112 \rangle}, m_{\langle 113 \rangle}, m_{\langle 122 \rangle}, m_{\langle 123 \rangle}, m_{\langle 222 \rangle}$, and $m_{\langle 223 \rangle}$.

With the help of Einstein summation convention and angular bracket notation, many of the terms can be represented in a compact form. For example, Einstein summation convention is implied in the term $\frac{\partial v_{ijkl}}{\partial x_l}$ which upon expansion is actually

$$\frac{\partial v_{ijkl}}{\partial x_l} = \frac{\partial v_{ijk1}}{\partial x_1} + \frac{\partial v_{ijk2}}{\partial x_2} + \frac{\partial v_{ijk3}}{\partial x_3}$$

The usefulness of the angular bracket notation can be illustrated by considering the term $\frac{\partial Q_{\langle ij}}{\partial x_{k\rangle}}$. Considering Q_{ij} to be symmetric, the symmetric part of the third order tensor obtained by taking gradient of Q_{ij} using Eq. (A.23) can be written as

$$\frac{\partial Q_{(ij}}{\partial x_{k)}} = \frac{1}{6}\left(\frac{\partial Q_{ij}}{\partial x_k} + \frac{\partial Q_{ki}}{\partial x_j} + \frac{\partial Q_{jk}}{\partial x_i} + \frac{\partial Q_{ik}}{\partial x_j} + \frac{\partial Q_{ji}}{\partial x_k} + \frac{\partial Q_{kj}}{\partial x_i}\right) \tag{A.44}$$

Since second order tensor Q_{lm} is symmetric, we can interchange the indices as $Q_{lm} = Q_{ml}$. Accordingly, we get

$$\begin{aligned}\frac{\partial Q_{(ij}}{\partial x_{k)}} &= \frac{1}{6}\left(2\frac{\partial Q_{ij}}{\partial x_k} + 2\frac{\partial Q_{ki}}{\partial x_j} + 2\frac{\partial Q_{jk}}{\partial x_i}\right) \\ &= \frac{1}{3}\left(\frac{\partial Q_{ij}}{\partial x_k} + \frac{\partial Q_{ki}}{\partial x_j} + \frac{\partial Q_{jk}}{\partial x_i}\right)\end{aligned} \tag{A.45}$$

Its trace-free and symmetric part is then obtained using Eq. (A.26) as

$$\frac{\partial Q_{\langle ij}}{\partial x_{k\rangle}} = \frac{1}{3}\left(\frac{\partial Q_{ij}}{\partial x_k} + \frac{\partial Q_{ki}}{\partial x_j} + \frac{\partial Q_{jk}}{\partial x_i}\right) - \frac{1}{15}\left\{\left(\frac{\partial Q_{il}}{\partial x_l} + \frac{\partial Q_{il}}{\partial x_l} + \frac{\partial Q_{ll}}{\partial x_i}\right)\delta_{jk} + \left(\frac{\partial Q_{jl}}{\partial x_l} + \frac{\partial Q_{jl}}{\partial x_l} + \frac{\partial Q_{ll}}{\partial x_j}\right)\delta_{ik} + \left(\frac{\partial Q_{kl}}{\partial x_l} + \frac{\partial Q_{kl}}{\partial x_l} + \frac{\partial Q_{ll}}{\partial x_k}\right)\delta_{ij}\right\} \tag{A.46}$$

Hence, with this example, it is clear that the Einstein summation convention and angular bracket notation greatly help to represent the equations in a compact form.

A.6 General Tensor Operations

In this section, several useful tensor operations are compiled. The most important for this book being the gradient, divergence, and Laplacian operations. The other operations are less frequently employed here and are given for the sake of completeness [116].

1. *Addition rule:* Two tensors strictly of the same order can be added or subtracted and the resulting tensor is of the same order.
2. *Product rule:* A product of mth order tensor with an nth order tensor gives a tensor of order $(m+n)$. For example, a tensor product of a first order tensor q_i with a second order tensor P_{ij} gives a third order tensor, say r_{ijk}.
   In terms of components, we can write

$$r_{ijk} = q_i P_{ij} \tag{A.47}$$

3. *Contraction:* In contraction operation, two indices of a tensor are equated, and a summation is then performed over this repeated index. For example, for a second order tensor P_{ij}, we equate $i = j$ and obtain P_{ii} as

$$P_{ii} = P_{11} + P_{22} + P_{33}$$

   The contraction operation decreases the order of the tensor by 2. In the above example, performing contraction of a second order tensor results in a scalar. Similarly, contracting a fourth order tensor gives a second order tensor.
4. *Inner product:* The inner product between a second order tensor A_{ij} and third order tensor B_{jkl} results in a third order tensor. In terms of components,

$$C_{ikl} = A_{ij} B_{jkl} \tag{A.48}$$

Generalizing the rule, the inner product of an nth order tensor and mth order tensor $(n, m \geq 1)$ results in a tensor of order $n+m-2$. The dot product of two vectors is probably a good example of inner product, where a scalar is obtained after the operation. Here, $n = m = 1$ and the inner product is of order zero $(1+1-2=0)$.

5. *Division:* Though addition, subtraction, and product operations are well defined in tensor algebra, there is no tensor operation corresponding to division.
6. *Gradient:* The gradient operator 'del' (a vector operator) is defined as

$$\nabla \equiv \mathbf{e_i}\frac{\partial}{\partial x_i} \tag{A.49}$$

The gradient of a tensor, say a second order tensor A_{ij}, is obtained by applying the gradient operator and gives a third order tensor as

$$B_{ijk} = \frac{\partial A_{ij}}{\partial x_k} \tag{A.50}$$

In general, the gradient operation on an nth order tensor gives a tensor of order $n+1$. It is easy to see that the repeated application of a gradient operation yields still higher derivatives.

7. *Divergence:* The inner product of a tensor and operator ∇ is termed as divergence. The divergence operation on an nth order tensor gives a tensor of order $n-1$. Thus, the divergence operation reduces the order of the tensor by 1. Probably, the best example is the divergence of the velocity vector which is given as

$$\nabla \cdot \mathbf{u} \equiv \frac{\partial u_k}{\partial x_k} = \frac{\partial u_1}{\partial x_1} + \frac{\partial u_2}{\partial x_2} + \frac{\partial u_3}{\partial x_3} \tag{A.51}$$

8. *Laplacian:* The Laplacian is obtained by taking the dot product of operator ∇ with itself as

$$\nabla^2 = \nabla \cdot \nabla = \mathbf{e_i}\frac{\partial}{\partial x_i} \cdot \mathbf{e_j}\frac{\partial}{\partial x_j} = \mathbf{e_i} \cdot \mathbf{e_j}\frac{\partial^2}{\partial x_i \partial x_j} = \delta_{ij}\frac{\partial^2}{\partial x_i \partial x_j} = \frac{\partial^2}{\partial x_i \partial x_i} \tag{A.52}$$

where we used Eqs. (A.14) and (A.12).

9. *Gauss's theorem:* This theorem relates a volume integral to a surface integral and is applicable to tensors of any order. According to this theorem,

$$\iiint_V \frac{\partial p_{ij}}{\partial x_k} \mathrm{d}V = \iint_A p_{ij} n_k \mathrm{d}A \tag{A.53}$$

where p_{ij} is a second order tensor, A is a smooth surface that encloses a volume V, and n_k denotes the unit outward-pointing normal on A. When the indices j and k are contracted, we obtain the divergence theorem.

10. *Doubly contracted product:* Consider a symmetric tensor p_{ij} and any tensor A_{ij}. The doubly contracted product which gives a scalar is defined as

$$P \equiv p_{ij} A_{ij} \tag{A.54}$$

According to the rule, the doubly contracted product of a symmetric tensor p_{ij} with any tensor A_{ij} is equal to p_{ij} times the symmetric part of A_{ij},

$$P \equiv p_{ij} A_{ij} = p_{ij} A_{(ij)} \tag{A.55}$$

Appendix B
Appendix to Burnett Equations

B.1 Derivation of $f^{(0)}$

The quantity $\log f$ is a summational invariant for collisions. Hence, it should be a linear combination of the three summational invariants $\psi^{(i)} = \{1, m\mathbf{c}, \frac{1}{2}mc^2\}$ and can be written as

$$\log f = \Sigma\alpha^{(i)}\psi^{(i)} = \alpha^{(1)} + \boldsymbol{\alpha}^{(2)}.m\mathbf{c} - \alpha^{(3)}\frac{1}{2}mc^2 \tag{B.1}$$

Since $\log f$ is a scalar, $\alpha^{(1)}$ and $\alpha^{(3)}$ are scalars and $\boldsymbol{\alpha}^{(2)}$ is a vector. Since the state of the gas is uniform and steady, all three must be independent of $\mathbf{x}$ and t.

$$\log f = \alpha^{(1)} + m(\alpha_x^{(2)}u + \alpha_y^{(2)}v + \alpha_z^{(2)}w) - \frac{1}{2}\alpha^{(3)}m(u^2 + v^2 + w^2) \tag{B.2}$$

Considering the following terms separately,

$$m\alpha_x^{(2)}u - \frac{1}{2}\alpha^{(3)}mu^2 = -\frac{1}{2}\alpha^{(3)}m\left[u^2 - 2u\frac{\alpha_x^{(2)}}{\alpha^{(3)}} + \left(\frac{\alpha_x^{(2)}}{\alpha^{(3)}}\right)^2 - \left(\frac{\alpha_x^{(2)}}{\alpha^{(3)}}\right)^2\right]$$

$$= -\frac{1}{2}\alpha^{(3)}m\left[\left(u - \frac{\alpha_x^{(2)}}{\alpha^{(3)}}\right)^2 - \left(\frac{\alpha_x^{(2)}}{\alpha^{(3)}}\right)^2\right]$$

Simplifying the other components in a similar way, (B.2) becomes

$$\log f = \log\alpha^{(0)} - \frac{1}{2}\alpha^{(3)}m\left[\left(u - \frac{\alpha_x^{(2)}}{\alpha^{(3)}}\right)^2 + \left(v - \frac{\alpha_y^{(2)}}{\alpha^{(3)}}\right)^2 + \left(w - \frac{\alpha_z^{(2)}}{\alpha^{(3)}}\right)^2\right]$$

A. Agrawal et al., *Microscale Flow and Heat Transfer*, Mechanical Engineering Series, https://doi.org/10.1007/978-3-030-10662-1

$$\text{where } \log \alpha^{(0)} = \alpha^{(1)} + \frac{1}{2}\alpha^{(3)} m \left[\left(\frac{\alpha_x^{(2)}}{\alpha^{(3)}} \right)^2 + \left(\frac{\alpha_y^{(2)}}{\alpha^{(3)}} \right)^2 + \left(\frac{\alpha_z^{(2)}}{\alpha^{(3)}} \right)^2 \right]$$

Defining $\mathbf{C}' \equiv \mathbf{c} - \boldsymbol{\alpha}^{(2)}/\alpha^{(3)}$, we get

$$f = \alpha^{(0)} e^{-\alpha^{(3)} \frac{1}{2} m C'^2} \tag{B.3}$$

This result was first obtained by Maxwell and the state of the system defined by (B.3) is said to be in the Maxwellian state.

The constants $\alpha^{(0)}$, $\boldsymbol{\alpha}^{(2)}$, and $\alpha^{(3)}$ need to be expressed in terms of the number density n, the mean velocity $\mathbf{c_0}$, and the absolute temperature T. According to the definition of number density,

$$n = \int f d\mathbf{c} = \alpha^{(0)} \int e^{-\alpha^{(3)} \frac{1}{2} m C'^2} d\mathbf{C}' \tag{B.4}$$

Expressing $\mathbf{C}'$ in terms of polar coordinates C', θ and ϕ where θ is the azimuthal angle and ϕ is the zenith angle,

$$\begin{aligned} n &= \alpha^{(0)} \int_0^{\infty} C'^2 e^{-\alpha^{(3)} \frac{1}{2} m C'^2} dC' \int_0^{\pi} \sin\phi d\phi \int_0^{2\pi} d\theta \\ &= \alpha^{(0)} \left(\frac{2\pi}{m\alpha^{(3)}} \right)^{\frac{3}{2}} \end{aligned} \tag{B.5}$$

Also,

$$\begin{aligned} n\mathbf{u} = \int \mathbf{c} f d\mathbf{c} &= \int \left(\frac{\boldsymbol{\alpha}^{(2)}}{\alpha^{(3)}} + \mathbf{C}' \right) f d\mathbf{C}' \\ &= n \frac{\boldsymbol{\alpha}^{(2)}}{\alpha^{(3)}} + \alpha^{(0)} \int e^{-\alpha^{(3)} \frac{1}{2} m C'^2} \mathbf{C}' d\mathbf{C}' \end{aligned} \tag{B.6}$$

Since the integrand is an odd function of the components $\mathbf{C}'$, the second term vanishes. Hence, we get $\mathbf{u} = \boldsymbol{\alpha}^{(2)}/\alpha^{(3)}$ and the parameter $\mathbf{C}'$ can be identified with the peculiar velocity. Using these results, (B.3) is written as

$$f = \alpha^{(0)} e^{-\alpha^{(3)} \frac{1}{2} m C'^2} = n \left(\frac{m\alpha^{(3)}}{2\pi} \right)^{\frac{3}{2}} e^{-\alpha^{(3)} \frac{1}{2} m C'^2} \tag{B.7}$$

Using the definition of temperature,

$$\frac{3}{2}k_B T = \frac{m}{2n}\int C^2 f d\mathbf{c}$$

$$= \frac{m}{2}\left(\frac{m\alpha^{(3)}}{2\pi}\right)^{\frac{3}{2}} \int C^2 e^{-\alpha^{(3)}\frac{1}{2}mC'^2} d\mathbf{C}$$

$$\therefore \quad \frac{3}{2}k_B T = \frac{3}{2\alpha^{(3)}} \quad \Rightarrow \alpha^{(3)} = \frac{1}{k_B T} \tag{B.8}$$

Hence, the final form of the distribution function in a steady and uniform state known as Maxwell's distribution function is given as

$$f^{(0)} = n\left(\frac{m}{2\pi kT}\right)^{\frac{3}{2}} \exp\left(-\frac{mC^2}{2k_B T}\right) \tag{B.9}$$

For a given density, mean velocity, and the absolute temperature, there appears to be only one possible permanent mode of distribution function. When the system is away from equilibrium, the actual distribution function will tend to approach this mode.

B.2 Derivation of $f^{(1)}$

B.2.1 Evaluation of Time Derivatives

In the evaluation of the distribution function at each approximation level, we will encounter the time derivatives of ρ, $\mathbf{u}$, and T. These time derivatives terms need to be replaced using the conservation laws (Eqs. (5.53)–(5.55)) as

$$\frac{\partial \rho}{\partial t} = -\frac{\partial(\rho u_k)}{\partial x_k},$$

$$\frac{\partial u_i}{\partial t} = -u_k\frac{\partial u_i}{\partial x_k} + F_i - \frac{1}{\rho}\sum \frac{\partial P_{ij}^{(r)}}{\partial x_j}$$

$$\frac{\partial T}{\partial t} = -u_k\frac{\partial T}{\partial x_k} - \frac{2}{3\rho R}\left\{\sum P_{ij}^{(r)}\frac{\partial u_i}{\partial x_j} - \sum \frac{\partial q_k^{(r)}}{\partial x_k}\right\}$$

Note that we have used $P_{ij} = p\delta_{ij} + p_{ij}$ in the above equations. The time derivatives of ρ, $\mathbf{u}$, and T are divided into parts as (subscript attached with ∂ symbol is not a tensorial index)

$$\frac{\partial n}{\partial t} = \sum \frac{\partial_r n}{\partial t}, \quad \frac{\partial u_k}{\partial t} = \sum \frac{\partial_r u_k}{\partial t}, \quad \frac{\partial T}{\partial t} = \sum \frac{\partial_r T}{\partial t} \tag{B.10}$$

The quantities on the right-hand side are not time-derivatives but they are defined as

$$\text{For } r = 0, \quad \frac{\partial_0 \rho}{\partial t} = -\frac{\partial \rho u_k}{\partial x_k} \tag{B.11}$$

$$\frac{\partial_0 u_i}{\partial t} = -u_k \frac{\partial u_i}{\partial x_k} + F_i - \frac{1}{\rho}\frac{\partial P_{ij}^{(0)}}{\partial x_j}, \tag{B.12}$$

$$\frac{\partial_0 T}{\partial t} = -u_k \frac{\partial T}{\partial x_k} - \frac{2}{3\rho R}\left\{P_{ij}^{(0)}\frac{\partial u_i}{\partial x_j} - \frac{\partial q_k^{(0)}}{\partial x_k}\right\}, \tag{B.13}$$

$$\text{For } r > 0, \quad \frac{\partial_r \rho}{\partial t} = 0, \tag{B.14}$$

$$\frac{\partial_r u_i}{\partial t} = -\frac{1}{\rho}\frac{\partial P_{ij}^{(r)}}{\partial x_j}, \tag{B.15}$$

$$\frac{\partial_r T}{\partial t} = -\frac{2}{3\rho R}\left\{P_{ij}^{(r)}\frac{\partial u_i}{\partial x_j} - \frac{\partial q_k^{(r)}}{\partial x_k}\right\} \tag{B.16}$$

At this stage, we have already solved for the zeroth approximation of the distribution function which turns out to be Maxwellian distribution (Eq. (B.9)). Utilizing this form of distribution function, expressions for zeroth approximation of pressure tensor and heat flux vector are obtained as

$$P_{ij}^{(0)} = \delta_{ij} p, \qquad q_i^{(0)} = 0 \tag{B.17}$$

Using these results, we have

$$\text{For } r = 0, \quad \frac{\partial_0 \rho}{\partial t} = -\frac{\partial \rho u_k}{\partial x_k} \tag{B.18}$$

$$\frac{\partial_0 u_i}{\partial t} = -u_k \frac{\partial u_i}{\partial x_k} + F_i - \frac{1}{\rho}\frac{\partial p}{\partial x_i}, \tag{B.19}$$

$$\frac{\partial_0 T}{\partial t} = -u_k \frac{\partial T}{\partial x_k} - \frac{2}{3}T\frac{\partial u_i}{\partial x_i}, \qquad [\because p = \rho R T] \tag{B.20}$$

$$\text{For } r > 0, \quad \frac{\partial_r \rho}{\partial t} = 0, \tag{B.21}$$

$$\frac{\partial_r u_i}{\partial t} = -\frac{1}{\rho}\frac{\partial P_{ij}^{(r)}}{\partial x_j}, \tag{B.22}$$

$$\frac{\partial_r T}{\partial t} = -\frac{2}{3\rho R}\left\{P_{ij}^{(r)}\frac{\partial u_i}{\partial x_j} - \frac{\partial q_k^{(r)}}{\partial x_k}\right\} \tag{B.23}$$

In terms of material derivative, we can write

$$\frac{D_0\rho}{Dt} = -\rho\frac{\partial u_k}{\partial x_k},\quad \frac{D_0 u_i}{Dt} = F_i - \frac{1}{\rho}\frac{\partial p}{\partial x_i},\quad \frac{D_0 T}{Dt} = -\frac{2T}{3}\frac{\partial u_k}{\partial x_k} \tag{B.24}$$

$$\frac{D_r\rho}{Dt} = 0,\quad \frac{D_r u_i}{Dt} = -\frac{1}{\rho}\frac{\partial P_{ij}^{(r)}}{\partial x_j},\quad \frac{D_r T}{Dt} = -\frac{2}{3\rho R}\left\{P_{ij}^{(r)}\frac{\partial u_i}{\partial x_j} - \frac{\partial q_k^{(r)}}{\partial x_k}\right\} \tag{B.25}$$

B.3 Expression for $\mathscr{D}^{(1)}$

As $f^{(0)}$ is a function of peculiar velocity C, it is more desirable to express $\mathscr{D}^{(1)}$ in terms of C. For this, the operator $\mathscr{D}$ of Boltzmann equation can be transformed as (see Appendix C.1),

$$\mathscr{D}^{(1)} = \frac{D_0 f^{(0)}}{Dt} + C_k\frac{\partial f^{(0)}}{\partial x_k} + \underline{\left(F_k - \frac{D_0 u_k}{Dt}\right)}\frac{\partial f^{(0)}}{\partial C_k} - \frac{\partial f^{(0)}}{\partial C_i}C_j\frac{\partial u_i}{\partial x_j}, \tag{B.26}$$

Using Eq. (B.24) (b) for the underlined term, and using $\frac{\partial f_0}{\partial t} = f^{(0)}\frac{\partial \log f^{(0)}}{\partial t}$, we have

$$\mathscr{D}^{(1)} = f^{(0)}\left\{\underbrace{\frac{D_0\log f^{(0)}}{Dt}}_{(a)} + \underbrace{C_k\frac{\partial \log f^{(0)}}{\partial x_k}}_{(b)} + \underbrace{\frac{1}{\rho}\frac{\partial p}{\partial x_k}\frac{\partial \log f^{(0)}}{\partial C_k}}_{(c)} - \underbrace{\frac{\partial \log f^{(0)}}{\partial C_i}C_j\frac{\partial u_i}{\partial x_j}}_{(d)}\right\}, \tag{B.27}$$

Taking the logarithm of the Maxwellian distribution,

$$\log f^{(0)} = \frac{3}{2}\log\frac{m}{2\pi k_B} + \log n - \frac{3}{2}\log T - \frac{mC^2}{2k_B T}$$

$$\Rightarrow \quad \frac{\partial \log f^{(0)}}{\partial C_i} = -\frac{mC_i}{k_B T} \tag{B.28}$$

$$(a) = \frac{D_0\log f^{(0)}}{Dt} = \frac{D_0\log n}{Dt} - \frac{3}{2}\frac{D_0\log T}{Dt} - \frac{mC^2}{2k_B}\frac{D_0}{Dt}\left(\frac{1}{T}\right) \tag{B.29}$$

Using Eq. (B.24) for material derivatives (ρ can be replaced by n in Eq. (B.24) (a)) the first two terms cancel out as

$$\frac{D_0 \log n}{Dt} - \frac{3}{2}\frac{D_0 \log T}{Dt} = \frac{1}{n}\frac{D_0 n}{Dt} - \frac{3}{2T}\frac{D_0 T}{Dt} = \frac{1}{n}n\frac{\partial u_k}{\partial x_k} - \frac{3}{2T}\left(\frac{-2T}{3}\right)\frac{\partial u_k}{\partial x_k} = 0$$

$$\Rightarrow (a) = \frac{D_0 \log f^{(0)}}{Dt} = \frac{mC^2}{2k_B T^2}\frac{D_0 T}{Dt} = -\frac{mC^2}{3k_B T}\frac{\partial u_k}{\partial x_k} \tag{B.30}$$

Thus, the addition of terms (a) and (d) is given by

$$\begin{aligned}
(a) + (d) &= \frac{D_0 \log f^{(0)}}{Dt} - \frac{\partial \log f^{(0)}}{\partial C_i} C_j \frac{\partial u_i}{\partial x_j} \\
&= -\frac{mC^2}{3k_B T}\frac{\partial u_i}{\partial x_i} + \frac{mC_i}{k_B T} C_j \frac{\partial u_i}{\partial x_j} \qquad \text{[Using (B.28)]} \\
&= \frac{m}{k_B T}\left\{C_i C_j \frac{\partial u_i}{\partial x_j} - \frac{1}{3}C^2 \delta_{ij}\frac{\partial u_i}{\partial x_j}\right\} \\
&= \frac{m}{k_B T} C_{\langle i} C_{j\rangle} \frac{\partial u_{\langle i}}{\partial x_{j\rangle}}
\end{aligned} \tag{B.31}$$

For middle terms, (b) and (c), we simplify as

$$\begin{aligned}
(b) + (c) &= C_k \frac{\partial \log f^{(0)}}{\partial x_k} + \frac{1}{\rho}\frac{\partial p}{\partial x_k}\frac{\partial \log f^{(0)}}{\partial C_k} \\
&= C_k \frac{\partial \log f^{(0)}}{\partial x_k} - \frac{m}{\rho k_B T} C_k \frac{\partial p}{\partial x_k} \qquad \text{[Using (B.28)]} \\
&= C_k \left(\frac{\partial \log f^{(0)}}{\partial x_k} - \frac{1}{\rho R T} p \frac{\partial \log p}{\partial x_k}\right) \\
&= C_k \left(\frac{\partial \log f^{(0)}}{\partial x_k} - \frac{\partial \log n k_B T}{\partial x_k}\right) \qquad \text{[Using } p = \rho R T = n k_B T\text{]} \\
&= C_k \left(\frac{\partial \log(f^{(0)}/n k_B T)}{\partial x_k}\right)
\end{aligned} \tag{B.32}$$

Dividing the Maxwellian distribution by $nk_B T$ and taking logarithm

$$\log \frac{f^{(0)}}{n k_B T} = \underbrace{\frac{3}{2}\log\frac{m}{2\pi} + \frac{5}{2}\log\frac{1}{k_B}}_{\text{constants}} - \frac{5}{2}\log T - \frac{mC^2}{2k_B T}$$

$$
\begin{aligned}
\therefore \qquad (b) + (c) &= C_k \frac{\partial \log(f^{(0)}/nk_B T)}{\partial x_k} \\
&= C_k \left[-\frac{5}{2} \frac{\partial \log T}{\partial x_k} - \frac{mC^2}{2k_B} \frac{\partial}{\partial x_k} \left(\frac{1}{T} \right) \right] \\
&= C_k \left[-\frac{5}{2} \frac{\partial \log T}{\partial x_k} + \frac{mC^2}{2k_B T} \frac{\partial \log T}{\partial x_k} \right] \\
&= \left(\frac{mC^2}{2k_B T} - \frac{5}{2} \right) C_k \frac{\partial \log T}{\partial x_k}
\end{aligned}
\tag{B.33}
$$

Hence, the final expression for $\mathscr{D}^{(1)}$ can be written as

$$
\mathscr{D}^{(1)} = f^{(0)} \left\{ \left(\frac{mC^2}{2k_B T} - \frac{5}{2} \right) C_k \frac{\partial \log T}{\partial x_k} + \frac{m}{k_B T} C_{\langle i} C_{j\rangle} \frac{\partial u_{\langle i}}{\partial x_{j\rangle}} \right\}
\tag{B.34}
$$

Appendix C
Appendix to Grad Equations

C.1 Transformation of Boltzmann Equation: $f(\mathbf{x}, \mathbf{c}, t) \Rightarrow f(\mathbf{x}, \mathbf{C}, t)$

The particle distribution function in the Boltzmann equation is expressed in terms of molecular velocity as $f(\mathbf{x}, \mathbf{c}, t)$. Sometimes, it is more convenient to express distribution function f in terms of peculiar velocity [$\mathbf{C} = \mathbf{c} - \mathbf{u}$]. In such a case, the Boltzmann equation has to be transformed into the new phase space coordinate system $f(\mathbf{x}, \mathbf{c}, t) \Rightarrow f(\mathbf{x}, \mathbf{C}(\mathbf{x}, \mathrm{t}), t)$. Care has to be taken while transforming the equation since the peculiar velocity $\mathbf{C}$ depends on $\mathbf{x}$ and t.

Starting with the original Boltzmann equation, we have

$$\frac{\partial f}{\partial t} + c_k \frac{\partial f}{\partial x_k} + F_k \frac{\partial f}{\partial c_k} = J(f, \ f_1) \tag{C.1}$$

The time derivative of distribution function transforms as

$$\begin{aligned} \frac{\partial f}{\partial t} &\Rightarrow \frac{\partial f}{\partial t} + \frac{\partial f}{\partial C_i}\frac{\partial C_i}{\partial t} = \frac{\partial f}{\partial t} + \frac{\partial f}{\partial C_i}\left(-\frac{\partial u_i}{\partial t}\right) \quad [C_i(\mathbf{x}, t) = c_i - u_i(\mathbf{x}, t),] \\ &= \frac{\partial f}{\partial t} - \frac{\partial u_i}{\partial t}\frac{\partial f}{\partial C_i} \quad \left[\text{Differentiating w.r.t time, } \frac{\partial C_i}{\partial t} = -\frac{\partial u_i}{\partial t}\right] \end{aligned} \tag{C.2}$$

The spatial derivative is transformed as

$$\begin{aligned} c_k \frac{\partial f}{\partial x_k} &\Rightarrow c_k \frac{\partial f}{\partial x_k} + c_k \frac{\partial f}{\partial C_j}\frac{\partial C_j}{\partial x_k} \qquad [C_j(\mathbf{x}, t) = c_j - u_j(\mathbf{x}, t),] \\ &= c_k \frac{\partial f}{\partial x_k} + c_k \frac{\partial f}{\partial C_j}\left(-\frac{\partial u_j}{\partial x_k}\right) \qquad \left[\because \frac{\partial C_j}{\partial x_k} = -\frac{\partial u_j}{\partial x_k}\right] \end{aligned}$$

A. Agrawal et al., *Microscale Flow and Heat Transfer*, Mechanical Engineering Series, https://doi.org/10.1007/978-3-030-10662-1

$$= C_k \frac{\partial f}{\partial x_k} + u_k \frac{\partial f}{\partial x_k} - C_k \frac{\partial f}{\partial C_j}\frac{\partial u_j}{\partial x_k} - u_k \frac{\partial f}{\partial C_j}\frac{\partial u_j}{\partial x_k} \tag{C.3}$$

and finally, the third term transforms as

$$\begin{aligned} F_k \frac{\partial f}{\partial c_k} \Rightarrow F_k \frac{\partial f}{\partial C_i}\frac{\partial C_i}{\partial c_k} &= F_k \frac{\partial f}{\partial C_i}\frac{\partial c_i}{\partial c_k} = F_k \frac{\partial f}{\partial C_i}\delta_{ik} \\ &= F_k \frac{\partial f}{\partial C_k} \qquad \left[\because \frac{\partial C_i}{\partial c_k} = \frac{\partial c_i}{\partial c_k} = \delta_{ik}\right] \end{aligned} \tag{C.4}$$

By combining all the transformations, we get

$$\begin{aligned} &\underline{\frac{\partial f}{\partial t}} - \underline{\frac{\partial u_j}{\partial t}\frac{\partial f}{\partial C_j}} + C_k \frac{\partial f}{\partial x_k} + \underline{u_k \frac{\partial f}{\partial x_k}} - C_k \frac{\partial f}{\partial C_j}\frac{\partial u_j}{\partial x_k} \\ &\quad - \underline{u_k \frac{\partial f}{\partial C_j}\frac{\partial u_j}{\partial x_k}} + F_k \frac{\partial f}{\partial C_k} = J(f,\ f_1) \end{aligned} \tag{C.5}$$

The underlined terms can be represented in terms of material derivative as

$$\frac{Df}{Dt} - \frac{\partial f}{\partial C_j}\frac{Du_j}{Dt} + C_k \frac{\partial f}{\partial x_k} - C_k \frac{\partial f}{\partial C_j}\frac{\partial u_j}{\partial x_k} + F_k \frac{\partial f}{\partial C_k} = J(f,\ f_1) \tag{C.6}$$

Rearranging some of the terms, we get the final expression as (dummy index j in the second term is changed to k)

$$\frac{Df}{Dt} + C_k \frac{\partial f}{\partial x_k} + \left(F_k - \frac{Du_k}{Dt}\right)\frac{\partial f}{\partial C_k} - C_k \frac{\partial f}{\partial C_j}\frac{\partial u_j}{\partial x_k} = J(f,\ f_1) \tag{C.7}$$

In the subsequent use of this equation (in Chaps. 5 and 6), the readers will notice that $\mathbf{x}$, $\mathbf{C}$, t are treated as independent variables. This differential treatment of $\mathbf{C}$ is referred to as mixed phase space treatment in the literature (see for example, [45]).

C.2 Note on Hermite Polynomials

Just like we expand a given vector in terms of a set of orthogonal base vectors, a sufficiently well-behaved function $f(x)$ can be expanded in terms of orthogonal functions like Legendre functions, Bessel functions, etc. The only difficulty lies in the fact that we are working in an infinite dimensional functional space.

Grad chose the orthogonal tensorial Hermite polynomials $\mathscr{H}_i^{(n)}$ [61] to expand the single particle distribution function because of its desirable mathematical properties. $\mathscr{H}_i^{(n)}$ is an n-th order tensor with subscripts, $i = (i_1, i_2, \ldots, i_n)$ as well as a polynomial of n-th degree. A complete set of orthonormal multi-dimensional Hermite polynomials $\mathscr{H}$ can be obtained by using products of such polynomials in a single variable.

In terms of Hermite polynomials, any function $f(x)$ can be represented as

$$f(\mathbf{x}) = \omega^{1/2} \sum_{n=0}^{\infty} \sum_{i=1}^{N} a_i^{(n)} \mathscr{H}_i^{(n)} \tag{C.8}$$

$$\text{where weight function,} \quad \omega(\mathbf{x}) = \frac{1}{(2\pi)^{N/2}} \exp\left(-\frac{x^2}{2}\right) \tag{C.9}$$

and Einstein summation convention is assumed for the inner summation as

$$\sum_{i=1}^{N} = \sum_{i_1=1}^{N} \sum_{i_2=1}^{N} \cdots \sum_{i_n=1}^{N} \tag{C.10}$$

The weight function $\omega(x)$ is normalized so that

$$\int \omega(\mathbf{x}) d\mathbf{x} = 1. \tag{C.11}$$

where $\int(\ldots)d\mathbf{x}$ represents an N-fold integration over $-\infty < x_i < +\infty$.

In expanded form, function $f(x)$ is represented as

$$f(\mathbf{x}) = \omega^{1/2} \left\{ a^{(0)} \mathscr{H}^{(0)} + \sum_{i=1}^{N} a_i^{(1)} \mathscr{H}_i^{(1)} + \sum_{i,j=1}^{N} a_{ij}^{(2)} \mathscr{H}_{ij}^{(2)} + \sum_{i,j,k=1}^{N} a_{ijk}^{(3)} \mathscr{H}_{ijk}^{(3)} + \cdots \right\} \tag{C.12}$$

Since $\mathscr{H}_i^{(n)}$ is symmetric in its subscripts, the scalar product of any tensor of order n, $a_i^{(n)}$ and $\mathscr{H}_i^{(n)}$ is equal to the product of symmetric part of $a_i^{(n)}$ and $\mathscr{H}_i^{(n)}$.

The orthogonality of Hermite polynomials can be established by proving the condition,

$$\int \omega \mathscr{H}_i^{(n)} \mathscr{H}_j^{(n)} d\mathbf{x} = \boldsymbol{\delta}_{ij}{}^n \tag{C.13}$$

where $\boldsymbol{\delta}_{ij}{}^n$ is a $2n$-th order tensor.

To represent a function $f(x)$ completely, we still need to evaluate the coefficients $a_i^{(n)}$. Just like we proceed in Fourier series expansion, we multiply both sides of Eq. (C.8) by $\omega^{(1/2)} \mathscr{H}_j^{(m)}$ and integrate as

$$\int \omega^{1/2} \mathscr{H}_j^{(m)} f(\mathbf{x}) d\mathbf{x} = \int \omega \sum_{n=0}^{\infty} \sum_{i=1}^{N} a_i^{(n)} \mathscr{H}_i^{(n)} \mathscr{H}_j^{(m)} d\mathbf{x} \tag{C.14}$$

Using orthogonality condition of Hermite polynomials (Eq. (C.13)), only one term in summation $\sum_{n=0}^{\infty}$ survives where $n = m$.

$$\begin{aligned}
\therefore \quad \int \omega^{1/2} \mathscr{H}_j^{(m)} f(\mathbf{x}) d\mathbf{x} &= \int \sum_{i=1}^{N} a_i^{(m)} \omega \mathscr{H}_i^{(m)} \mathscr{H}_j^{(m)} d\mathbf{x} \\
&= \sum_{i=1}^{N} a_i^{(m)} \int \omega \mathscr{H}_i^{(m)} \mathscr{H}_j^{(m)} d\mathbf{x} \\
&= \sum_{i=1}^{N} a_i^{(m)} \boldsymbol{\delta_{ij}}^m \qquad \text{[Using Eq. (C.13)]}
\end{aligned}$$

Choosing one term of the $n!$ contained in $\boldsymbol{\delta_{ij}}^m$ and summing over i gives the result $a_j^{(m)}$. Consequently, we can write as

$$\int \omega^{1/2} \mathscr{H}_j^{(m)} f(\mathbf{x}) d\mathbf{x} = m! a_j^{(m)}$$

$$\therefore \quad a_j^{(n)} = \frac{1}{n!} \int \omega^{1/2} \mathscr{H}_j^{(n)} f(\mathbf{x}) d\mathbf{x} \tag{C.15}$$

where dummy index m is changed to n.

In the process of generating Hermite polynomials, we need to take the gradients of the weight function frequently as

$$\begin{aligned}
\nabla_i \omega &= \frac{1}{(2\pi)^{N/2}} \frac{\partial}{\partial x_i} \left(\exp\left[-\frac{x_k x_k}{2} \right] \right) \\
&= \frac{1}{(2\pi)^{N/2}} \exp\left[-\frac{x_k x_k}{2} \right] \frac{\partial}{\partial x_i} \left(-\frac{x_k x_k}{2} \right) \\
&= \omega \left(-\frac{1}{2} \right) 2 x_k \frac{\partial x_k}{\partial x_i} \qquad \text{[Using Eq. (C.9)]} \\
&= -\omega x_k \delta_{ik} \qquad \left[\because \frac{\partial x_k}{\partial x_i} = \delta_{ik} \right]
\end{aligned}$$

$$\therefore \nabla_i \omega = -x_i \omega, \tag{C.16}$$

Similarly, we can prove

$$\nabla_i \left(\frac{1}{\omega}\right) = \frac{x_i}{\omega} \tag{C.17}$$

The Hermite polynomial of order n ($n \geq 0$ is an integer) can be generated using the definition,

$$\mathscr{H}^{(n)} = \frac{(-1)^{(n)}}{\omega} \nabla^{(n)} \omega \tag{C.18}$$

First few Hermite polynomials are obtained as

$$\begin{aligned}
\mathscr{H}^{(0)} &= \frac{1}{\omega}\omega = 1 \\
\mathscr{H}_i^{(1)} &= \frac{-1}{\omega}\nabla_i \omega = \frac{-1}{\omega}(-x_i\omega) = x_i \\
\mathscr{H}_{ij}^{(2)} &= \frac{1}{\omega}\left[\nabla_i \nabla_j \omega\right] = \frac{1}{\omega}\left[\nabla_i(-x_j\omega)\right] = \frac{-1}{\omega}\left[x_j \nabla_i \omega + \omega \nabla_i x_j\right] \\
&= \frac{-1}{\omega}\left[x_j(-x_i\omega) + \omega\delta_{ij}\right] \\
&= x_i x_j - \delta_{ij} \\
\mathscr{H}_{ijk}^{(3)} &= \frac{-1}{\omega}\left[\nabla_i \nabla_j \nabla_k \omega\right] = \frac{-1}{\omega}\left[\nabla_i \nabla_j(-x_k\omega)\right] = \frac{1}{\omega}\left[\nabla_i(x_k \nabla_j \omega + \omega \nabla_j x_k)\right] \\
&= \frac{1}{\omega}\left[\nabla_i(-x_k x_j \omega + \omega\delta_{jk})\right] = \frac{1}{\omega}\left[-\nabla_i(x_k x_j \omega) + \nabla_i(\omega\delta_{jk})\right] \\
&= \frac{-1}{\omega}\left[x_k x_j \nabla_i \omega + x_k \omega \nabla_i x_j + x_j \omega \nabla_i x_k - \delta_{jk}\nabla_i \omega\right] \\
&= \frac{-1}{\omega}\left[x_k x_j(-x_i\omega) + x_k\omega\delta_{ij} + x_j\omega\delta_{ik} - \delta_{jk}(-x_i\omega)\right] \\
&= x_i x_j x_k - \left(x_i\delta_{jk} + x_j\delta_{ik} + x_k\delta_{ij}\right)
\end{aligned} \tag{C.19}$$

Similarly, we can obtain an expression for fourth order Hermite polynomial $\mathscr{H}_{ijkl}^{(4)}$ as

$$\begin{aligned}
\mathscr{H}_{ijkl}^{(4)} = x_i x_j x_k x_l &- \left[x_i x_j \delta_{kl} + x_i x_k \delta_{jl} + x_i x_l \delta_{jk} + x_j x_k \delta_{il} + x_j x_l \delta_{ik} + x_k x_l \delta_{ij}\right] \\
&+ \left[\delta_{ij}\delta_{kl} + \delta_{ik}\delta_{jl} + \delta_{il}\delta_{jk}\right]
\end{aligned} \tag{C.20}$$

References

1. Acosta R, Muller R, Tobias C (1985) Transport processes in narrow (capillary) channels. AIChE J 31(3):473–482
2. Agarwal RK, Yun KY, Balakrishnan R (2001) Beyond Navier–Stokes: Burnett equations for flows in the continuum-transition regime. Phys Fluids 13(10):3061–3085; see also: Agarwal RK, Yun KY, Balakrishnan R (2002) Erratum: "Beyond Navier–Stokes: Burnett equations for flows in the continuum-transition regime" [Phys. Fluids 13, 3061 (2001)]. Phys Fluids 14:1818
3. Agrawal A (2016) Higher-order continuum equation based heat conduction law. INAE Lett 1(2):35–39
4. Agrawal A, Prabhu SV (2008) Deduction of slip coefficient in slip and transition regimes from existing cylindrical Couette flow data. Exp Thermal Fluid Sci 32(4):991–996
5. Agrawal A, Prabhu SV (2008) Survey on measurement of tangential momentum accommodation coefficient. J Vac Sci Technol A 26(4):634–645
6. Agrawal A, Djenidi L, Agrawal A (2009) Simulation of gas flow in microchannels with a single 90° bend. Comput Fluids 38(8):1629–1637
7. Ai DK (1960) Cylindrical Couette flow in a rarefied gas according to Grad's equations. PhD thesis, California Institute of Technology
8. Akintunde A, Petculescu A (2014) Infrasonic attenuation in the upper mesosphere–lower thermosphere: a comparison between Navier–Stokes and Burnett predictions. J Acoust Soc Am 136(4):1483–1486
9. Albertoni S, Cercignani C, Gotusso L (1963) Numerical evaluation of the slip coefficient. Phys Fluids 6(7):993–996
10. Alexeenko AA, Fedosov DA, Gimelshein SF, Levin DA, Collins RJ (2006) Transient heat transfer and gas flow in a MEMS-based thruster. J Microelectromech Syst 15(1):181–194
11. Arkilic E, Schmidt M, Breuer K (1997) Gaseous slip flow in long microchannels. J Microelectromech Syst 6(2):167–178
12. Arkilic EB, Breuer KS, Schmidt MA (2001) Mass flow and tangential momentum accommodation in silicon micromachined channels. J Fluid Mech 437:29–43
13. Avcı M, Aydın O (2008) Laminar forced convection slip-flow in a micro-annulus between two concentric cylinders. Int J Heat Mass Transf 51(13–14):3460–3467
14. Aydın O, Avcı M (2006) Heat and fluid flow characteristics of gases in micropipes. Int J Heat Mass Transf 49(9–10):1723–1730
15. Aydın O, Avcı M (2007) Analysis of laminar heat transfer in micro-Poiseuille flow. Int J Thermal Sci 46(1):30–37

A. Agrawal et al., *Microscale Flow and Heat Transfer*, Mechanical Engineering Series, https://doi.org/10.1007/978-3-030-10662-1

16. Balaj M, Roohi E, Mohammadzadeh A (2017) Regulation of anti-Fourier heat transfer for non-equilibrium gas flows through micro/nanochannels. Int J Thermal Sci 118:24–39
17. Balakrishnan R, Agarwal RK (1997) Numerical simulation of Bhatnagar-Gross-Krook-Burnett equations for hypersonic flows. J Thermophys Heat Transf 11(3):391–399
18. Barber RW, Emerson DR (2001) A numerical investigation of low Reynolds number gaseous slip flow at the entrance of circular and parallel plate micro-channels. In: ECCOMAS computational fluid dynamics conference, Swansea, Wales
19. Bentz JA, Tompson R, Loyalka S (1997) The spinning rotor gauge: measurements of viscosity, velocity slip coefficients, and tangential momentum accommodation coefficients for N_2 and CH_4. Vacuum 48(10):817–824
20. Bhatnagar PL, Gross EP, Krook M (1954) A model for collision processes in gases. I. Small amplitude processes in charged and neutral one-component systems. Phys Rev 94:511–525
21. Bird G (2013) The DSMC method. CreateSpace Independent Publishing Platform, California
22. Bobylev A (1982) The Chapman–Enskog and Grad methods for solving the Boltzmann equation. Akad Nauk SSSR Dokl 262:71–75
23. Bobylev A (2008) Generalized Burnett hydrodynamics. J Stat Phys 132(3):569–580
24. Bobylev A, Bisi M, Cassinari M, Spiga G (2011) Shock wave structure for generalized Burnett equations. Phys Fluids 23(3):030607
25. Brenner H (2005) Navier–Stokes revisited. Phys A Stat Mech Appl 349(1–2):60–132
26. Burnett D (1936) The distribution of molecular velocities and the mean motion in a non-uniform gas. Proc Lond Math Soc 2(1):382–435
27. Cercignani C (1988) The Boltzmann equation and its applications, 1st edn. Applied mathematical sciences, vol 67. Springer, New York
28. Cercignani C (2000) Rarefied gas dynamics: from basic concepts to actual calculations, vol 21. Cambridge University Press, Cambridge
29. Cercignani C, Daneri A (1963) Flow of a rarefied gas between two parallel plates. J Appl Phys 34(12):3509–3513
30. Cercignani C, Lampis M, Lorenzani S (2004) Variational approach to gas flows in microchannels. Phys Fluids 16:3426–3437
31. Chapman S, Cowling TG (1970) The mathematical theory of non-uniform gases: an account of the kinetic theory of viscosity, thermal conduction and diffusion in gases. Cambridge University Press, Cambridge
32. Chen RY (1973) Flow in the entrance region at low Reynolds numbers. J Fluids Eng 95(1):153–158
33. Chen S, Tian Z (2009) Simulation of microchannel flow using the lattice Boltzmann method. Phys A Stat Mech Appl 388(23):4803–4810
34. Choi S (1991) Fluid flow and heat transfer in microtubes. Micromechanical sensors, actuators, and systems. ASME, New York, pp 123–134
35. Colin S, Lalonde P, Caen R (2004) Validation of a second-order slip flow model in rectangular microchannels. Heat Transf Eng 25(3):23–30
36. Comeaux KA, Chapman DR, MacCormack RW (1995) An analysis of the Burnett equations based on the second law of thermodynamics. In: 33rd Aerospace sciences meeting and exhibit, p 415
37. Dadzie SK (2013) A thermo-mechanically consistent Burnett regime continuum flow equation without Chapman–Enskog expansion. J Fluid Mech 716:R6
38. Date AW (2005) Introduction to computational fluid dynamics. Cambridge University Press, Cambridge
39. De Groot SR, Mazur P (2013) Non-equilibrium thermodynamics. Courier Dover Publications, New York
40. Deissler R (1964) An analysis of second-order slip flow and temperature-jump boundary conditions for rarefied gases. Int J Heat Mass Transf 7(6):681–694
41. Demsis A, Verma B, Prabhu SV, Agrawal A (2009) Experimental determination of heat transfer coefficient in the slip regime and its anomalously low value. Phys Rev E 80(1):016311

42. Demsis A, Prabhu SV, Agrawal A (2010) Influence of wall conditions on friction factor for flow of gases under slip condition. Exp Thermal Fluid Sci 34(8):1448–1455
43. Dombrowski N, Foumeny E, Ookawara S, Riza A (1993) The influence of Reynolds number on the entry length and pressure drop for laminar pipe flow. Can J Chem Eng 71(3):472–476
44. Dongari N, Agrawal A, Agrawal A (2007) Analytical solution of gaseous slip flow in long microchannels. Int J Heat Mass Transf 50(17):3411–3421
45. Dosch A, Zank GP (2016) Transport processes in space physics and astrophysics. Springer, Berlin
46. Duan Z (2012) Second-order gaseous slip flow models in long circular and noncircular microchannels and nanochannels. Microfluid Nanofluid 12(5):805–820
47. Duan Z, Muzychka Y (2010) Slip flow in the hydrodynamic entrance region of circular and noncircular microchannels. J Fluids Eng 132(1):011201
48. Durst F, Ray S, Ünsal B, Bayoumi O (2005) The development lengths of laminar pipe and channel flows. J Fluids Eng 127(6):1154–1160
49. Duryodhan VS, Singh SG, Agrawal A (2013) Liquid flow through a diverging microchannel. Microfluid Nanofluid 14(1–2):53–67
50. Duryodhan VS, Singh A, Singh SG, Agrawal A (2015) Convective heat transfer in diverging and converging microchannels. Int J Heat Mass Transf 80:424–438
51. Duryodhan VS, Singh SG, Agrawal A (2017) Effect of cross aspect ratio on flow in diverging and converging microchannels. J Fluids Eng 139(6):061203
52. Eu CB (1980) A modified moment method and irreversible thermodynamics. J Chem Phys 73(6):2958–2969
53. Ewart T, Perrier P, Graur IA, Méolans JG (2007) Mass flow rate measurements in a microchannel, from hydrodynamic to near free molecular regimes. J Fluid Mech 584:337–356
54. Ewart T, Perrier P, Graur IA, Méolans JG (2007) Tangential momentum accommodation in microtube. Microfluid Nanofluid 3(6):689–695
55. Fan H, Xue Q (2000) A new analytic solution of the Navier–Stokes equations for microchannel flows. Microscale Thermophys Eng 4(2):125–143
56. Fiscko KA, Chapman DR (1989) Comparison of Burnett, super-Burnett, and Monte Carlo solutions for hypersonic shock structure. Progress in astronautics and aeronautics (AIAA), pp 374–395
57. Gad-el Hak M (1999) The fluid mechanics of microdevices—the Freeman scholar lecture. J Fluids Eng 121(1):5–33
58. Gad-el Hak M (2001) The MEMS handbook. CRC Press, Boca Raton
59. Gavasane A, Sachdev SS, Mittal BK, Bhandarkar UV, Agrawal A (2011) A critical assessment of the Maxwell slip boundary condition for rarefied wall bounded flows. Int J Micro-Nano Scale Transp 2:109–116
60. Gembarovič J, Majernik V (1988) Non-Fourier propagation of heat pulses in finite medium. Int J Heat Mass Transf 31(5):1073–1080
61. Grad H (1949) Note on N-dimensional Hermite polynomials. Commun Pure Appl Math 2(4):325–330
62. Grad H (1949) On the kinetic theory of rarefied gases. Commun Pure Appl Math 2(4):331–407
63. Grad H (1952) The profile of a steady plane shock wave. Commun Pure Appl Math 5(3):257–300
64. Grad H (1958) Principles of the kinetic theory of gases. Springer, Berlin, pp 205–294
65. Graur I, Veltzke T, Méolans J, Ho M, Thöming J (2015) The gas flow diode effect: theoretical and experimental analysis of moderately rarefied gas flows through a microchannel with varying cross section. Microfluid Nanofluid 18(3):391–402
66. Greenshields CJ, Reese JM (2007) The structure of shock waves as a test of Brenner's modifications to the Navier–Stokes equations. J Fluid Mech 580:407–429
67. Gross EP, Jackson EA (1958) Kinetic theory of the impulsive motion of an infinite plane. Phys Fluids 1(4):318–328

68. Gu XJ, Emerson DR (2009) A high-order moment approach for capturing non-equilibrium phenomena in the transition regime. J Fluid Mech 636:177–216
69. Gu XJ, Emerson DR, Tang GH (2009) Kramer's problem and the Knudsen minimum: a theoretical analysis using a linearized 26-moment approach. Contin Mech Thermodyn 21(5):345
70. Hemadri V, Varade VV, Agrawal A, Bhandarkar UV (2016) Investigation of rarefied gas flow in microchannels of non-uniform cross section. Phys Fluids 28(2):022007
71. Hemadri V, Varade VV, Agrawal A, Bhandarkar UV (2017) Rarefied gas flow in converging microchannel in slip and early transition regimes. Phys Fluids 29:032002
72. Hemadri V, Agrawal A, Bhandarkar UV (2018) Determination of tangential momentum accommodation coefficient and slip coefficients for rarefied gas flow in a microchannel. Sādhanā 43(10):164
73. Hemadri V, Biradar GS, Shah N, Garg R, Bhandarkar UV, Agrawal A (2018) Experimental study of heat transfer in rarefied gas flow in a circular tube with constant wall temperature. Exp Thermal Fluid Sci 93:326–333
74. Hooman K (2008) A superposition approach to study slip-flow forced convection in straight microchannels of uniform but arbitrary cross-section. Int J Heat Mass Transf 51(15–16):3753–3762
75. Hsia YT, Domoto G (1983) An experimental investigation of molecular rarefaction effects in gas lubricated bearings at ultra-low clearances. J Lubr Technol 105(1):120–129
76. Huang C, Gregory JW, Sullivan JP (2007) Microchannel pressure measurements using molecular sensors. J Microelectromech Syst 16(4):777–785
77. Inman RM (1964) Laminar slip flow heat transfer in a parallel-plate channel or a round tube with uniform wall heating. NASA Technical note D-2393. National Aeronautics and Space Administration, Washington, DC
78. Jadhav RS, Singh N, Agrawal A (2017) Force-driven compressible plane Poiseuille flow by Onsager-Burnett equations. Phys Fluids 29(10):102002
79. Jang J, Wereley ST (2006) Gaseous slip flow analysis of a micromachined flow sensor for ultra small flow applications. J Micromech Microeng 17(2):229
80. Jie D, Diao X, Cheong KB, Yong LK (2000) Navier–Stokes simulations of gas flow in micro devices. J Micromech Microeng 10(3):372
81. Karniadakis GE, Beskok A, Aluru N (2006) Microflows and nanoflows: fundamentals and simulation, vol 29. Springer, Berlin
82. Kennard EH (1938) Kinetic theory of gases with an introduction to statistical mechanics. McGraw-Hill, New York
83. Kogan MN (1969) Rarefied gas dynamics. Plenum Press, New York
84. Kremer GM (2010) An introduction to the Boltzmann equation and transport processes in gases. Springer, Berlin
85. Kundu PK (1990) Fluid mechanics. Academic Press, New York
86. Lee WY, Wong M, Zohar Y (2002) Microchannels in series connected via a contraction/expansion section. J Fluid Mech 459:187–206
87. Lihnaropoulos J, Valougeorgis D (2011) Unsteady vacuum gas flow in cylindrical tubes. Fusion Eng Des 86(9–11):2139–2142
88. Liu J, Tai YC, Pong CMHC (1995) MEMS for pressure distribution studies of gaseous flows in microchannels. In: Proceedings of the IEEE Micro Electro Mechanical Systems. IEEE, Piscataway, pp 209–215
89. Lockerby DA, Reese JM, Emerson DR, Barber RW (2004) Velocity boundary condition at solid walls in rarefied gas calculations. Phys Rev E 70(1):017303
90. Loyalka S (1996) Theory of the spinning rotor gauge in the slip regime. J Vac Sci Technol A 14(5):2940–2945
91. Mahendra AK, Singh RK (2013) Onsager reciprocity principle for kinetic models and kinetic schemes. arXiv preprint arXiv:13084119
92. Maranzana G, Perry I, Maillet D (2004) Mini-and micro-channels: influence of axial conduction in the walls. Int J Heat Mass Transf 47(17–18):3993–4004

93. Maurer J, Tabeling P, Joseph P, Willaime H (2003) Second-order slip laws in microchannels for helium and nitrogen. Phys Fluids 15(9):2613–2621
94. Maxwell JC (1879) On stresses in rarified gases arising from inequalities of temperature. Philos Trans R Soc Lond A 170:231–256
95. McLennan JA (1974) Onsager's theorem and higher-order hydrodynamic equations. Phys Rev A 10(4):1272
96. McNamara S, Gianchandani YB (2005) On-chip vacuum generated by a micromachined Knudsen pump. J Microelectromech Syst 14(4):741–746
97. Mitsuya Y (1993) Modified Reynolds equation for ultra-thin film gas lubrication using 1.5-order slip-flow model and considering surface accommodation coefficient. J Tribol 115:289–289
98. Miyamoto M, Shi W, Katoh Y, Kurima J (2003) Choked flow and heat transfer of low density gas in a narrow parallel-plate channel with uniformly heating walls. Int J Heat Mass Transf 46(14):2685–2693
99. Mohammadzadeh A, Rana AS, Struchtrup H (2015) Thermal stress vs. thermal transpiration: a competition in thermally driven cavity flows. Phys Fluids 27(11):112001
100. Moharana MK, Khandekar S (2013) Effect of aspect ratio of rectangular microchannels on the axial back conduction in its solid substrate. Int J Microscale Nanoscale Thermal Fluid Transp Phenom 4(3/4):211
101. Morini GL, Spiga M, Tartarini P (2004) The rarefaction effect on the friction factor of gas flow in microchannels. Superlattices Microstruct 35(3):587–599
102. Muralidhar K, Biswas G (2005) Advanced engineering fluid mechanics. Narosa Publishing House, New Delhi
103. Myong RS (1999) Thermodynamically consistent hydrodynamic computational models for high-Knudsen-number gas flows. Phys Fluids 11(9):2788–2802
104. Myong R (2011) A full analytical solution for the force-driven compressible Poiseuille gas flow based on a nonlinear coupled constitutive relation. Phys Fluids 23(1):012002
105. Niazmand H, Renksizbulut M, Saeedi E (2008) Developing slip-flow and heat transfer in trapezoidal microchannels. Int J Heat Mass Transf 51(25–26):6126–6135
106. Ohira SI, Toda K (2008) Micro gas analyzers for environmental and medical applications. Anal Chim Acta 619(2):143–156
107. Onsager L (1931) Reciprocal relations in irreversible processes. I. Phys Rev 37(4):405
108. Onsager L (1931) Reciprocal relations in irreversible processes. II. Phys Rev 38(12):2265
109. Özişik M, Vick B (1984) Propagation and reflection of thermal waves in a finite medium. Int J Heat Mass Transf 27(10):1845–1854
110. Panigrahi PK (2016) Transport phenomena in microfluidic systems. Wiley, New York
111. Panigrahi AK, Singh V, Singh SG (2017) Nonlithographic fabrication of plastic-based nanofibers integrated microfluidic biochip for sensitive detection of infectious biomarker. Appl Mater Interfaces 9(46):39994–40005
112. Petculescu A (2016) Acoustic properties in the low and middle atmospheres of Mars and Venus. J Acoust Soc Am 140(2):1439–1446
113. Pfahler J, Harley J, Bau H, Zemel JN (1991) Gas and liquid flow in small channels. Am Soc Mech Eng Dyn Syst Control Div 32:49–60
114. Pollard W, Present RD (1948) On gaseous self-diffusion in long capillary tubes. Phys Rev 73(7):762
115. Pong K, Ho CM, Liu J, Tai YC (1994) Non-linear pressure distribution in uniform microchannels. In: Proceedings of application of microfabrication to fluid mechanics, ASME Winter annual meeting, Chicago, pp 51–56
116. Pope SB (2001) Turbulent flows. IOP Publishing, Bristol
117. Porodnov B, Suetin P, Borisov S, Akinshin V (1974) Experimental investigation of rarefied gas flow in different channels. J Fluid Mech 64(3):417–438
118. Prabhakar A, Kumar YBV, Tripathi S, Agrawal A (2015) A novel, compact and efficient microchannel arrangement with multiple hydrodynamic effects for blood plasma separation. Microfluid Nanofluid 18(5–6):995–1006

119. Prabhu SV, Mahulikar SP (2014) Effects of density and thermal conductivity variations on entropy generation in gas micro-flows. Int J Heat Mass Transf 79:472–485
120. Rath A, Singh N, Agrawal A (2018) A perturbation-based solution of Burnett equations for gaseous flow in a long microchannel. J Fluid Mech 844:1038–1051
121. Rohsenow WM, Hartnett JP, Cho YI (1998) Handbook of heat transfer, vol 3. McGraw-Hill, New York
122. Romero M, Velasco R (1995) Onsager's symmetry in the Burnett regime. Phys A Stat Mech Appl 222(1–4):161–172
123. Sambasivam R (2012) Extended Navier–Stokes equations: derivations and applications to fluid flow problems. PhD thesis, Universität Erlangen-Nürnberg
124. Satapathy AK (2010) Slip flow heat transfer in an infinite microtube with axial conduction. Int J Thermal Sci 49(1):153–160
125. Schamberg R (1947) The fundamental differential equations and the boundary conditions for high speed slip-flow, and their application to several specific problems. PhD thesis, California Institute of Technology
126. Schlichting H (1960) Boundary layer theory. McGraw-Hill, New York
127. Sharipov F, Graur I (2014) General approach to transient flows of rarefied gases through long capillaries. Vacuum 100:22–25
128. Sharma A (2016) Introduction to computational fluid dynamics: development, application and analysis. Wiley, New York
129. Shavaliyev MS (1993) Super-Burnett corrections to the stress tensor and the heat flux in a gas of Maxwellian molecules. J Appl Math Mech 57(3):573–576
130. Singh N, Agrawal A (2014) The Burnett equations in cylindrical coordinates and their solution for flow in a microtube. J Fluid Mech 751:121–141
131. Singh N, Agrawal A (2016) Onsager's-principle-consistent 13-moment transport equations. Phys Rev E 93(6):063111
132. Singh N, Dongari N, Agrawal A (2014) Analytical solution of plane Poiseuille flow within Burnett hydrodynamics. Microfluid Nanofluid 16(1–2):403–412
133. Singh N, Gavasane A, Agrawal A (2014) Analytical solution of plane Couette flow in the transition regime and comparison with direct simulation Monte Carlo data. Comput Fluids 97:177–187
134. Singh SG, Jain A, Sridharan A, Duttagupta S, Agrawal A (2009) Flow map and measurement of void fraction and heat transfer coefficient using an image analysis technique for flow boiling of water in a silicon microchannel. J Micromech Microeng 19(7):075004
135. Singh SG, Bhide R, Duttagupta S, Sridharan A, Agrawal A (2010) Experimental study of water boiling in microchannel. Adv Multiphase Flow Heat Transf 3:246–303
136. Singh N, Jadhav RS, Agrawal A (2017) Derivation of stable Burnett equations for rarefied gas flows. Phys Rev E 96(1):013106
137. Sreekanth A (1969) Slip flow through long circular tubes. In: Trilling L, Wachman HY (eds) Proceedings of the sixth international symposium on rarefied gas dynamics. Academic Press, New York, pp 667–680
138. Steckelmacher W (1986) Knudsen flow 75 years on: the current state of the art for flow of rarefied gases in tubes and systems. Rep Prog Phys 49(10):1083
139. Stemme E, Stemme G (1993) A valveless diffuser/nozzle-based fluid pump. Sens. Actuators A 39(2):159–167
140. Struchtrup H (2004) Stable transport equations for rarefied gases at high orders in the Knudsen number. Phys Fluids 16(11):3921–3934
141. Struchtrup H (2005) Macroscopic transport equations for rarefied gas flows. Springer, Berlin
142. Struchtrup H, Torrilhon M (2003) Regularization of Grad's 13 moment equations: derivation and linear analysis. Phys Fluids 15(9):2668–2680
143. Suetin P, Porodnov B, Chernjak V, Borisov S (1973) Poiseuille flow at arbitrary Knudsen numbers and tangential momentum accommodation. J Fluid Mech 60(3):581–592
144. Tang G, Li Z, He Y, Tao W (2007) Experimental study of compressibility, roughness and rarefaction influences on microchannel flow. Int J Heat Mass Transf 50(11–12):2282–2295

145. Tekasakul P, Bentz J, Tompson R, Loyalka S (1996) The spinning rotor gauge: measurements of viscosity, velocity slip coefficients, and tangential momentum accommodation coefficients. J Vac Sci Technol A 14(5):2946–2952
146. Timokhin MY, Struchtrup H, Kokhanchik A, Bondar YA (2017) Different variants of R13 moment equations applied to the shock-wave structure. Phys Fluids 29(3):037,105
147. Torrilhon M (2016) Modeling nonequilibrium gas flow based on moment equations. Ann Rev Fluid Mech 48(1):429–458
148. Torrilhon M, Struchtrup H (2004) Regularized 13-moment equations: shock structure calculations and comparison to Burnett models. J Fluid Mech 513:171–198
149. Truesdell C, Muncaster RG (1980) Fundamentals of Maxwel's kinetic theory of a simple monatomic gas: treated as a branch of rational mechanics, vol 83. Academic Press, New York
150. Tunc G, Bayazitoglu Y (2001) Heat transfer in microtubes with viscous dissipation. Int J Heat Mass Transf 44(13):2395–2403
151. Turner SE, Sun H, Faghri M, Gregory OJ (1999) Local pressure measurement of gaseous flow through microchannels. ASME-Publications-HTD 364:71–80
152. Tzou D, Chiu K (2001) Temperature-dependent thermal lagging in ultrafast laser heating. Int J Heat Mass Transf 44(9):1725–1734
153. Uribe FJ (2011) The shock wave problem revisited: the Navier–Stokes equations and Brenner's two velocity hydrodynamics. In: Coping with complexity: model reduction and data analysis. Springer, Berlin, pp 207–229
154. Uribe F, Garcia A (1999) Burnett description for plane Poiseuille flow. Phys Rev E 60(4A):4063–4078
155. Uribe F, Velasco R, Garcia-Colin L (2000) Bobylev's instability. Phys Rev E 62(4):5835
156. Varade V, Agrawal A, Pradeep AM (2014) Behaviour of rarefied gas flow near the junction of a suddenly expanding tube. J Fluid Mech 739:363–391
157. Varade V, Agrawal A, Pradeep AM (2014) Experimental study of rarefied gas flow near sudden contraction junction of a tube. Phys Fluids 26(6):062002
158. Varade V, Agrawal A, Prabhu SV, Pradeep AM (2015) Early onset of flow separation with rarefied gas flowing in a 90° bend tube. Exp Thermal Fluid Sci 66:221–234
159. Varade V, Agrawal A, Prabhu SV, Pradeep AM (2015) Velocity measurement in low Reynolds and low Mach number slip flow through a tube. Exp Thermal Fluid Sci 60:284–289
160. Varade V, Duryodhan VS, Agrawal A, Pradeep AM, Ebrahimi A, Roohi E (2015) Low Mach number slip flow through diverging microchannel. Comput Fluids 111:46–61
161. Veltzke T, Baune M, Thöming J (2012) The contribution of diffusion to gas microflow: an experimental study. Phys Fluids 24(8):082004
162. Verma B, Demsis A, Agrawal A, Prabhu SV (2009) Semiempirical correlation for the friction factor of gas flowing through smooth microtubes. J Vac Sci Technol A 27(3):584–590
163. Wallace MAG, Pleil JD (2018) Evolution of clinical and environmental health applications of exhaled breath research: review of methods and instrumentation for gas-phase, condensate, and aerosols. Anal Chim Acta 1024:18–38
164. Weiss W (1995) Continuous shock structure in extended thermodynamics. Phys Rev E 52:R5760–R5763
165. White C, Borg MK, Scanlon TJ, Reese JM (2013) A DSMC investigation of gas flows in micro-channels with bends. Comput Fluids 71:261–271
166. Woods L (1979) Transport processes in dilute gases over the whole range of Knudsen numbers. Part 1. General theory. J Fluid Mech 93(3):585–607
167. Woods L, Troughton H (1980) Transport processes in dilute gases over the whole range of Knudsen numbers. Part 2. Ultrasonic sound waves. J Fluid Mech 100(2):321–331
168. Wu L (2008) A slip model for rarefied gas flows at arbitrary Knudsen number. Appl Phys Lett 93(25):253103
169. Wu P, Little W (1984) Measurement of the heat transfer characteristics of gas flow in fine channel heat exchangers used for microminiature refrigerators. Cryogenics 24(8):415–420
170. Xu K (2003) Super-Burnett solutions for Poiseuille flow. Phys Fluids 15(7):2077–2080

171. Yamaguchi H, Hanawa T, Yamamoto O, Matsuda Y, Egami Y, Niimi T (2011) Experimental measurement on tangential momentum accommodation coefficient in a single microtube. Microfluid Nanofluid 11(1):57–64
172. Yang HT (1955) Rayleigh's problem at low Mach number according to the kinetic theory of gases. Doctoral dissertation, California Institute of Technology
173. Yang R, Chen G, Laroche M, Taur Y (2005) Simulation of nanoscale multidimensional transient heat conduction problems using ballistic-diffusive equations and phonon Boltzmann equation. J Heat Transf 127(3):298–306
174. Yu D (1995) An experimental and theoretical investigation of fluid flow and heat transfer in microtubes. In: ASME/JSME thermal engineering conference, vol 1
175. Zhao S, Povitsky A (2009) A hybrid molecular and continuum method for low-Reynolds-number flows. Nonlinear Anal Theory Methods Appl 71(12):e2551–e2564
176. Zheng Y, Garcia AL, Alder BJ (2002) Comparison of kinetic theory and hydrodynamics for Poiseuille flow. J Stat Phys 109(3–4):495–505
177. Zhong X, MacCormack RW, Chapman DR (1993) Stabilization of the Burnett equations and application to hypersonic flows. AIAA J 31(6):1036–1043

Index

A. Agrawal et al., *Microscale Flow and Heat Transfer*, Mechanical Engineering Series, https://doi.org/10.1007/978-3-030-10662-1

O

P

R

S

Printed by Printforce, the Netherlands